AF722819

MOLECULAR STRUCTURE AND ENERGETICS

Volume 2

Physical Measurements

MOLECULAR STRUCTURE AND ENERGETICS

Volume 2

Series Editors

Joel F. Liebman
University of Maryland Baltimore County

Arthur Greenberg
New Jersey Institute of Technology

Advisory Board

Israel Agranat, The Hebrew University of Jerusalem (Israel)
Marye Anne Fox, University of Texas at Austin
Sharon G. Lias, National Bureau of Standards
Alan P. Marchand, North Texas State University
Eiji Ōsawa, Hokkaido University (Japan)
Heinz D. Roth, AT&T Bell Laboratories
Othmar Stelzer, University of Wuppertal (FRG)
Ronald D. Topsom, La Trobe University (Australia)
Joan Selverstone Valentine, University of California, Los Angeles
Deborah Van Vechten, Sesqui Science Services
Kenneth B. Wiberg, Yale University

© 1987 VCH Publishers, Inc. New York

Distribution: VCH Verlagsgesellschaft mbH, P.O. Box 1260/1280, D-6940 Weinheim, Federal Republic of Germany

USA and Canada: VCH Publishers, Inc., 303 N.W. 12th Avenue, Deerfield Beach, FL 33442-1705, USA

MOLECULAR STRUCTURE AND ENERGETICS

Volume 2

Physical Measurements

Edited by

Joel F. Liebman
and
Arthur Greenberg

VCH

Joel F. Liebman
Department of Chemistry
University of Maryland Baltimore County
Catonsville, Maryland 21228

Arthur Greenberg
Division of Chemistry
New Jersey Institute of Technology
Newark, New Jersey 07102

Library of Congress Cataloging-in-Publication Data

Molecular structure and energetics.

Includes bibliographies and indexes.
Contents: v. 1. Chemical bonding models—v. 2. Physical measurements—v. 3. Studies of organic molecules.
1. Molecular structure—Collected works. 2. Molecular dynamics—Collected works. I. Liebman, Joel F. II. Greenberg, Arthur.
QD461.M629 1986 541.2′2 86-19250
ISBN 0-89573-140-1 (v. 2)

© 1987 VCH Publishers, Inc.

This work is subject to copyright.

All rights are reserved, whether the whole or part of the material is concerned, specifically those of translation, reprinting, re-use of illustrations, broadcasting, reproduction by photocopying machine or similar means, and storage in data banks.

Registered names, trademarks, etc. used in this book, even when not specifically marked as such, are not to be considered unprotected by law.

Printed in the United States of America.

ISBN 0-89573-140-1 VCH Publishers
ISBN 3-527-26476-0 VCH Verlagsgesellschaft

SERIES FOREWORD

Molecular structure and energetics are two of the most ubiquitous, fundamental and, therefore, important concepts in chemistry. The concept of molecular structure arises as soon as even two atoms are said to be bound together since one naturally thinks of the binding in terms of bond length and interatomic separation. The addition of a third atom introduces the concept of bond angles. These concepts of bond length and bond angle remain useful in describing molecular phenomena in more complex species whether it be the degree of pyramidality of a nitrogen in a hydrazine, the twisting of an olefin, the planarity of a benzene ring, or the orientation of a bioactive substance when binding to an enzyme. The concept of energetics arises as soon as one considers nuclei and electrons and their assemblages, atoms and molecules. Indeed, knowledge of some of the simplest processes, eg, the loss of an electron or the gain of a proton, has proven useful for the understanding of atomic and molecular hydrogen, of amino acids in solution, and of the activation of aromatic hydrocarbons on airborne particulates.

Molecular structure and energetics have been studied by a variety of methods ranging from rigorous theory to precise experiment, from intuitive models to casual observation. Some theorists and experimentalists will talk about bond distances measured to an accuracy of 0.001 Å, bond angles to 0.1°, and energies to 0.1 kcal/mol and will emphasize the necessity of such precision for their understanding. Yet other theorists and experimentalists will make equally active and valid use of such seemingly ill-defined sources of information as relative yields of products, vapor pressures, and toxicity. The various chapters in this book series use as their theme "Molecular Structure and Energetics", and it has been the individual authors' choice as to the mix of theory and of experiment, of rigor and of intuition that they have wished to combine.

As editors, we have asked the authors to explain not only "what" they know but "how" they know it and explicitly encouraged a thorough blending of data and of concepts in each chapter. Many of the authors have told us that writing their chapters have provided them with a useful and enjoyable (re)education. The chapters have had much the same effect on us and we trust readers will share our enthusiasm. Each chapter stands autonomously as a combined review and tutorial of a major research area. Yet clearly there are interrelations between them and to emphasize this coherence we have tried to have a single theme in each volume. Indeed the first four volumes of this series were written in parallel, and so for these there is an even higher degree of unity. It is this underlying unity of molecular structure and energetics with all of chemistry that marks the series and our efforts.

Another underlying unity we wish to emphasize is that of the emotions and of the intellect. We thus enthusiastically thank Alan Marchand for the opportunity to write a volume for his book series, which grew first to multiple volumes, and then became the current, autonomous series for which this essay is the foreword. We likewise thank Marie Stilkind of VCH Publishers for her versatility, cooperation and humor. We also wish to emphasize the support, the counsel, the tolerance and the encouragement we have long received from our respective parents, Murray and Lucille, Murray and Bella; spouses, Deborah and Susan; parents-in-law, Jo and Van, Wilbert and Rena; and children, David and Rachel. Indeed, it is this latter unity, that of the intellect and of emotions, that provides the motivation for the dedication for this series:

"To Life, to Love, and to Learning".

Joel F. Liebman
Baltimore, Maryland

Arthur Greenberg
Newark, New Jersey

PREFACE

The eight chapters of the present volume provide the theoretical underpinnings and empirical results for important classes of experimental probes of molecular structure and energetics. All the essays interweave modern theory and modern experiment.

Chapter 1, by the Hargittais, discusses experimental determinations of molecular structure in which the bond lengths are precise to 0.001 Å and the interatomic angles to 0.1°. At this level of accuracy, comparison of results measured on the solid and gaseous phases of a given compound generally proffers useful insights into the magnitude and nature of intra- and inter-molecular interactions. Application is made to a wide variety of inorganic and organic compounds from gaseous ion pairs to the molecular solids of polyfunctionalized ring systems.

Chapter 2, by Stein and Brown, provides an interpolation between the gaseous and solid states of organic compounds by systematically discussing the accretion and annelation of benzene rings in the process that eventually forms the "infinite" sheets that constitute graphite. These authors make extensive use of the admittedly sparse experimental literature on the species that constitute the conceptually intermediate steps in this process. By so using the available experimental data and systematically employing a variety of semiempirical methods developed for "small" molecules, a coherent understanding arises.

Chapter 3, by Chickos, likewise interrelates the gaseous and solid states by discussing the energetics of sublimation. The strengths, weaknesses and principles associated with the primary techniques for its experimental measurement and some of the more successful predictive methods are presented. Thermochemical distinctions between pure enantiomers and racemates are also made. Complementing these experimental and theoretical approaches is a tabulated data compendium of the experimental measurements reported of heat of sublimation.

Chapter 4, by Wiberg, surveys the field of organic thermochemistry and discusses the major techniques of combustion and reaction calorimetry and emphasizes how high accuracy is much more easily achieved for the latter. He explicitly demonstrates how knowledge of the experimental properties of one class of compounds allows for the better understanding of others, eg, the synthetic, structural and thermochemical relationships between alkanes, alcohols, esters, alkenes, ketones and acetals. The complementary use of high accuracy quantum mechanical calculations and high accuracy experimental measurements is also demonstrated.

Chapter 5, by Pihlaja, discusses a variety of techniques useful in obtaining qualitative and quantitative information about the energetics of organic compounds. These methods include bond energy analyses and molecular mechanics, measurements of appearance and ionization potentials, and isomer and conformer equilibration. The author also explicitly discusses the heats of formation of ca. 200 organic compounds measured in the past decade and generally interrelates these numbers with other thermochemical and structural results.

Chapter 6, by Skinner and Connor, presents both the special experimental techniques and conceptual approaches designed to study the energetics of organometallic compounds. Many of these compounds form the explicit interface of organic and inorganic chemistry and have molecular features that are rarely found elsewhere, eg, unicoordinate carbon and metal-metal multiple bonds. A set of extensive tables of measured and of derived thermochemical quantities accompanies the text wherein species as simple as diatomic ions and as complicated as polynuclear metal complexes are to be found.

Chapter 7, by Dixon and Lias, discusses state-of-the-art quantum mechanical calculations and experimental determinations of the proton affinities of gas phase molecules. The theoretical results on comparatively few species are discussed because these authors have explicitly limited their attention to those studies for which identically well-defined, high accuracy calculations on both the neutral and ion have been reported. Similarly, the experimental results on comparatively few species are discussed because the authors explicitly limit their attention to "benchmark" molecules to which all other measured proton affinities are ultimately related. By so doing, quantitative, as well as qualitative, understanding is achieved.

Chapter 8, by Liebman and Van Vechten, delineates the single logical framework which underlies the various concepts of heats of formation, strain energy and resonance energy. This framework is defined in terms of their "conceptual vectors" and "diagonal" logic. Explicit application thereof is made to unsubstituted alicyclic and aromatic hydrocarbons and "polyhedral" *closo*-borane ions. Some of the subtle but crucial facets of the definition of the reaction of interest are pointed out in reference to a seemingly simple question based on a restricted universe composed solely of hydrogen, sodium and chlorine. Limitations and interrelations of various quantum and thermochemical approaches are also given, as are brief applications to polymer chemistry, biochemistry and solid state chemistry. The universality of the logical framework is the central theme.

CONTRIBUTORS

Robert L. Brown
Chemical Kinetics Division
National Bureau of Standards
Gaithersburg, Maryland 20899

James S. Chickos
Department of Chemistry
University of Missouri-St. Louis
St. Louis, Missouri 63121

Joseph A. Connor
Chemistry Department
University of Kent
Canterbury
England CT2 7NH

David H. Dixon
E.I. duPont de Nemours & Company
Central Research and Development
Experimental Station 328/106C
Wilmington, Delaware 19898

István Hargittai
Hungarian Academy of Science
Budapest Pf 117
H-1431 Hungary

Magdolna Hargittai
Hungarian Academy of Sciences
Budapest Pf 117
H-1431 Hungary

Sharon G. Lias
Ion Kinetics and Energetics Data Center
National Measurement Laboratory
National Bureau of Standards
Gaithersburg, Maryland 20899

Joel F. Liebman
Department of Chemistry
University of Maryland Baltimore County
Catonsville, Maryland 21228

Kalevi Pihlaja
Department of Chemistry
University of Turku
SF-20500 Turku 50
Finland

Henry A. Skinner
Chemistry Department
University of Manchester
Manchester M13 9PL
England

Stephen E. Stein
Chemical Kinetics Division
National Bureau of Standards
Gaithersburg, Maryland 20899

Deborah Van Vechten
Sachs/Freeman Associates
Code 4671
Naval Research Laboratory
Washington, DC 20375

Kenneth B. Wiberg
Department of Chemistry
Yale University
New Haven, Connecticut 06500

CONTENTS

CHAPTER 1

The Importance of Small Structural Differences

István Hargittai* and Magdolna Hargittai**

University of Connecticut, Storrs, Connecticut

CONTENTS

1. INTRODUCTION

Molecular structure has three major aspects: the *geometry,* referring to the relative positions of the atomic nuclei in the molecule and often described by bond lengths, bond angles, and angles of internal rotation; the intramolecular *motion,* meaning primarily the relative displacements of the atomic nuclei with respect to their equilibrium positions; and the *electron density* distribution. This chapter is primarily concerned with molecular geometry and its variations. The geometrical aspects of a molecular structure are, of course, strongly interrelated with the other two aspects.

*Visiting Professor of Physics (1983/84) and Chemistry (1984/85), and **Visiting Scientist (1983/85). Permanent address: Hungarian Academy of Sciences, Budapest Pf 117, H-1431 Hungary.

© 1987 VCH Publishers, Inc.
MOLECULAR STRUCTURES AND ENERGETICS, Vol. 2

There is no definition for *small* structural difference, and there is no need for it. A structural difference need not be small to be important; on the contrary. However, even very small differences (eg, 0.01 Å in bond length or a couple of degrees in bond angle), if reliably determined, may reveal important structural information.

It is stressed that real structural differences should be considered and apparent differences should be eliminated. What may such apparent differences be? Generally speaking, there may be differences due to the different meaning of the structural parameters used in various comparisons: differences that may be as much as an order of magnitude larger than the real structural differences that are relevant to discuss or than the experimental errors of their determination, for that matter. The apparent differences are usually not very large, but they may be, so it is important to understand them.

Apparent differences may occur between parameters determined by various techniques, due to the difference in the nature of interaction between radiation and matter in the experiment. Thus, for example, the interatomic distance determined by X-ray diffraction represents the distance between charge centroids of the electron density distribution, whereas the interatomic distance is truly internuclear when determined by electron diffraction, neutron diffraction, or microwave spectroscopy. The larger the asphericity of the electron density distribution, the larger the difference between the X-ray diffraction interatomic distance and the internuclear distance. In recent studies the small-angle experimental data from X-ray diffraction are often excluded from the final refinement of the structural parameters to decrease this effect.

The consequences of intramolecular vibrations are present in the results from practically all experimental techniques used to determine molecular geometry. The best representation of an internuclear distance is the equilibrium distance r_e, corresponding to the minimum position of the potential energy function. This parameter results from quantum chemical calculations, at least in principle, while it is not directly attainable from the experiment. The experimentally determined geometrical parameters are always the result of averaging over the molecular vibrations. In crystals, molecular librations also influence the determined parameters. In any case, direct, uncritical comparison of calculated and measured geometries should be treated with caution.

There is an abundance of internuclear distance parameter types in the literature, and the most important ones are as follows.[1]

- r_e equilibrium distance between the positions of atomic nuclei corresponding to the minimum of the potential energy
- r_α^0 and r_z distance between average nuclear positions in the ground vibrational state
- r_α distance between average nuclear positions in the thermal equilibrium at temperature T

r_g average bond length (or generally, average internuclear distance) at temperature T

r_a effective internuclear distance in the expression of the molecular contribution to electron scattering intensities

r_0 effective internuclear distance obtained from the rotational constants (usually, though not always, refers to the ground vibrational state)

r_s effective internuclear distance determined from the isotopic substitution coordinates of the respective atoms

r_{av} average internuclear distance corrected for the effects of intramolecular motion (usually, though not always, obtained from a combined analysis of rotational spectroscopic and electron diffraction data)

The apparent differences, sometimes also called operational differences, which are due to the differences in the interaction between radiation and matter and to the consequences of motion, must be eliminated before the remaining differences, especially when they are small, can be ascribed to real structural effects. A clear-cut separation of the operational and real structural differences in not always possible, as will be seen later, so this problem has to be treated with caution, also.

In determining small structural differences it is especially instrumental to utilize information from as many techniques as are available. One of the most telling examples of the importance of small structural differences is the ring distortions in substituted benzene derivatives. The first report on the ring distortion in a benzene derivative was by Keidel and Bauer[2] from an early (1956) gas electron diffraction study of the phenylsilane structure. Today there are extensive studies of this question. The determination of the ring deformations is a challenging task and each technique has its merits and difficulties (see Table 1-1).

In the remainder of this chapter, we discuss selected examples primarily from structural differences between free molecules and molecules in crystal and between molecules with different substituents. It is hoped that our largely phenomenological approach will contribute to a better understanding of the structural phenomena discussed.

2. CONSEQUENCES OF INTERMOLECULAR INTERACTIONS

Elucidating the effects of intermolecular interactions may greatly facilitate our understanding the structure and energetics of crystals. The geometrical changes of the molecular structures cover a wide range in energy. Molecular shape, symmetry, and conformation change more easily when the molecule becomes part of a crystal than do bond angles and especially bond lengths. Kitaigorodskii[4] suggested four approaches to the investigation of the effect

TABLE 1-1. Advantages and Disadvantages of Five Techniques for Determining Benzene Ring Deformations[a]

Merits	Difficulties
X-ray crystallography	
Vast amount of structural data accumulated and accessible by computer retrieval Accurate bond angles Differences between bond lengths may often be accurately determined	Intermolecular interactions may influence structure Not very accurate bond lengths; their physical meaning does not correspond to internuclear distance
Neutron Crystallography	
Accurate bond angles and internuclear distances, including those involving hydrogen atoms	Intermolecular interactions may influence structure
Electron Diffraction	
Yields structure of free molecules Increased accuracy for symmetrically substituted derivatives Also gives information on intramolecular motion	Not very accurate for unsymmetrically substituted derivatives
Microwave Spectroscopy	
Yields structure of free molecules Extremely accurate geometrical parameters from substitution coordinates	Poor for determining position of atoms near center of gravity Not suitable for most symmetrically substituted derivatives
Quantum Chemical Calculations	
Predict directions of ring distortions and small structural differences Can be used to interpret structural peculiarities May give geometry for experimentally inaccessible molecules	Decreased reliability in determining absolute values of parameters, especially with semiempirical techniques Even ab initio results may depend on basis sets and other factors (eg, geometry optimization, polarization, correlation effects)

[a]Reference 3.

of the crystal field on molecular structure: (a) comparison of gaseous (ie, free) and crystalline molecules, (b) comparison of symmetrically (ie, crystallographically) independent molecules in the crystal, (c) analysis of the structure of molecules whose symmetry in the crystal is lower than their free molecular symmetry, and (d) comparison of the molecular structure in different polymorphic modifications. Bernstein and Hagler[5] called attention to the fact that the energetics of intramolecular deformations may be such that some molecules will be able to achieve a more symmetrical conformation under crystal field effects than when they are isolated in the vapor. Some of their results will be mentioned later. Similarly, biphenyl has higher molecular symmetry (a coplanar structure) in the crystal[6] than in the vapor[7] where the two benzene rings are rotated by about 45° relative to each other. Thus more general formulation for item (c) above can be suggested as the analysis of the structure of molecules whose crystal symmetry is different from their free molecular symmetry.

A. Gas/Crystal Comparisons

Both comparisons (a) and (c) above refer to the structures of free and crystalline molecules. Also, they provide, perhaps, the most straightforward information, since the structure of the free molecule is determined exclusively by intramolecular interactions. Thus any difference that is reliably detected will carry information on the effects of the crystal field on the molecular structure. However, before discussing more subtle structural differences in molecular crystals as compared with free molecules, it is appropriate to point out some striking differences between ionic crystals and the corresponding vapor phase molecules.

B. Ionic Crystals and Vapor Phase Molecules

Although molecules cannot be singled out as building blocks of ionic crystals, the free molecules of *some* compounds may be considered as if they were taken out of the crystal. A nice example is sodium chloride, whose main vapor components are monomeric and dimeric molecules. They are indicated in the crystal structure in Figure 1-1, as is a tetrameric species. Recent mass spectrometric studies of cluster formation of cesium chloride[8] determined a great relative abundance of species with 27 atoms in the cluster. The corresponding $3 \times 3 \times 3$ cube may again be considered as a piece

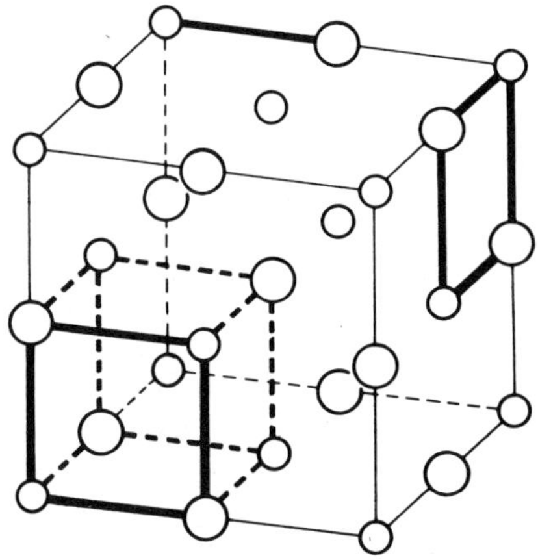

Figure 1-1. Part of sodium chloride crystal structure, indicating NaCl, $(NaCl)_2$, and $(NaCl)_4$ units.

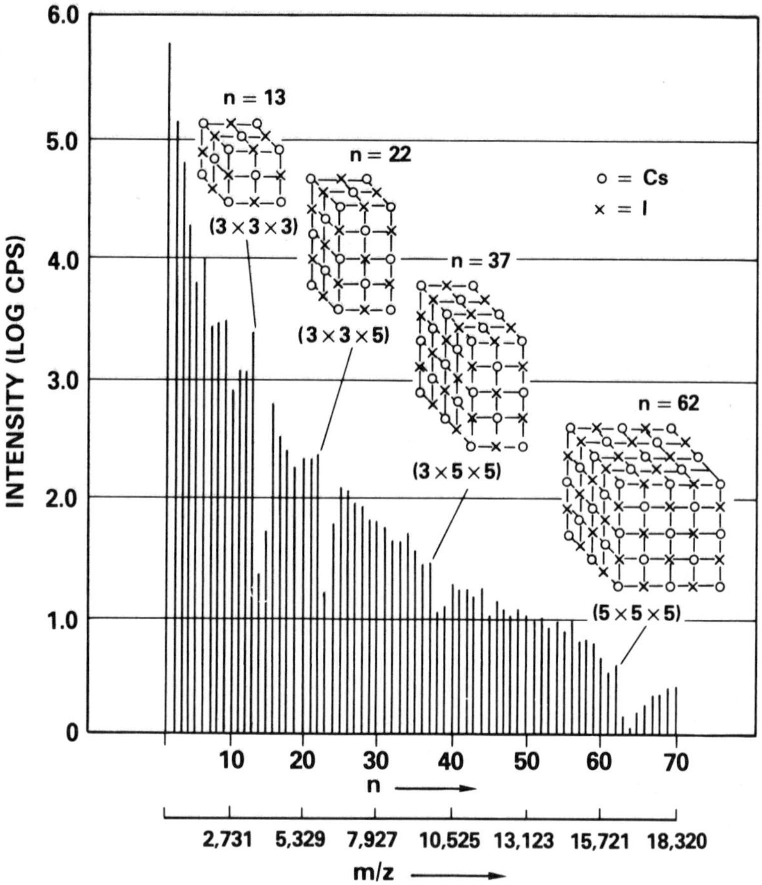

Figure 1-2. SIMS spectrum of CsI cluster ions. Spectral evidence for several stable "cubiclike" structures is apparent at n = 13, 22, 37, and 62, where m/Z is the mass/charge ratio. (Reproduced with permission from Reference 9.)

directly from the crystal. Secondary ion mass spectrometry (SIMS) of cesium iodide[9] and other alkali halides[10] has given evidence for relatively high abundance of several stable "cubelike" structures as clusters. Figure 1-2 depicts the SIMS spectrum of cesium iodide cluster ions after Barlak and co-workers.[9]

Another series of simple molecules whose structure may easily be traced back to the crystal structure is shown in Figure 1-3. It is clear how, for example, various MX_2 and MX_3 molecules may derive different shapes and symmetries from the same kind of crystal structure. The crystal structure is represented by the octahedral arrangement of six "ligands" around the "central atom."

The unit corresponding to the copper chloride trimeric molecules also seems to be already present in the crystal. The molecules have a ring struc-

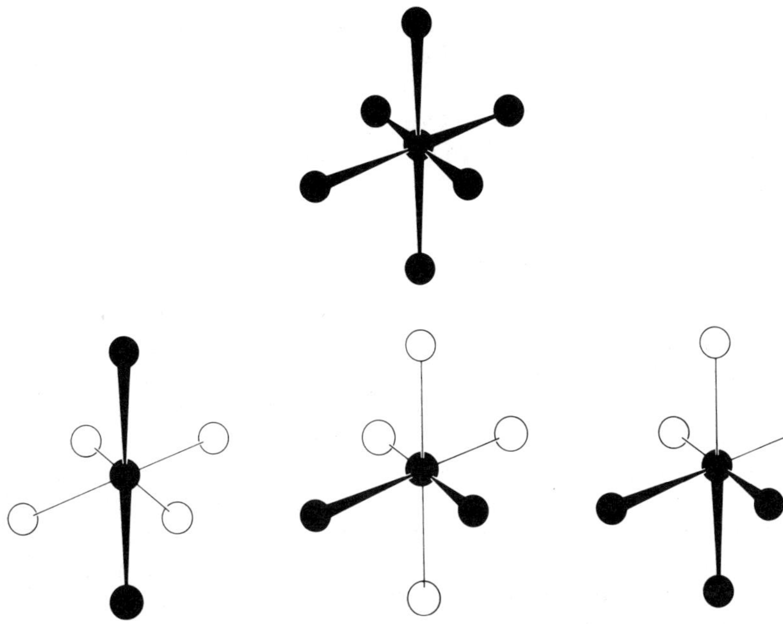

Figure 1-3. Different shapes of MX_2 and MX_3 molecules derived from the crystal structure in which the central atom has an octahedral environment.

ture with alternating copper and chloride atoms. The low-temperature phase of copper chloride has a zinc-blende structure that transforms on heating into a wurtzite structure.[11] In both arrangements the six-membered ring is the smallest unit that can be isolated from the structure upon evaporation.[12]

Even though the shapes of the free molecules mentioned above may be said to have originated from the crystal structure, the interatomic distances change considerably. They are collected for some lithium halides in Table 1-2.[13–15] As we go from the crystal to the free dimers and then to the monomers, the interatomic distances drastically decrease. The extreme accuracy of the microwave spectroscopic data[15] for the monomers is noteworthy. On the other hand, the large uncertainties on the dimers call for the reinvestigation of these compounds by up-to-date techniques.

TABLE 1-2. Interatomic Distances (Å) of Lithium Halides

Phase	Lithium chloride	Lithium bromide	Lithium iodide
Crystal[a]	2.57	2.75	3.00
Vapor			
Dimer[b]	2.23 ± 0.03	2.35 ± 0.02	2.54 ± 0.02
Monomer[c]	2.02067 ± 0.00006	2.17042 ± 0.00004	2.39191 ± 0.00004

[a]Reference 13.
[b]References 13 and 14.
[c]Reference 15.

There is even less structural similarity for many other metal halides as the crystalline molecules are compared with the molecules in the vapor phase. Aluminum trichloride and iron trichloride crystallize in hexagonal layer structures. Upon melting followed by evaporation at relatively low temperatures, dimeric molecules are formed. At higher temperatures they dissociate into monomers.

C. Molecular Crystals and Free Molecules

The vast majority of molecular crystals are formed by organic substances. In some cases their vapor contains more rotational isomers than the crystal. Thus, for example, the vapor of ethane-1,2-dithiol consists of a conformational equilibrium of anti and gauche forms with respect to the rotation about the central bond.[16] This is illustrated by the electron diffraction radial distributions in Figure 1-4. Only the anti form was found in the crystal.[17]

Bis(cyclopentadienyl)titanium borohydride has also been studied both in the crystal[18] and the vapor.[19] The rings seem to take an orientation of high symmetry relative to each other and to the Ti . . . B axis in the vapor as seen in Figure 1-5. The structure is less symmetrical in the crystal.

For some time beryllocene was thought to be a striking example of differ-

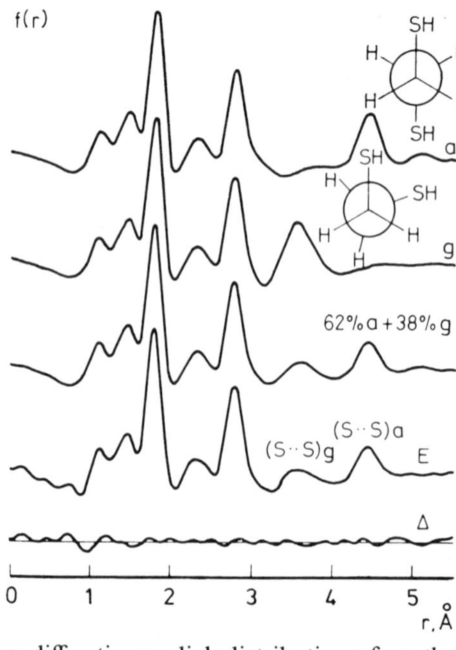

Figure 1-4. Electron diffraction radial distributions for ethane-1,2-dithiol: E = experimental; theoretical distributions, *a* and *g* for the anti and gauche forms, and 62% *a* + 38% *g* for their mixture. (From Reference 16.)

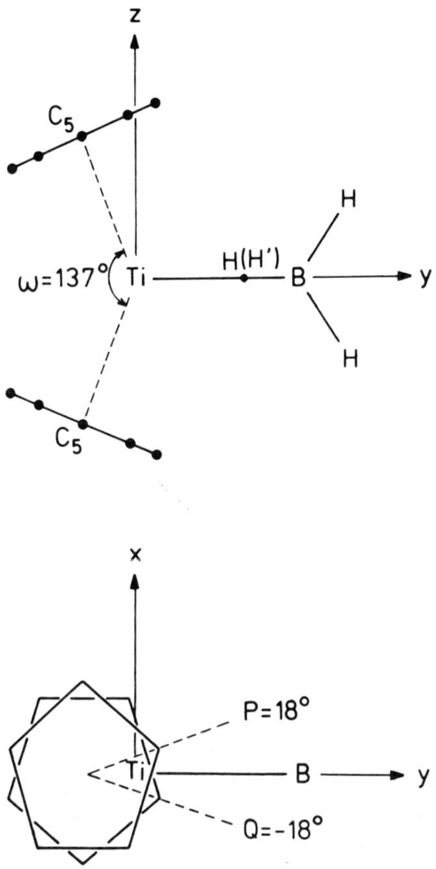

Figure 1-5. The projection models of the vapor phase bis(cyclopentadienyl)-titanium borohydride molecule in two different planes. The angles *P* and *Q* are 0° when one of the apexes of the five-membered rings coincides with the Ti . . . B axis. (From Reference 19.)

ent structures in vapor and crystal. The model originally suggested for the free molecule[20] had C_{5v} symmetry and parallel, staggered cyclopentadienyl rings with the beryllium on their coinciding five-fold axis. The metal atom was thought to oscillate between alternative positions along this axis. Another model represented the crystal molecular structure[21] at -120°C. One of the two parallel rings slipped by 1.2Å in this model. This behavior was ascribed to the influence of intermolecular interactions. Subsequent quantum chemical calculations, however, provided the following information on the structure of the free molecule[22]: it has rather low symmetry with one σ-bonded and one π-bonded ring, as in the crystal; even a ferrocenelike D_{5h} or D_{5d} model appeared to be more stable than the C_{5v} model. The dynamic and fluxional character of the beryllocene structure has been repeatedly emphasized.

Until recently it has been argued that the crystal field may influence molecular symmetry and conformation, but scarcely if at all the bond configuration. With increasing accuracy of the experimentally determined structural parameters and with detailed understanding of their physical meaning, new possibilities are opening up to investigate even subtle effects caused by intermolecular interactions in the crystal. At the same time, these considerations can save us from ascribing apparent differences to crystal field effects when they are not real structural differences indeed.

In the comparison of some structural features of *p*-bis(trimethylsilyl) benzene in gas and crystal below, α is the ipso angle at the substitutent in the ring, $\langle CC \rangle$ is the mean C—C bond length of the ring, and Δ(SiC) is the Si—C(methyl)—Si—C(benzene) bond length difference:

Parameter	Gas from electron diffraction[23]	Crystal from X-ray diffraction[24]
α	115.7(6)°	115.8(5)°
$\langle CC \rangle$	1.408(3) Å	1.388 Å
Δ(SiC)	0.033(7) Å	−0.021 Å

The angular deformation is the same in the two phases; the ring itself, however, seems to be smaller in the crystal than in the gas. This observation is consistent with general experience with similar data. Figure 1-6 presents some correlation between the α angles and the $C_1 \ldots C_4$ distances of the benzene rings in a series of paradisubstituted benzene derivatives.[3] The X-ray crystallography results on the one hand, and the gas electron diffraction and neutron crystallography results on the other, tend to be separated. Apparently the $C_1 \ldots C_4$ interatomic distances from X-ray diffraction are a little too short. This may originate from the benzene ring electron density distribution being somewhat shifted toward the center of the benzene ring. It is consistent with this notion that the ring-to-substituent distances appear to be longer in X-ray crystallography than the analogous internuclear distances determined by electron diffraction in, for example, *p*-bis-(trimethylsilyl)benzene.

The molecular structure of naphthalene has been repeatedly investigated in gas[26] and in crystal.[25] The results of the experimental studies as well as calculations[27] indicate no appreciable crystal field effects, as seen in the data of Table 1-3.

The internal angles of the benzene ring in fluorobenzene are also in excellent agreement as determined by various techniques in different phases,[28–32] as demonstrated[29] by the data of Table 1-4. The most pronounced deformation occurs in the ipso angle of the ring. Ab initio molecular orbital calculations[33] using the gradient force method seem to somewhat underestimate this deformation. Similar discrepancies occurred in the experimental and calculated angles of 1,3,5-trifluorobenzene, namely: C—C(F)—C =

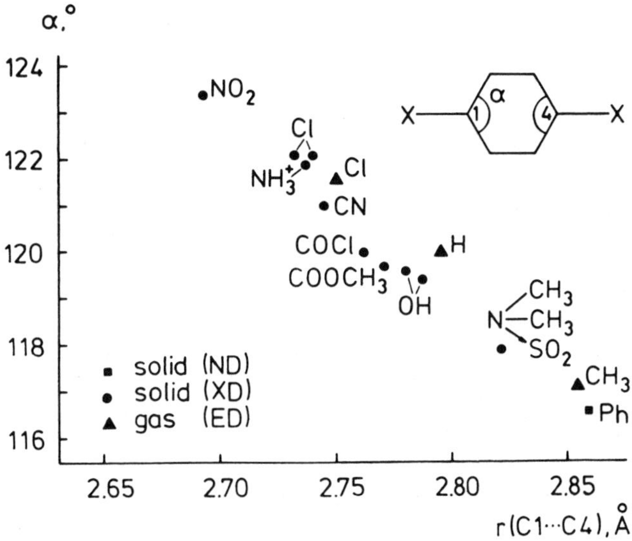

Figure 1-6. Correlation between the ring ipso angles α and the $C_1 \ldots C_4$ distances of para-disubstituted benzene derivatives mainly from X-ray crystallography (XD) and also from gas electron diffraction (ED) and neutron crystallography (ND) (After Reference 3.)

Table 1-3. Bond Lengths (Å) and Bond Angles (°) of the Naphthalene Carbon Skeleton (D_{2h} symmetry)

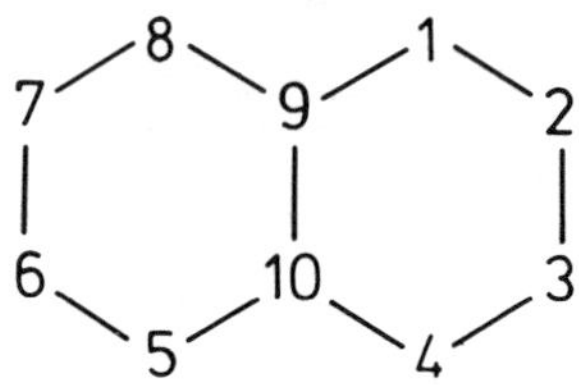

Parameter	Crystal experiment[a]	Free molecule: Experiment[b]	Free molecule: Calculation[c]
Bond length			
C_1-C_2	1.377(2)	1.381(2)	1.371
C_2-C_3	1.411(2)	1.417(4)	1.420
C_1-C_9	1.424(2)	1.422(3)	1.425
C_9-C_{10}	1.421(2)	1.412(8)	1.415
Bond angle			
$C_9-C_1-C_2$	120.2(1)		120.8
$C_1-C_2-C_3$	120.5(4)		120.4
$C_1-C_9-C_{10}$	119.0(1)	119.5(3)	118.9

[a]X-Ray diffraction (−150°C),[25a] in good agreement with earlier results (room temperature)[25b].
[b]Electron diffraction,[26a] in good agreement with earlier results.[26b]
[c]Ab initio, corrected.[27]

Table 1-4. Internal Ring Angles (°) in Fluorobenzene

	Experimental						
	Free molecule		Nematic liquid crystal		Crystal		MO calculations, ab initio[33]
	Microwave	Electron	NMR[32]		X-ray diffraction		
Angle	spectroscopy[28]	diffraction[29]	1	2	*p*-FBA[a]	*p*-FAHCl[b]	
α	123.4	123.4 ± 0.2	124.2(5)	123.0(2)	123.8	123.3	122.3_5
β	117.9	118.0 ± 0.2	117.6(3)	118.2(1)	117.9	118.0	118.5_8
γ	120.5	120.2 ± 0.3	120.3(1)	120.3(1)	120.3	120.3	120.3_2
δ	119.8	120.2 ± 0.4	120.0(1)	120.0(1)	119.8	120.0	119.8_5

[a]From the structure of *p*-fluorobenzoic acid[30] after subtracting the angular distortions caused by the COOH substituent.[34] Estimated standard deviations are 0.2°.
[b]From the structure of *p*-fluoroaniline hydrochloride[31] after subtracting the angular distortions caused by the NH_3^+ substituent.[34] Estimated standard deviations are 0.2°.

123.7 ± 0.2° from electron diffraction[35] and 122.7° from quantum chemical calculations.[33]

It was interesting to compare[36] the molecular geometries of *p*-methylsulfonylbenzoic acid ($CH_3SO_2C_6H_4COOH$) determined by X-ray diffraction in crystal[37] and methylphenylsulfone ($CH_3SO_2C_6H_5$) determined by electron diffraction in gas.[38] Although these are two different substances, the structural

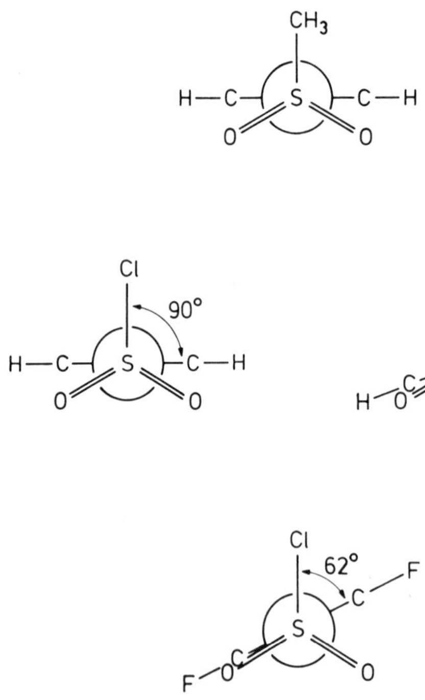

Figure 1-7. The conformations of methylphenylsulfone ($C_6H_5SO_2CH_3$), benzenesulfonyl chloride ($C_6H_5SO_2Cl$), and pentafluorobenzenesulfonyl chloride ($C_6F_5SO_2Cl$). The projections are viewed along the S—C (Ph) bond.

effect of the COOH substituent on the benzene ring is known to be negligible.[34] Since the same is assumed to hold beyond the ring, the comparison is meaningful. The conformation of the two molecules is essentially the same, with the C−S−C plane being perpendicular to the plane of the ring as shown in Figure 1-7. Incidentally, the conformation of the analogous benzene sulfonyl chloride ($C_6H_5SO_2Cl$) could not be determined so unambiguously.[39,40] The "perpendicular" model and an asymmetric one, both depicted in Figure 1-7, could be brought equally well into agreement with the electron diffraction data. The "perpendicular" model implied large-amplitude torsional motion about the S−C (Ph) bond. The root-mean-square amplitude of this torsion was estimated to be 8.5°. If the asymmetric form is supposed to be a consequence of averaging over the torsional motion, it then corresponds to an estimated 4 kcal/mol barrier. The rotational form of pentafluorobenzene sulfonyl chloride ($C_6F_5SO_2Cl$) displays even greater deviation from the perpendicular model as determined by electron diffraction[41] and shown also in Figure 1-7.

The geometrical parameters of free methylphenylsulfone and crystalline *p*-methylsulfonylbenzoic acid are compared in Table 1-5. Except for the dis-

TABLE 1-5. Bond Lengths (Å) and Bond Angles (°) in Free Methylphenylsulfone and Crystalline *p*-methylsulfonylbenzoic Acid

Parameter	Methylphenylsulfone (by gas electron diffraction)[a]	*p*-Methylsulfonylbenzoic acid (by X-ray crystallography)[b]		
		Conventional refinement	High-θ refinement	High-θ refinement, corrected for libration
Bond length				
C−C (mean)	1.399 ± 0.003	1.389	1.392	1.398
S−Ph	1.771 ± 0.012	1.770	1.769	1.771
S−Me	1.785 ± 0.012	1.757	1.756	1.768
S=O	1.445 ± 0.003	1.436	1.442	1.452
		1.440	1.443	1.452
Bond angle				
Ph−S−Me	103.9 ± 0.7	103.5	103.7	103.6
O=S=O	118.4 ± 0.6	118.3	118.6	118.7
Ph−S=O	108.2 ± 0.4	108.7	108.6	108.5
		108.7	108.7	108.6
Me−S=O	108.6 ± 0.6	108.3	108.1	108.1
		108.4	108.1	108.2
α	121.9 ± 1.3	121.6	121.7	122.0
β	118.9 ± 0.8	118.9	118.9	118.7
		119.2	119.2	119.1
γ	119.6 ± 0.7[c]	119.9	119.9	119.7
		119.8	119.7	119.5
δ	121.2 ± 0.7[c]	120.6	120.6	120.9

[a]Estimated total errors are given as error limits; bond lengths are r_g parameters.
[b]Least-squares standard deviations are 0.001–0.002 Å and 0.06–0.14°.
[c]Derived parameter.

Table 1-6. Selected Bond Lengths (Å) in *p*-Dicyanobenzene[42] and *p*-Diisocyanobenzene[43]

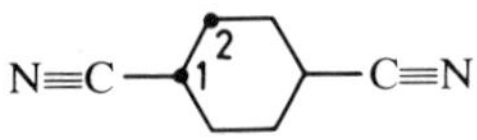		C_1–C_2	CN	
X-Ray diffraction	Conventional	1.388(1)	1.145(2)	
	High-θ refinement	1.395(2)	1.149(2)	
	Corrected for thermal motion	1.400(2)	1.169(2)	
Electron diffraction		1.398(3)	1.167(2)	
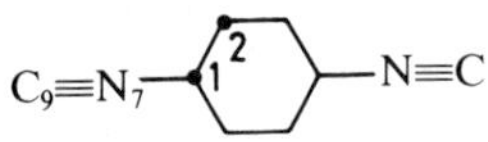		⟨CC⟩	C_1–N_7	N_7–C_9
X-Ray diffraction	Conventional	1.385(2)	1.399(2)	1.156(3)
	High-θ refinement	1.389(2)	1.393(3)	1.163(4)
	Corrected for thermal motion	1.396(2)	1.396(2)	1.166(4)
Electron diffraction		1.400(3)	1.388(3)	1.176(2)

crepancies in the bond lengths S=O and S–CH_3 (cf Reference 36), the agreement is impressive. Thus both the molecular conformation and the benzene ring deformation suggest no appreciable intermolecular interactions in the crystal of *p*-methylsulfonylbenzoic acid.

Variations in the ring angular deformations of gaseous and crystalline substituted benzene derivatives may be a sensitive indicator of intermolecular interactions. Recent parallel gas electron diffraction and X-ray crystallographic studies of *p*-dicyanobenzene[42] and *p*-diisocyanobenzene[43] have led to interesting results in this respect. Table 1-6 contains some of the bond lengths of the two molecules from various structure refinements. These data show the significance of operational differences and the ways to eliminate them. On the other hand, these data do not indicate appreciable changes in the gas versus crystal structures. It is very revealing, however, to examine the variations of the ipso ring angle (from the best refinements):

	Free molecule	Crystal molecule	Difference
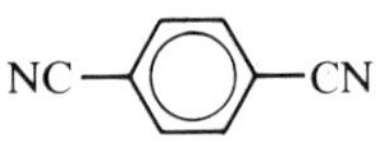	122.1(2)°	121.6(1)°	−0.5°
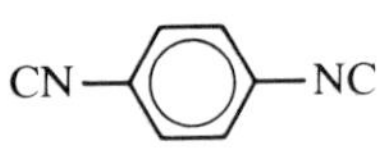	121.7(2)°	122.2(3)°	+0.5°

The changes are marginal, but their trend can be interpreted in a reasonable way invoking the marked difference in the molecular packing in the two crystals. For *p*-dicyanobenzene, the smaller deformation in the crystal versus the free molecule indicates an attenuation of the polar character of the cyano substituent under the impact of intermolecular interactions. The *p*-

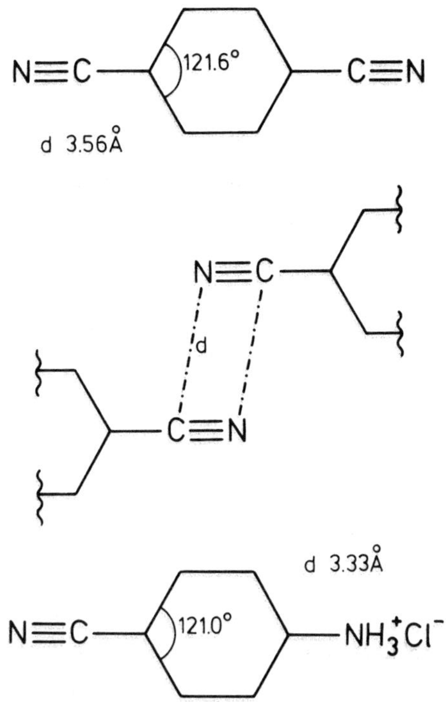

Figure 1-8. Schematic representation of the possible intermolecular interaction between antiparallel cyano groups of *p*-dicyanobenzene and also those of *p*-cyanoaniline hydrochloride in the crystal.

dicyanobenzene molecules are parallel to each other in the crystal, and they form layers.[44,45] Each layer is displaced with respect to the next, leading to antiparallel contacts between neighboring cyano groups in addition to parallel contacts. The shortest distance between two antiparallel cyano groups is 3.56 Å as seen in Figure 1-8. These interactions are also the origin of the relatively high melting point of this compound (viz, 223–226°C). A similar pattern of interactions occurs in the crystal of *p*-cyanoaniline hydrochloride.[31] Since no parallel cyano groups are present in this structure, the antiparallel cyano groups can get even closer, as indicated in Figure 1-8. The ipso ring angle in this molecule is further, if only slightly, decreased to 121.0(2)°. In this particular case again, the change in the parasubstituent fortunately did not change the conditions of the comparison.

The gas/crystal difference in the ring angular deformation is just the opposite for *p*-diisocyanobenzene.[43] Here the ring is more deformed in the crystal than in the free molecule. The crystal structure of *p*-diisocyanobenzene is strikingly different from that of *p*-dicyanobenzene. The intermolecular interactions that occur between the isocyano groups and the benzene rings, as shown schematically in Figure 1-9, apparently lead to an increase in the polar character of the isocyano substituent, and accordingly to somewhat

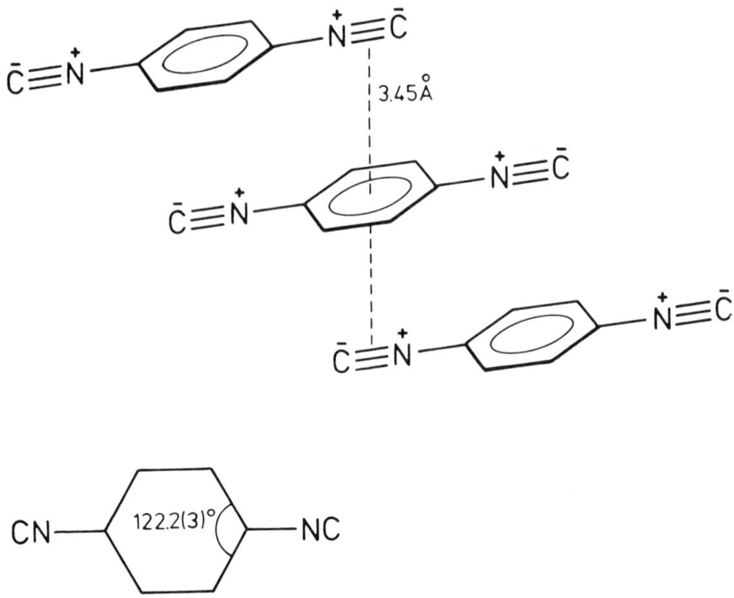

Figure 1-9. Schematic representation of the possible intermolecular interactions between the isocyano groups and benzene rings of *p*-diisocyanobenzene in the crystal.

enhanced ring deformation in the crystal as compared with the free *p*-diisocyanobenzene molecule, or for that matter, with the crystal *p*-dicyanobenzene molecule. As marginal as these findings may be, they could serve to stimulate structural chemists to search for gas/crystal differences by means of accurately determined systems, and theoretical chemists to build models and perform calculations on them in which both the intramolecular and intermolecular interactions are adequately represented.

The gas/crystal structural changes obviously depend on the relative strengths of the intramolecular versus intermolecular interactions. More pronounced changes are expected, for example, in relatively weak coordination linkages under the influence of the crystal field than in stronger bonds.

D. Weak Intramolecular Interactions

The N—B bond of the trimethylamine boron trichloride $(CH_3)_3N \cdot BCl_3$ donor–acceptor complex is considerably longer in gas[46] than in crystal.[47] The two sets of structural parameters are compared in Figure 1-10. The N—Si dative bond of 1-methylsilatrane is so much shorter in crystal[48] than in gas[49] that in the latter there is hardly any such bond according to the geometrical data cited in Figure 1-11. It may be supposed that the intermolecular forces somewhat compress the molecules along the coordination linkage in the

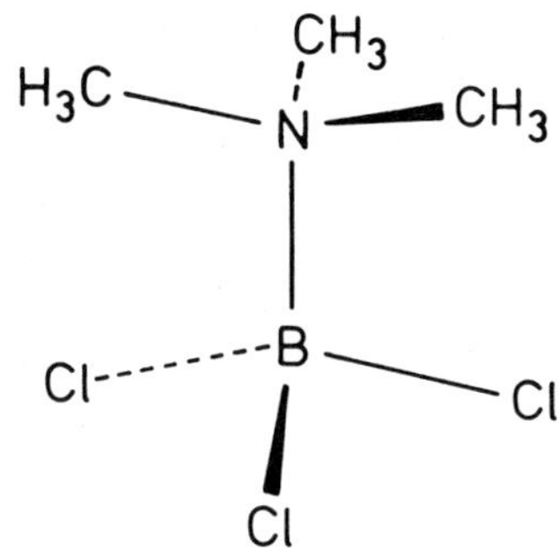

gas (ED)		crystal (XD)
1.659(6) Å	B-N	1.609(6) Å
108.7(5)°	C-N-C	108.6(14)°
110.8(3)°	Cl-B-Cl	109.5(2)°

Figure 1-10. Selected geometrical parameters of the trimethylamine boron trichloride molecule in gas[46] and in crystal.[47]

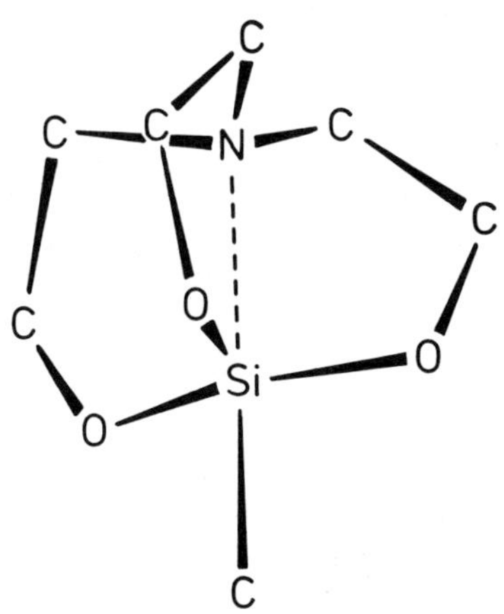

gas (ED)		crystal (XD)
2.45(5) Å	N···Si	2.175(4) Å
79.0(15)°	N-Si-O	82.7(2)°
116.5(9)°	O-Si-O	118.4(4)°
101.0(15)°	C-Si-O	97.2(5)°

Figure 1-11. Selected geometrical parameters of the 1-methylsilatrane molecule in gas[49] and in crystal.[48]

crystal. Part of the gas/crystal bond length differences in these cases originates, of course, from the differing physical meanings of the gas electron diffraction and crystal X-ray diffraction results, since these coordination linkages entail considerable charge transfer. The other part of the differences, however, is supposed to be real structural, and the geometrical changes in the rest of the molecule are fully consistent with this notion.

The bond angles of the donor part of $(CH_3)_3N \cdot BCl_3$ are the same in gas and in crystal within experimental error (Figure 1-10). It is generally observed[50] that the donor geometry is not very sensitive to complex formation. On the other hand, the acceptor configuration usually changes considerably upon complex formation.[50] The free BCl_3 molecule has D_{3h} symmetry with 120° bond angles. The greater angular deformation of the BCl_3 moiety of the complex in the crystal than in the gas is consistent with a stronger coordination linkage in the crystal.

Consider now the bond angles in 1-methylsilatrane (Figure 1-11). Sup-

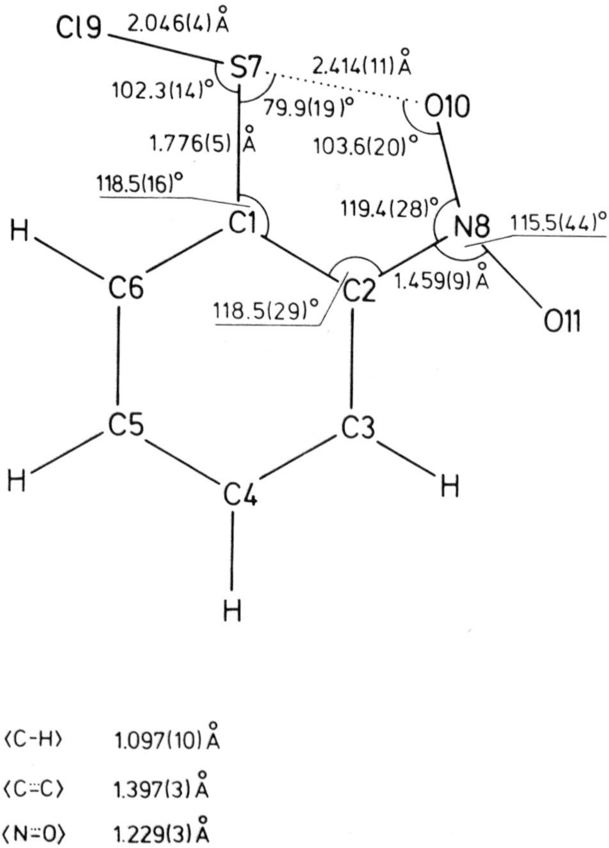

Figure 1-12. The molecular geometry of 2-nitrobenzenesulfenyl chloride.[51]

posing pentacoordination of silicon, the bond angles determined for crystal are closer to the ideal trigonal bipyramidal configuration than those for gas. However, if one ignores the N—Si linkage and supposes tetra coordination of silicon, the structure in gas is nearer the ideal tetrahedral configuration than that in crystal. Thus the bond angle differences are consistent with the notion of the N—Si bond being stronger in crystal than in gas. Part of the bond length difference, however, originates from the difference in the meaning of the two bond lengths, since the N—Si linkage involves considerable charge transfer.

Recently the molecular geometry of 2-nitrobenzenesulfenyl chloride has been determined by gas electron diffraction.[51] The whole molecule is planar; the S—Cl bond is anti with respect to the nitro group. The geometrical parameters are shown in Figure 1-12. There is a short 2.41Å distance between the nonbonded sulfur and oxygen atoms being a part of an essentially linear Cl—S . . . O chain. This intramolecular nonbonded interaction seems to determine the conformation of the free molecule, making it similar to that of analogous systems in the crystal. The conformations of the free benzenesulfenyl chloride[52] and nitrobenzene[53] molecules have been determined earlier. The close S . . . O contact is supposed to hinder the addition reactions at sulfur, and this feature is invoked to explain the reduced reactivity of orthonitrosulfenyl chlorides as compared with analogous para derivatives.

E. Influence of Hydrogen Bond Formation

Even strong intramolecular interactions may be influenced by the crystal structure. Thus the C=S bond of thioacetamide is considerably shorter in gas[54] than in crystal[55] (Figure 1-13). The changes in the bond angles are consistent with this difference, considering simple models of molecular geometry. The intramolecular structural changes are ascribed to the formation of intermolecular hydrogen bridges in thioacetamide as well as in similar systems [eg, acetamide, carbamide (urea), thiocarbamide (thiourea)]. Molecular motion[56] and electron density deformation[57] are thought to account for part of the differences observed.

The structural differences of gaseous[58] and crystalline[59] sulfuric acid molecules (Figure 1-14) have also been ascribed to hydrogen bond formation in the latter. In spite of the considerable differences in the bond lengths and bond angles referring to the two phases as well as to the same molecule, in particular the S—O versus the S=O bonds, the four oxygen atoms are situated in four apexes of a nearly regular tetrahedron in both phases.[60]

Unfortunately, the data available for gas/crystal molecular structure comparison are scarce and most of the pertinent X-ray crystallography information does not originate from modern low-temperature experiments. It is envisioned that this kind of work will be intensified in the near future.

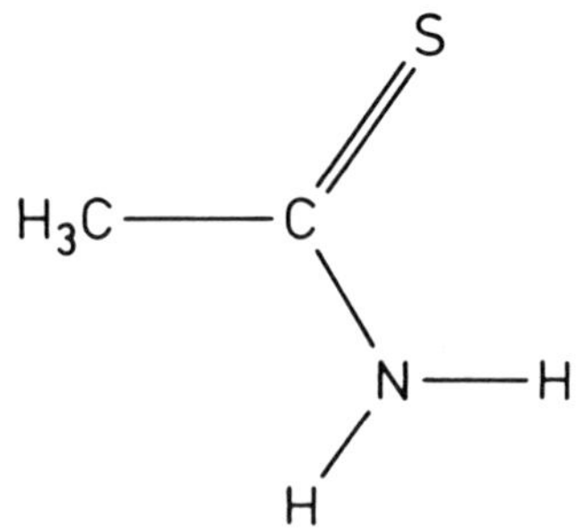

gas (ED)		crystal (XD)
1.356(3) Å	C-N	1.324(8) Å
1.647(3) Å	C=S	1.713(6) Å
122.9(3)°	C-C=S	120.7(4)°
114.8(4)°	C-C-N	117.7(6)°
122.3°	N-C=S	121.6(4)°

Figure 1-13. Geometrical parameters of thioacetamide in gas[54] and in crystal.[55]

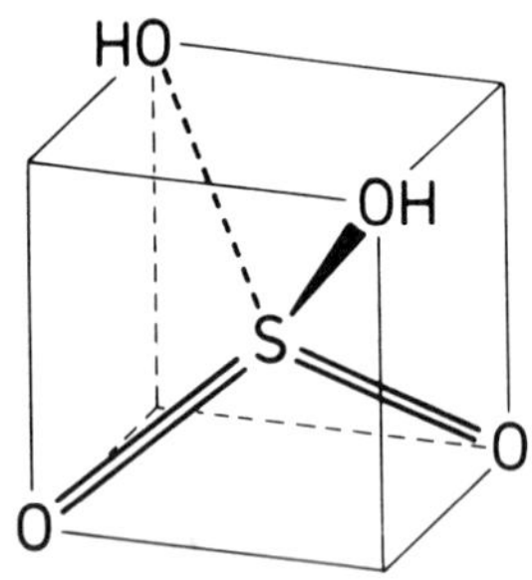

gas (MW)		crystal (XD)
1.422(10) Å	S=O	1.426(15) Å
1.574(10) Å	S-O	1.535(15) Å
123.3(10)°	=S=	118.6(1)°
107.5°	=S-	108.3°
101.3(10)°	-S-	104(1)°
2.50 Å	=O···O=	2.45 Å
2.43 Å	=O···O-	2.40 Å
2.43 Å	-O···O-	2.42 Å

Figure 1-14. The molecular geometry of sulfuric acid in gas[58] and in crystal.[59]

F. Symmetrically Independent Molecules in Crystals

Structural variations may occur also in molecules that build the same crystal and have the same structural formula but are not related by symmetry operations of the space group of their crystal structure. A special case occurs when such molecules are nevertheless related by some symmetry operations that are *not* part of the crystal space group. This phenomenon is known as hypersymmetry.[61] For example, the phenol molecules, connected by hydrogen bonds, form spirals with three-fold screw axis in the crystal as shown in Figure 1-15. This screw axis is not extended, however, to the whole crystal. Tolane (diphenylacetylene) is another example. As shown in Figure 1-16, molecules A and B are related by two-fold rotation plus translation, which is not part of the crystal space group.

The validity of the hypersymmetry operations depends on whether the molecules involved have indeed the same structure, implying both conformation and bond configuration. Since the molecules have somewhat different environments, there may be structural changes to various extents. Moreover, structures that were considered to be identical before may no longer be so considered as more accurate data become available.

According to recent statistics by Zorkii and Razumaeva,[62] about 8% of all homomolecular crystals contain symmetrically independent molecules. "Homomolecular" is the term applied to crystals built of chemically identical molecules.[63] Recognizing the importance of structural differences in the symmetrically independent molecules, even a special name, "quasi-heteromolecular," has been proposed for their crystals.[62]

According to quantitative criteria for defining the extent of differences between two different structures, some 20% of all quasi-heteromolecular

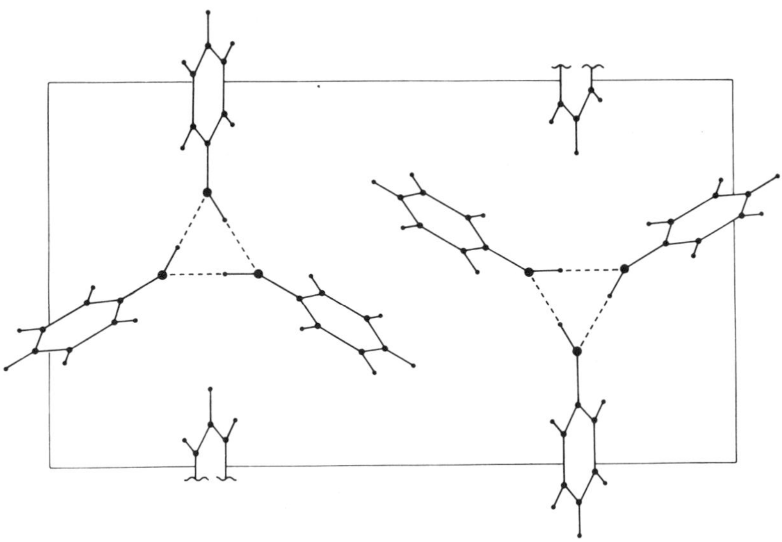

Figure 1-15. To the discussion of the hypersymmetry of phenol crystals.[61]

Figure 1-16. To the discussion of the hypersymmetry of tolane crystals.[61]

organic crystals were found to have appreciable structural differences in their symmetrically independent molecules.[62] This percentage was even higher among quasi-heteromolecular coordination compounds of transition metals.[64] The symmetrically independent molecules in fact showed a great variety of structural differences.[65]

G. Conformational Polymorphism

The energy differences between the polymorphs of organic crystals are estimated to be similar to the free energy differences of rotational isomers of many organic molecules, namely, a few kilocalories per mole. When the molecules adopt different conformations in the different polymorphs, the change in rotational isomerism is attributed to the influence of the crystal field, since the difference in the intermolecular forces is the single variable in the polymorphic systems. Bernstein and co-workers[5,66] have extensively studied conformational polymorphism of various organic compounds with a variety of techniques in addition to X-ray crystallography. Among the molecules investigated were *N*-(*p*-chlorobenzylidine)-*p*-chloroaniline (**1**),

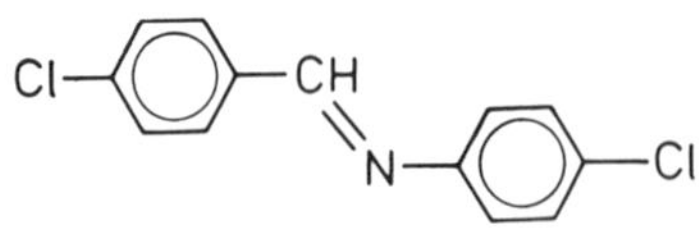

1

which exists in at least two forms, and *p*-methyl-*N*-(*p*-methylbenzylidene)aniline (**2**),

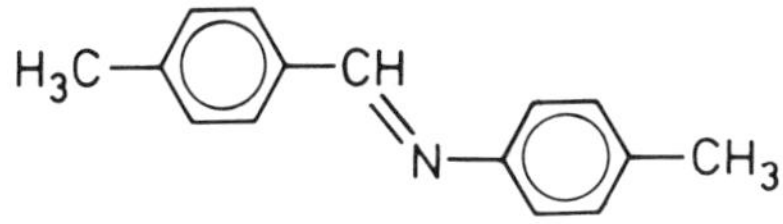

2

which exists in at least three forms. For (**1**) a high-energy planar conformation was shown to occur in triclinic lattice. A lower energy form with normal exocyclic angles was found in the orthorhombic compound. The question of why molecule (**1**) would not always pack with its lowest energy conformation was intriguing.

The X-ray diffraction work has been augmented by lattice energy calculations employing different potential functions. The results did not depend on the choice of the potential function, and they showed that the crystal packing and the (intra)molecular structure together adopt an optimal compromise. The minimized lattice energies were analyzed in terms of partial atomic contributions to the total energy. Even for the trimorphic molecule (**2**), the relative energy contributions of various groups to the total energy were the same for all polymorphs. This indicated that the various groups in all polymorphs had similar energy environments. However, this obviously could be achieved in some lattices only by adopting a different from the most favorable conformation with respect to the structure of the isolated molecule. The investigation of conformational polymorphism proved to be a suitable and promising tool for investigating the nature of crystal forces influencing molecular conformation, or possibly molecular structure in a broader sense.

Possible variations in bond angles and bond lengths have been ignored in the considerations above. The energy requirements for changing bond angles and bond lengths are certainly higher than those for conformational changes, and, accordingly, higher than what may be available in polymorphic transitions. However, some relaxation of the bond configuration may take place, especially considering the (intra)molecular structure also as a compromise adopted by the bond configurations and the rotational forms. Although it is often assumed in conformational analysis of free molecules, especially by electron diffraction, that the various rotational isomers differ in the angles of rotation only, this is not an intrinsic requirement. On the contrary, not only may such an assumption conceal appreciable differences in geometrical parameters, it may even be the origin of hidden errors in the determination of the rotational angles themselves.

H. Bond Configuration Relaxation During Internal Rotation

Bond configuration relaxation during internal rotation has been investigated by quantum chemical calculations for a series of 1,2-dihaloethanes.[67] The bond angle CCX may change as much as 4° during internal rotation accord-

TABLE 1-7. Calculated Bond Angle Changes During Internal Rotation in 1,2-Dihaloethanes[67]

XCH_2CH_2X	ΔCCX^a	$\Delta\phi^b$
X = F	1°	3°
X = Cl	3°	9°
X = Br	4°	11°
X = I	4°	11°

[a]Calculated differences of bond angles CCX of the gauche and anti conformers.

[b]Calculated differences of gauche rotational angles referring to ignoring versus including bond configuration relaxation during internal rotation.

ing to these calculations. If there is then a mixture of, say, anti and gauche forms, as is often the case, and the relaxation of the bond configuration is ignored, this may lead to considerable errors in the determination of the gauche angle of rotation. The error may amount to as much as 11° according to calculations based on experimentally established relative abundances. The results are summarized in Table 1-7.

3. SUBSTITUENT EFFECTS

A. Substituted Benzene Derivatives

The impact of substituents on the ring geometry in substituted benzene derivatives has already been mentioned. In monosubstituted derivatives the ipso ring angle α is controlled mainly by the electronegativity χ of the substituent,[3,68] but separate correlations have been established for elements belonging to different rows of the periodic table as shown in Figure 1-17. These correlations may be used to estimate the electronegativity of functional groups from the observed ring distortions, or alternatively, to predict the ring distortions for functional groups of known electronegativity. Thus, for example, the α angle in $C_6H_5CF_3$ was predicted[3] to be 120.3° from the estimated electronegativity of the trifluoromethyl group 3.2 and was eventually determined by electron diffraction to be 119.5 ± 0.4°.[69] The geometries of about 20 monosubstituted benzene derivatives having at least a symmetry plane perpendicular to the ring plane have been determined by semiempirical complete neglect of differential overlap (CNDO/2) calculations.[70] The results were consistent with a large body of experimental data.[34] In a statistical analysis of the CNDO/2 results, the geometrical parameters of the shaded triangles of Figure 1-18 proved to have the smallest standard deviations. Hence a simplified model of ring distortion was proposed in which these parts of the molecule are unchanged and transferred from benzene. Table 1-8 presents the electron diffraction results for toluene and ethyl-

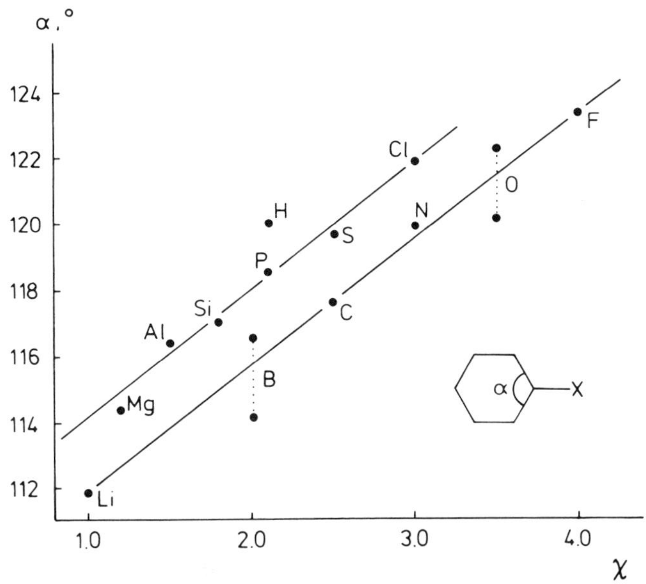

Figure 1-17. Empirical relationships between the ring ipso angles α and substituent electronegativities χ in monosubstituted benzene derivatives. (After Reference 3.)

benzene from an analysis[70] based on the proposed model. The results for toluene from a less constrained electron diffraction structure refinement[71] are also presented. Currently George, Bock, and Trachtman[72] are carrying out large basis set, ab initio calculations on monosubstituted and symmetrically para-disubstituted benzene derivatives. These calculations produce additional data, and moreover, facilitate our understanding of the origins of ring deformation.

The available X-ray crystallography data on selenium–phenyl derivatives indicate some correlation between the ring-to-selenium bond length and the ipso ring angle as shown in Figure 1-19. The data originate from the crystallographically independent phenyl groups in $(C_6H_5)_3Se(NCS)$,[74] $(C_6H_5)_3SeCl \cdot H_2O$,[74] and $(C_6H_5)_2SeC(C(O)CH_3)_2$.[75] It is seen that with increas-

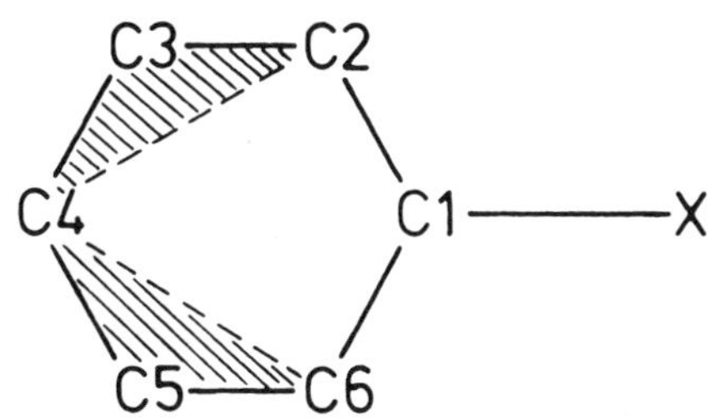

Figure 1-18. Illustration of the model for simplified description of the ring deformations in substituted benzene derivatives. The parameters related to the shaded areas may be assumed from benzene itself. (After Reference 70.)

Table 1-8. Ring Bond Lengths (Å) and Angles (°) in Toluene and Ethylbenzene from Electron Diffraction

Parameter	Toluene: Complete analysis[71]	Toluene: Simplified model[70] (see text)	Ethylbenzene: Simplified model[70] (see text)
Bond length			
C_1-C_2	1.408(5)	1.402(3)	1.400(4)
C_2-C_3	1.396(4)[a]	1.399[b]	1.399[b]
C_3-C_4	1.396(4)[a]	1.399[b]	1.399[b]
Bond angle			
C_6-C_1-C_2	118.7(4)	118.6(6)	118.3(10)
C_1-C_2-C_3	120.4(4)	121.0	121.1(6)
C_2-C_3-C_4	120.6(5)	120[b]	120[b]
C_3-C_4-C_5	119.4(6)	119.5	119.5(3)

[a]Assumed to be equal.
[b]Assumed from benzene.

ing Se—C (Ph) bond length there may be a tendency for increased ipso ring angle. If the bond lengthening is associated with increased electron-withdrawing ability of the substituent, then this phenomenon is consistent with the other structural changes observed in the substituted benzene derivatives.

Electron diffraction is especially suitable to determine the structures of symmetrically substituted benzene derivatives. Figure 1-20 presents the angular deformations and mean ring C—C bond lengths in symmetrically

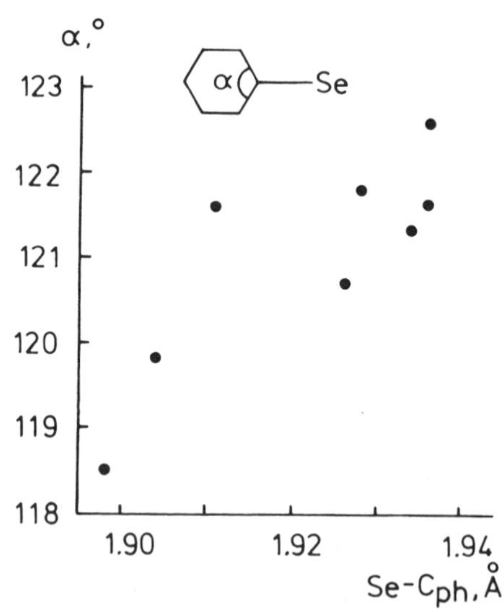

Figure 1-19. Se—C (Ph) bond lengths and the ring ipso angles α in selenium-substituted benzene derivatives. (After Reference 73; for sources, see text.)

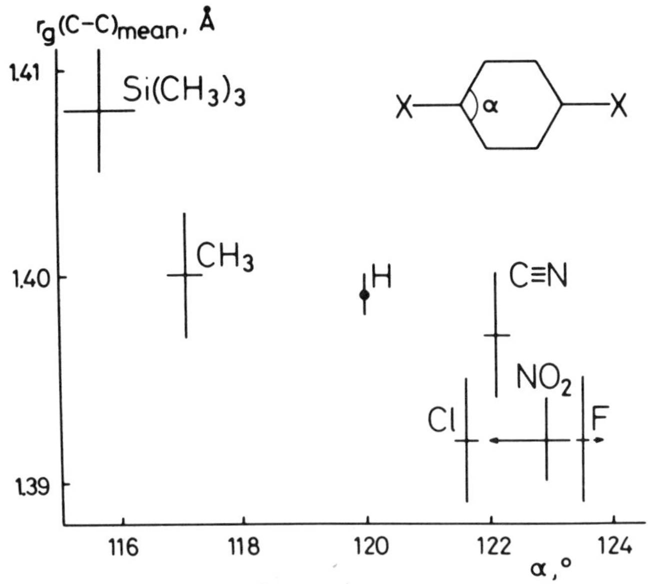

Figure 1-20. Ipso ring angles α and mean ring C–C bond lengths in symmetrically para-disubstituted benzene derivatives. (After Reference 23.)

para-disubstituted benzene derivatives. It is clear that the electronegative substituents somewhat compress the ring, the ipso angle increases, and the overall ring size decreases. The electropositive ligands somewhat elongate the ring, the ipso angle decreases, and the ring expands. The angular changes are considerable, but the bond length changes are at the limit of the present sensitivities of electron diffraction. The original reports for the XC_6H_4X para-disubstituted derivatives, where X = $Si(CH_3)_3$, CH_3, Cl, CN, NO_2, and F, appeared in References 23, 71, 76, 42, 77, and 78, respectively. Figure 1-21 presents a correlation between the ipso ring angles and the $C_1 \ldots C_4$ nonbonded distances in the same series of the para-disubstituted benzene derivatives.[23] The dashed least-squares line was obtained from equally weighted data points. However, the $\alpha/C_1 \ldots C_4$ interdependence reflects in part purely geometric relations represented by the solid line with a good approximation. The remaining difference seems to be rather characteristic.

The structures of 1,3,5-trisubstituted benzene derivatives investigated so far are in qualitative agreement with an additivity scheme based on available information on other substituted derivatives. The experimentally determined C–C bond lengths and CCC ipso angles are as follows:

1,3,5-$C_6H_3X_3$	C–C (Å)	C–C(X)–C (°)
X = CH_3 (Ref. 79)	1.401 ± 0.002	118.2 ± 0.2
X = Cl (Ref. 35)	1.392 ± 0.002	122.0 ± 0.2
X = F (Ref. 35)	1.390 ± 0.002	123.7 ± 0.2

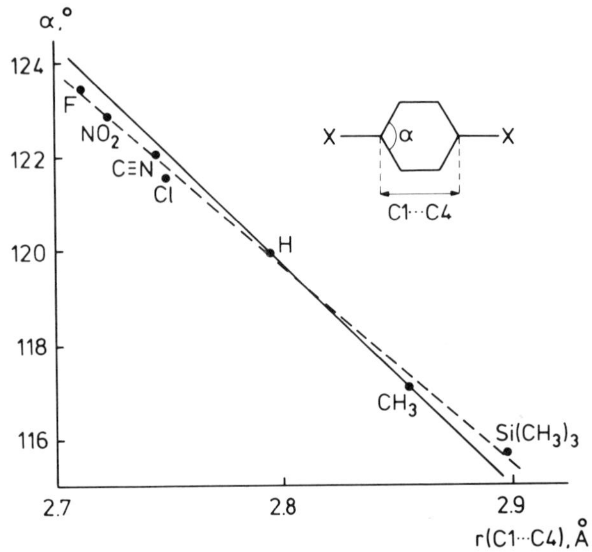

Figure 1-21. Correlation between the ring ipso angles α and the $C_1 \ldots C_4$ distances of symmetrically para-disubstituted benzene derivatives exclusively determined by electron diffraction in the vapor phase. (After Reference 23.)

B. Examples from Sulfur Stereochemistry

The molecular structure of a relatively large series of simple sulfur compounds has been determined in the vapor phase.[60] Substituent effects thus can be examined from several points of view: structural variations in analogous sulfides, sulfoxides, sulfones or in XSO_2Cl sulfonyl chlorides with changing X ligands, or the variations in S—C bond lengths alone under the influence of various substituents on the carbon atom,[80] and so on.

Comparison of SX_2, SOX_2, and SO_2X_2 molecular geometries with X = $(CH_3)_2N$, CH_3, CF_3, Cl, or F shows nonmonotonic angular changes in the XSX bond angles as one; then both lone pairs of electrons of sulfur are gradually replaced by S=O double bonds. The smallest bond angles occur in the sulfoxides, and they are accompanied by the longest bonds in the series.[60]

Ligand electronegativities χ_X and χ_Y can nicely be correlated with the structural changes in XSO_2Y sulfones. The variations in the geometrical parameters provided a possibility for estimating group electronegativities.[81,82]

Essentially all structural variations could be accounted for by the valence shell electron pair repulsion (VSEPR) model,[83] whose applicability to these systems was reinforced by quantum chemical calculations.[84] In addition to the electron pair repulsions, the importance of nonbonded interactions has been indicated by the constancy of 1,3-nonbonded separations (eg, O . . . O, F . . . F).[60] In fact, in going from the oxygen stereochemistry to sulfur and then to selenium and tellurium,[73] the relative importance of nonbonded

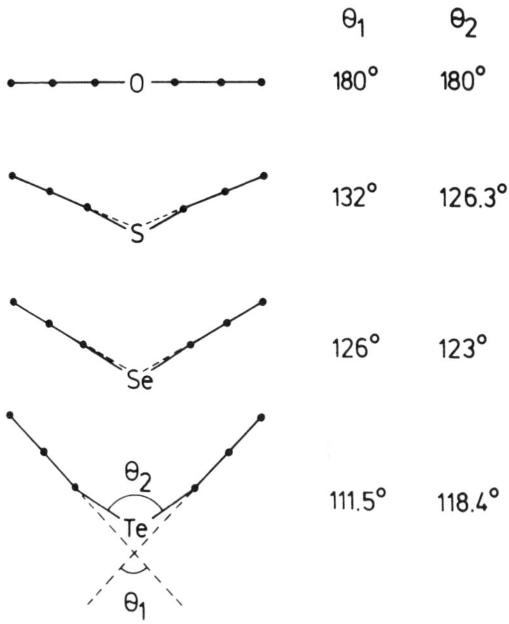

Figure 1-22. The dihedral angles between the mean plane of the phenylene rings in perfluoro-9,10-diheteroanthracene molecules. (After Reference 73; for sources, see text.)

interactions decreases as compared with that of electron pair repulsions. Figure 1-22 shows the dihedral angles between the mean planes of the phenylene rings in perfluoro-9,10-diheteroanthracene molecules. These are all crystal structure results,[85–88] so the conformational results should be interpreted with caution in terms of consequences of exclusively intramolecular interactions. The pattern of structural changes, however, seems to be very consistent. The molecule of dibenzo-*p*-dioxin is practically planar; the others have the so-called butterfly configuration (Figure 1-23). With larger chalcogen atoms and diminishing nonbonded repulsions, the dihedral angles

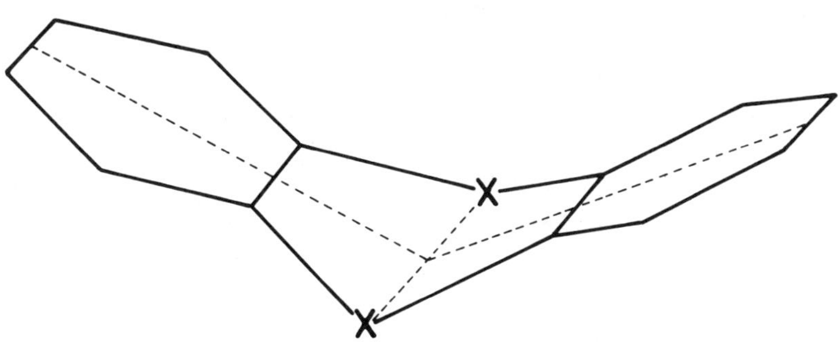

Figure 1-23. The "butterfly" model of the skeleton of perfluoro-9,10-diheteroanthracene molecules.

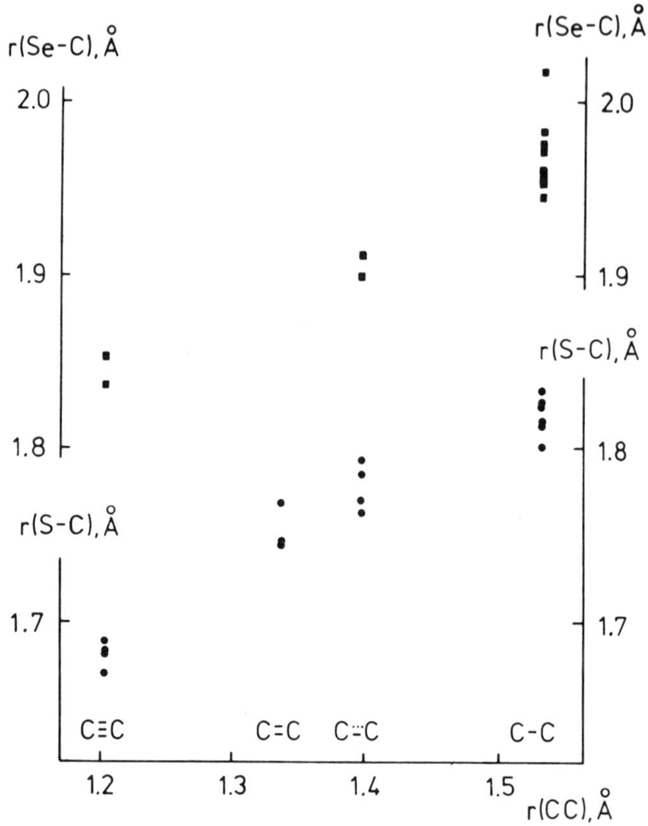

Figure 1-24. Variations of the S−C and Se−C bond lengths with the carbon bonding situation. The abscissa represents the C−C bond lengths in acetylene, ethylene, benzene, and ethane. (After Reference 73.)

decrease. The 2,6(3,5) C . . . C nonbonded distances of the central ring increase as follows: 2.31, 2.69, 2.85, and 3.06 Å.

The length of the sulfur–carbon bond changes sensitively with its environment,[80] and the fewer data on the selenium–carbon bond show very much the same, as illustrated in Figure 1-24 for a series of selected compounds. Of the available data on free molecules, the shortest sulfur–carbon bonds are in the vicinity of 1.55 Å (eg, 1.554(3) Å in $H_2C{=}C{=}S$[89]), and the longest near 1.90 Å (eg, 1.894(5) Å in $(CCl_3)_2SO_2$).[90] Even within the sulfide and sulfinyl series or within the sulfone and sulfonyl series, the variations are rather large depending on the environment. The S−C bond of CH_3SO_2Cl is 1.763(5) Å,[91] or 0.13 Å shorter than that of $(CCl_3)_2SO_2$ and almost that much shorter than this bond in CF_3SO_2Cl, namely, 1.856(6) Å.[92] Similarly large differences have been observed in the two kinds of S−C single bond of molecules with $H_3C{-}S{-}C{\equiv}$ bonding situation:

	$CH_3SC{\equiv}CH$ (Ref. 93)	$CH_3SC{\equiv}CSCH_3$ (Ref. 94)	CH_3SCN (Ref. 95)	CH_3CH_2SCN (Ref. 96)
$S-CH_3$ (Å)	1.814(3)	1.806(2)	1.824(2)	1.820
$S-C{\equiv}$ (Å)	1.683(3)	1.671(2)	1.684(3)	1.690

Ab initio and semiempirical calculations on the sulfonyl chloride structures paralleled and interpreted the variations in their S—Ç bond lengths.[97,98] While the inclusion of sulfur d orbitals in the basis set was imperative to account for the S=O bonds, the d-orbital participation was found to be much smaller in the S—C bonding. Significantly, however, even minor changes in the d-orbital involvement caused considerable changes in the S—C bond length.

Wherever electron pair repulsions are a dominating effect, the VSEPR model[83] has found wide applicability in predicting and interpreting the structural changes that occur upon substitution. On the other hand, the experimental observation upon substitution of some structural changes that were in seeming contradiction with this model facilitated a better understanding and even some generalization of the VSEPR model.[84,99,100]

4. CONCLUDING REMARKS

The reliable determination and interpretation of small structural differences are matters of far-reaching importance in structural chemistry. Furthermore they enhance the possibilities of structural chemistry in applications to and interactions with other branches of chemistry. An example for the utilization of accurate geometrical data consists of the bond length/bond strength–dissociation energy correlations by Szabó and Konkoly Thege.[101] These investigators have shown that a 1-kcal/mol change in the dissociation energy of a C—C bond is accompanied by an 0.0025 Å change in bond length. The corresponding change for inorganic bonds may be as much as 0.026 Å. Szabó and Konkoly Thege[101] have also demonstrated how the correlations involving geometry and energetics have improved over the years[101–103] as more accurate data have become available.

Data banks and bibliographies greatly facilitate the utilization of structural data. The Cambridge Structural Database[104] is being used as a search tool[105] for systematic analysis of crystal molecular structures. For vapor phase structural data, critical compilations[106] and annotated bibliographies[107,108] are available from the Ulm Center for Structural Documentation. Extensive compilations of the quantum chemical calculations augment the experimental information.

The ever-increasing accuracy of structural parameters is to be accompanied by an understanding of their physical meaning, as emphasized in Section 1. Molecules exhibiting low-frequency, large-amplitude motion, especially when studied under high-temperature conditions, may pose basic questions even as to their shape and symmetry. A recent discussion of the linearity of iron dichloride and other first-row transition metal dihalides illustrates this point.[109]

We conclude our discussion of molecular structure by citing a recent statement of Hoffmann:

> There is no more basic enterprise in chemistry than the determination of the geometrical structure of a molecule. Such a determination, when it is well done, ends all speculation as to the structure and provides us with the starting point for the understanding of every physical, chemical and biological property of the molecule.[110]

ACKNOWLEDGMENTS

The authors, who were on leave from the Hungarian Academy of Sciences in Budapest, express their appreciation to the University of Connecticut and to their colleagues for hospitality extended to them.

REFERENCES

1. See, eg, Kuchitsu, K. In "Diffraction Studies on Non-Crystalline Substances", Hargittai, I.; and Orville-Thomas, W.J. Eds.; Akadémiai Kiadó: Budapest, and Elsevier: Amsterdam, 1981.
2. Keidel, F.A.; Bauer, S.H. *J. Chem. Phys.* **1956,** *25,* 1218.
3. Domenicano, A.; Hargittai, I.; Schultz, G. Eighth Austin Symposium on Molecular Structure. Abstracts, MA7, Austin, Tex., 1980.
4. Kitaigorodskii, A.I. In "Advances in Structure Research by Diffraction Methods", Vol. 3, Brill, R.; and Mason, R. Eds.; Pergamon Press: Oxford, and Vieweg & Sohn: Braunschweig, 1970, p. 173. Kitaigorodskii, A.I. "Molekulyarnie Kristalli". Nauka: Moscow, 1971.
5. Bernstein, J.; Hagler, A.T. *J. Am. Chem. Soc.* **1978,** *100,* 673.
6. Charbonneau, G.-P.; Delugeard, Y. *Acta Crystallogr. Sect. B* **1977,** *33,* 1586. Cailleau, H.; Bauduor, J.L.; Zeyen, C.M.E. ibid, **1979,** *35,* 426. Takeuchi, H.; Suzuki, S.; Dianoux, A.J.; Allen, G. *Chem. Phys.* **1981,** *55,* 153.
7. Bastiansen, O. *Kristallografiya* **1981,** *26(5),* 935. First investigation: Bastiansen, O. *Acta Chem. Scand.* **1952,** *6,* 205.
8. Martin, T.P. *Phys. Rep.* **1983,** *95,* 167.
9. Barlak, T.M.; Campana, J.E.; Wyatt, J.R.; Dunlap, B.I.; Colton, R.J. *Int. J. Mass Spectrom. Ion Phys.* **1983,** *46,* 523.
10. Barlak, T.M.; Campana, J.E.; Wyatt, J.R.; Colton, R.J. *J. Phys. Chem.* **1983,** *87,* 3441. Campana, J.E.; Dunlap, B.I. *Int. J. Mass Spectrom. Ion Phys.* **1984,** *57,* 103.
11. Wells, A.F. "Structural Inorganic Chemistry", 5th ed.; Clarendon Press: Oxford, 1984.
12. Shelton, R.A.J. *Trans. Faraday Soc.* **1961,** *57,* 2113.
13. Bauer, S.H.; Ino, T.; Porter, R.F. *J. Chem. Phys.* **1960,** *33,* 685.
14. Akishin, P.A.; Rambidi, N.G. *Zh. Strukt. Khim.* **1960,** *5,* 23.
15. Lide, D.R.; Cahill, P.; Gold, L.P. *J. Chem. Phys.* **1964,** *40,* 156.
16. Schultz, G.; Hargittai, I. *Acta Chim. Hung.* **1973,** *75,* 381.
17. Hayashi, M.; Shiro, Y.; Oshima, T.; Murata, H. *Bull. Chem. Soc. Japan* **1965,** *38,* 1734.
18. Melmed, K.M.; Coucouvanis, D.; Lippard, S.J. *Inorg. Chem.* **1973,** *12,* 232.
19. Mamaeva, G.I.; Hargittai, I.; Spiridonov, V.P. *Inorg. Chim. Acta* **1973** *25,* L123.

20. Almenningen, A.; Bastiansen, O.; Haaland, A. *J. Chem. Phys.* **1964,** *40,* 3434.
21. Wong, C.-H.; Lee, T.-Y.; Chao, K.-J.; Lee, S. *Acta Crystallogr. Sect. B* **1972,** *28,* 1662.
22. Marynick, D.S. *J. Am. Chem. Soc.* **1977,** *99,* 1436. Chiu, N.S.; Schäfer, L. ibid, **1978,** *100,* 2604. Jemmis, E.D.; Alexandratos, S.; Schleyer, P.v.R.; Streitwieser, A.; Schaefer, H.F. ibid, **1978,** *100,* 5695. Demuynck, J.; Rohmer, M.M. *Chem. Phys. Lett.* **1978,** *54,* 567. Gleiter, R.,; Böhm, M.C.; Haaland, A.; Lusztyk, J. *J. Organometal. Chem.* **1979,** *170,* 285. Boldyrev, A.I.; Charkin, O.P. *Zh. Strukt. Khim.* **1977,** *18,* 783 (contains information on the beryllocene structure in its appendix). Skancke, P.N. *Int. J. Quant. Chem.* **1984,** *26,* 729.
23. Rozsondai, R.; Zelei, B.; Hargittai, I. *J. Mol. Struct.* **1982,** *95,* 187.
24. Menczel, G.; Kiss, J. *Acta Crystallogr. Sect. B* **1975,** *31,* 1787.
25. (a) Ponomarev, V.I.; Filipenko, O.S.; Atovmyan, L.O. *Kristallografiya* **1976,** *21,* 392. (b) Cruickshank, D.W.J.; Sparks, R.A. *Proc. R. Soc. London, Ser. A* **1960,** *258,* 270.
26. (a) Ketkar, S.N.; Fink, M. *J. Mol. Struct.* **1981,** *77,* 139. (b) Almenningen, A.; Bastiansen, O.; Dyvik, F. *Acta Crystallogr.* **1961,** *14,* 1056.
27. Sellers, H.; Boggs, J.E. *J. Mol. Struct.* **1981,** *74,* 137.
28. Nygaard, L.; Bojesen, I.; Pedersen, T.; Rastrup-Andersen, J. *J. Mol. Struct.* **1968,** *2,* 209.
29. Portalone, G.; Schultz, G.; Domenicano, A.; Hargittai, I., *J. Mol. Struct.* **1984,** *118,* 53.
30. Colapietro, M.; Domenicano, A.; Pela Ceccarini, G. *Acta Crystallogr. Sect. B* **1979,** *35,* 890.
31. Colapietro, M.; Domenicano, A.; Marciante, C.; Portalone, G. *Acta Crystallogr. Sect. B* **1981,** *37,* 387.
32. Lounila, J.; Väänänen, T. *Mol. Phys.* **1983,** *49,* 859.
33. Boggs, J.E.; Pang, F.; Pulay, P. *J. Comput. Chem.* **1982,** *3,* 344.
34. Domenicano, A.; Murray-Rust, P. *Tetrahedron Lett.* **1979,** 2283.
35. Almenningen, A.; Hargittai, I.; Brunvoll, J.; Domenicano, A.; Samdal, S. *J. Mol. Struct.* **1984,** *116,* 199.
36. Brunvoll, J.; Colapietro, M.; Domenicano, A.; Marciante, C.; Portalone, G.; Hargittai, I. *Z. Naturforsch. B,* **1984,** *39,* 607.
37. Colapietro, M.; Domenicano, A.; Marciante, C.; Portalone, G. To be published.
38. Brunvoll, J.; Exner, O.; Hargittai, I.; Kolonits, M.; Scharfenberg, P. *J. Mol. Struct.* **1984,** *117,* 317.
39. Brunvoll, J.; Hargittai, I. *J. Mol. Struct.* **1976,** *30,* 361.
40. Van Eijck, B.P.; Hargittai, I.; Mayer, I. *J. Mol. Struct.* **1980,** *69,* 301.
41. Vajda, E.; Hargittai, I. *Z. Naturforsch. A* **1983,** *38,* 765.
42. Colapietro, M.; Domenicano,A.; Portalone, G.; Schultz, G.; Hargittai, I. *J. Mol. Struct.* **1984,** *112,* 141.
43. Colapietro, M.; Domenicano, A.; Portalone, G.; Torrini, I.; Hargittai, I.; Schultz, G. *J. Mol. Struct.* **1984,** *125,* 19.
44. Van Rij, C.; Britton, D. *Acta Crystallogr. Sect. B,* **1977,** *33,* 1301.
45. Drück, U.; Littke, W. *Acta Crystallogr. Sect B,* **1978,** *34,* 3095.
46. Hargittai, M.; Hargittai, I. *J. Mol. Struct.* **1977,** *39,* 79.
47. Clippard, P.H.; Hanson, J.C.; Taylor, R.C. *J. Cryst. Mol. Struct.* **1971,** *1,* 363.
48. Párkányi, L.; Bihátsi, L.; Hencsei, P. *Cryst. Struct. Commun.* **1978,** *7,* 435.
49. Shen, Q.; Hilderbrandt, R.L. *J. Mol. Struct.* **1980,** *64,* 257.
50. Hargittai, M.; Hargittai, I. "The Molecular Geometries of Coordination Compounds in the Vapour Phase". Akadémiai Kiadó: Budapest, and Elsevier: Amsterdam, 1977.
51. Schultz, G.; Hargittai, I.; Kapovits, I.; Kucsman, A. *J. Chem. Soc. Faraday Trans. 2* **1984,** *80,* 1273.
52. Zaripov, N.M.; Popik, M.V.; Vilkov, L.V.; Mannafov, T.G. *Zh. Strukt. Khim.* **1980,** *21,* 37.
53. Shishkov, I.F. Thesis, Moscow State University, 1980. Cited in Batyukhnova, O.G.; Sadova, N.I.; Vilkov, L.V.; Pankrushev, Yu.A. *J. Mol. Struct.* **1983,** *97,* 153., Høg, J.H. Thesis, University of Copenhagen, 1971.
54. Hargittai, M.; Samdal, S.; Seip, R. *J. Mol. Struct.* **1981,** *71,* 147.
55. Truter, M. *J. Chem. Soc.* **1960,** 997.
56. Sheringer, C. *Acta Crystallogr. Sect. A,* **1980,** *36,* 814.
57. Kutoglu, A.; Sheringer, C.; Meyer, H.; Schweig, A. *Acta Crystallogr. Sect. B* **1982,** *38,* 2626.
58. Kuczkowski, R.L.; Suenram, R.D.; Lovas, F.J. *J. Am. Chem. Soc.* **1981,** *103,* 2561.

59. Pascard-Billy, C. *Acta Crystallogr.* **1965,** *18,* 827.
60. Hargittai, I. "The Structure of Volatile Sulphur Compounds". Akadémiai Kiadó: Budapest, and Reidel: Dordrecht, Holland, 1985.
61. Zorkii, P.M.; Koptsik, V.A. In "Sovremennie Problemi Fizicheskoi Khimii", Gerasimov, Ya.I.; and Akishin, P.A. Eds.; Izd. Moscov. Univ.: Moscow, 1979.
62. Zorkii, P.M.; Razumaeva, A.E. *Zh. Strukt. Khim.* **1979,** *20,* 463.
63. Belsky, V.K.; Zorkii, P.M. *Acta Crystallogr. Sect. A* **1977,** *33,* 1004.
64. Chernikova, N.Yu.; Lavut, E.E.; Zorkii, P.M. *Koord. Khim.* **1979,** *5,* 1266.
65. Lavut, E.E.; Zorkii, P.M.; Chernikova, N.Yu. *Zh. Strukt. Khim.* **1981,** *22,* 89.
66. Hagler, A.T.; Bernstein, *J. Am. Chem. Soc.* **1978** *100,* 6349. Bernstein, J.; Anderson, T.E.; Eckhardt, C.J. ibid. **1979,** *101,* 541. Bernstein, J. *Acta Crystallogr. Sect. B* **1979,** *35,* 360. Bernstein, J.; Engel, Y.M.; Hagler, A.T. *J. Chem. Phys.* **1981,** *75,* 2346. Bar, I.; Bernstein, J. *Acta Crystallogr. Sect. B* **1981,** *37,* 569; *J. Phys. Chem.* **1982,** *86,* 3223.
67. Scharfenberg, P.; Hargittai, I. *J. Mol. Struct.* **1984,** *112,* 65.
68. Domenicano, A.; Vaciago, A.; Coulson, C.A. *Acta Crystallogr. Sect. B,* **1975,** *31,* 1630; See also Domenicano, A.; Vaciago, A.; Coulson, C.A. ibid, **1975,** *31,* 221; Domenicano, A.; Vaciago, A. ibid, **1979,** *35,* 1382; Domenicano, A.; Murray-Rust, P.; Vaciago, A. ibid, **1983,** *39,* 457.
69. Schultz, G.; Hargittai, I.; Seip, R. *Z. Naturforsch. A.* **1981,** *36,* 669.
70. Scharfenberg, P.; Hargittai, I.; Rozsondai, B. Ninth Austin Symposium on Molecular Structure. Abstracts A16. Austin, Tex., 1982.
71. Domenicano, A.; Schultz, G.; Kolonits, M.; Hargittai, I. *J. Mol. Struct.* **1979,** *53,* 197.
72. Bock, C.-W.; Trachtman, M.; George,P. *J. Mol. Struct.* (THEOCHEM), In press, 1985.
73. Hargittai, I.; Rozsondai, B. In "Organic Chemistry of Se and Te Containing Functional Groups", Patai, S.; and Rappoport, Z. Eds.; Wiley: New York, 1985.
74. Ash, P.; Lee, J.-S.; Titus, D.D.; Mertes, K.B.; Ziolo, R.F. *J. Organomet. Chem.* **1977,** *135,* 91. Lee, J.S.; Titus, D.D. *J. Cryst. Mol. Struct.* **1976,** *6,* 279.
75. Wei, K.-T.H.; Paul, I.C.; Chang, M.-M.Y.; Musher, J.I. *J. Am. Chem. Soc.* **1974,** 96, 4099.
76. Schultz, G.; Hargittai, I.; Domenicano, A. *J. Mol. Struct.* **1980,** *68,* 281.
77. Penionzhkevich, N.P.; Sadova, N.I.; Popik, M.I.; Vilkov, L.V.; Pankrushev, Yu.A. *Zh. Strukt. Khim.* **1979,** *20,* 603.
78. Domenicano, A.; Schultz, G.; Hargittai, I. *J. Mol. Struct.* **1982,** *78,* 97.
79. Almenningen, A.; Hargittai, I.; Samdal, S.; Brunvoll, J.; Domenicano, A.; Lowrey, A. *J. Mol. Struct.* **1983,** *96,* 373.
80. Hargittai, I. In "Organic Sulfur Chemistry: Theoretical and Experimental Advances", Csizmadia, I.G.; Mangini, A.; and Bernardi, F., (Eds).; Elsevier: Amsterdam, 1985.
81. Hargittai, I. *Z. Naturforsch. A* **1979,** *34,* 755.
82. Hargittai, I.; Bliefert, C. *Z. Naturforsch. B* **1980,** *35,* 1953.
83. Gillespie, R.J. "Molecular Geometry". Van Nostrand-Reinhold: London, 1972.
84. Schmiedekamp, A.; Cruickshank, D.W.J.; Skaarup, S.; Pulay, P.; Hargittai, I.; Boggs, J.E. *J. Am. Chem. Soc.* **1979,** *101,* 2002.
85. Rainville, D.P.; Zingaro, R.A.; Meyers, E.A. *J. Fluorine Chem.* **1980,** *16,* 245.
86. Rainville, D.P.; Zingaro, R.A.; Meyers, E.A. *Cryst. Struct. Commun.* **1980,** *9,* 771.
87. Rainville, D.P.; Zingaro, R.A.; Meyers, E.A. *Cryst. Struct. Commun.* **1980,** *9,* 909.
88. Rainville, D.P.; Zingaro, R.A.; Meyers, E.A. *Cryst. Struct. Commun.* **1980,** *9,* 291.
89. Landsberg, B.M. *Chem. Phys. Lett.* **1979,** *60,* 265.
90. Hargittai, M.; Vajda, E.; Nielsen, C.J.; Klaeboe, P.; Seip, R.; Brunvoll, J. *Acta Chem. Scand. A* **1983,** *37,* 341.
91. Hargittai, M.; Hargittai, I. *J. Chem. Phys.* **1973,** *59,* 2513.
92. Brunvoll, J.; Hargittai, I.; Kolonits, M. *Z. Naturforsch. A* **1978,** *33,* 1236.
93. Van Eijck, B.P. Dissertation, University of Utrecht, 1971.
94. Beagley, B.; Ulbrecht, V.; Katsumata, S.; Lloyd, D.R.; Connor, J.A.; Hudson, G.A. *J. Chem. Soc. Faraday Trans. 2* **1977,** 73, 1278.
95. Dreizler, H.; Rudolph, H.D.; Schleser, H. *Z. Naturforsch. A* **1970,** *25,* 1643.
96. Bjørseth, A.; Marstokk, K.M. *J. Mol. Struct.* **1972,** *11,* 15.
97. Mayer, I.; Hargittai, I. *Z. Naturforsch. A* **1979,** *34,* 911.
98. Mayer, I.; Révész, M.; Hargittai, I. *Acta Chim. Hung.* **1983,** *114,* 159.
99. Hargittai, I. *J. Mol. Struct.* **1979,** *56,* 287.
100. Hargittai, I. *Inorg. Chem.* **1982,** *21,* 4334.
101. Szabó, Z.G.; Konkoly Thege, I. *Acta Chim. Hung.* **1975,** *86,* 127.

102. Szabó, Z.G. *Z. Elektrochem.* **1957,** 61, 1083.
103. Szabó, Z.G. Special Publication of the Chemical Society, No. 16, pp. 113–116, London, **1962.** Szabó, Z.G. *Z. Phys. Chem. Neue Folge* **1967,** *55,* 1.
104. Allen, F.H.; Bellard, S.; Brice, M.D.; Cartwright, B.A.; Doubleday, A.; Higgs, H.; Hummelink, T.; Hummelink-Peters, B.G.; Kennard, O.; Motherwell, W.D.S.; Rodgers, J.R.; Watson, D.G. *Acta Crystallogr. Sect. B* **1979,** *35,* 2331.
105. See, eg, Allen, F.H.; Kennard, O.; Taylor, R. *Acc. Chem. Res.* **1983,** *16,* 146.
106. Landolt-Börnstein, Numerical Data and Functional Relationships in Science and Technology (new series), Vol. 7. "Structure Data of Free Polyatomic Molecules," Hellwege, K.-H.; and Hellwege, A.M., Eds.; Springer-Verlag: Berlin, 1976.
107. Starck, B.; Mutter, R.; Spreter, C.; Kettermann, K.; Boggs, A.; Botskor, M.; Jones, M. "Bibliography of Microwave Spectroscopy, 1945–1975", Physics Data No. 9-1. Fachinformationszentrum: Karlsruhe, 1977.
108. Buck, I.; Maier, E.; Mutter, R.; Seiter, U.; Spreter, C.; Starck, B.; Hargittai, I.; Kennard, O.; Watson, D.G.; Lohr, A.; Pirzadeh, T.; Schirdewahn, H.G.; Majer, Z. "Bibliography of Gas Phase Electron Diffraction, 1930–1979," Physics Data No. 21-1. Fachinformationszentrum: Karlsruhe, 1981.
109. Hargittai, M.; Hargittai, I. *J. Mol. Spectrosc.* **1984,** *108,* 155.
110. Hoffmann, R. Foreword, In Vilkov, L.V.; Mastryukov, V.S.; Sadova, N.I. "Determination of the Geometrical Structure of Free Molecules". MIR Publishers: Moscow, 1983.

CHAPTER 2

Pathways to Graphite: Properties of Very Large Polybenzenoid Hydrocarbons

Stephen E. Stein and Robert L. Brown

National Bureau of Standards, Gaithersburg, Maryland

CONTENTS

© 1987 VCH Publishers, Inc.
MOLECULAR STRUCTURE AND ENERGETICS, Vol. 2

1. INTRODUCTION

Graphite and many other carbonaceous materials are composed of highly condensed, largely planar polyaromatic molecules, referred to as polynuclear aromatic hydrocarbons or polycyclic aromatic hydrocarbons. There is no reason to believe that reactions of these structures cannot be described in conventional chemical terms. However, the inability to isolate such structures for individual study as well as for determining their structural variability has resulted in these reactions being described in very general, phenomenological terms. Reactivity is thought to depend on "active site" density, although the chemical nature of these sites is a matter of speculation.

Because of our interest in the chemistry of carbonization and our view that reactions of carbonaceous materials are likely to follow straightforward organic reaction mechanisms, though perhaps with unique energetic and steric features, we have been applying semiempirical organic theories to very large polyaromatic species. Some of these theories seem very well suited for this purpose, since they are computationally facile (very large molecules can be treated), have been very reliable for polybenzenoid structures, and can be tested and parameterized by a wide range of background data for ordinary polyaromatic hydrocarbon molecules. By using these theories to develop structure–reactivity relations, we hope eventually to develop well-grounded working models for reactions of carbonaceous materials.

Some 30 years ago, there was considerable interest in applying molecular orbital (MO) theories to very large polyaromatic molecules. Coulson and coworkers were particularly concerned with this problem.[1] A number of the papers at a symposium entitled "From Benzene to Graphite" examined the reasonableness of this approach as well as the importance of edge structures in reactivity.[2,3] However, at the time of this symposium high-speed computers were not available and the scope and reliability of semiempirical calculations were not fully known. The largest molecule studied contained 60 atoms, a size too small to have well-developed edges.

We have proceeded in the spirit of this symposium, but with the advantage of having modern computers and a range of more thoroughly tested organic theories. In addition, we have found an atom-labeling scheme that greatly facilitates the computations,[4] thereby enabling calculations to be done on even larger molecules.

Initial results of our calculations using the simplest theories have just been reported.[5] Calculations were done on hexagonally symmetric molecules using structure–resonance theory (SRT)[6] to evaluate resonance energies, perturbation molecular orbital (PMO) theory[7a,b] to obtain electron localization energies and Hückel molecular orbital (HMO) theory[8] to derive approximate electron densities in free radicals. The key general finding of these studies was that the edge structures can have a dramatic effect on the properties of even very large, highly condensed aromatic molecules (> 200 C atoms). In accord with expectations based on known reactivities of polyaromatic molecules, phenanthrenelike edges were the most stable and anthracenelike edges were the least.

This chapter is devoted to a more thorough application of SRT and PMO theories to these large polyaromatic molecules. Resonance energies, Pauling bond orders,[16] ring aromaticities,[11] and electron localization energies are examined. In addition to providing structure–reactivity relations and testing and reasonableness of predictions of the simpler theories, this work is intended to lay a logical framework on which to build future calculations using more sophisticated theories.

As in our previous analysis,[5] this work treats four series of hexagonally symmetric polybenzenoid molecules, each with one of the following edge structures:

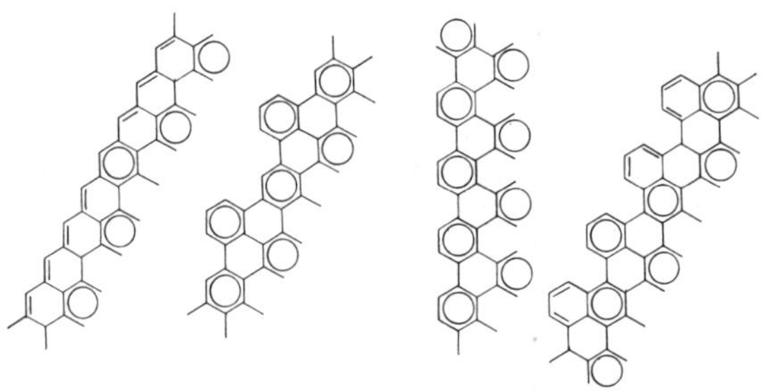

1 **2** **3** **4**

This treatment enables the separation of edge and size effects. Early members of these series are shown in Figure 2-1. At first glance, it might appear that a large number of other arrangements of rings at the edges are possible. However, we find that the structures given represent most of the imaginable local arrangements.

As suggested by the double bonds and aromatic circles shown in structures **1–4,** molecules having edge type 2 or 3 can be logically viewed as being created by the fusion of parasubstituted polyphenyl chains. Molecules with edge types 1 or 4 cannot be created in this way; they contain "extra" double bonds. According to the ideas of Clar,[9] molecules in series 2 or 3 are "fully benzenoid" and should therefore be more stable than those in 1 or 4.

The two theories used in this chapter are mathematically related, both using counts of Kékulé structures as their primary input. Structure–resonance theory is used to compute resonance energies, Pauling bond orders, and ring aromaticities. Its estimates of the first two of these properties have been shown to be in excellent agreement with both experimental data and results of more sophisticated self-consistent field–molecular orbital

Figure 2-1. Homologous series of hexagonally symmetric molecules. In each series, a particular molecule can be identified by a generating index K. Benzene constitutes the first member of each series and corresponds to $K = 0$. Various molecular properties can be expressed in terms of this index. As an example the following formulas give the numbers of carbon atoms N_C in each molecule:

Series 1:	$N_C = 6(1 + 2K + K^2)$	
Series 2:	$N_C = 6(1 + 3K + K^2)$,	$K = 0, 3, 6, \ldots$
	$N_C = 6(3 + 3K + K^2)$,	$K = 1, 2, 4, 5, 7, 8, \ldots$
Series 3:	$N_C = 6(1 + 3K + 3K^2)$	
Series 4:	$N_C = 6(1 + 3K + 4K^2)$	

In each series except 2, all the molecules have the same corner structure. Molecules in series 2 have three different corner structures: one for $K = 0, 3, 6, \ldots$, a second for $K = 1, 4, 7, \ldots$, and a third for $K = 2, 5, 8, \ldots$.

(SCFMO) calculations.[10,11] Ring aromaticity estimates are less easily compared to either experiment or MO calculations.

PMO theory is used for electron localization energies, a measure of the energy required to fix a delocalized π electron at a specific position in a molecule. SRT is not used for this purpose, since it yields values that diverge with increasing molecular size.[5]

2. RESONANCE ENERGY

A useful comparative measure of the thermodynamic stability (or aromaticity) of a π-electron network is the resonance energy per π electron (REPE).

In this chapter, two forms of SRT are used to obtain REPE values. The first form (SRT-1) simply assumes that the total resonance energy of a polybenzenoid molecule is proportional to the logarithm of its number of Kékulé structures.[6] This is a quantitative expression of the organic chemist's rule of thumb that as the number of Kékulé structures of a species increases, so does its stability. REPE is simply proportional to the number of Kékulé structures per π electron, KSPE, which is proportional to the log of the number of Kékulé structures divided by the number of π electrons. The second form of SRT (SRT-2) is more complex.[12] It finds resonance energies by adding contributions from individual parts of the molecule (see Appendix A for details). REPE values obtained by SRT-1 and SRT-2 for our four homologous series are plotted as a function of molecular size in Figure 2-2. The reciprocal of the square root of the number N_C of carbons was used for the abscissa. This was done to expand the scale near the origin (ie, as $N_C \rightarrow \infty$). The largest species examined by SRT-1 contained roughly 20,000 C atoms; for SRT-2 the largest had about 2000 C atoms.

Both forms of SRT predict the same relative stability, namely, series 1 $<$ series 4 $<$ series 2 $\simeq$ series 3. The intersection of each curve with the y-axis corresponds to REPE for an arbitrarily large molecule. Although the rate of approach to these limiting values differs for the two forms of SRT, relative limiting values are nearly the same. For series 2 (or 3), 4, and 1 these values are 1:0.94:0.81 for SRT-1 and 1:0.92:0.78 for SRT-2. The limit for series 4 lies one-third of the way between the limits for series 2 and series 1.

Both these forms of SRT make the physically unreasonable prediction that in an arbitrarily large layer, the average energy per electron depends on the structures at the edges. We view this as an artifact of SRT that casts doubt on its quantitative accuracy for predicting resonance energies of large, highly condensed layers. Moreover, we have not been able to devise a reasonable "correction" to SRT to remedy this problem. Since SRT resonance energies closely parallel those of SCFMO calculations, it is of interest to find whether both methods fail in this way, or the point at which these two theories begin to diverge. We are now pursuing this problem.

Group additivity is an alternative method for estimating stabilities of polyaromatic molecules.[13] The heat of formation of a polyaromatic molecule is estimated with this method by summing up the contributions from the different groups composing the molecule. These groups are defined in terms of two C atom types: a benzenelike C atom C_B, which belongs to only one ring, and a polyaromatic C atom C_{BF}, which is a member of two or three fused rings. Group values for $C_B-(H)$ were derived from benzene; $C_{BF}-(C_B)_2(C_{BF})$ from naphthalene; $C_{BF}-(C_B)(C_{BF})_2$ from phenanthrene, and $C_{BF}-(C_{BF})_3$ from the heat of sublimation of graphite. Since group additivity accounts for nearest neighbor effects only and aromaticity is generally thought to be a long-range property, it is perhaps surprising that it should work at all for polyaromatic molecules. It has been shown,[13] however, to

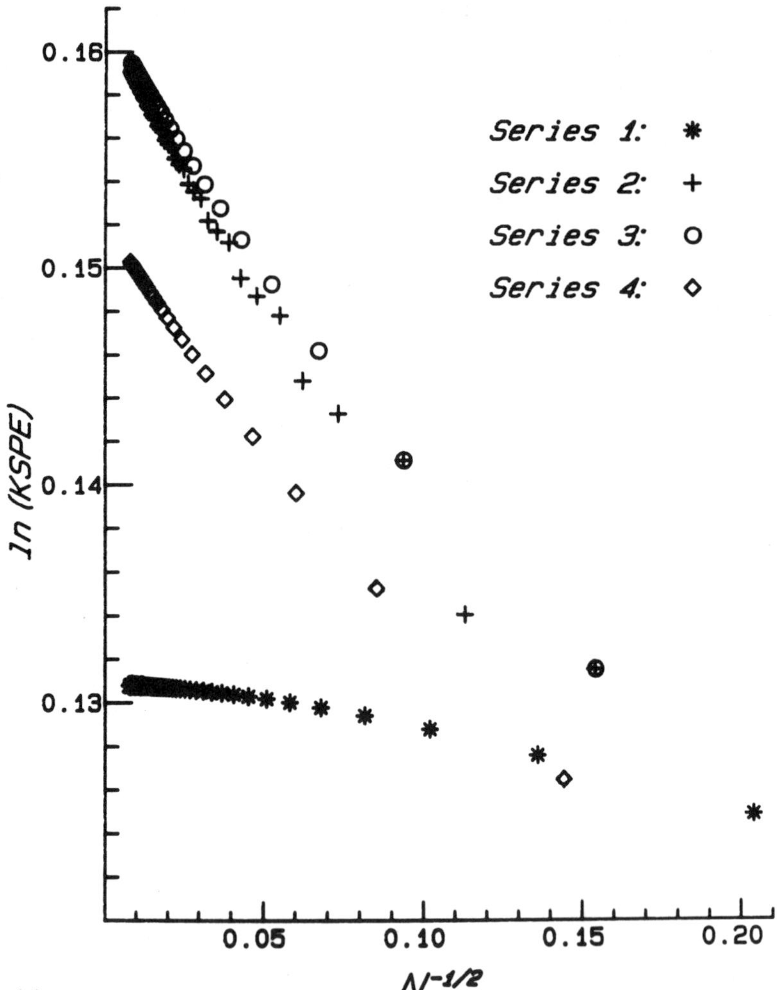

(a)

Figure 2-2. (*a*) Resonance energies per π electron from SRT-1. The logarithm of the number of Kékulé structures divided by the number of π electrons is plotted as a function of the reciprocal of the square root of the number N_C of carbon atoms. The origin corresponds to an infinitely large molecule. (*b*) Resonance energies per π electron from SRT-2, plotted as a function of reciprocal molecular size were calculated from the formula given in Appendix A.

predict available heats of formation as well as does molecular orbital and related theories using the same number of "adjustable parameters." (For a further discussion see Chapter 1 in Volume 4 of this series.) However, it cannot really predict aromaticity. This is evident from the fact that in the absence of strain it predicts the following two isomers to be nearly equally

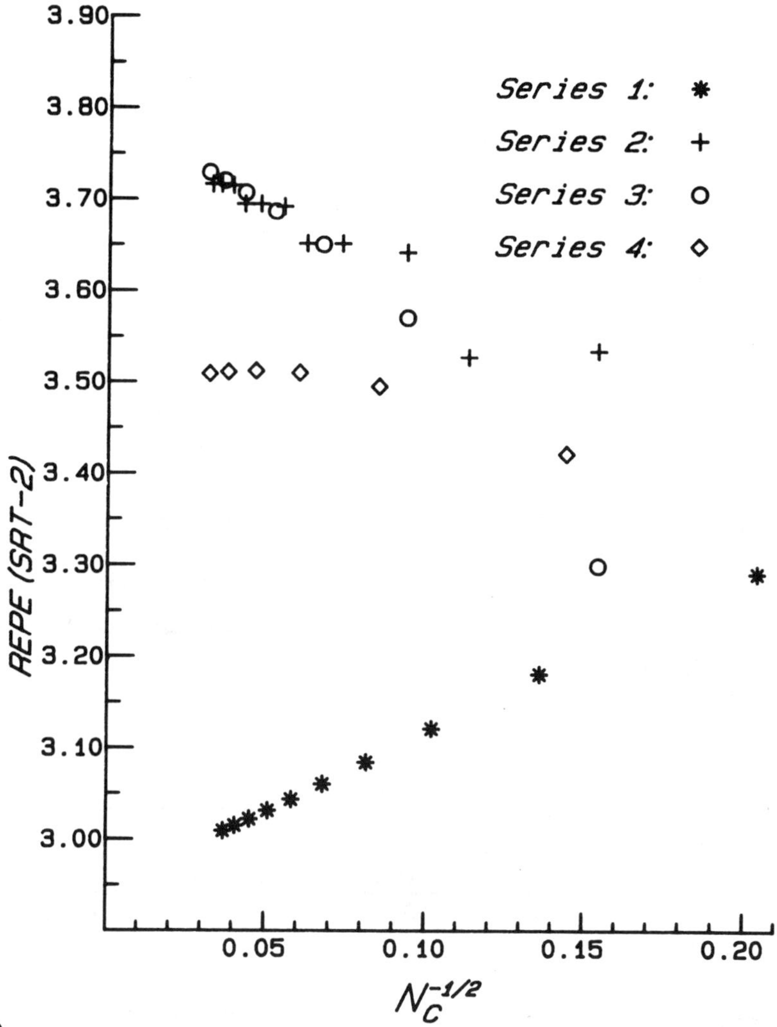

(b)

stable, although MO theory and experiment show the second to be nonaromatic and far less stable than the first.[14]

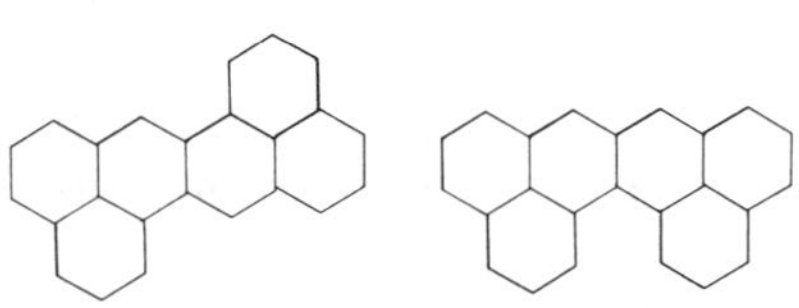

Heats of formation per π electron predicted by group additivity (GA) and SRT-2 are given in Figure 2-3 for the molecules in the different homologous

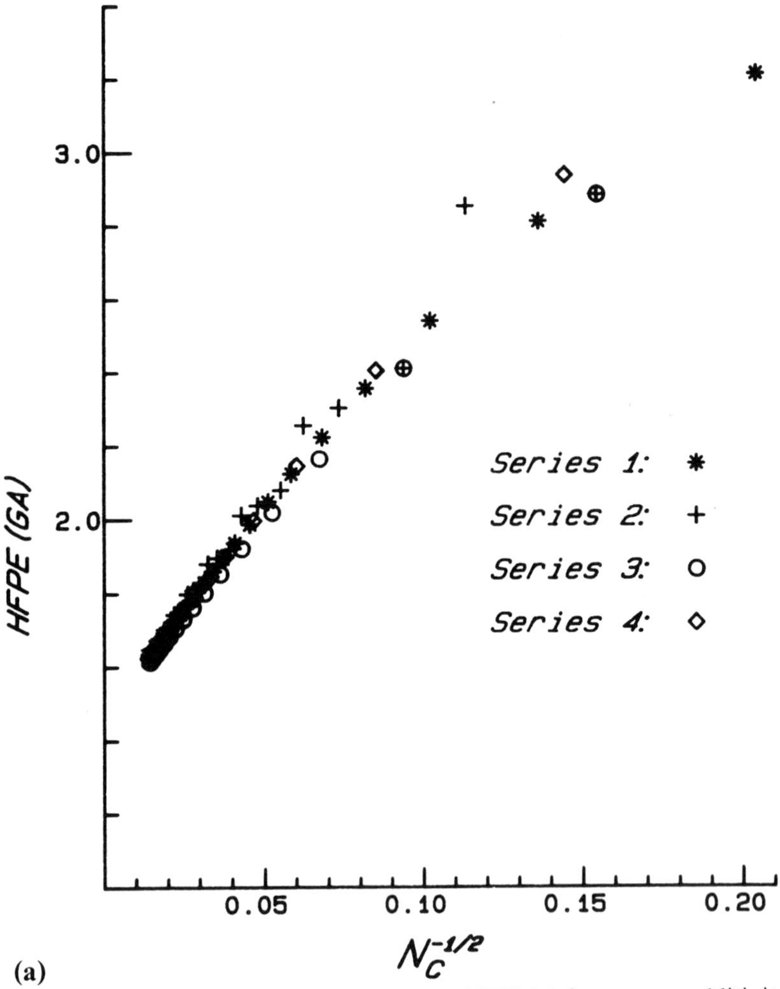

(a)

Figure 2-3. Heats of formation per π electron (HFPE) (*a*) from group additivity and (*b*) from SRT-2.

series. Group additivity predicts very little difference in the stabilities of molecules of comparable size. The much larger differences predicted by SRT-2 arise from the resonance energy terms shown in Figure 2-2*b* and in steric factors introduced[15] to account for H atom interactions around the edges of the molecules. In view of the abovementioned inability of GA to account for aromaticity, we feel that the relative stabilities it predicts are not reliable. On the other hand, unlike SRT, GA values for REPE converge to a unique value regardless of edge structure. It appears that if one is interested in relative stabilities of isomeric molecules, GA is not as reliable as SRT (and MO theories in general), although for equilibrium calculations

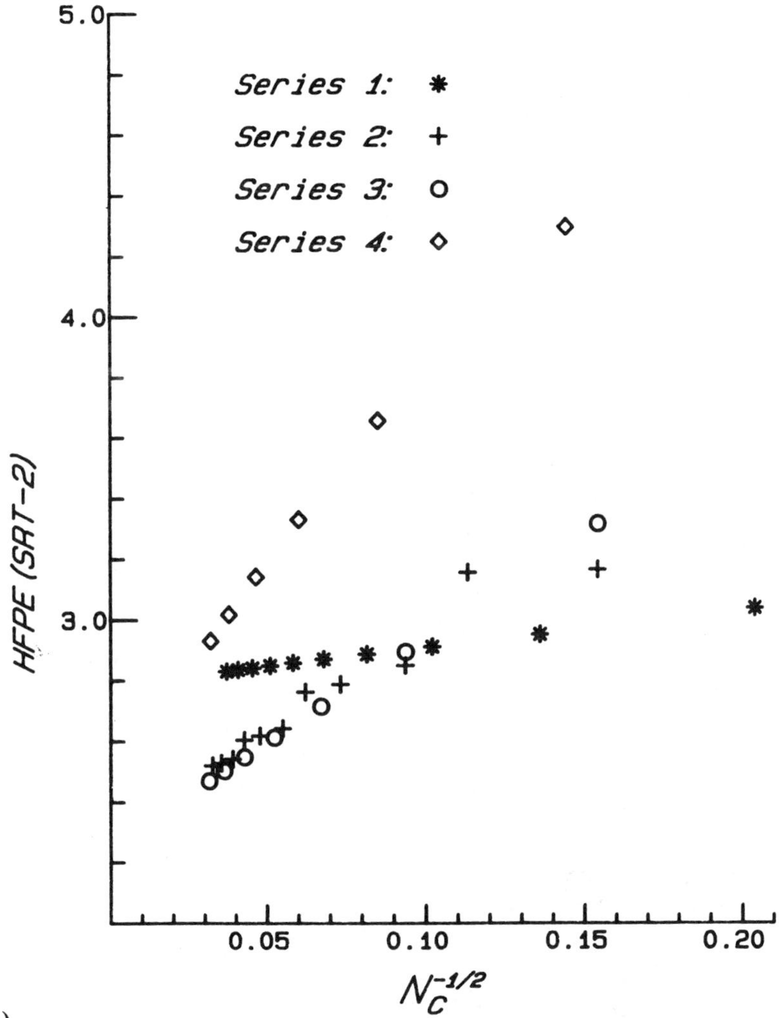

(b)

involving large species, we prefer GA over SRT because the former cannot fail drastically.

3. BOND ORDERS

Bond order is a measure of the effective number of covalent bonds joining two atoms. Since our calculations are concerned only with π electrons in aromatic molecules, bond order ranges from zero (no π-bonding) to one (an isolated double bond). Symmetry arguments imply a bond order of ½ in benzene and ⅓ in graphite.

According to SRT, the order of a bond is simply the ratio of the number of Kékulé structures in the molecule in which the bond of interest is double to the total number of Kékulé structures in the molecule (see Appendix B). These are the so-called Pauling bond orders, not Hückel bond orders. Through the relation between bond order and bond length, this procedure predicts bond lengths in excellent agreement with experiment.[16] In addition, because the bond order is a measure of π-electron density between two carbon centers, we expect that it can be used to identify the bonds most susceptible to electrophilic attack. This should produce correlations between bond order and reactivity.

It is convenient to present certain results of calculations of local properties (bond orders, ring aromaticities, and localization energies) as one-twelfth "slices" of selected hexagonal molecules. An example of such a "slice" is shown in Figure 2-4. Bond orders for "slices" from four molecules of comparable size, each belonging to a different series, are shown in Figure 2-5. The width of each bond is a measure of the calculated bond order, with the thinnest bond having order near zero and the broadest near unity. In all cases, bonds more than three or four rings away from the edge approach the graphitic value of ⅓.

Series 1 is clearly the most exceptional (pathological?). Near the corner along the direction parallel to the edge, bond orders alternate between zero and unity, resembling linear polyene molecules. Near the midpoint of the edge, bond orders become uniform (order = ½). This general behavior becomes damped, but it is still noticeable for the second, third, and even fourth interior layer of atoms. The bonds perpendicular to the edge

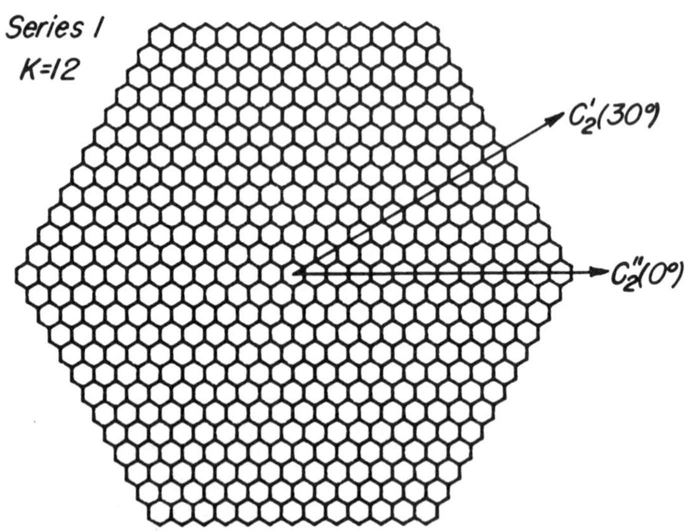

Figure 2-4. Hexagonal molecule showing representative section used to display results.

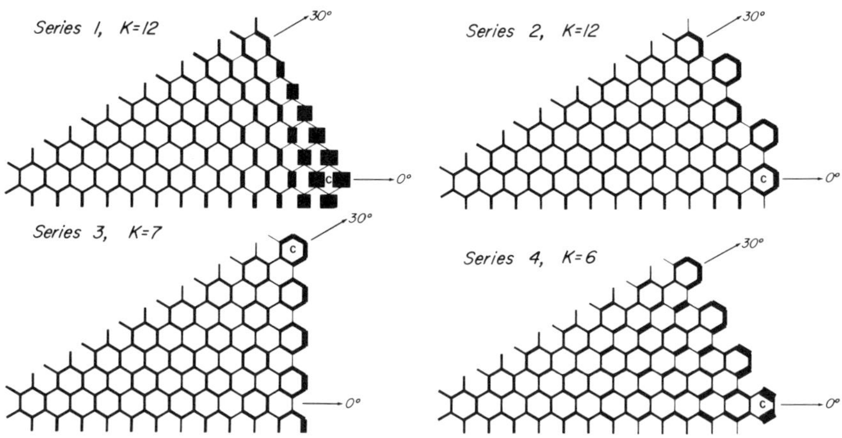

Figure 2-5. Bond orders in the 30° wedge indicated in Figure 2-4. Four molecules of comparable size are shown, one from each of the four series depicted in Figure 2-1. They are identified by the value of the generating parameter *K*. Corners on the molecules are denoted by the letter *c*. The widths of the bonds are proportional to the squares of the bond orders. Squared values were used to accentuate differences.

approach a value of zero near the edge, giving the outer structure an "onion-skin" arrangement.

Next to series 1, series 4 shows the greatest nonuniformity of bond order near the edges. In contrast to series 1, bond alternation is most apparent in the direction perpendicular to the edges and bonds of lower order are, more or less, parallel to these edges. This creates an "orange-section" arrangement of bonds.

Series 2 and 3 are closely related. The outermost aromatic rings contain the bonds of highest order, and bond alternation is damped only several C atoms away from the edge.

The arrangements of edge bond orders shown in Figure 2-5 resemble those of noncondensed molecules containing similar structural features. This is illustrated in Figure 2-6, where molecules with edge types 1 and 3 are compared to related chainlike polyaromatic molecules. For molecules with edge type 1 and their linear polyacene analogues, alternant edge bonds vary smoothly from unity at the corner to 0.5 at the center. The rate of this change for these two cases is, however, quite different. This change is linear for the chain molecule, but for the condensed species the decline from unity is slow near the corner and rapid near the center. Edges of series 3 molecules are also rather similar to corresponding chainlike molecules. However, condensed structures have a somewhat more uniform edge bond order arrangement.

The behavior of bond orders in the limit of arbitrarily large molecular size has been examined for both exterior and interior bonds. Plausible extrapolations were obtained for all series except the first. Although we examined molecules in series 1 containing as many as 6000 C atoms, we were unable

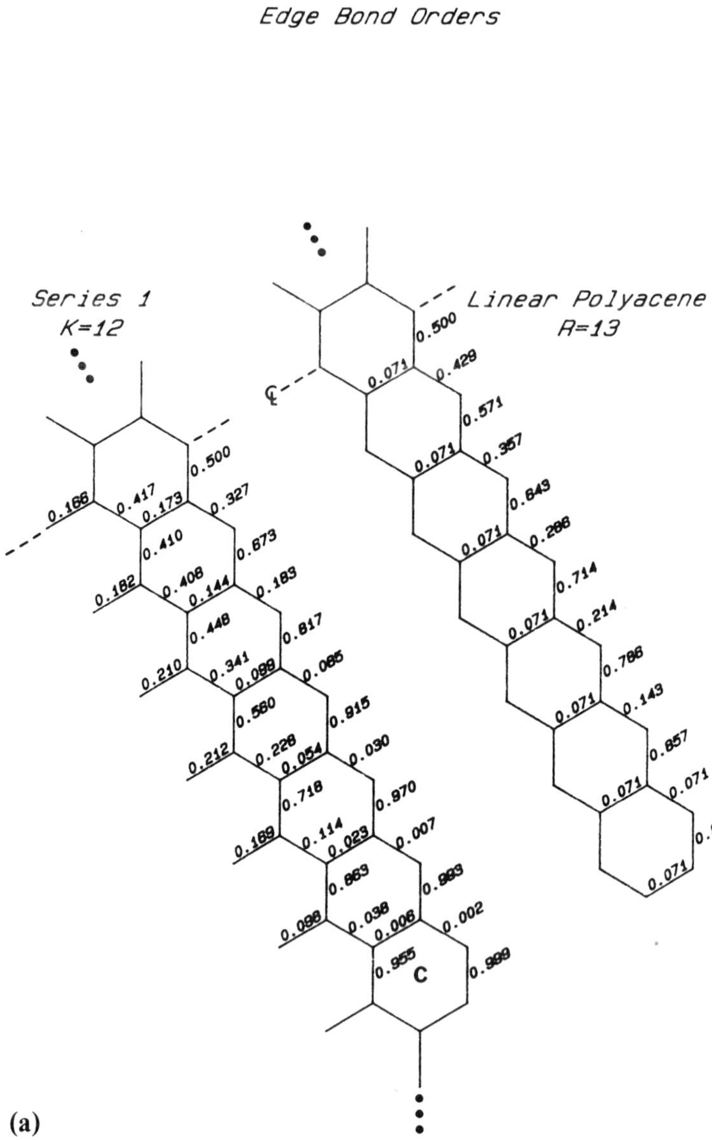

(a)

Figure 2-6. (*a*) Comparison of the bond orders along the edge of a series 1 molecule with those in a linear polyacene chain. The number of benzene rings *R* in the chain is equal to the number rings along one side of the hexagonal molecule. For this series, $R = K + 1$, where *K* is the series generating parameter. The centerline denoted here corresponds to the 30° axis as illustrated in Figure 2-4. The ring at a corner is denoted by the letter *c*. (*b*) Comparison of the bond orders along the edge of a series 3 molecule with those in a phenanthrene chain. The number of rings along the side of the hexagonal molecule is the same as the number of rings *R* in the chain. For this series, $R = 2K + 1$, where *K* is the generating parameter. The centerline corresponds to the 0° axis as denoted in Figure 2-4. A corner ring is denoted by the letter *c*.

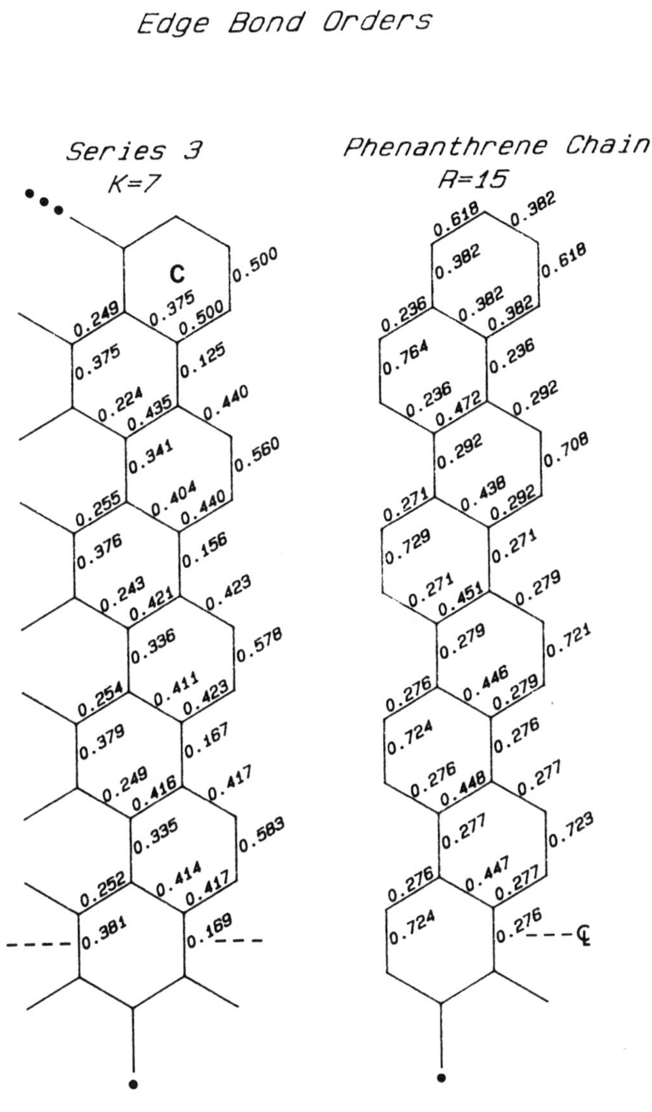

(b)

---

to conclude whether in the limit of infinite size, edge bonds become uniform (0.5) or alternate. The details of this problem are discussed in Appendix C. Results of the edge bond order extrapolations as N_C goes to infinity for series 3 are shown in Figure 2-7*a,* along with the corresponding limits for an isolated phenanthrene chain. As can be seen by comparing Figure 2-7*a* with Figure 2-6*b,* limiting values are being closely approached even for relatively

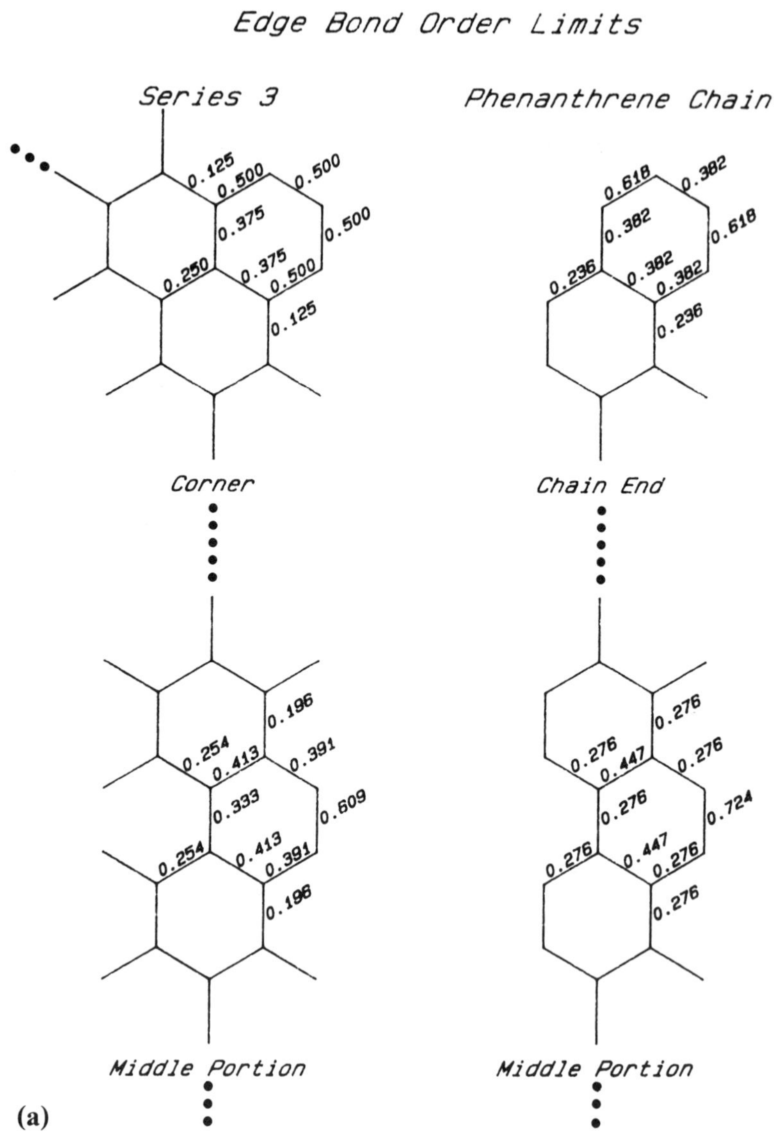

Figure 2-7. Limits as N_C approaches infinity for bond orders at the edges (*a*) of series 3 molecules and (*b*) of series 2 and 4 molecules.

small molecules. Furthermore, the variation along the edge is small, becoming negligible several rings away from the corners in the larger molecules. The limits as N_C goes to infinity for bond orders at the edges of series 2 and series 4 molecules are shown in Figure 2-7*b*. Only limits for bonds in the middle regions along the edge are shown. As in the case of series 3 molecules, the variation with increasing N_C is slow and should produce an accu-

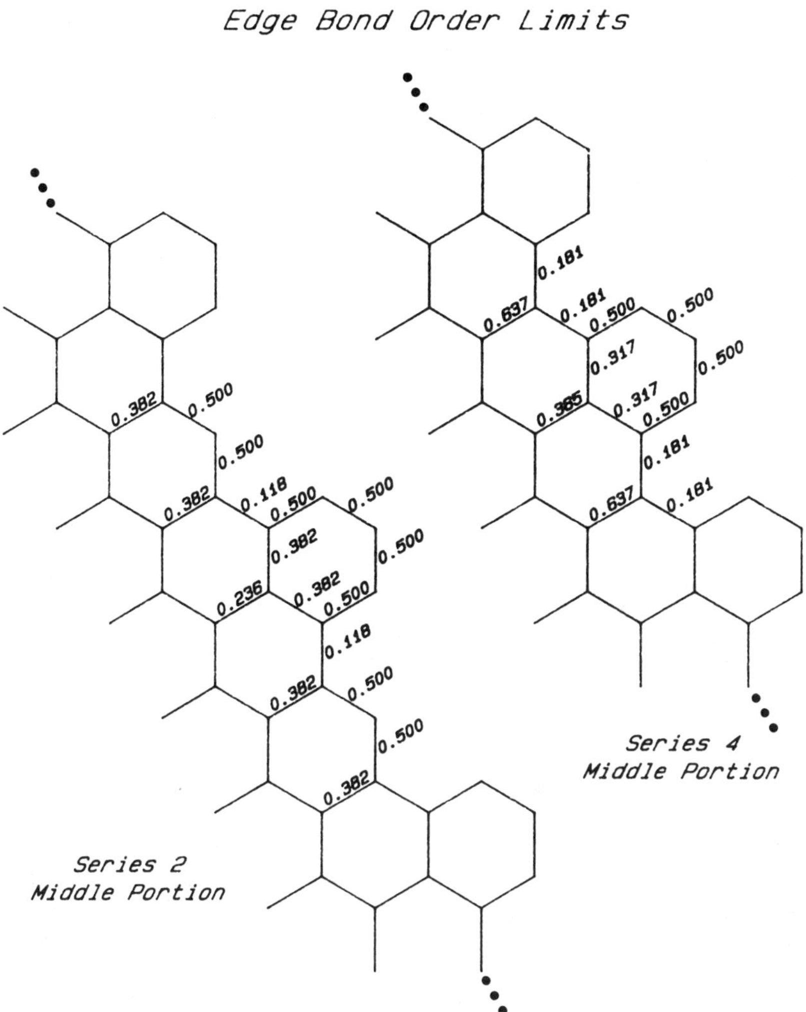

(b)

rate extrapolation. Likewise the variation along the edge is small, as can be seen from the relatively small molecules shown in Figure 2-5.

The convergence of interior C—C bonds toward their limiting value of ⅓ is shown in Figure 2-8. The values shown are for the six bonds forming the center-most ring and the six adjacent bonds. The rate of convergence to the limiting value is not the same throughout the different series. For series 1, these central bond orders are within 10% of the limit for molecules containing more than 250 carbon atoms; series 2 and 3 molecules must have more than 750 carbon atoms to satisfy this condition, and series 4 molecules must contain more than 1750 carbons.

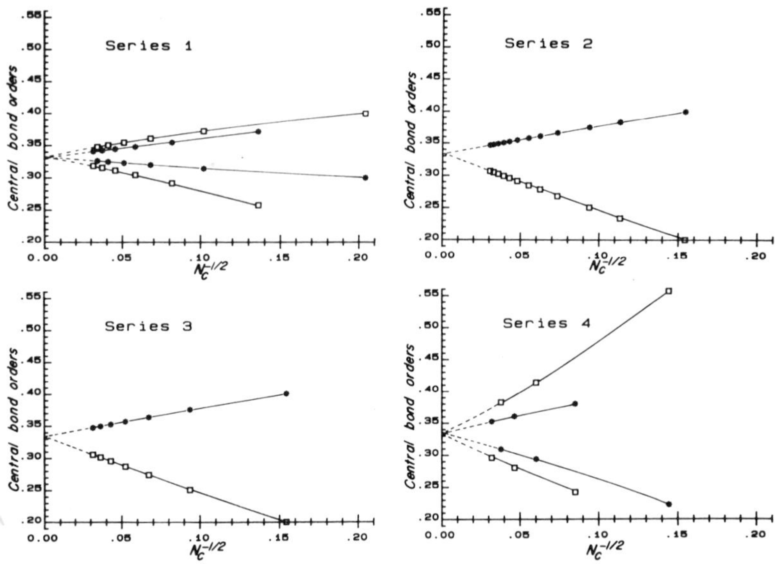

Figure 2-8. Limits as N_C approaches infinity of bond orders in the centers of the series molecules. The asterisks denote the six bonds forming the center-most ring in each molecule. The squares denote the six bonds adjacent to these and positioned like the C—H bonds in benzene. Molecules in series 1 and 4 show effects of even/odd alternation of the *K* values.

4. RING AROMATICITIES

A ring aromaticity index for a particular ring in a molecule can be defined as the ratio of the resonance energy due to that ring to the resonance energy of benzene. This index is calculated according to SRT-2 as described in Appendix A. It can be used to identify localized aromatic regions within a molecule. Since such regions would be those least subject to chemical attack, we expect this index to correlate with reactions involving participation of individual aromatic rings (eg, intercalation). In a manner analogous to that used for displaying bond orders, Figure 2-9 shows ring aromaticities for members of each of the four series. The diameters of the circles are proportional to the ring aromaticities.

Analogous to the case of the bond order values, graphitic aromaticity values of 0.4 are approached in rings near the center of the molecules for all cases. However, the effects of edge structure on aromaticity indices appear to penetrate further into the molecules than is the case for bond order. In other respects, the situation in similar to the bond order results. Series 1 shows an "onion-skin" structure for the aromaticities, while series 4 indicates an "orange-section" arrangement. Series 2 and 3 exhibit rings of high aromaticity at the edges. The variation in aromaticities across these molecules is shown more quantitatively in Figure 2-10, which plots aromaticities

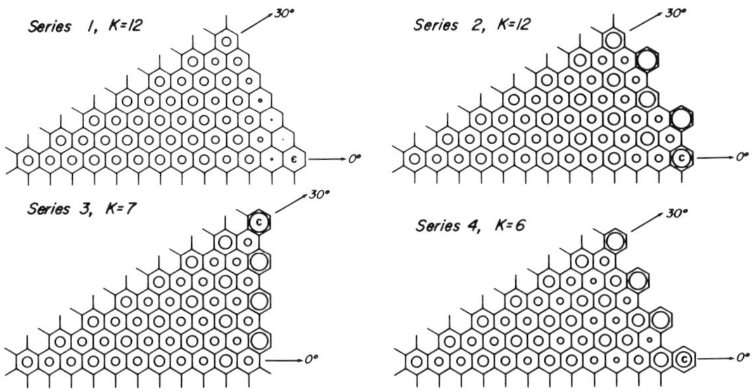

Figure 2-9. Ring aromaticities in the 30° wedge indicated in Figure 2-4. These are the same molecules used to depict the bond order distributions shown in Figure 2-5. The diameters of the circles in the benzene rings are proportional to the ring aromaticities. The largest circles correspond to an aromaticity index of 0.85.

for the rings falling along the 0° and 30° molecular axes (shown in Figure 2-4) as a function of distance from the center of the molecule.

The relationship between aromaticities at the edges of series 1 and 3 molecules and related fused-ring chains is shown in Figure 2-11. Aromaticities in the linear analogue of series 1 are uniform for the nonterminal rings, but there is in contrast a rather large variation along the edge of the series 1

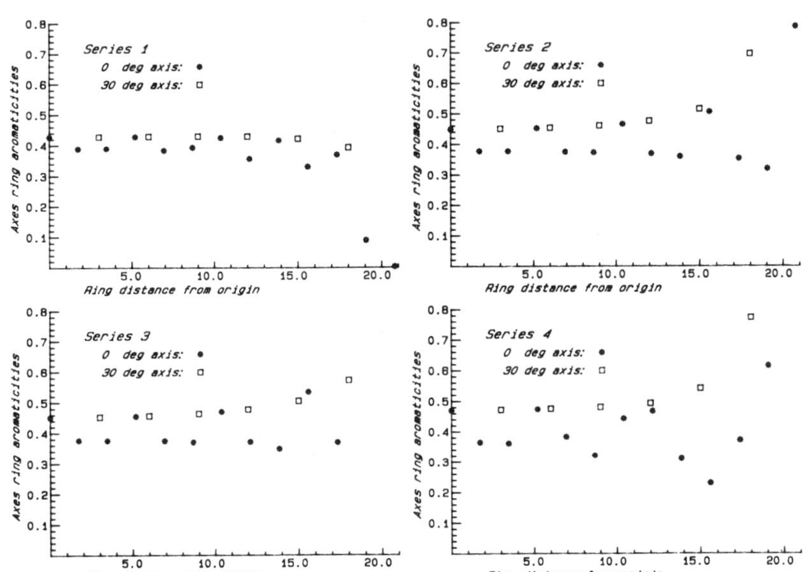

Figure 2-10. Ring aromaticities for rings falling along the 0° and 30° molecular axes. The ring distance is the distance from the center of a ring to the center of the molecule in units of carbon–carbon bond distance.

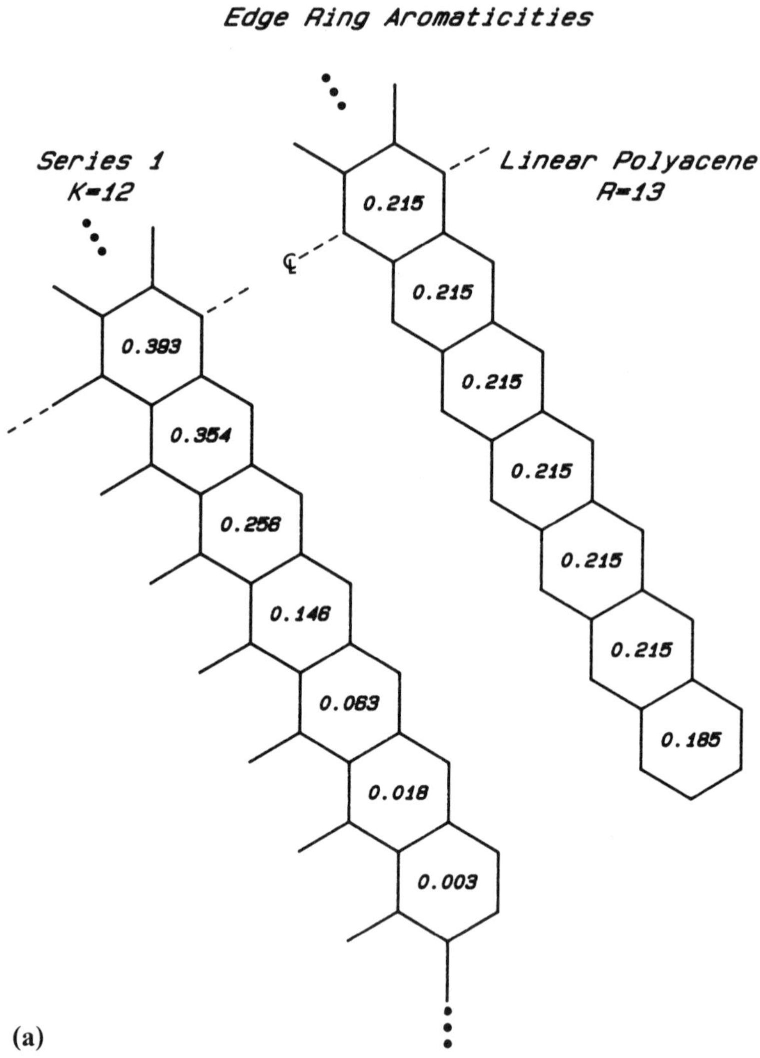

Figure 2-11. (*a*) Comparison of edge ring aromaticities in a series 1 molecule with the corresponding ones in a linear polyacene chain. The centerline corresponds to the 30° axis illustrated in Figure 2-4. (*b*) Comparison of aromaticities of rings along the edge of a series 3 molecule with those in a phenanthrene chain. The centerline corresponds to the 0° axis as depicted in Figure 2-4.

polyaromatic molecule. For series 3, aromaticities at the edge more closely parallel those in its linear analogue, but we still see a much larger variation in the condensed structure.

We consider now how these edge ring aromaticities behave in the limit of arbitrarily large molecules. Series 1 is the only one whose molecules show

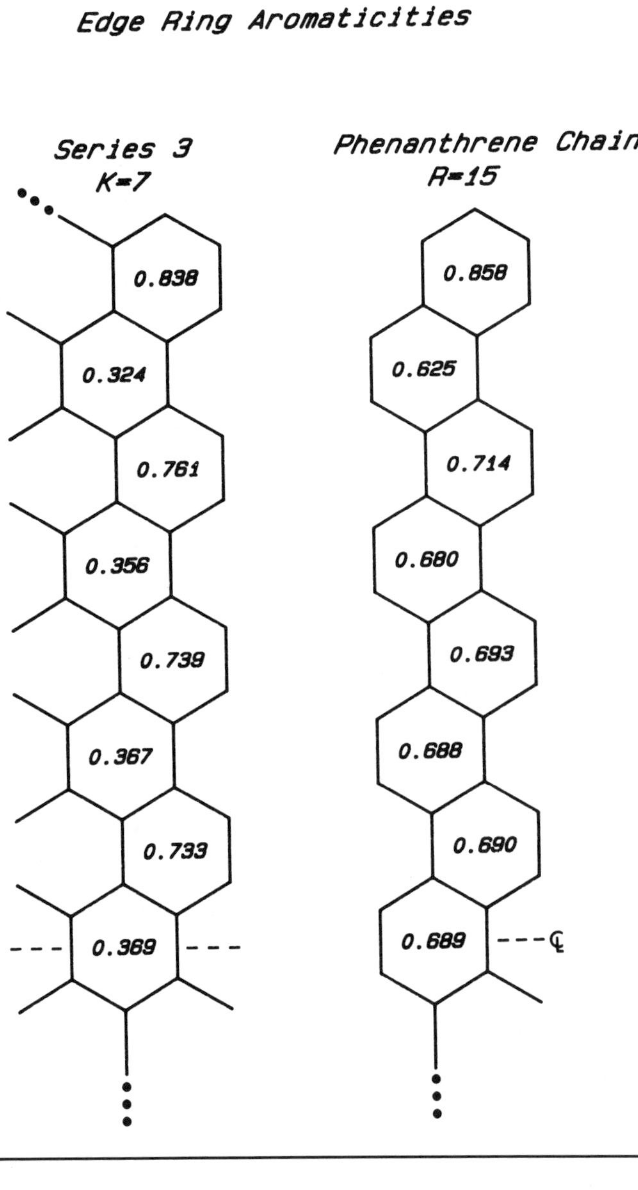

(b)

large variations in aromaticities along the edges. To get some indication of how the limiting edge aromaticity distribution might appear, it would be necessary to examine series 1 molecules larger than that shown in Figure 2-9. Unfortunately, this was not feasible because of the time-consuming nature of the aromaticity calculation in comparison with that for bond orders. For the remaining three series, however, changes in edge aromatic-

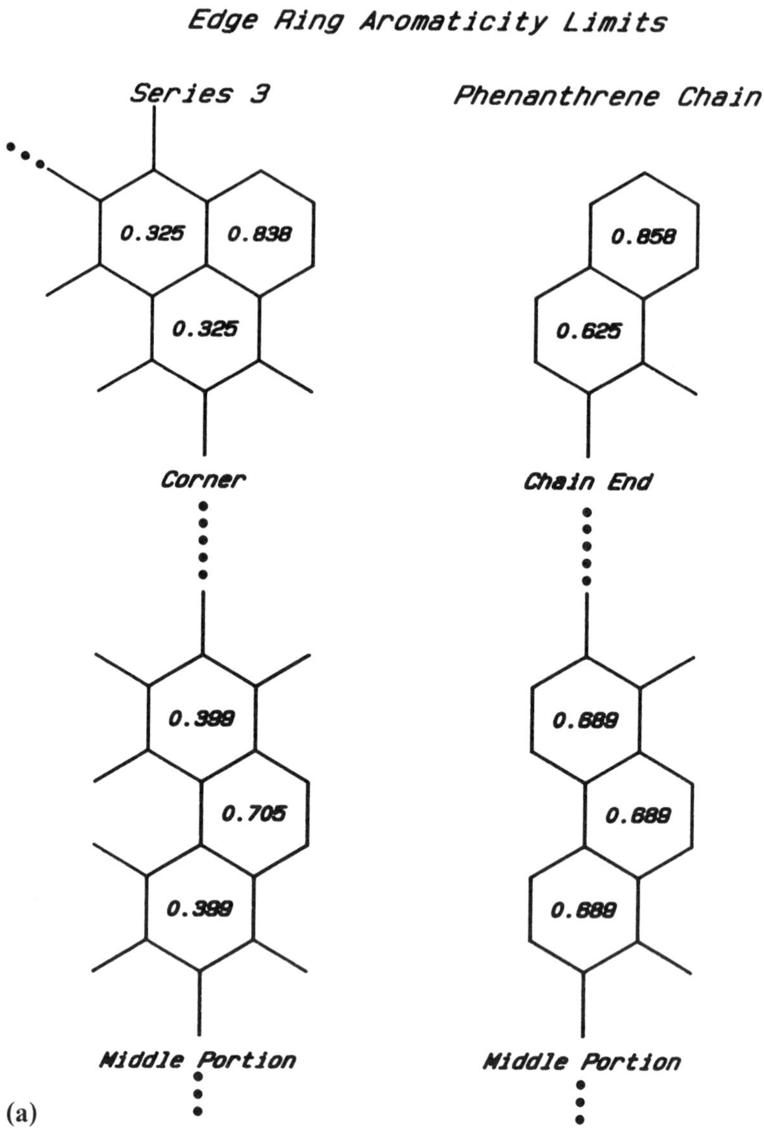

(a)

Figure 2-12. Limits as N_C approaches infinity for ring aromaticities at the edges of (*a*) series 3 molecules, and their comparison with corresponding limiting values for a phenanthrene chain, and (*b*) series 2 and 4 molecules.

ities with increasing molecular size become slow by the time the sizes shown in Figure 2-9 are reached. Thus, extrapolations to infinite size were feasible. The results are shown in Figure 2-12.

We examine finally the limiting values of the aromaticity in the central ring of these molecules. For each series, changes in the values with increasing molecular size were sufficiently regular to permit extrapolation to infi-

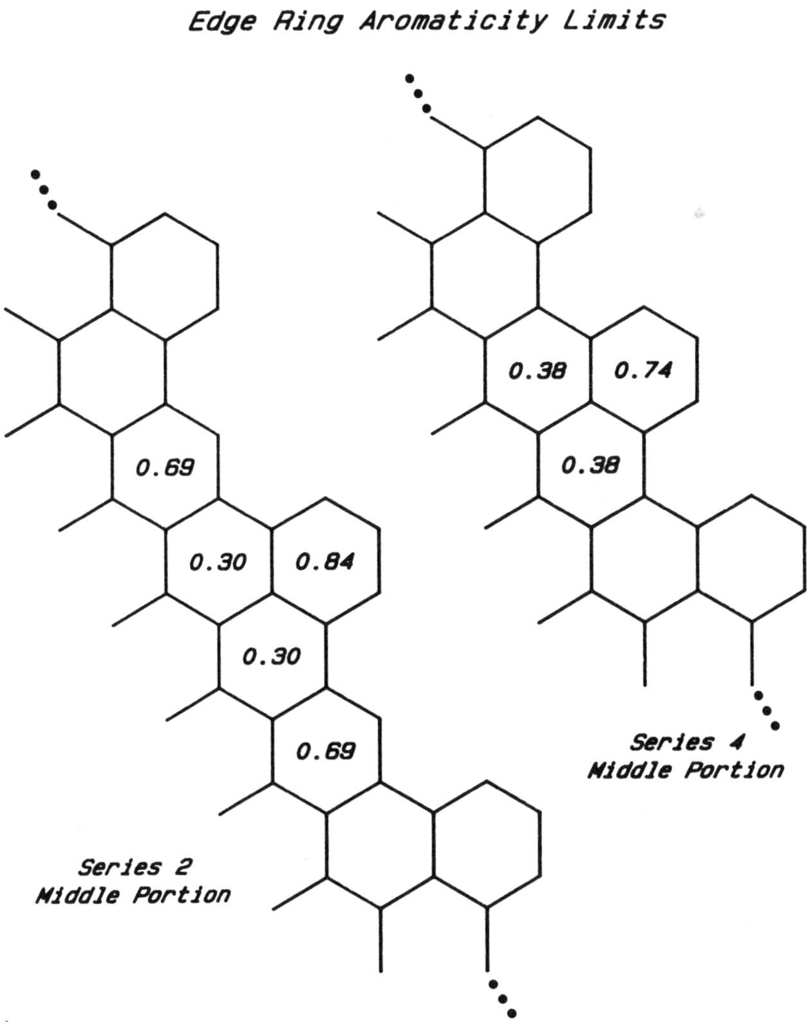

(b)

nite size in all cases. The results are shown in Figure 2-13. For all series, the aromaticity of the central ring extrapolates to a value of 0.4.

5. LOCALIZATION ENERGIES DETERMINED BY PERTURBATION MOLECULAR ORBITAL THEORY

Localization energy is the change in resonance energy resulting from the localization of a π electron at a particular carbon center in a delocalized structure. Therefore, differences in localization energy between different

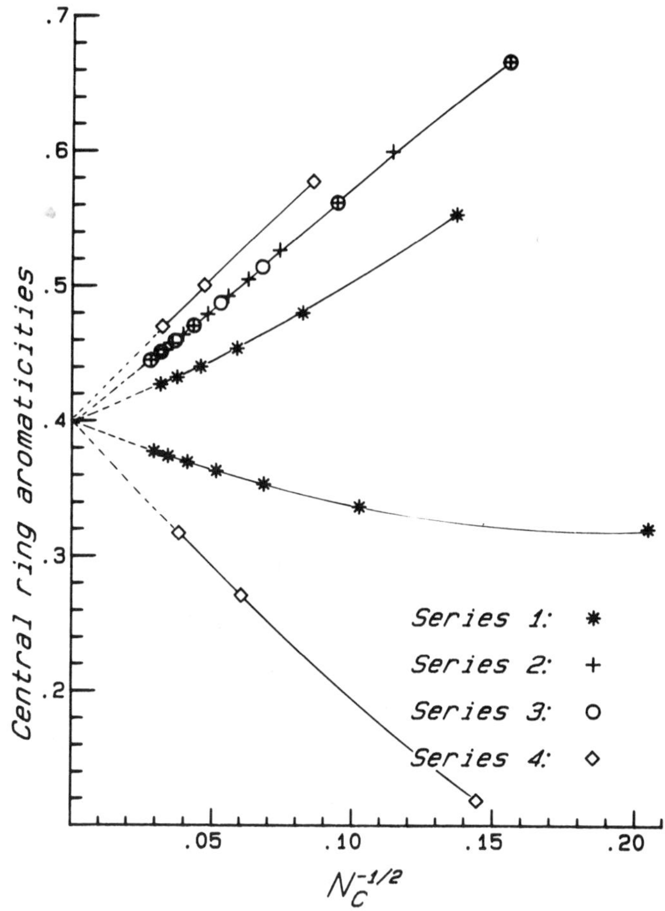

Figure 2-13. Limits as N_C approaches infinity of ring aromaticities for the center ring in series 1–4 molecules.

sites are equal to differences in reaction enthalpy for free radical and electrophilic reactions that localize a π electron at these sites. Differences in activation energy and reaction enthalpy are expected to be related; therefore, localization energy provides a measure of relative reactivity, since the lower the localization energy the lower the activation energy. This approach of course ignores steric (σ-bond) effects.

According to PMO theory, localization energies are evaluated by applying first-order perturbation theory to the Hückel molecular orbitals[7,8] (see Appendix D for details). Localization of one electron in the even-parent system results in a system that has a nonbonding MO (NBMO). The localization energy is taken to be 2β times the sum of the NBMO coefficients on the atoms to which the localization center is directly linked; β is the Hückel resonance integral. Localization energies are shown in Figures 2-14 through

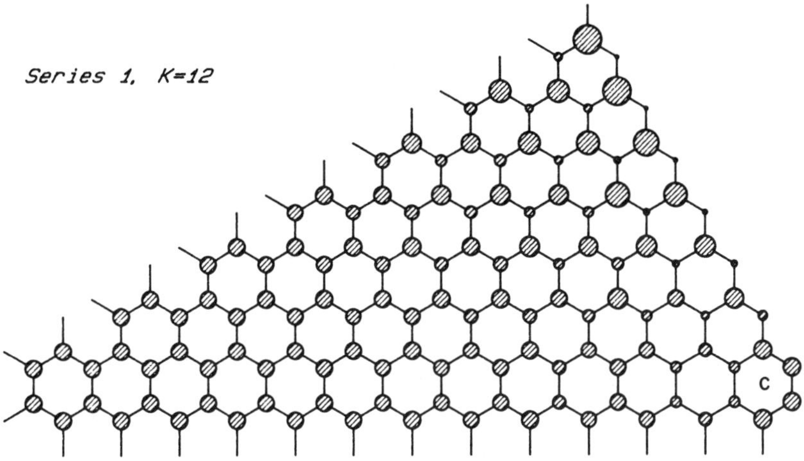

Figure 2-14. PMO localization energies for a series 1 molecule. Areas of the shaded circles are proportional to the change in π-electron energy required to localize a π electron at a particular carbon center.

2-17 for the same molecules displayed in Figures 2-5 and 2-9. The areas of the shaded circles at each center are proportional to their localization energies.

Series 1 molecules are by far the most reactive. Edge atoms in larger species have virtually no localization energy, and even their interior atoms are quite reactive. The most and least reactive position in the series 1 molecule shown in Figure 2-14 have the values 0.03β and 2.27β, respectively, and the average of all positions is 0.69β. For the sake of comparison, localization energies for benzene are 2.31β, the 1- and 2-positions in naphthalene have

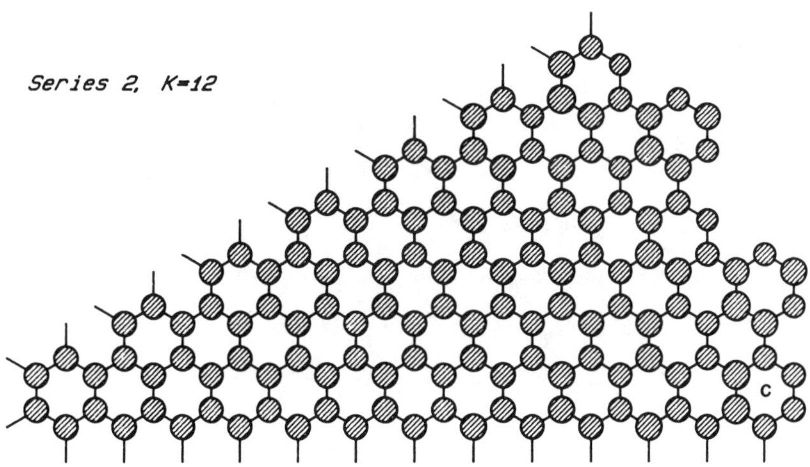

Figure 2-15. PMO localization energies for a series 2 molecule.

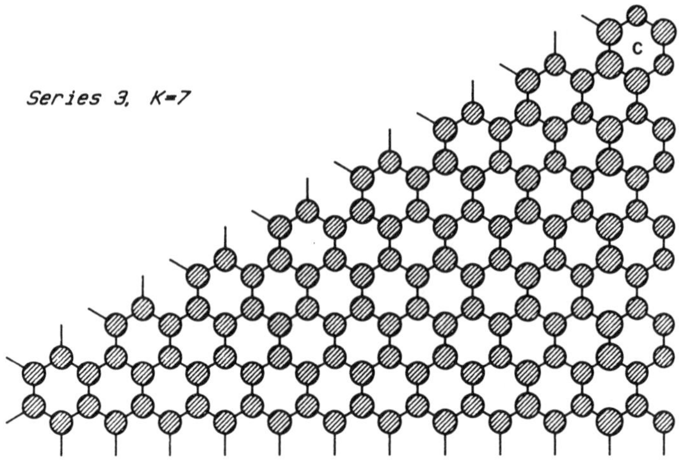

Figure 2-16. PMO localization energies for a series 3 molecule.

values 1.81β and 2.12β, and the most reactive site (the 9-position) in anthracene has the value 1.26β.

Series 4 molecules are the next most reactive. The molecule in Figure 2-17 has minimum, maximum, and average values of 0.26β, 2.23β, and 1.19β. The most exposed C atoms in this molecule are the least reactive.

Series 2 and 3 have relatively uniform reactivities. The minimum, maximum, and average values for the series 2 molecule in Figure 2-15 are 1.20β, 2.29β, and 1.66β. For the series 3 molecule in Figure 2-16, corresponding values are 1.23β, 2.31β, and 1.66β. These calculations indicate, quite remarkably, that even very large, highly condensed polyaromatic molecules need not be highly reactive. Edge structure is far more important than

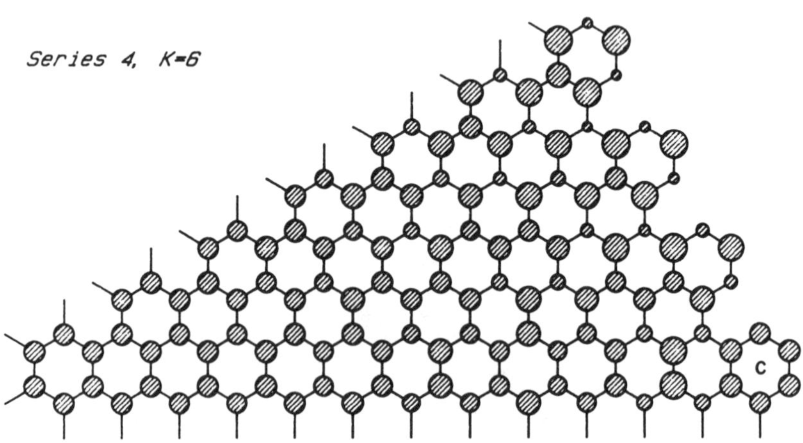

Figure 2-17. PMO localization energies for a series 4 molecule.

size as a determining factor for localization energies, at least for molecules of the size studied in this work.

As the molecules in each series become larger, extrapolations indicate that average localization energies should eventually approach zero. However, the rate at which this happens varies widely among the different series. For example, the average localization energy decreases rapidly in series 1 molecules, becoming less than 1.26β (the value for the 9-position in anthracene) for molecules having more than 350 carbons. In contrast, the decrease is extremely slow for the more stable series 3, with the average becoming less than 1.26β only for species containing more than 120,000 carbons. A molecule of that size would measure 650 Å across.

Although PMO theory may be a useful indicator of relative reactivity for different sites in a single molecule, it is not clear that it can give reliable comparisons of reactivity between different molecules. Therefore, quantitative results of the calculations above, while intriguing, are still open to question and must be tested against more exact theories.

SUMMARY

Our applications of simple theories to very large molecules suggest that the properties near the edge depend very strongly on the specific edge structures. All calculations show that the "zig-zag" edge structure (series 1) is thermodynamically unstable and exceedingly reactive. This edge type, in effect, blocks π-electron conjugation and is therefore nonaromatic. The outermost layers are more closely related to polyolefins than to polyaromatic molecules.

The edges of series 2 and 3 are composed of the most stable structures possible. They lead to highly efficient conjugation between edge and interior atoms and are the only two edge types permissible for "fully benzenoid" molecules. Even in irregular graphitelike molecules, these edge structures are expected to stabilize the part of the molecules to which they are attached. Simple calculations suggest that molecules containing these edges are not highly reactive.

Edges of series 4, like those of series 1, are fundamentally unstable, though their blockage of π-electron conjugation is not nearly as severe.

These results also show that the simple theories used usually yield values for each series that converge in a reasonable manner in the limit of infinite size. There is, however, one notable failure, namely, the prediction by SRT of different limiting resonance energies per carbon atom for different series.

When π-electron effects control the reactions of large polyaromatic structures, the relations given in this chapter can assist the analysis of reaction pathways. We expect that use of more advanced theories will refine, not overturn the conclusions of the present calculations.

APPENDIX A:
Resonance Energies and Ring Aromaticities Determined by SRT-2

The total resonance energy E_T of a molecule is given in SRT-2 as the sum of individual ring resonance energies E_B.

$$E_T = \sum_{\mathrm{B}} E_{\mathrm{B}} = 2 \sum_{\mathrm{B}} \frac{C_B\gamma_1 + 0.5 \sum_{N=1}^{6} C_{\mathrm{BN}}\, \gamma_2}{C_T}$$

Where C_T is the total number of Kékulé structures for the molecule, C_{B} is the number of structures with ring B deleted, and C_{BN} is the number resulting from deletion of a naphthalene unit consisting of ring B and one of its six possible adjacent neighbors N. Note that one must take half the value for the double sum in the expression above; otherwise in the summation over B, each naphthalene unit would be counted twice. The resonance energy for benzene is equal to the value of the parameter γ_1.

The ring aromaticity index A_{B} for a particular ring in a molecule is the ratio of E_{B} to the resonance energy of benzene, γ_1. We have:

$$A_{\mathrm{B}} = \frac{E_{\mathrm{B}}}{\gamma_1} = \frac{2[C_{\mathrm{B}} + 0.5(\gamma_2/\gamma_1) \sum_{N=1}^{6} C_{\mathrm{BN}}]}{C_T}$$

Herndon[11] uses a value of 0.4 for the ratio γ_2/γ_1.

APPENDIX B:
Bond Order Calculation with Structure–Resonance Theory

The bond orders according to SRT are calculated as follows. Let i denote a particular carbon center in the even-parent molecule. There are a maximum of three adjacent centers, j, k, and l, to which it can be bound. The bond orders of the three possible bonds are given by the quantities:

$$\frac{\beta_{ij}}{|B|} \qquad \frac{\beta_{ik}}{|B|} \qquad \frac{\beta_{il}}{|B|}$$

where $|B|$ is the determinant of the bond matrix B and equals the number of Kékulé structures in the parent molecule.[4] The β_{ij} are the elements of the cofactor matrix of B; they give the number of Kékulé structures in the parent system in which the bond between i and j is a double bond.

The sum $\beta_{ij} + \beta_{ik} + \beta_{il}$ is equal to the number of Kékulé structures for the parent. Thus, the bond orders for bonds connecting a particular carbon center to adjacent centers must sum to unity.

APPENDIX C:
Limits as N_C Approaches Infinity for Pauling Bond Orders at the Edges of Series 1 Molecules

For comparison, we look first at the limits for the bond orders in a linear polyacene chain. Starting from one end of a finite chain, we number the rings sequentially from 1 up to R, the total number. If i is the number of a particular ring in the chain, the bond orders for this ring are denoted by:

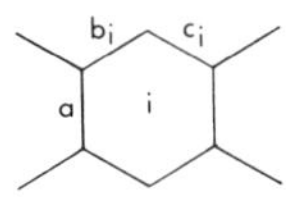

where $a = 1/(R + 1)$, $b_i = i/(R + 1)$, $c_i = 1 - b_i$. Note that the order of the cross-linking bond does not depend on the ring number i. For convenience, consider the center ring i_c of an odd-member chain. We have $i_c = (R + 1)/2$. Pick another ring k, an arbitrary number, j, of rings away on either side of i_c. Thus, $k = i_c \pm j = [(R + 1)/2] \pm j$. The bond order b_k is consequently $b_k = ½ \pm [j/(R + 1)]$. For whatever value of j chosen, no matter how large, and thus no matter how far on either side of the center of the chain, we get $b_k \rightarrow 0.5$ as $R \rightarrow \infty$. The limiting bond orders are thus $a \rightarrow 0$, $b_i = c_i \rightarrow 0.5$.

We now examine the situation in the series 1 molecules. As can be seen in Figure 2-6*a*, the bond orders along the edges of these molecules strongly resemble those in a polyacene chain. We wish to see whether, in analogy with the isolated chain, the bond orders on rings an arbitrarily large distance away from the center of a side approach the value 0.5 as N_C approaches infinity. To examine this question, it is convenient to tabulate for a particular molecule the difference between the bond order of a given edge bond and the value for the same bond on the corresponding isolated chain. This is done in Table C-1 for the example shown in Figure 2-6*a*. We give here the bond orders, their differences Δ_K, and an edge bond index defined as $i' = i - (R + 1)/2$, where $R = K + 1$. Plots of Δ_K versus i' are antisymmetric about $i' = 0$ for all values of K. A number of such plots are shown in Figure C-1 for odd K values from 1 to 29. Plots for even K fall between those shown. To indicate the general nature of the relationship, spline functions passing through the discrete values of Δ_K are displayed instead of individual points. As indicated above, in an arbitrarily large region around $i' = 0$, the edge bond orders for isolated chains approach the value 0.5 as R approaches infinity. For this situation also to occur for series 1 edge bonds, the slopes of Δ_K versus i' would have to approach zero as N_C approached infinity. Figure C-1 shows that although the slopes at $i' = 0$ are decreasing with increas-

TABLE C-1. Differences in Edge Bond Orders Between Series 1, $K = 12$ and $R = 13$ Polyacene Chain

Ring number i	Series 1 edge bond order	Polyacene chain bond order	Δ_K	$i' = i - (R + 1)/2$
1	0.00096	0.07143	−0.07047	−6
2	0.00745	0.14286	−0.13541	−5
3	0.03017	0.21429	−0.18412	−4
4	0.08451	0.28571	−0.20120	−3
5	0.18330	0.35714	−0.17384	−2
6	0.32726	0.42857	−0.10131	−1
7	0.50000	0.50000	0.00000	0
8	0.67274	0.57143	0.10131	1
9	0.81670	0.64286	0.17384	2
10	0.91549	0.71429	0.20120	3
11	0.96983	0.78571	0.18412	4
12	0.99255	0.85714	0.13541	5
13	0.99904	0.92857	0.07047	6

ing K, this decrease is a very slow one. A plot of these slopes against the reciprocal of the square root of the total numbers of carbon centers in the series 1 molecules is shown in Figure C-2. We see that the extrapolation to zero is a large one. To extend this curve appreciably would require computational resources beyond our present capabilities. We therefore feel unable to reach any conclusion regarding the limiting edge bond order distribution.

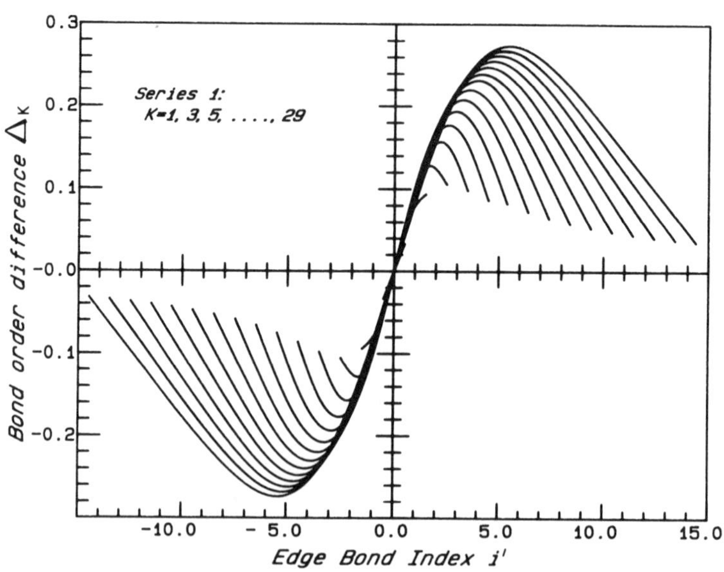

Figure C-1. Differences between bond orders of polyacene chains and series 1 edges.

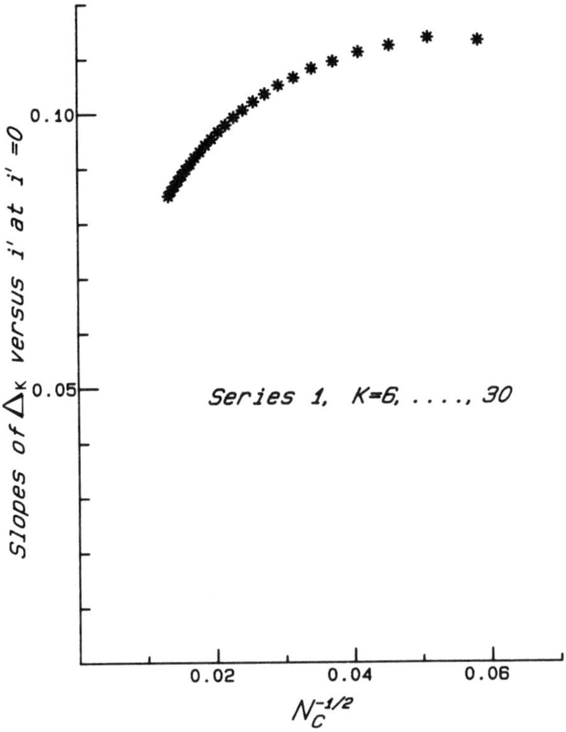

Figure C-2. Rate of change in series 1 edge bond orders at the center of an edge.

APPENDIX D: Localization Energies Determined by Perturbation Molecular Orbital Theory

The localization energy is the change in π-electron energy resulting from the localization of a π electron at a particular carbon center. Localization of one electron in the even-parent system results in an odd system that has a nonbonding MO. The localization energy is taken to be 2β times the sum of the normalized NBMO coefficients to which the localization center is linked; (β is the Hückel resonance integral). In terms of the cofactor matrix elements of the bond matrix B, the normalized NBMO coefficients are $\beta_{ij}(\sum_j \beta_{ij}^2)^{-1/2}$.

The localization energy at center i is thus:

$$\delta E_i = 2\beta(\sum_j \beta_{ij}^2)^{-1/2} \sum_k \beta_{ik}$$

The k index is summed over the centers adjacent to center i. But this sum is equal to the absolute value of the determinant of B. Thus we have:

$$\delta E_i = 2\beta(\sum_j \beta_{ij}^2 \, |B|^{-2})^{-1/2}$$

These localization energies are shown in Figures 2-14 through 2-17 for the same molecules displayed in Figures 2-5 or 2-9. The areas of the shaded circles at each center are proportional to the localization energies.

ACKNOWLEDGMENT

This work was supported by grants from the Gas Research Institute (No. 5081-261-0556) and the Air Force Office of Scientific Research (ISSA-84-00033).

REFERENCES

1. Coulson, C.A. *Nature* **1947,** *159,* 265. Coulson, C.A.; Taylor, R. *Proc. Phys. Soc.* **1952,** *A65,* 815. Coulson, C.A.; Schaad, L.J.; Burnelle, L. Proceedings of the Third Biennial Carbon Conference, 1958, p. 27.
2. Pullman, B. Proceedings of the Third Biennial Carbon Conferences, p. 3.
3. Matsen, F.A. Proceedings of the Third Biennial Carbon Conferences, p. 21.
4. Brown, R.L. *J. Comput. Chem.* **1983,** *4,* 556.
5. Stein, S.E.; Brown, R.L. *Carbon* **1985,** *23,* 105.
6. Herndon, W.C. *Isr. J. Chem.* **1980,** *20,* 270.
7. (a) Dewar, M.J.S. "The Molecular Orbital Theory of Organic Chemistry". New York: McGraw-Hill, 1969. (b) Dewar, M.J.S., Dougherty, R.C. "The PMO Theory of Organic Chemistry." New York: Plenum Press, 1975.
8. Heilbronner, E.; Bock, H. "The HMO Model and Its Application". New York: Wiley-Interscience, 1976.
9. Clar, E.J. "Polycyclic Hydrocarbons", Vol. 1. New York: Academic Press, 1964, pp. 37–38.
10. Swinborne-Sheldrake, R.; Herndon, W.C. *Tetrahedron Lett.* **1975,** *10,* 755.
11. Herndon, W.C.; Ellzey, M.L., Jr. *J. Am. Chem. Soc.* **1974,** *96,* 6631.
12. Herndon, W.C. *J. Org. Chem.* **1981,** *46,* 2119.
13. Stein, S.E.; Golden, D.M.; Benson, S.W. *J. Phys. Chem.* **1977,** *81,* 314.
14. Clar, E.J. "Polycyclic Hydrocarbons", Vol. 1. New York: Academic Press, 1964, pp. 26–27.
15. Herndon, W.C. *Thermochim. Acta* **1974,** *8,* 225.
16. Herndon, W.C. *J. Am. Chem. Soc.* **1974,** *96,* 7605.

CHAPTER 3

Heats of Sublimation

James S. Chickos

University of Missouri–St. Louis, St. Louis, Missouri

CONTENTS

1. INTRODUCTION

Evaporation enthalpies for both the solid and liquid phases, (ΔH_s, ΔH_v) are important thermodynamic properties of the condensed phase. Most frequently, evaporation enthalpies are used to correct heats of formation data to a standard state,[1-3] but since these quantities are a measure of the magnitude of the interactions in the condensed phase, studies of various other types desire an evaluation or estimation of ΔH_s and ΔH_v.[4-8] Some examples include the use of ΔH_s in studies of intermolecular interactions in the solid state such as for estimating lattice energies via atom–atom potentials[4,7] and studies of molecular motion in solids[7c]; as a measure of the differences in interaction between chiral solids and their racemic modifications[5]; and for estimating transport properties in the environment of various chemicals such as pesticides and other commercial preparations.[6,8]

© 1987 VCH Publishers, Inc.
MOLECULAR STRUCTURE AND ENERGETICS, Vol. 2

We became interested in measuring latent heats nearly a decade ago to correct heats of combustion data to the gas phase. Since most of our interests at that time centered around highly strained molecules, the generally low long-term stability of many of these molecules precluded the possibility of interesting other research groups in making these measurements for us. Furthermore, as we learned subsequently, very few groups in North America were active in this area at that time. We therefore decided to explore some potential methods of making these measurements ourselves.

Our initial attempts were focused on the construction of a mass effusion apparatus. After some preliminary effort, we realized the extent of commitment necessary in both resources and time to make reliable measurements and abandoned this approach.

The sublimation apparatus described by Melia and Merrifield[9] and used to measure vapor pressures and heat of sublimation of tris(acetylacetonato) complexes of Sc(III), V(III), and Cr(III) appeared to be, from our perspective, an attractive alternative. Our initial efforts in this area were focused on demonstrating the applicability and versatility of this technique and on defining this method's inherent limitations.[10] As we have pursued our own interests and have responded to the requests of colleagues for specific measurements, it has been necessary to continually modify the apparatus. The modifications in general have reflected the need to measure increasingly lower vapor pressures. We have been able to extend the range of this technique to measure vapor pressures over a range of pressures from 10^5 to 10^{-3} Pa. The description that follows briefly describes the basic principles of head space analysis (HSA) as we have applied it to measure vapor pressures and heats of sublimation and reports on the results of our efforts to expand the dynamic range of this technique. A portion of this work has not yet been reported. Section 3 deals with our application of this technique to study differences in solid state interactions that occur between chiral and racemic solid modifications. Section 4 contains a comprehensive summary of the heats of sublimation that have been reported for organic compounds in the period 1910–1984.

2. THE MEASUREMENT OF VAPOR PRESSURE BY HEAD SPACE ANALYSIS

A. Basic Principles

Head space analysis has been used by a number of research groups in recent years to measure equilibrium vapor pressures.[9–13] A typical apparatus is shown in Figure 3-1. The experiment in general can be performed either isothermally or under steady state conditions. By keeping the ballast and

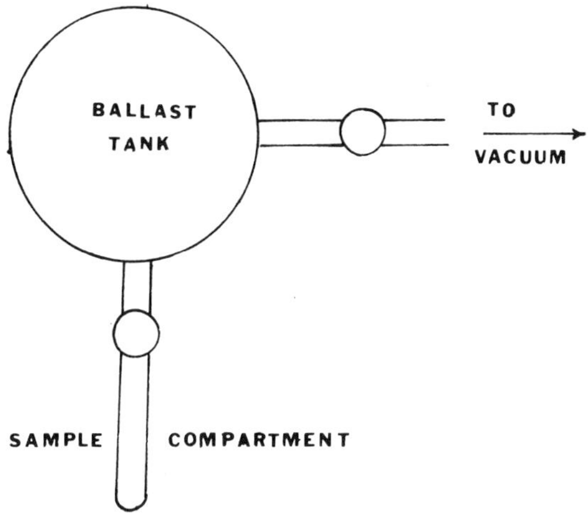

Figure 3-1. Apparatus for isothermal indirect head space analysis.

sample compartment at the same temperature, equilibrium will be achieved when

$$P_{\text{condensed phase, } T_1} \text{ in the sample compartment} = P_{\text{vapor, } T_1} \text{ in the ballast chamber} \qquad (3\text{-}1a)$$

where P represents the vapor pressure of the sample at temperature T_1. This is basically the experimental procedure described by Melia and Merrifield.[9] An analysis of the contents in the ballast tank of known volume followed by treatment of this quantity by the ideal gas law yields the experimental vapor pressure of the compound in question. Although satisfactory results can be obtained by performing these experiments isothermally, we have found that small temperature fluctuations in the constant temperature bath often are sufficient to cause condensation of the vapor in the ballast tank. Thus a potential source of error can be introduced at this point if the experiments are performed in this manner. Additional limitations will be described below. In our experience, the probability of condensation increased when the volatility of the sample decreased and the working temperature increased.

An alternative procedure is to maintain the ballast chamber at a higher temperature than the sample compartment. A steady state will be achieved when the pressures in the two compartments are equal.

$$P_{\text{condensed phase, } T_1} \text{ in the sample compartment} = P_{\text{vapor, } T_2} \text{ in the ballast chamber} \qquad (3\text{-}1b)$$

where $T_2 > T_1$.

By an appropriate design of the apparatus (Figure 3-2), heat flow between the two compartments can be restricted and thermal equilibrium main-

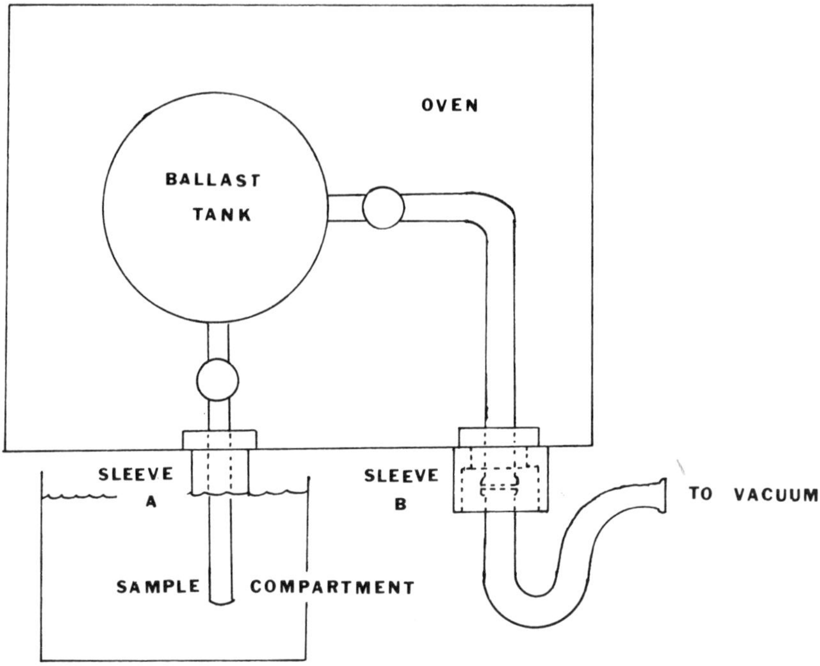

Figure 3-2. Apparatus for indirect steady state head space analysis (not drawn to scale).

tained in each. In this mode, condensation of the vapor in the ballast tank is improbable and, as will be described below, adsorption of sample on the surface of the vessel can be minimized and/or corrected.

Analysis of the vapor density in the ballast compartment can be achieved by direct or indirect methods. When the vapor pressure is high, or the analytical technique sufficiently sensitive, direct measurement as a function of temperature is possible. The heats of sublimation of anthracene (by fluorescence[14,15]) and uracil (by infrared[16]) have been determined in this manner; for a more detailed listing, the reader is referred to Section 4. The major advantage of direct head space analysis is the ease with which this experiment can be performed. Disadvantages often include the limited temperature range in which the detector response remains linear with concentration, the impossibility of obtaining vapor pressures directly from these results, and perhaps most significantly, the propensity to employ excessively high temperatures to obtain a vapor density suitable for spectroscopic detection.

Vapor in the ballast tank can be analyzed indirectly by transfer of the contents of ballast compartment and/or condensation of the vapor followed by quantitative analysis of the contents by a variety of spectroscopic techniques. Advantages of indirect analysis include more flexibility in the choice of analytical tools, increased sensitivity and sample size (because the entire contents of the ballast tank is analyzed, rather than a small cross section), the ability to use a lower temperature (as a result of the increase in effective sample size), access to a broader temperature range for measurement, and the capability of correcting for adsorbed substrate (see below). The major disadvantage to the indirect method is the increased time and effort that must be expended as a consequence of the additional steps required.

Use of the gas laws under ideal conditions (low pressure, high temperature) affords a relationship for converting directly to vapor pressure. When the experiment is performed in the isothermal mode, the contents and volume of the ballast tank, the temperature of the ballast tank, and molecular weight all serve to define the sample's vapor pressure. In the steady state mode these quantities serve to define the vapor pressure in the ballast tank. The condition of steady state requires this pressure to equal the vapor pressure of the sample at the lower sample temperature.[10]

B. Description of the Apparatus

Our apparatus at its current stage of development (Figure 3-2) can be built with only a modest investment in time and equipment. The ballast tank consists of a 12-L flask connected to the sample compartment and vacuum system by means of 10-mm Teflon stopcocks. The apparatus is housed in a plywood oven lined with aluminum foil and is heated by four 250-W infrared lamps. These heaters are sufficient to maintain oven temperatures up to 300 K and provide uniform heating of the ballast tank. The lamps are controlled by a variable transformer. A fan further ensures temperature uniformity within ± 2 K.

The interfaces between oven and sample compartment and oven and collection tube are fitted with aluminum sleeves and heated with 100-W cartridge heaters. Their temperatures are controlled electronically at a temperature intermediate between the temperatures of the oven and sample compartments. Sleeve A prevents the condensation of vapor at the interface between the constant temperature bath and oven, while sleeve B assures that vapor condenses only after it has entered the collection tube. An O-ring seal connects the collection tube to the ballast tank and can be partially heated by sleeve B if necessary. The collection U tube, which is easily removed by means of the O-ring seals, is generally cooled in a liquid nitrogen trap to ensure complete condensation.

The vacuum system employed depends in part on the vapor pressure range to be measured. For pressures above 10^{-1} Pa, only an efficient forepump is necessary. For measurements below 10^{-1} Pa, pumping speed becomes increasingly important, as will be discussed below.

C. The Measurement of Vapor Pressures: A Description of the Procedure

The measurement of vapor pressures by the method of head space analysis requires time to establish equilibrium (isothermal experiment) or a steady

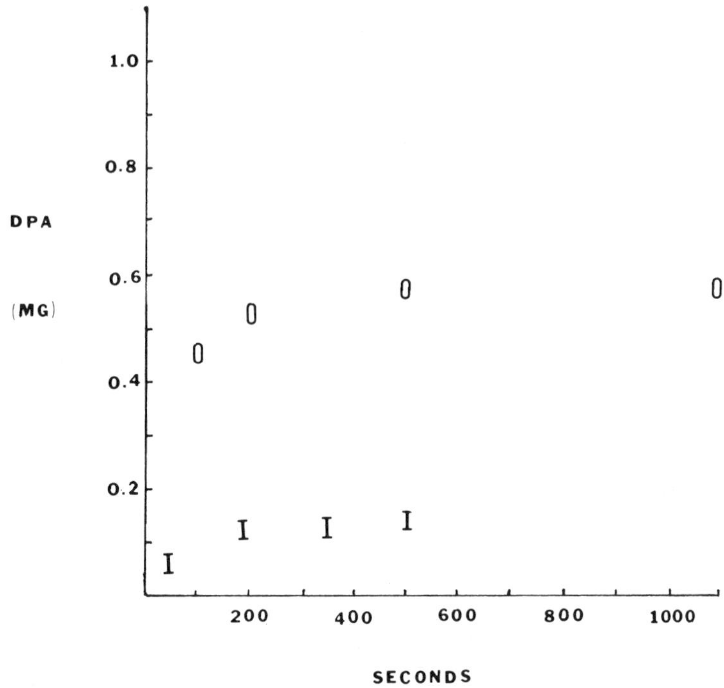

Figure 3-3. Rate of equilibration of diphenylacetylene (DPA, 500 mg) with a 12 L ballast tank. Sample temperature (a) 317 K, **0**; (b) 287 K, **I.**

state. Although a period of hours has been allowed to achieve this result,[9] it is remarkable how quickly it is achieved. Figure 3-3 describes the results of allowing an evacuated sample of diphenylacetylene (DPA: $\backsimeq$ 500 mg) to reach steady state with its vapor in a 12-L ballast tank at a background pressure of approximately 0.2 Pa. Two sets of experiments are reported, which correspond to sample temperatures of 287 and 317.5 K reaching steady state vapor pressures with a ballast tank maintained at 383 K. These results indicate that a steady state is reached after only a few hundred seconds. Similar results have been obtained with a variety of other substances. In general, we have used 600 s as a standard time interval to establish the steady state.

A similar type of experiment can be performed to evaluate the time necessary for complete transfer of the contents from the ballast tank to the collection tube. In this instance, however, the time necessary for transfer depends on a variety of factors including the geometry of the vacuum system, the diameter of the vacuum line and the pumping speed of the forepump and diffusion pumps used. When measuring vapor pressures above 0.1 Pa at a ballast temperature of 350 K (1 Pa at a ballast temperature of 300 K), comparison of the total amount of sample collected as a function of time also becomes invariant with time. In our system this has generally occurred after 500 s of collection. However if this experiment is performed

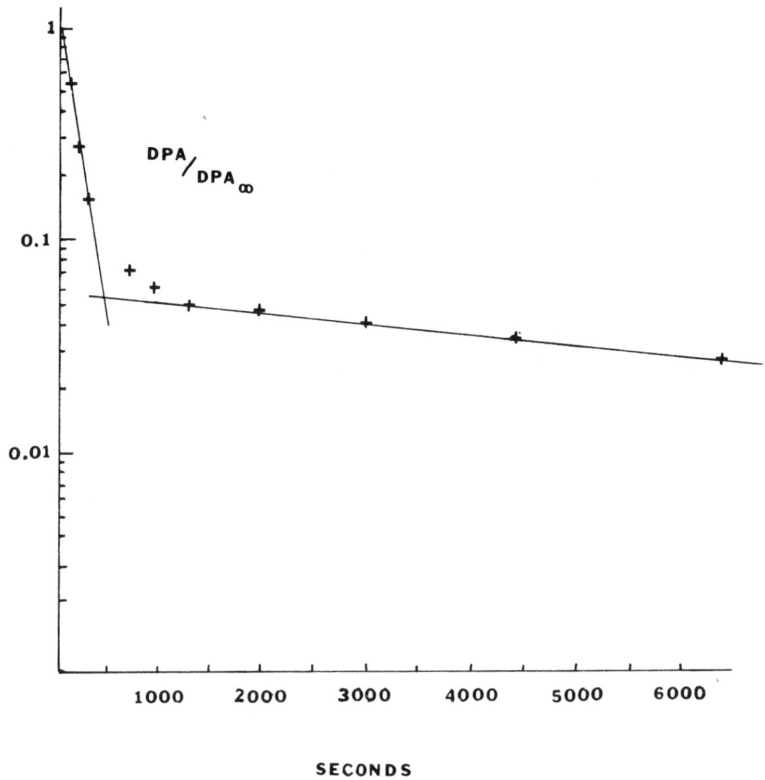

Figure 3-4. The fraction of diphenylacetylene collected from a 12-L ballast tank as a function of time.

differently or if the sample size is small, corresponding to a vapor pressure below the limiting pressures indicated above, sample size does continue to vary with time.

By monitoring the amount of sample collected over designated time intervals, results such as those shown in Figure 3-4 are obtained for diphenylacetylene. The ordinate represents the fraction of the total amount of DPA collected as a function of time for a single "steady state equilibration" between a ballast tank at 385 K and a sample at 317.5 K. The amount of recovered DPA was monitored sequentially at various times over a period of 2×10^4 s. These results can be interpreted as two completing first-order processes characterized by rate constants that differ roughly by an order of magnitude. The rapid process can be described as the transfer of vapor from the ballast compartment and is characterized by a rate constant that is a measure of the pumping speed of the vacuum system. The slower process is best described as the deadsorption of DPA off the surface of the ballast tank and subsequent transfer to the collection tube. Both processes obey first-order kinetics.

When this experiment was repeated using the same sample temperature but at a higher ballast tank temperature, the amount of material collected in the first 600 s remained essentially constant, but the amount of material collected in subsequent time intervals decreased. At a constant ballast temperature but higher sample temperature (higher steady state pressure), the quantities of sample collected both during the first 600 s and subsequently increased, but the former contributed more significantly to the increase.

These results are consistent with the interpretation cited above and offer an explanation of the curvature observed in log P versus $1/T$ plots when measuring vapor pressures around 0.1 Pa.[10] Even when curvature is not observed, the proportional increase in contribution of adsorbed sample to the total as the vapor pressure decreases explains why the heats of sublimation measured by this technique tend to be a few percentage points lower than those measured by other methods. In addition, some simple modifications to the experimental procedure for improving accuracy and for possible expansion of the range of this method into the subpascal region become apparent from these results.

The results in Figure 3-4 indicate that quantitative transfer (99%) of the vapor can be achieved after 600 s. Since this period of time reflects the pumping speed of vacuum system, not of the sample, it should remain invariant to the nature of the sample being measured and need be determined only once for each system. By limiting the period of collection to 600 s, the contribution of adsorbed substrate to the total collected can be minimized. This is demonstrated for diphenylacetylene, whose vapor pressure as a function of temperature is reported in Figure 3-5. The data collected in the 10^{1}–10^{-1} Pa range serves to establish the slope of the line in the Clapeyron relationship of log vapor pressure versus $1/T$ (K^{-1}). As can be seen in Figure 3-5, the apparent vapor pressure levels off below 10^{-1} Pa at approximately 4×10^{-2} Pa if the time for collection of the sample exceeds 1200 s (circles). By restricting the collection period to 600 s in the 10^{-2}-Pa region and correcting the data for the contribution of adsorbed substrate in the 10^{-3}-Pa region, linearity can be observed from 1 to 10^{-3} Pa. Estimation of the amount of adsorbed substrate contributing to the first 600 s was achieved in this instance by two consecutive 600-s collections. The difference between the quantities obtained in these two collections then serves as a more accurate estimate of diphenylacetylene's true vapor pressure at the corresponding sample temperature. Such a correction appears reasonable as long as the contribution of adsorbed material is small.

Extension of the range of this technique much below 10^{-3} Pa would require extensive modification of the vacuum system. An increase in pumping speed by at least an order of magnitude is technologically possible and would be an additional improvement necessary to extend the range further. As currently set up, however, this technique is capable of measuring vapor pressures from 10^{5} to 10^{-3} Pa, and it covers a range of vapor pressures broad

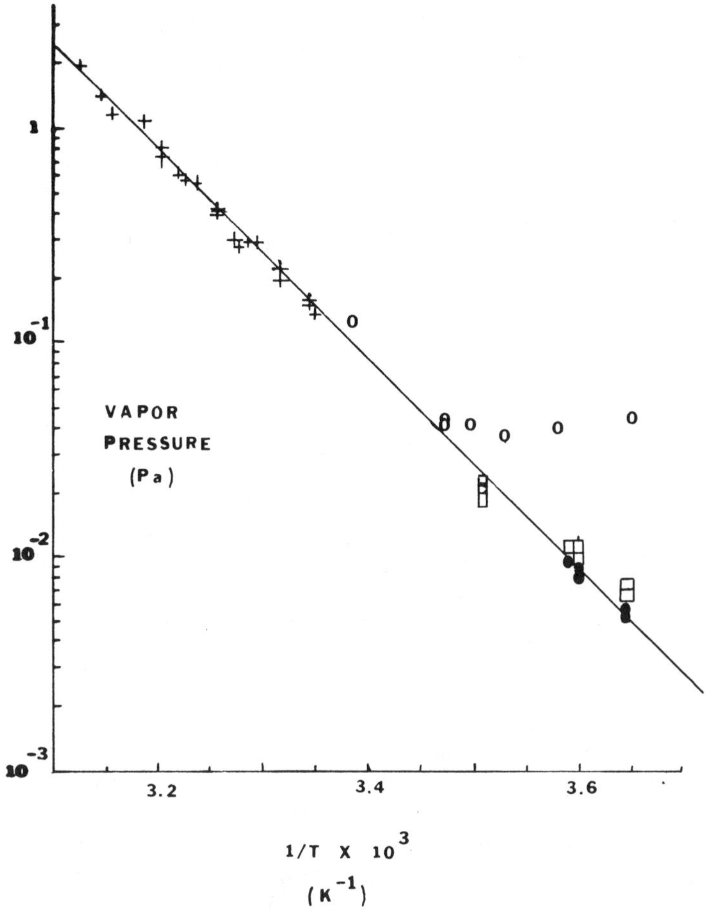

Figure 3-5. Log P vs 1/T for diphenylacetylene: (a) +, collection time: variable, > 600s; (b) ○, collection time: 1200 s; (c) □, collection time: 600 s; (d) ●, difference between two sequential 600 s collections.

enough to permit measurement of the vapor pressures and heats of sublimation of most typical organic molecules.

D. Reliability of the Method

The usefulness of any technique used to measure vapor pressures and heats of sublimation depends ultimately on the reliability of the measurements obtained. Comparisons of previous heats of sublimation obtained by indirect head space analysis, as described above, are provided Table 3-1. Agreement obtained by this method is generally within 5% of the literature values. As noted above, these results are usually a few percent lower than accepted

TABLE 3-1. A Comparison of Heats of Sublimation Determined by Indirect Head Space Analysis with Literature Values

	Head space analysis		Literature[a]	
Compound	ΔH_s (kJ/mol)	(Ref).	ΔH_s (kJ/mol)	(Ref.)
dl-Camphor	51.5 ± 2.6	(10)	51.8 ± 8	(84)
			53.6	(85)
			50.7	(86)
Biphenyl	76. ± 4	(10)	77.9 ± .3	(87)
			80.4 ± 1.6	(88)
			83.6 ± 2.5	(89)
Phenol	65.3 ± 3.3	(10)	69.7 ± .9	(90)
			68.7 ± .5	(91)
			67.8	(92)
Naphthalene	67.8 ± 3.5	(10)	72.6 ± .1	(35)
			72.8 ± .3	(37)
			72.5 ± .1	(36)
2,2,4,4-Tetramethyl-1,3-cyclobutanedione	70.3 ± 3.5	(10)	72.2 ± .6	(93)
Thymol	89.1 ± 4.5	(10)	91.2 ± 4.1	(92)
Tetrathiafulvalene	92 ± 6.3	(4)	95.3 ± 1	(94)
dl-Dimethyl tartrate	112 ± 5.6	(5)	113.8	(33)

[a]Typical values are included for comparison; for a complete list see Tables 3-5 and 3-6.

values. A reasonable explanation of the origin of this error is based on the contributions of adsorbed substrate to the total sample collected. Previous measurements did not correct for these contributions, which are proportionately higher at the lower vapor pressures; hence the slightly lower heats of sublimation. By a more careful timing of the collection period, it is likely that improvement in the accuracy of this method can be realized.

3. HEAT OF SUBLIMATION OF CHIRAL SYSTEMS

A. Introduction

Some years ago we became aware of a report in which fractional sublimation of a partially resolved, optically active material led to optical fractionation of the mixture.[17] A search revealed a few other similar reports scattered in the literature.[18–20] Since fractional sublimation can be a potentially useful technique for enhancing "optical purity," we decided to employ the vapor pressure apparatus we had developed in an effort to gain a better understanding of the effect that was being observed. We embarked on a program of systematically investigating the vapor pressure and heats of sublimation of a variety of chiral solids. A parallel course of study was initiated by a colleague who examined the composition of the sublimate as a function of the optical purity of the initial mixture for $d - l$ mixtures whose phase diagrams were known.[21]

B. Origin of the Effect: Heats of Vaporization and Fusion

Enthalpies of sublimation can be directly related to the corresponding enthalpies of fusion ΔH_{fus} and vaporization ΔH_v, by Equation 3-2. This relationship is exact for enthalpies obtained at the same temperature (eg, the fusion temperature) or when corrected for heat capacity differences between the condensed and gas phases. The enthalpy of sublimation of a solid at temperature T is given by:[22]

$$\Delta H_s(T) = \Delta H_{\text{fus}}(T_M) + \Delta H_V(T_V) + \int_T^{T_M} Cp(\text{c})\, dT + \int_T^{T_V} Cp(\text{l})\, dT + \int_{T_v}^{T} Cp(\text{g})\, dT \qquad (3\text{-}2)$$

where Cp represents the heat capacity of the solid (c), the liquid (l), and the vapor (g), and T_M and T_V represents the normal melting and boiling points, respectively. Neglecting for the moment the heat capacity contributions, which are small (see Subsection B of Section 4), the major contributors to $\Delta H_s(T)$ are $\Delta H_{\text{fus}}(T_M)$ and $\Delta H_v(T_V)$. Comparison of the heats of sublimation of the racemic pair $\Delta H_S^R(T)$ to one of the enantiomers $\Delta H_S^E(T)$, leads to the conclusion that any differences between $\Delta H_S^R(T)$ and $\Delta H_S^E(T)$ are likely to be the result of differences in the corresponding heats of evaporization and/or fusion, Equation 3-3.

$$\Delta H_s^R(T) - \Delta H_s^E(T) = (\Delta H_v^R(T_V) - \Delta H_v^E(T_V)) + (\Delta H_{\text{fus}}^R(T_M) - \Delta H_{\text{fus}}^E(T_M)) \qquad (3\text{-}3)$$

Heats of vaporization usually make up the largest contribution to $\Delta H_s(T)$, and differences between enantiomer and racemic mixture in the liquid state might be expected. However most available evidence suggests that this term contributes only minor differences to $\Delta H_S^R(T) - \Delta H_S^E(T)$. Studies that have tried to identify the nature and magnitude of the differences in interactions between liquid enantiomers and their racemic counterparts include: identical densities and dielectric constants for 2-octanol[23] and a heat of mixing ($d + l \rightarrow dl$) of only 13 J/mol[24,25]; identical vapor pressures and boiling points of (d) and (dl)-2-butanol[24–26]; heats of mixing for 2-(p-nitrophenyl)butane and α-methylbenzylamine of 1.88 and 9.6 J/mol,[24] respectively; excess molar volumes for limonene, fenchone, and α-methylbenzylamine which are small[27]; and heats of mixing of solutions of enantiomers of α-methyl-α-ethylsuccinic acid ($CHCl_3$, 298 K),[25] α-methyl-α-isopropylsuccinic acid ($CHCl_3$, 298 K),[25] tartaric acid (H_2O, 298 K),[28] and threonine (H_2O, 298 K)[28] of 140, 209, 2.0, and 5.5 J/mol, respectively. In contrast, a standard free energy difference of 2.76 kJ/mol at 296 K has been recently reported for

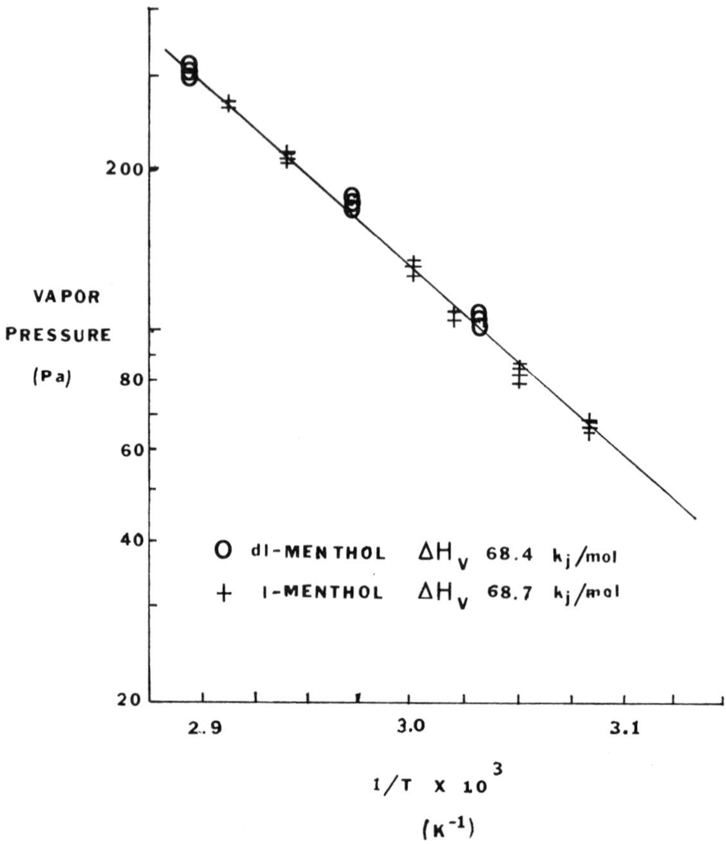

Figure 3-6. Log *P* versus $1/T$ for (*dl*), (*l*)-menthol in the liquid state.

hydrogen bonding between homodimers ($R \rightarrow R$; $S \rightarrow S$) and heterodimers ($R \rightarrow S$; $S \rightarrow R$) in 2-butanol.[29] Studies of hydrogen bonding in racemic and (*R*)-2,2,4-trimethylpentane-3-ol, however, show differences of only 200 J/mol.[30]

Although some of these results do not necessarily preclude differences in $\Delta H_v(T)$ and in vapor pressure, they are indicative of similarities in $\Delta H_v(T_V)$ for both the *d* and *dl*, particularly when strong intermolecular forces are not involved. Experimental results obtained in our laboratory support this contention. The vapor pressure of *d*- and *dl*-ephedrine (mp 312, 352 K) measured at 354.4 K gave results of 71.9 ± 1 and 69.1 ± 2.5 Pa, respectively.[31] The vapor pressures of liquid *l*- and *dl*-menthol (mp 316, 301 K, respectively) have been measured from 324 to 343 K. The results shown in Figure 3-6 do not suggest any substantial difference in either vapor pressure or enthalpy of vaporization, $\Delta H_v^R(T_V) - \Delta H_v^E(T_V)$.[31]

Heats of fusion of enantiomers and their racemates generally make up only about 25% of the total contribution to $\Delta H_s(T)$ for typical organic solids.

TABLE 3-2A. Heats of Fusion of Some Representative Chiral Solids

Compound	Reference	Melting point (K)	ΔH_{fus} (kJ/mol)	Phase diagram case (Fig. 3-7)
Mandelic acid	22, 79			
Racemic		392	25.5	B
Active		406	26.4	
Menthyl 3,5-dinitrobenzoate	5			
Racemic		427	30.6	A
Active		401	34.5	
Carvone oxime	5			
Racemic		364	16.8	C
Active		344	15.0	
3-(*m*-Chlorophenyl)hydracrylic acid	22, 79			
Racemic		340	23.8	A
Active		368	28.0	
2-(*o*-Chlorophenoxy)propionic acid	22, 79			
Racemic		388	32.1	B
Active		369	26.8	

TABLE 3-2B. Differences in the Heats of Solution of Racemic and Optically Active Solids[a]

Compound	$\Delta H_{sol}(dl) - \Delta H_{sol(active)}$ 298 K (kJ/mol solute)	Phase diagram case (Fig. 3-7)
Alanine	$+2.0 \pm 0.2$	
Glutamic acid[b]	$+4.1 \pm 0.2$	A
Histidine	$+1.5 \pm 0.2$	
Threonine[b]	-0.17 ± 0.13	A
Tartaric acid	$+9.4 \pm 0.3$	B
Valine	$+2.3 \pm 0.01$	

[a]From Reference 95.
[b]It is interesting and perhaps surprising that although glutamic acid and threonine both form conglomerates,[22] a reversal in sign is observed between the two.

Here, however, differences are observed between the two forms as demonstrated in Tables 3-2 except in certain circumstances. Chiral molecules crystallize in one of three general categories (see Figure 3-7). Molecules of a given chirality prefer to cohabit the crystal with other molecules of either the same (case A, conglomerate) or opposite chirality (case B, racemic compound formation), or the crystal lattice does not differentiate chirality (case C, solid solution). Approximately 5–10% of all organic substances crystallize in the form of conglomerates. Formation of solid solutions also appears to be rare. The bulk of organic molecules prefer to form racemic compounds in which the *d* and *l* are present in a 1:1 ratio, although a few cases are known with different stoichiometry.[22] Examples exist in each category, as

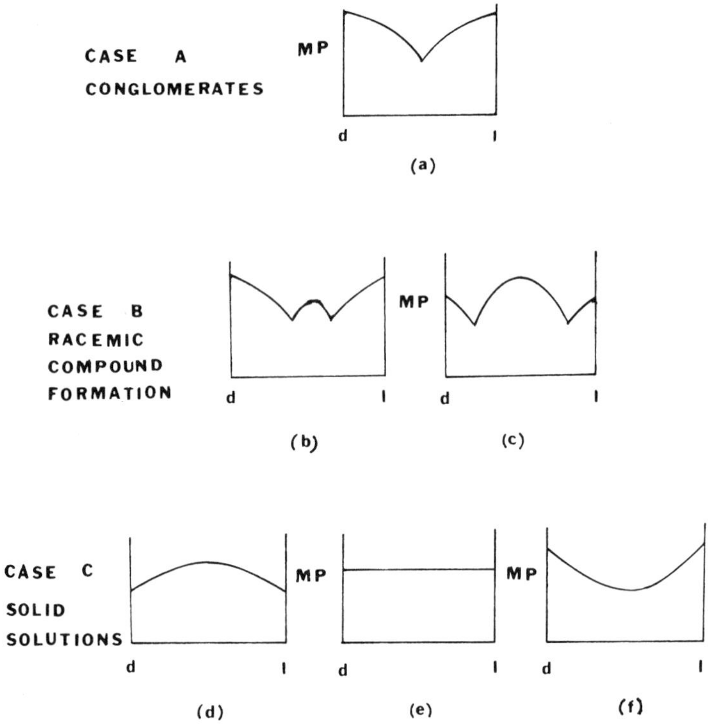

Figure 3-7. Typical phase diagrams for enantiomers and their racemates. (*a*) Case A: conglomerates. (*b*) Case B: racemic compound formation. (*c*) Case C: solid solutions.

demonstrated in Tables 3-2 and 3-3, which clearly document the dependence of a variety of physical properties such as melting point, density and heat of fusion on the *d-l* composition of the solid, even in solid solutions (see below). Consequently, referring back to Equation 3-3, if differences in ΔH_s and vapor pressure are to be expected, these differences for the most part should reflect any differences observed between heats of fusion for enantiomer and racemate. Of significance to this discussion is whether these differences can be correlated with other physical properties and whether they can be of use in resolution studies.

C. Heats of Sublimation

Until our own work, we are aware of only two studies of the heat of sublimation on both enantiomer and racemate. Both studies measured the heat of sublimation of (*d*) and (*dl*)-dimethyl tartrate. The conclusions reached by examination of these results were inconclusive. One study found a difference of 6.7 ± .6 kJ/mol between (*d*) and (*dl*)-dimethyl tartrate,[32] while the

TABLE 3-3. Some Physical Constants of Enantiomers and Their Racemates[a]

Compounds	Melting point (K)	ΔH_s (kJ/mol)	ΔH_{fus} (kJ/mol)	ΔH_v (kJ/mol)	ΔH_s(calc) (kJ/mol)	VP[b] (Pa)	Temperature (K)	Case (Fig. 3-7)
l-Menthol	316	95.8 ± 5	11.9	68.7[c]	80.6	6.5	298	
						0.4	278	
dl-Menthol	301	78.7 ± 4	10.3	68.4[c]	78.7	6.4	298	B
						0.7	278	
d-Dimethyl tartrate	322	77.4 ± 8	17.4	64.3[d]	81.7	1.4	320	
						0.5	310	
dl-Dimethyl tartrate	360	112.4 ± 6	27.0	62.8[d]	89.8	0.4	320	B
						0.1	310	
d-Carvone oxime	344	90.8 ± 5	15.0			12.9	345	
						1.8	325	
dl-Carvone oxime	364	101.7 ± 5	16.8			8.6	345	C
						1.0	325	

[a] Data from Reference 5 unless otherwise noted.
[b] Vapor pressure at the corresponding temperature, column 8.
[c] From Reference 96.
[d] See Figure 3-6.

other found a difference of only 0.75 kJ.[33] Substantial differences in the vapor pressures between the two compounds were observed in the latter case.

The crystalline systems we chose to investigate are representative of the most common phase diagrams observed in chiral organic solids. (*d*),(*dl*)-Dimethyl tartrate was chosen for reasons mentioned above as an example of a system whose phase diagram exhibits compound formation (case B). The melting point at the *dl* composition is greater than for pure enantiomer (Figure 3-7*c*). (*dl*),(*l*)-Menthol is an example of a similar case in which the melting of pure enantiomer is greater than for the (*dl*) compound (Figure 3-7*b*). (*d*),(*dl*)-Carvone oxime was chosen as an example of a system that forms a solid solution (case C, Figure 3-7*d*). This system is unusual, however. Although the crystal lattice does not differentiate chirality, the physical properties are reflective of the (*d*)-(*l*) composition of the solid, with the racemic composition exhibiting a maximum in the melting point. Other important crystalline systems that have not yet been examined include conglomerates (case A, Figure 3-7*a*), ideal solid solutions, and solid solutions exhibiting a minimum in the melting point (case C, Figure 3-7*e*,*f*).

Table 3-3 summarizes a variety of physical constants measured for the materials mentioned above. First it should be noted that the difference in ΔH_s and vapor pressure observed between enantiomer and racemate have nearly the same magnitude as the experimental uncertainties associated with these numbers. This is particularly true of the results of *d*-dimethyl tartrate, for which data were collected over a temperature span of only 10 K. However, since comparisons between enantiomer and racemate are relative numbers, comparison of ΔH_s and vapor pressures are believed to be qualitatively significant. This conclusion is further reinforced by the agreement obtained from heats of sublimation calculated from the heats of fusion and vaporization (column 6) according to the first two terms of Equation 3-2. Heats of sublimation estimated in this manner by ignoring heat capacity corrections for enthalpies obtained at higher temperatures can generally be considered to be a lower limit for ΔH_s (see below).

Several trends become evident upon examination of the data in Table 3-3. Differences in heats of sublimation between enantiomer and racemate are paralleled by similar differences in the heats of fusion. The crystalline form with the highest melting point exhibits the greatest ΔH_s, ΔH_{fus} and generally the lowest vapor pressure.

Thus although there are not enough data to clearly establish this trend, it does suggest that fractionation of optically pure materials can occur and that predictions about the composition of the vapor are possible from examination of the appropriate phase diagram. Eutectic compositions generally exhibit the lowest heats of fusion (see results for menthyl 3,5-dinitrobenzoate, Table 3-2A) and therefore should be the most volatile. Depending on the initial optical purity relative to the composition of the closest eutectic, fractional sublimation can lead to optical enrichment of either the sublimate or the residual sample. This has been demonstrated for *d*, *dl*-mandelic acid (phase diagram, Figure 3-7*b*) by starting at optical purities on each side of the eutectic composition.[21] In both cases the eutectic composition sublimed preferentially.

In summary, heats of sublimation and heats of fusion are expected to vary depending on the phase diagram. Fractional sublimation is not expected in general to lead to optically pure material except perhaps when conglomerates and certain types of solid solution are involved. However if the eutectic composition lies close to either the pure enantiomer or the *dl* pair, substantial optical enrichment of the sublimate or residual substrate is possible, as has been reported.[17–21] Additional experimentation is necessary to confirm the trends noted in this work. Demonstration of the feasibility of optical purification or enrichment of racemates by fractional sublimation relative to more conventional techniques such as recrystallization would also be worthwhile and could prove to be of considerable practical utility in many studies involving resolutions.

D. Vapor Pressures of Conglomerates

As noted previously, differences in vapor pressures have been observed between enantiomer and racemic compound. In the case of conglomerates in which the eutectic composition corresponds to a 1:1 ratio of *d* to *l*, the vapor pressure of the eutectic should correspond to twice the vapor pressure of a single enantiomer ($v_d + v_l = 2v_{dl}$), since a conglomerate is a physical mixture of *dd* and *ll*. The inverse of this statement suggests that if at a particular temperature the vapor pressure of a racemic compound equals twice the vapor pressure of pure enantiomer, this value should correspond to the transition temperature for the racemic compound to conglomerate transformation. Whether this is a necessary and sufficient condition for the existence of polymorphism of the type just described has not yet been experimentally demonstrated. The foregoing implies that a chamber consisting of physically distinct crystals of *d* and *l* should grow at the expense of racemic compound (*dl*) at a temperature in which the vapor pressure of the *dl* just exceeds the sum of the vapor pressures of each respective enantiomer.

The relatively rare occurrence of conglomerates coupled with their potential usefulness in resolutions—conglomerates often can be resolved by physical methods[22]—has prompted the development of methods to aid in their detection. Ironically it was just such considerations of vapor pressure that led Meyerhoff in 1904 to propose the "double-solubility" rule, which is of considerable value in detecting conglomerates.[22,34] The rule predicts that the solubility of conglomerates will be greater than or equal to the sum of the solubilities of each enantiomer. The success of the rule is based on the similarities between vapor pressure and solubility and on the mathematic relationships that describe their temperature behavior under ideal conditions, the Clapeyron and Schroder–Van Larr equations. This topic has recently been reviewed.[5,22]

4. HEATS OF SUBLIMATION

A. Heat of Sublimation Standards

An examination of the literature (Tables 3-5 and 3-6, in subsection D), reveals that a number of different substances have been used as standards for heat of sublimation and vapor pressure measurements. The most frequently used standards include materials such as benzoic acid, naphthalene, benzophenone, ferrocene, and anthracene. A more detailed listing can be found in the tables. These compounds have been chosen as standards for a variety of reasons, including availability in pure form, long-term thermal stability, and ease of measurement. Recent work has afforded very accurate measurements for naphthalene[35–37] and benzoic acid.[36–38] Both compounds have been recommended as standards for both vapor pressure and ΔH_s measurements.[39] Some objections to the use of benzoic acid, however, have recently been raised.[36]

From our perspective, selection of standards is long overdue and more standards are needed. The measurement of vapor pressures or heats of sublimation of substances with vapor pressures in the 1-Pa range is relatively straightforward. The experiment becomes progressively more difficult as the vapor pressure decreases. The ability of a technique to accurately measure vapor pressures or ΔH_s in one pressure region does not in itself guarantee the same degree of accuracy in a different pressure region.[81] An examination of the data in Tables 3-5 and 3-6 supports such a conclusion. The precision in heats of sublimation reported for many compounds often exceeds their accuracy. This suggests that the systematic errors revealed by comparison of the results obtained by different methods are a function of pressure and/or temperature. Only by establishing a series of standards whose vapor pressures cover the entire vapor pressure region experimentally accessible and the entire temperature range commonly employed, can the accuracy of a technique be assessed within a given pressure and temperature range. This series of standards would serve the same general purpose as the series of melting point standards that are used to standardize thermometers.

B. Heats of Sublimation: Corrections to a Standard State

Enthalpies of sublimation at 298 or 0 K are often required for a variety of studies, including correcting heats of formation to a standard state and for standardizing atom–atom potentials used in the calculations of lattice energies. Enthalpies measured at other temperatures can be corrected to the reference temperature T_R by correcting for the differences in heat capacities at constant pressure ΔCp between the gas and condensed phases, Cp(g) and Cp(c), respectively. Since many enthalpies are measured over a range of

$$\Delta H(298, 0\text{ K}) = \Delta H_s(\overline{T}_M) + \Delta H(T_T) - \int_{298,0}^{T_M} Cp\, dT \tag{3-4}$$

temperatures, corrections are customarily from the mean temperature of this range, $\overline{T}_M$. If a phase transition occurs between $\overline{T}_M$ and the reference temperature, the enthalpy change of this transition $\Delta H(T_T)$ must also be included.

Since heat capacity corrections are generally small and often comparable in magnitude to the uncertainty associated with the heat of sublimation measured, varying degrees of approximation have been used to correct to 298 or 0 K. Recent technological advancements in the measurement of heats of sublimation have afforded data for which these corrections are more significant.

If the heat of sublimation is accurately known and the heat capacity data are available, corrections to a standard state T_R simply involves integrating the $\Delta Cp\, dT$ term. Assuming ΔCp to be independent of temperature, yields:

$$\Delta H(T_R) = \Delta H(\overline{T}_M) + \Delta H(T_T) + \Delta Cp(\overline{T}_M - T_R) \tag{3-5}$$

Since Cp does vary with temperature, mean values can be used for Cp(g) and Cp(c) in evaluating ΔCp in Equation 3-5 for the temperature range $\overline{T}_M$–T_R.

Often heat capacities for both the gas and condensed phases are not known. The heat capacity of the condensed phase can be measured directly, for example, by differential scanning calorimetry (DSC). Reasonable estimates of Cp(g) can be obtained by bond additivity or group additivity methods and by structural methods employing statistical mechanics. Specific details on these estimation techniques for Cp(g) have been described in detail by Benson.[40]

Vapor pressures that have been measured over a significantly large temperature range show some curvature in plots of ln P versus $1/T$. In this instance ΔCp can be extracted by fitting the results to the equation:[41,42]

$$R \ln \frac{P_2}{P_1} = -\frac{\Delta G^\circ(T_R)}{T_R} + \frac{\Delta H^\circ(T_R)}{T_R} + \Delta Cp(T_R)\left(\frac{T_R}{T} - 1 + \ln \frac{T}{T_R}\right) \tag{3-6}$$

where T_R is a reference temperature, P_0 is a standard pressure (1 Pa), $\Delta Cp(T_R)$ is the value of $Cp(\text{g}) - Cp(\text{c})$ at T_R, and $\Delta G^\circ(T_R)$ and $\Delta H^\circ(T_R)$ are the standard thermodynamic functions evaluated at the reference temperature. Generally the temperature range studied is not large enough to permit the observation of curvature; the only parameters that can be evaluated from Equation 3-6 in such circumstances are $\Delta G^\circ(T_R)$ and $\Delta H^\circ(T_R)$.

An alternative and more approximate method that has been used to correct heats of sublimation is to treat the gas as ideal and to assume that the Dulong–Petit value holds for the solid.[4,43,44] The molar heat capacity of the gas at constant volume $Cv(\mathrm{g})$ can be partitioned into translational, rotational, vibrational, and electronic components:

$$Cv(\mathrm{g}) = C_{v(\mathrm{trans})} + C_{v(\mathrm{rot})} + C_{v(\mathrm{vib})} + C_{v(\mathrm{elec})}$$

where $C_{v(\mathrm{trans})} = 3/2\,R$, $C_{v(\mathrm{rot})} = 3/2\,R$ (R for linear molecules), $C_{v(\mathrm{vib})} = (3N - 6)R$ ($(3N - 5)R$ for linear molecules), where N is the number of atoms in the molecule and $C_{v(\mathrm{elec})} \simeq 0$.

Treating a mole of gas containing N atoms per molecule as a solid of M vibrating atoms, where $M = N \times A$ (A = Avogadro's number), the total vibrational degree of freedom for the solid is $3M$. The molar heat capacity according to the Dulong–Petit treatment is $3R$, which for M atoms gives a total heat capacity of $3RM/A = 3RN$. Substituting into Equation 3-4 for the heat capacities of solid and gas and noting that:

$$Cp(\mathrm{g}) = Cv(\mathrm{g}) + R, \qquad \text{and} \qquad Cv(\mathrm{s}) \simeq Cp(\mathrm{s})$$

yields:

$$\Delta H_s(T_R) = \Delta H_s(\overline{T}_M) + \Delta H(T_T) - \int_{T_R}^{\overline{T}_M} [R + 3R + (3N - 6)R - 3RN]\, dT \tag{3-7}$$

which upon simplification:

$$\Delta H_s(T_R) = \Delta H_s(T_M) + \Delta H(T_T) + \int_{T_R}^{\overline{T}_M} 2R\, dT \tag{3-8}$$

and integration results in:

$$\Delta H_s(T_R) = \Delta H_s(\overline{T}_M) + 2R(\overline{T}_M - T_R) + \Delta H(T_T) \tag{3-9}$$

Equation 3-9 states that heats of sublimation will always increase with decreasing temperature. For this condition to hold, according to Equation 3-4, the heat capacity of the solid must be greater than for the corresponding gas. Examination of Tables 3-5 and 3-6 shows that when ΔH_s is reported at two temperatures, an increase is observed with decreasing temperature. Other methods of estimating heat capacities are described in subsection C, below.

C. Estimations of Heats of Sublimation

The difficulty in obtaining reliable heats of sublimation data has prompted the development of techniques that can be used to estimate ΔH_s when such data are not readily available. One of the earliest methods was application of Walden's rule[45] (Equation 3-11), and Trouton's Constant[46] (Equation 3-

12) to evaluate the heats of fusion and vaporization, where T_M and T_B represent the melting and boiling points, respectively.

$$\Delta H_s(298) \approx \Delta H_{\text{fus}}(T_M) + \Delta H_v(T) \tag{3-10}$$

$$\frac{\Delta H_{\text{fus}}(T_M)}{T_M} \approx 56.5 \text{ J/K} \tag{3-11}$$

$$\frac{\Delta H_v(T_B)}{T_B} \approx 92 \text{ J/K} \tag{3-12}$$

An alternative approach to estimating ΔH_v in Equation 3-10 was developed by Klages,[47] who proposed that for nonpolar and slightly polar compounds,

$$\Delta H_v(298) = 22.6 + 0.151(T_B - 298) \text{ kJ/mol} \tag{3-13}$$

and for hydrogen-bonded compounds,

$$\Delta H_v(298) = 28.5 + 0.188(T_B - 298) \text{ kJ/mol} \tag{3-14}$$

The major limitation of Equations 3-12 to 3-14 is that normal boiling points for solids are often unknown or inaccessible.

Results reported for acridine (mp 384, bp 619 K) and phenazine (mp 445, bp 612 K) according to Equations 3-10 to 3-12 are 78.7 and 81.4 kJ/mol,[48] whereas Equations 3-11 and 3-13 give 92.8 and 95.2 kJ/mol, respectively, for ΔH_s. These results can be contrasted to the experimentally determined values of 91.6 ± 2.5 (Reference 49) and 99.9 ± 2.5 (Reference 50) (90.4 ± 4.7: Reference 49) kJ/mol for each, respectively.

A similar method has been used to determine ΔH_s for diphenyldisulfide and dibenzyldisulfide.[51] In this instance ΔH_v (298) was estimated by the additivity method of Laidler.[52] Heats of fusion were estimated by Walden's rule and adjusted to 298 K using Sidgewick's approximation[53]:

$$\Delta H_{fus}(298) = \Delta H_{fus}(T_M) + 0.013(298 - T_M) \tag{3-15}$$

Experimental comparisons are not available for these compounds.

Other equations used to correct $\Delta H_{\text{fus}}(T_M)$ and $\Delta H_v(T)$ to 298 K include:[54]

$$\Delta H_{\text{fus}}(298) = \Delta H_{\text{fus}}(T_M) - \Delta Cp(\text{lc})(T_M - 298) \tag{3-16}$$

$$\Delta H_v(298) = \Delta H_v(T) - \Delta Cp(\text{gl})(T - 298) \tag{3-17}$$

$$\Delta H_v(298) = \Delta H_v(T)\left[\frac{T_C - T}{T_C - T_B}\right]^{0.38} \tag{3-18}$$

where ΔCp(lc) is the molar heat capacity difference between liquid (l) and solid (c), ΔCp(gl) is the same difference between gas and liquid, and T_C is the critical temperature. The heat capacities required in Equations 3-16 and 3-17 for solids and liquids can be estimated using a form of Kopp's law. The heat capacity of a solid or liquid can be approximated by the sum of the heat capacities of the constituent atoms. Values for the following atoms in solids and liquids have been assigned:[55] C, 7.5 (11.7); H, 9.6 (18); B, 11.3

(19.7); Si, 15.9 (24.3); O, 16.7 (25.1); F, 20.9 (29.3); P, S, 22.6 (31.0); and 25.9 (34.3) J/K for heavier atoms. Values in parentheses are for liquids. Heat capacities of the corresponding gas phase can be estimated as described in subsection B above or by other methods.[56] Methods of estimating the critical temperature required in Watson's equation (3-18)[57] are described by Thomson.[58] All required knowledge of the normal boiling point. Other semiempirical rules have been developed for estimating heat capacities of a variety of organic compounds.[59]

Numerous successful methods have been developed for estimating $\Delta H_{v(298)}$ of liquids, particularly hydrocarbons.[52,53,56–58,60–70] These methods vary in the empirical parameters required. Some methods depend on experimental parameters (eg, critical properties); others require only knowledge of molecular structure. The relative magnitudes of ΔH_v and ΔH_{fus} observed for many organic compounds suggest that the better the estimation of ΔH_v, the closer will Equation 3-10 approximate experimental values of ΔH_s. Application of these methods to estimate $\Delta H_v(298)$ of solids together with a better assessment of $\Delta H_{\text{fus}}(298)$ should improve the accuracy of Equation 3-10. The obstacles imposed by crude estimations such as Walden's rule of $\Delta H_{\text{fus}}(T_M)$[56,71] can be circumvented by direct experimental measurement. Using current instrumentation, this measurement can be accomplished quickly and accurately and is experimentally accessible for many organic solids. When fusion cannot be observed directly, ΔH_{fus} may be available by other techniques such as the temperature dependence on solubility (Schroder–Van Larr equation, Section 3).[22]

An alternate method of estimating heats of sublimation by molecular structure increments has been developed by Bondi.[72] The method assumes additivity in group increments to yield the total sublimation energy,

$$\Delta H_s(\text{RX}) = \Delta H_s(\text{R}) + \Delta H(\text{X}) \tag{3-19}$$

Group increments have been tabulated, and the method is quite useful for rough estimates of ΔH_s, particularly when molecular instability is involved.[6]

The vaporization energy $\Delta E(s)$, where

$$\Delta E_{(s)} = \Delta H_s - RT$$

of condensed polynuclear aromatic hydrocarbons was found to vary linearly with the number of carbon atoms in the molecule.[73] The average contribution was found to be 6.3 kJ/mol per carbon atom. A similar additivity was observed by Aihara for a variety of substituted aromatics, and group contributions were derived.[74] Although the accuracy of the sublimation enthalpies used to derive the two relationships described above has been questioned,[61] these results do support the concept of group additivity at a qualitative level. The sublimation enthalpies of planar aromatic hydrocarbons have been correlated with molecular structure most recently by Morawetz,[61] and bond contributions by type have been tabulated.

Other accurate but also more restrictive relationships for some simple homologous series have been developed.[75] The relationships are of the form:

$$\Delta H_s = A + B(n) \tag{3-20}$$

where A is a measure of the contribution to ΔH_s of the functional group and B is a constant that reflects the sensitivity of ΔH_s to homologating an alkyl side chain. Some of the systems investigated include molecules of the types ROH, RCO_2H, RCO_2CH_3, and $RCONH_2$, where $R = C_nH_{2n+1}$ and n is incremented by 2. Relationships have been observed for both the even and odd series but are restricted to the members of the series that are isostructural (ie, members of the same polymorphic form).[75] Similar relationships have been established for even and odd alkanes,[63]

$$\Delta H(298) \approx -8.58 + 8.95\,(n_C), \quad \text{(even } n_C\text{; kJ/mol)} \tag{3-21}$$

$$\Delta H(298) \approx 26.6 + 8.95\,(n_C), \quad \text{(odd } n_C\text{; kJ/mol)} \tag{3-22}$$

$$\Delta H(298) \approx 4.1 + 8.24\,(n_C), \quad \text{(even } n_C\text{; kJ/mol)} \tag{3-23}$$

where N_C is the number of carbon atoms in the molecule. The differences observed for even and odd members of these series appears to result from differences in crystal structure, particularly from packing differences in the end group layers in the crystal.[62,78] Equations 3-21 to 3-23 appear to have been established by two independent but parallel studies of lattice energies of normal alkanes. Equations 3-21 and 3-22 were derived from experimental heats of sublimation of even normal alkanes from C_{12} to C_{20} and odd alkanes C_{13} to C_{19}, respectively (ΔH_s calculated for liquids from Equation 3-2).[61] Equation 3-23 was derived from a relationship similar to Equation 3-20 established for lattice energies of even n-alkanes up to C_{16} by Mnyukh (for triclinic structures, $A = -12.45$, $B = -8.39$ kJ/mol).[76] Using crystal structure data for $C_{18}H_{38}$ and $C_{32}H_{66}$, together with a Lennard–Jones 6–12 potential, Equation 3-23 was derived and compared to ΔH_s values obtained by an alternative estimation technique.[77]

The approximate nature of the assumption of group additivity in heats of sublimation is particularly dramatized in the ΔH_{fus} and ΔH_s results reported in Tables 3-2 and 3-3. Differences of several kilojoules per mole are routinely observed in ΔH_{fus} between racemic and active forms of the same chiral molecule, and differences up to 16 kJ/mol have been measured.[79] These differences are clearly due to differences in packing. Similar deviations from group additivity also have been observed in the chlorobenzenes.[80] The dependence of ΔH_s on molecular packing in the crystal suggests that estimations of ΔH_s by the methods of group additivity are likely to remain either qualitative or highly restrictive.

D. Heats of Sublimation Compendium: Tables 3-5 and 3-6

The most frequent use of heat of sublimation data is to correct heats of formation to a standard state. Compendia on heats of formation usually include references for heats of sublimation and are a valuable reference source for this information.[1–3] However these compendia are sources for heats of formation; the available information on heats of sublimation is not exhaustive, nor does the coverage include all available data. Heat of sublimation data had not been compiled since 1960,[82] and because results are diffusely scattered in the literature, an exhaustive search was initiated to update this information.

This compendium has resulted in the collection of some 900 organic compounds whose heats of sublimation had been reported between 1910 and 1984 and represents an attempt to list all determinations reported for each substance. A few organometallic compounds are also included, but the list here is far from comprehensive. Coverage for organics is comprehensive for 1910–1984 and includes data reported in the sublimation compendium by Jones (1960),[82] and thermochemical compendia of Cox and Pilcher (1970),[2] Pedley and Rylance (1977),[1] and the chemical literature. With a few exceptions, estimates of heats of sublimation are not included. The conventions described below apply to both tables (3-5 and 3-6).

The compounds are arranged in alphabetical order in each table, with the usual exception of carbon and hydrogen. The sequencing within a given elemental composition is in terms of monoatomic increments in reverse alphabetical order. Column 2 lists the chemical name of each substance. Occasionally, a trivial name is also included for ease of identification. Other identifying information is included, if available, such as the space group, or melting point if the compound was known to exhibit polymorphism at the time of measurement. Otherwise, reference to column 3, which refers to the temperature range in which the measurement were made, serves to experimentally define the most probable crystalline modification used. Absolute configurations and/or sign of rotation are included for chiral molecules when possible.

In a few cases of reports of ΔH_s for a molecular complex, such complexes are listed under the molecular formula of the member that would be listed first. Thus the molecular complex formed from tetrathiafulvalene—7,7,8,8-tetracyanoquinodimethane ($C_6H_4S_4 \cdot C_{12}H_4N_4$)—is listed under $C_6H_4S_4$.

The experimental heats of sublimation are listed in column 4 in kilojoules per mole, along with the experimental uncertainty if reported. Column 5 lists the temperature to which this heat of sublimation refers. In many cases the temperature reported in column 5 is simply the mean temperature of the experiments. In other cases, heat capacity corrections to the standard state (298 K) are included. The reader is referred to the original literature for specific details regarding the experimental or approximate nature of these corrections.

In a few instances, the heats of sublimation reported in column 4 are prefaced with a U. This implies that in our estimation these results are unreliable. Criteria used in this evaluation include comparisons with other literature values reported for the same substance; where comparisons were not directly available, it was determined how well a given technique reproduced literature values when such comparisons could be made.

Column 6 abbreviates the technique used in the reported measurement; see Table 3-4 for a definition of the abbreviations used. The final column lists the reference(s) to the original work, which are arranged chronologically in the reference section following the table. Recent improvement in equipment and design, as well as the development of techniques for simultaneous analysis, generally afford results that are more reliable than earlier ones. Consequently the entries in the table for each substance are arranged in reverse chronological order. In addition, reference to one of the compendia previously mentioned is included when possible as an additional source to the original literature, thereby significantly reducing the chances of citation error. When the original literature was not consulted, the *Chemical Abstracts* reference is included in the reference.

Table 3-4. Abbreviations Used in Tables 3-5 and 3-6

Abbreviation	Instrument or technique[a]
A	Calculated from the vapor pressure data reported by the method of least squares
B	Calculated from the sum of the heats of vaporization and fusion
BG	Bourdon gage
C	Calorimetric determination
DBM	Dibutylphthalate manometer
DM	Diaphragm manometer
E	Estimated
GS	Gas saturation, transpiration
HSA	Head space analysis
I	Isoteniscope
IPM	Inclined piston manometry
KG	Knudsen gage
LE	Langmuir evaporation
ME	Mass effusion—Knudsen effusion
MG	McLeod gage
MM	Mercury manometer
NA	Not available at the time of publication
QF	Quartz fiber
QR	Quartz resonator
RG	Rodebush gage
T	Tensimeter
TCM	Thermal conductivity manometer
TE	Torsion effusion
TGA	Thermal gravimetric analysis
TSGC	Temperature scanning gas chromatography
V	Viscosity gage
VG	MKS Baratron vacuum gauge

[a]A description of many of these instruments and techniques can be found in Reference 83.

TABLE 3-5. Reported Heats of Sublimation of Organic Compounds, 1910–1983

Molecular formula	Compound	Temp. range (K)	ΔH_s (kJ 1 kcal = 4.184 kJ)	T (K)	Method (see Table 3-4)	Reference
CBrN	Cyanogen bromide	(273–308)	45.2±4.2		MM	[54/7] [70/1]
		(356–308)	47		GS	[20/1]
CBr_4	Carbon tetrabromide					
	Monoclinic	(295–319)	54.4±1.3	(307)	BG	[59/5]
	Cubic	(321–329)	49.4±1.3	(325)	BG	[59/5]
CClN	Cyanogen chloride	(196–259)	35.7	(228)	A	[47/2]
CCl_4	Carbon tetrachloride		43.3	(226)	B	[63/6]
		(209–225)	38.8	(217)		[60/1] [48/1]
		(227–248)	37.9			[48/1]
CFN	Cyanogen fluoride	(139–192)	24.4	(166)	A	[47/2]
CF_4	Tetrafluoromethane		17.0	(76)		[63/6]
ClN	Cyanogen iodide	(298–414)	58.6	(356)	A	[47/2]
			59.8±0.4			[43/2] [70/1]
		(278–374)	58.3	(326)		[33/2]
CHI_3	Iodoform		69.9			[43/1]
CHN	Hydrogen cyanide	(202–254)	37.6	(228)	A	[47/2]
CHN_3O_6	Trinitromethane		54.8±4.2			[70/7] [77/1]
			46.7±0.4			[67/4] [70/1] [77/1]
CH_2N_2	Cyanamide	(227–289)	75.9	(290)	TE,ME	[83/7]
			75.2	(298)		[83/7]
CH_2N_4	Tetrazole	(333–363)	97.5±4.2	(348)	ME	[51/3] [70/1]
CH_2O_2	Formic acid	(203–218)	62.1±1	(213)	TE,ME	[78/16]
		(265–268)	60.7	(266)		[30/1] [60/1]
		(253–275)	60.1	(264)	A	[47/2]
$(CH_2O_2)\cdot(CH_2O_2)$	Formic acid dimer	(203–218)	64.1±1	(213)	TE,ME	[78/16]
CH_3Cl	Methyl chloride		28.0		B	[40/3]
CH_3Cl_2OP	Methylphosphonic dichloride		62.3±4.2			[55/4] [70/1]
CH_3I	Methyl iodide	(176–227)	40.2±0.4	(191)	VG	[82/6]
			U 69.9			[48/2] [60/1]
CH_3NO	Formamide	(251–273)	72.4	(264)	TE,ME	[83/7]
			71.7	(298)		[83/7]
			71.7	(276)		[79/11]
CH_4	Methane	(67–88)	9.62	(77)	A	[47/2]
		(54–90)	9.2	(72)		[63/6] [55/2]

Molecular formula	Compound	Temp. range (K)	ΔH_s (kJ; 1 kcal = 4.184 kJ)	T (K)	Method (see Table 3-4)	Reference
		(79-89)	10.0	(199)		[60/1]
CH_4N_2O	Urea	(338-362)	96.9	(351)	TE,ME	[83/7]
			98.6	(298)		[83/7]
			95.4	(361)		[78/19]
		(345-368)	87.9±2.1	(356)		[56/6] [60/1] [70/1]
			88.2	(357)		[53/1]
CH_4N_2S	Thiourea	(368-395)	106.6	(384)	TE,ME	[83/7]
			107.6	(298)		[83/7]
			112±1.5	(298)	C	[82/8]
$CH_4N_4O_2$	Nitroguanidine	(402-473)	142.7±2.	(298)	ME	[78/15]
CH_5N_3S	Thiosemicarbazide		125.8±1.5	(298)	C	[82/8]
CH_5O_3P	Methylphosphonic acid		48.1±4.2			[55/4] [70/1]
CH_6N_4S	Thiocarbohydrazide		152.1±3.0	(298)	C	[82/8]
CO	Carbon monoxide	(51-68)	8.1	(60)	A	[47/2]
CO_2	Carbon dioxide	(139-195)	26.3	(167)	A	[47/2]
$C_2Cl_3F_3$	1,1,2-Trichloro-1,2,2-trifluoroethane	(205-233)	32.9	(219)	A	[47/2]
$C_2Cl_4F_2$	1,1,2-Tetrachloro-1,2-difluoroethane	(237-293)	38.2	(265)	A	[47/2]
C_2Cl_6	Hexachloroethane (mp 186.6)	(306-459)	48.8	(382)	A	[47/2]
	Triclinic	(286-447)	59±1.7	(367)		[47/3] [60/1] [70/1]
	Cubic	(286-447)	51	(367)		[47/3] [60/1]
			NA			[41/1]
C_2F_6	Hexafluoroethane		26	(103)		[63/6] [48/6]
C_2N_2	Cyanogen	(177-230)	33.6	(204)	A	[47/2]
		(198-240)	32.4	(224)		[75/10] [25/2]
$C_2N_6O_{12}$	Hexanitroethane	(293-313)	70.7±1.7	(303)	ME	[69/10] [77/1]
C_2H_2	Acetylene	(133-191)	21.8	(162)		[60/1]
		(130-189)	22.7	(160)	A	[47/2]
$C_2H_2F_3NO$	Trifluoroacetamide	(288-308)	77.7±1.4	(298)	I	[78/6]
$C_2H_2I_2$	_trans_-Diiodoethylene	(253-265)	40.7	(258)	ME	[33/1] [60/1]
$C_2H_2O_4$	Oxalic acid (anhydrous)					
	Orthorhombic	(311-325)	97.9±2.2	(318)		[53/4] [60/1]
	Monoclinic	(311-323)	93.3	(317)		[53/4] [60/1]
$C_2H_3NO_2$	Oxamic acid	(355-363)	107.9	(359)	ME	[53/1] [60/1]
$C_2H_3ClO_2$	Chloroacetic acid		75.3±4.2			[28/1] [49/4] [70/1]
$C_2H_3Cl_3O_2$	Trichloroacetaldehyde hydrate	(263-319)	50.9	(291)	A	[47/2]

Molecular formula	Compound	Temp. range (K)	ΔH_S (kJ 1 kcal = 4.184 kJ)	T (K)	Method (see Table 3-4)	Reference
$C_2H_3FN_2O_5$	2-Fluoro-2,2-dinitroethanol		55.6±2.1			[68/7] [77/1]
C_2H_3NS	Methyl isothiocyanate	(238–293)	31.5	(266)	A	[47/2]
$C_2H_4Br_2$	1,2-Dibromoethane	(229–248)	54.8	(239)		[48/1]
		(251–281)	49.8	(258)	A	[48/1] [47/2]
C_2H_4BrCl	1-Bromo-1-chloro-ethane	(237–283)	44.3	(260)	A	[47/2]
$C_2H_4I_2$	1,2-Diiodoethane		65.7±4.1			[54/6] [70/1]
$C_2H_4N_2O_2$	Diformylhydrazine	(340–373)	205.1±0.7	(356)	ME	[80/11]
		(370–403)	100.8			[56/6] [60/1]
$C_2H_4N_2O_2$	Oxamide	(370–398)	115.8	(387)	TE,ME	[83/7]
		(353–369)	113.0	(361)	ME	[53/1] [60/1] [70/1]
$C_2H_4N_2S_2$	Dithio-oxamide (rubeanic acid)	(345–372)	105.1	(361)	TE,ME	[83/7]
		(360–378)	105.4	(369)	ME	[53/1] [60/1]
$C_2H_4N_4$	Dicyandiamide	(420–450)	128.7	(436)	TE,ME	[83/7]
$C_2H_4O_2$	Acetic acid	(213–230)	67.3±1	(223)	TE,ME	[78/16]
		(238–283)	U 41.4			[60/1]
$(C_2H_4O_2)_2$	Acetic acid dimer	(213–230)	70±1	(213)	TE,ME	[78/16]
$C_2H_4O_3$	Methyl hydrogen carbonate	(204–237)	18.2±1.6	(220)		[73/9]
C_2H_5NO	Acetamide	(273–293)	77.8	(284)	TE,ME	[83/7]
			77.2	(298)		[83/7]
			78.7±0.3			[75/18] [77/1]
			80.3±1	(298)		[71/16]
			80.3±1.3	(298)	C	[65/8]
		(298–349)	77.4±0.4	(353)	GS	[59/3]
$C_2H_5NO_2$	Methyl carbamate	(287–305)	74.5±0.8	(296)	GS	[59/4]
$C_2H_5NO_2$	Glycine	(408–431)	136.5±2	(419)	TE,ME	[79/1]
		(325–425)	U 96.2±4	(375)	LE	[77/2]
		(413–450)	138.1±4.6	(298)		[77/3]
		(453–471)	136.4±0.4	(462)	ME	[65/1] [70/1]
		(412–417)	130.5±2	(414)	ME	[59/1]
C_2H_5NS	Thioacetamide		83.3±0.3	(298)	C	[82/7]
C_2H_6	Ethane		20.5	(90)	B	[63/6]
$C_2H_6N_2O_2$	N-Methyl-N-nitro methanamine		41.4±2.1			[67/6] [77/1]
		(315–324)	69.9±2.1			[52/2] [70/1] [77/1]
$C_2H_6O_2S$	Dimethyl sulfone		77.±2.9			[70/1] [U/3]
$C_2H_6O_4$	*Bis*-hydroxymethyl-peroxide		94.1±4.2		ME	[53/3] [70/1]
$C_2H_7O_3P$	Ethylphosphonic acid		50.6±4.2			[55/4] [70/1]

Molecular formula	Compound	Temp. range (K)	ΔH_S (kJ 1 kcal = 4.184 kJ)	T (K)	Method (see Table 3-4)	Reference
$C_3H_2N_2$	Malononitrile	(255–281)	79.1±8		ME	[67/3] [70/1]
$C_3H_2OS_2$	1,3-Dithiol-2-one		73.6±0.8	(298)		[73/16] [77/1]
$C_3H_2OS_3$	1,3-Dithiole-2-thione		75.4±0.4	(298)		[73/16] [77/1]
$C_3H_3N_3$	1,3,5-Triazine	(212–229)	58.2	(222)	TE,ME	[83/7]
			NA			[82/9]
		(242–264)	U 43.1	(253)	ME	[68/8]
$C_3H_3N_3O_2$	6-Azauracil		141		LE	[74/8]
$C_3H_3N_3O_3$	Cyanuric acid	(440–473)	131	(458)	ME,TE	[83/7]
			133	(298)		[83/7]
$C_3H_4N_2$	Imidazole	(288–310)	80.8	(301)	ME,TE	[83/7]
			75.4	(298)	C	[80/5]
			85.3	(298)		[61/3]
$C_3H_4N_2O$	2-Cyanoacetamide	(325–348)	99.7	(336)	TE,ME	[83/7]
$C_3H_4N_2$	Pyrazole	(253–273)	72.7	(265)	TE,ME	[83/7]
			69.2±3	(298)	C	[80/5]
			71.8			[79/11]
			67.7			[61/3]
$C_3H_4N_4O$	6-Azacytosine		NA			[75/14]
$C_3H_4OS_2$	1,3-Dithiolan-2-one		80.3±0.4			[73/16] [77/1]
$C_3H_4O_2S$	Thiete sulfone		83.7±2.5		B	[69/11] [77/1]
$C_3H_4O_3$	Ethylene carbonate		78.5±4.2			[71/12] [77/1]
			73.2±2.5			[70/1] [58/3]
$C_3H_4N_2O_4$	3-Nitro-2-isoxazoline-2-oxide		71.1±8.4			[69/10] [77/1]
$C_3H_4S_3$	1,3-Dithiolan-2-thione	(294–303)	81.8±0.8	(298)		[67/5] [70/1]
C_3H_6	Cyclopropane		29.2	(145)	B	[63/6]
$C_3H_6N_6$	2,4,6-Triamino-s-triazine (melamine)	(417–614)	121.3±4.2	(515)	GS	[60/6] [70/1]
$C_3H_6N_6O_3$	n,n'n"-Trinitro-hexahydrotriazine	(325–360)	112.5±0.8		ME	[74/11]
$C_3H_6N_6O_3$	n,n'n"-Trinitroso-hexahydrotriazine	(343–447)	134.3±0.7	(298)	ME	[78/15]
		(383–411)	112.1		ME	[74/11]
			112.1			[53/6] [60/1]
$C_3H_6O_2$	Propionic acid	(225–238)	74.1±1	(233)	TE	[78/16]
			73.2±1	(233)	ME	[78/16]
$C_3H_6O_2 \cdot C_3H_6O_2$	Propionic acid dimer	(225–238)	81.3±1	(233)	TE	[78/16]
			79.4±1	(233)	ME	[78/16]
$C_3H_6O_3$	1,3,5-Trioxane	(212–231)	57.9	(223)	TE,ME	[83/7]
			55.6	(298)		[83/7]

Molecular formula	Compound	Temp. range (K)		ΔH_s (kJ 1 kcal = 4.184 kJ)	T (K)	Method (see Table 3-4)	Reference
				56.5	(298)	C	[75/17]
				56.2±0.2	(298)	C	[69/8] [77/1]
$C_3H_6S_3$	1,3,5-Trithiane	(320-339)		91.5	(331)	TE,ME	[83/7]
				93.9	(298)		[83/7]
C_3H_7NO	n-Methyl acetamide	(288-303)		54.0			[52/4] [60/1]
C_3H_7NO	Propanamide			79.2±0.3			[75/18] [77/1]
		(318-346)		79.1±0.4		GS	[59/3]
$C_3H_7NO_2$	L-(d)-Alanine			132.8±1	(414)	TE,ME	[79/1]
		(413-450)		132.4±1	(431)		[77/3]
				144.8±4.2	(298)		[77/3]
	D-(l)-Alanine	(342-442)	U	105±8	(392)	LE	[77/2]
		(453-469)		138.3±8	(461)	ME	[65/1] [70/1]
$C_3H_7NO_2$	β-Alanine	(318-418)	U	105±4	(368)	LE	[77/2]
$C_3H_7NO_2$	Ethyl carbamate	(256-273)		77.7	(265)	TE,ME	[83/7]
				76.3	(298)		[83/7]
				71.9	(322)		[76/10]
		(292-307)		89.1±0.8	(299)	GS	[59/4]
$C_3H_7NO_2$	Sarcosine (N-methylglycine)	(380-413)		146±1	(298)	C	[78/4]
$C_3H_7NO_3$	DL-Serine	(354-454)	U	83.7±4	(404)	LE	[77/2]
$C_3H_7NO_2S$	L-Cysteine	(337-437)	U	96.2±4.2	(387)	LE	[77/2]
C_3H_8	Propane			28.5	(86)	B	[63/6]
$C_3H_8SO_2$	Ethyl methyl sulfone			77.8±2.9			[U/3] [70/1]
$(C_3H_9B) \cdot (C_7H_{13}N)$	Trimethylboron-aza-bicyclo[2.2.2]octane	(273-388)		79.6			[48/3]
C_4N_2	Dicyanoacetylene	(263-273)		44.3	(265)	I	[57/4] [75/10]
C_4H_2	Butadiyne	(190-232)		36.2	(211)	A	[47/2]
		(188-234)		36.3			[33/3]
$C_4H_2N_2$	Fumaronitrile	(250-269)		69.6	(260)	TE,ME	[83/7]
				68.6	(298)		[83/7]
		(245-281)		72±0.8	(263)	ME	[67/3] [70/1]
$C_4H_2N_2S$	4-Cyanothiazole			73.9±0.4	(298)	C	[66/5] [70/1]
$C_4H_2O_3$	Maleic anhydride	(247-267)		68.8	(258)	TE,ME	[83/7]
				70.0	(298)		[78/10]
		(208-325)		71.5±5			[49/2] [70/1]
$C_4H_2O_4$	Butyndioic acid			NA			[72/7]
$C_4H_2O_4$	3,4-Dihydroxy-3-cyclobutene-1,2-dione (squaric acid)	(469-499)		152.5	(486)		[83/7]
				154.3	(298)		[83/7]
				83.7±16.7	(298)	E	[71/5] [77/1]

Molecular formula	Compound	Temp. range (K)	ΔH_S (kJ 1 kcal = 4.184 kJ)	T (K)	Method (see Table 3-4)	Reference
$C_4H_3BrN_2O_2$	5-Bromouracil		128.4		LE	[74/8]
$C_4H_4BrN_3O$	5-Bromocytosine	(403-468)	148.4±1.5	(435)		[75/4]
$C_4H_4IN_3O$	5-Iodocytosine	(413-463)	144±1.5	(438)		[75/4]
$C_4H_4N_2$	Pyrazine		56.3±0.5		C	[62/4] [70/1]
$C_4H_4N_2$	Succinonitrile	(279-298)	70±0.3	(289)		[60/7] [71/13] [77/1]
$C_4H_4N_2O_2$	Uracil	(452-587)	131±5	(298)	TE,GS	[80/18]
		(378-428)	120.5±1.3	(403)	QR	[80/19]
		(500-545)	133.9±8	(523)	HSA	[78/17]
			126.5±2.2	(298)	C	[77/13]
		(393-458)	120.5±5.2	(426)	LE	[75/16] [74/8]
			U 83.7			[65/2]
$C_4H_4N_2OS$	2-Thiouracil		129.3		LE	[74/8]
$C_4H_4N_2OS$	4-Thiouracil		125.5		LE	[74/8]
$C_4H_4N_2S_2$	2,4-Dithiouracil	(393-443)	119.7±2.4	(418)		[75/14]
$C_4H_4N_6$	8-Azaadenine	(418-463)	128.4±1.3	(440)		[75/14]
$C_4H_4N_8O_{13}$	Bis(2,2,2-trinitro-ethyl)-N-nitrosoamine	(333-354)	97.9±0.8	(343)	ME	[73/1]
$C_4H_4N_8O_{14}$	Bis(2,2,2-trinitro-ethyl)-N-nitroamine	(340-356)	117.6±0.8	(348)	ME	[73/1]
$C_4H_4O_2$	Cyclobutane-1,2-dione	(251-289)	69.1±3.5	(270)	HSA	[U/5]
$C_4H_4O_2$	Cyclobutane-1,3-dione	(274-322)	73.6±3.7	(298)	HSA	[78/14]
$C_4H_4O_3$	Succinic anhydride	(290-311)	82.2	(302)	ME,TE	[83/7]
$C_4H_4O_4$	*cis*-Butenedioic acid	(348-389)	105.4±1.7	(368)	ME	[74/5]
		(357-367)	110±2.5			[38/1] [60/1] [70/1]
		(356-371)	109±4.2			[34/1]
$C_4H_4O_4$	*trans*-Butenedioic acid	(371-391)	123.6±2.0	(381)	TE,ME	[77/4]
			136±6.3	(365)	QF	[38/1] [35/1] [60/1]
		(358-371)	134.3±4.2			[34/1]
$C_4H_5N_3O$	Cytosine		167±10	(298)	TE	[80/9]
		(450-470)	176±10	(298)	C	[80/21]
			150.6		ME	[74/8] [75/16]
			NA			[77/15]
$C_4H_5N_3O_2$	5-Aminouracil		145.6		LE	[74/8]
$C_4H_5N_3O_2$	6-Azathymine	(358-403)	112.5±2.3	(380)		[74/14]
$C_4H_5N_3S$	2-Thiocytosine	(408-458)	158±1.6	(433)		[75/14]
$C_4H_5N_7O_{12}$	Bis(2,2,2-trinitro-ethyl)amine	(338-349)	80.8±0.4		ME	[73/1] [77/1]

Molecular formula	Compound	Temp. range (K)	ΔH_s (kJ 1 kcal = 4.184 kJ)	T (K)	Method (see Table 3-4)	Reference
C_4H_6	2-Butyne	(200–239)	37.4	(220)	A	[47/2]
$C_4H_6NO_2$	3-Amino-5-methyl-isoxazole		81.6±2.5			[73/20] [77/1]
$C_4H_6O_4$	Dimethyl oxalate		47.4±0.5		BG	[76/5] [75/13]
$C_4H_6O_4$	Succinic acid	(356–376)	120.5	(368)	TE,ME	[83/7]
			123.1	(298)	TE,ME	[83/7]
		(333–387)	120.9±1.7		ME	[74/15]
		(372–401)	117.6±3.3	(298)	ME	[70/1] [60/4]
$C_4H_6S_3$	1,3-Dithian-2-thione	(321–348)	88.6	(335)		[67/5]
			91.4±2.5	(298)		[67/5] [70/1]
$C_4H_7NO_2$	Diacetimide		73.2±0.8	(298)	C	[65/8]
$C_4H_7NO_3$	N-Acetylglycine		127.0±1.0	(389)	TE,ME	[79/1]
$C_4H_7NO_4$	L-Aspartic acid	(370–470)	U 96±4.2	(420)	LE	[77/2]
C_4H_8	cyclobutane		36.4	(145)	B	[63/6]
		(181–214)	29.0	(198)	A	[47/2]
$C_4H_8Cl_2S$	2,2'-Dichloroethyl sulfide		84.5		B	[63/6] [47/4]
$C_4H_8N_2O_2$	Dimethylglyoxime		97.1±2.1			[56/7] [70/1] [60/1]
$C_4H_8N_4O_2$	1,4-Dinitroso-piperazine	(325–360)	101.3±8	(343)		[74/11] [77/1]
$C_4H_8N_4O_4$	1,4-Dinitropiperazine	(325–360)	111.3±8	(343)		[74/11] [77/1]
$C_4H_8N_8O_8$	1,3,5,7-Tetranitro-1,3,5,7-tetrazacyclo-octane					
	δ form	(461–487)	161.9	(474)		[76/8]
	δ form	(415–479)	161±0.3	(447)		[78/15]
	β form	(371–403)				[69/12]
$C_4H_8O_2$	Butanoic acid	(238–255)	76.0±1.5	(248)	TE,ME	[78/16]
$(C_4H_8O_2)\cdot(C_4H_8O_2)$	Butanoic acid dimer	(238–255)	85±1.5	(248)	TE,ME	[78/16]
$C_4H_8O_2$	1,4-Dioxane	(237–272)	35.6	(255)	A	[47/2]
$C_4H_8O_4$	1,3,5,7-Tetroxane		79.6±0.2	(298)	C	[69/8] [77/1]
			79.5		C	[75/17]
$C_4H_8S_2$	1,3-Dithiane	(250–271)	72.6	(263)	TE,ME	[83/7]
$C_4H_8S_2$	1,4-Dithiane	(253–276)	72.4	(268)	TE,ME	[83/7]
C_4H_9NO	Butanamide		86.4±0.4			[75/18] [77/1]
		(292–304)	85.4±1.7	(298)	ME	[73/19] [77/1]
		(353–373)	87	(363)		[60/1]
		(336–382)	86.4±0.4	(359)	GS	[59/3]
		(298–341)	87.0±0.8	(320)	ME	[59/3]

Molecular formula	Compound	Temp. range (K)	ΔH_s (kJ 1 kcal = 4.184 kJ)	T (K)	Method (see Table 3-4)	Reference
$C_4H_9NO_2$	2-Aminoisobutyric acid		129.5±0.4	(455)	ME	[65/1]
$C_4H_9NO_2$	DL-2-Aminobutanoic acid		132±2	(409)	TE,ME	[79/1]
$C_4H_9NO_2$	(1)-2-Aminobutanoic acid		162.8±0.8	(455)	ME	[65/1]
$C_4H_9NO_3$	DL-Threonine	(341-441)	U 96±8	(391)	LE	[77/2]
C_4H_{10}	n-Butane		35.9	(107)	B	[66/3]
$C_4H_{10}O$	t-Butyl alcohol	(253-298)	51.3	(275)	A	[47/2]
$C_4H_{10}O_2S$	Diethyl sulfone		86.2±2.5			[U/3] [70/1]
$C_4H_{10}O_4$	<u>meso</u>-Erythritol		135.1±2.2			[50/1] [60/1] [70/1]
$C_4H_{11}NO_2$	Diethanolamine		105.9±2.2	(298)	C	[82/5]
C_5Cl_6	Hexachlorocyclo-pentadiene		73.6	(283)	B	[63/6] [58/5]
C_5F_{10}	Decafluorocyclo-pentane		38.2	(115)		[63/6] [51/8] [56/9]
C_5F_{12}	n-Dodecafluoropentane		43.7	(145)		[63/6] [51/8] [56/9]
C_5N_4	Tetracyanomethane		61.1±8.8	(298)	C	[73/3]
$C_5H_4N_4$	Purine		NA			[74/7]
$C_5H_4N_4O$	Hypoxanthine	(423-473)	158.1±1.6	(448)		[75/14]
$C_5H_4N_4S$	6-Mercaptopurine	(413-458)	148.5±1.5	(435)		[75/14]
$C_5H_4O_2S$	2-Thenoic acid	(315-323)	97.1	(319)	E	[53/12] [60/1]
$C_5H_4O_3$	2-Furoic acid	(317-328)	108.4±2.2		ME	[53/1] [60/1] [70/1]
$C_5H_5FN_2O_2$	1-Methyl-5-fluoro-uracil	(480-515)	125.5±8		HSA	[78/17]
$C_5H_5FN_2O_2$	3-Methyl-5-fluoro-uracil	(465-487)	79.5±17		HSA	[78/17]
$C_5H_5ClN_2O$	1-Methyl-6-chloro-uracil	(417-465)	108.8±8		HSA	[78/17]
$C_5H_5FN_2O_2$	3-Methyl-6-chloro-uracil	(444-493)	104.6±8		HSA	[78/17]
$C_5H_5NO_2$	Pyrrole-2-carboxylic acid	(350-354)	126.8	(352)	ME	[53/12] [60/1]
$C_5H_5N_5$	Adenine		126.3		LE	[75/16] [74/8]
			108.7±8		ME	[65/2] [70/1]
$C_5H_5N_5O$	Guanine		186.2		LE	[75/16] [74/8]
$C_5H_6N_2O_2$	1-Methyluracil	(378-418)	112.5±2.6	(398)	QR	[80/19]
		(435-480)	104.6±8	(457)	HSA	[78/17]
$C_5H_6N_2O_2$	3-Methyluracil	(438-498)	75.3±8	(463)	HSA	[78/17]
$C_5H_6N_2O_2$	5-Methyluracil (thymine)	(378-428)	124.4±1.3	(403)	QR	[80/19]
			138±10	(298)	TE	[80/9]
			134.1±4.2	(298)	C	[77/13]

Molecular formula	Compound	Temp. range (K)		ΔH_S (kJ 1 kcal = 4.184 kJ)	T (K)	Method (see Table 3-4)	Reference
				124.3		LE	[75/16] [74/8]
$C_5H_6N_2O_2$	6-Methyluracil	(426-503)		131	(298)		[80/13]
$C_5H_7NO_3$	DL-5-Oxoproline			133.2±1	(405)	TE,ME	[79/1]
$C_5H_8Br_4$	Pentaerythitol tetrabromide			NA			[41/1]
$C_5H_8NO_2$	5-Amino-3,4-dimethyl-isoxazole			87.9±2.5			[73/21] [77/1]
$C_5H_8N_2$	2,3-Diazabicyclo[2.2.1]-hept-2-ene			43.9±2.1			[74/17] [77/1]
$C_5H_8N_4O_{12}$	Pentaerythritol-tetranitrate	(328-405)		150.4±1.3	(298)	ME	[78/15]
			U	121.3		ME	[69/3]
		(370-411)		151.9±2.1			[53/6] [60/1] [70/1]
C_5H_8OS	Tetrahydro-4-H-thiopyran-4-one			71.7±1.7	(317)	I	[72/15]
				72.6±1.7	(298)		[72/15] [77/1]
$C_5H_8O_2$	Methyl methacrylate	(194-223)		60.7	(205)		[52/5] [60/1]
$C_5H_8O_2S$	2,5-Dihydro-2-methyl-thiophene-1,1-dioxide			60.7±2.5			[69/11] [77/1]
$C_5H_8O_2S$	2,5-Dihydro-3-methyl thiophene-1,1-dioxide			64.±2.5			[69/11] [77/1]
C_5H_9NO	δ-Valerolactam	(293-312)		74.5	(303)		[53/5] [60/1]
$C_5H_9NO_2$	L-(1)-proline			127.4±1	(406)	TE,ME	[79/1]
		(380-420)		149±4	(400)	C	[78/4]
		(323-423)	U	50±8	(373)	LE	[77/2]
				96.7±0.8	(454)	ME	[65/1]
$C_5H_9NO_3$	4-Hydroxy-L-proline			162.6±2	(471)	TE,ME	[79/1]
$C_5H_9NO_4$	L-Glutamic acid	(353-453)	U	121±34	(403)	LE	[77/2]
C_5H_{10}	Cyclopentane			42.6	(122)	B	[63/6]
$C_5H_{10}O_5$	1,3,5,7,9-Pentoxecane			87.9±0.5	(298)	C	[74/16]
$C_5H_{11}NO$	Pentanamide	(333-374)		89.3±0.4		GS	[59/3] [70/1]
		(353-373)		89.1			[60/1]
$C_5H_{11}NO_2$	Butyl carbamate	(292-316)		94.1±8		GS	[59/4]
$C_5H_{11}NO_2$	(dl)-2-Amino-2-methyl-butanoic acid			134.2±1	(413)	TE,ME	[79/1]
		(439-469)		125.8±0.4	(454)		[65/1]
$C_5H_{11}NO_2$	DL-Norvaline			121.1±0.4	(455)	ME	[65/1]
$C_5H_{11}NO_2$	DL-Valine	(320-420)	U	79.5±8	(370)	LE	[77/2]
$C_5H_{11}NO_2$	l-Valine			162.8±8	(455)	ME	[65/1]
$C_5H_{11}NO_2S$	DL-Methionine	(363-463)	U	134±8	(413)	LE	[77/2]
	L-(d)-Methionine			164±4	(298)	C	[81/5]
	D-(1)-Methionine			125.1±0.8	(455)	ME	[65/1]

Molecular formula	Compound	Temp. range (K)	ΔH_S (kJ 1 kcal = 4.184 kJ)	T (K)	Method (see Table 3-4)	Reference
C_5H_{12}	2,2-Dimethylpropane		33.2			[63/6] [36/1]
		(171-249)	23.9	(210)	A	[47/2]
C_5H_{12}	n-Pentane		42.0	(143)	B	[63/6]
$C_5H_{12}NS$	Tetramethylthiourea		83.0±0.2	(298)	C	[82/7]
$C_5H_{12}O_2S$	t-butyl methyl sulfone		82.4±2.5			[U/3] [70/1]
$C_5H_{12}O_4$	Pentaerythritol (tetragonal)	(397-410)	131.4		ME	[51/3] [60/1]
		(379-408)	143.9±0.8		ME	[53/4] [60/1]
$C_6Cl_4O_2$	Tetrachlorobenzo-quinone	(333-356)	98.7±8.3			[27/2] [60/1] [70/1]
C_6Cl_6	Hexachlorobenzene	(312-337)	79.5±12			[77/10]
		(369-397)	92±8.2		RG	[49/3] [70/1]
C_6F_{12}	Dodecafluorocyclo-hexane	(293-333)	36.2	(313)		[57/5]
C_6N_2	Dicyanodiacetylene	(295-335)	35.9	(315)	I	[57/4]
C_6N_4	Tetracyanoethylene		81.2±5.9	(350)		[63/4] [70/1]
		(290-312)	84.3	(302)	TE,ME	[83/7]
$C_6HCl_3O_2$	Trichloroquinone	(301-327)	88.7±8.3			[27/2] [60/1] [70/1]
C_6HCl_5O	Pentachlorophenol		67.4±2.1			[U/4] [70/1]
C_6HF_5O	Pentafluorophenol		67.4±1.7		GS	[69/5] [70/1]
$C_6H_2ClN_3O_6$	1,3,5-Trinitrochloro-benzene		NA			[50/2]
$C_6H_2Cl_2O_2$	2,6-Dichloro-benzoquinone	(274-315)	69.9±8.3			[27/2] [60/1] [70/1]
$C_6H_2Cl_4O_2$	Tetrachlorohydro-quinone	(333-356)	88.7			[27/2] [60/1]
$C_6H_3ClO_2$	Chlorobenzoquinone	(264-289)	69.0±8.3	(276)		[27/2] [60/1] [70/1]
$C_6H_3Cl_3$	1,2,3-Trichlorobenzene	(289-303)	65.7	(296)	RG	[49/6] [60/1]
$C_6H_3Cl_3$	1,2,4-Trichlorobenzene	(279-298)	62.3	(289)	RG	[49/6] [60/1]
$C_6H_3Cl_3$	1,3,5-Trichlorobenzene	(282-301)	56.5	(291)	RG	[49/6] [60/1]
$C_6H_3Cl_3O_2$	Trichlorohydroquinone	(314-355)	101.3	(324)		[27/2] [60/1]
$C_6H_3N_3O_6$	1,3,5-Trinitrobenzene	(313-395)	107.3±0.6	(298)	ME	[78/15]
			99.6±2.1			[50/2] [70/1]
$C_6H_3N_3O_7$	Picric acid	(314-406)	105.1±1.6	(298)	ME	[78/15]
$C_6H_3N_3O_8$	2,4,6-Trinitro-resorcinol	(325-436)	120.8±1.1	(298)	ME	[78/15]
C_6H_4BrCl	1,4-Bromochlorobenzene	(294-337)	67.9±0.8	(316)		[61/2]
$C_6H_4Br_2$	1,4-Dibromobenzene	(278-353)	73.3±0.4	(326)		[61/2]
		(228-347)	73.8	(288)		[59/5]
		(248-303)	59.8	(298)	ME,GS	[40/1] [60/1]

Molecular formula	Compound	Temp. range (K)	ΔH_s (kJ 1 kcal = 4.184 kJ)	T (K)	Method (see Table 3-4)	Reference
C_6H_4ClI	1,4-Chloroiodobenzene	(303-323)	61.1±0.6			[53/8] [60/1] [60/1]
$C_6H_4ClNO_2$	3-Chloro-1-nitrobenzene	(275-286)	74.7±1.7			[35/1] [38/1] [60/1]
$C_6H_4Cl_2$	1,4-Dichlorobenzene	(293-311)	64.8±0.8	(303)		[61/2] [70/1]
		(311-325)	63±0.4	(318)		[61/2]
		(248-303)	56.9	(275)	ME	[40/1] [60/1]
$C_6H_4Cl_2O_2$	2,6-Dichlorohydroquinone		92.0±8.3			[27/2] [60/1] [70/1]
$C_6H_4INO_2$	3-Iodo-1-nitrobenzene	(295-306)	83.2±1.2	(300)		[35/1] [38/1] [60/1]
$C_6H_4N_2O$	2,1,3-Benzoxadiazole		64.9±1.7	(298)	GS,ME	[80/6]
$C_6H_4N_2O_4$	1,2-Dinitrobenzene	(343-397)	87.9±2.1	(298)	TE	[76/1]
		(328-338)	86.6±1.2	(309)		[35/1] [38/1] [60/1]
$C_6H_4N_2O_4$	1,3-Dinitrobenzene	(332-383)	87.0±0.8	(298)	TE	[76/1]
		(315-329)	81.1±1.7	(323)		[35/1] [38/1] [60/1]
			81.2±1.7			[50/5] [70/1]
$C_6H_4N_2O_4$	1,4-Dinitrobenzene	(339-398)	96.2±2.5	(298)	TE	[76/1]
		(345-368)	89.1±1.7	(343)		[35/1] [38/1] [60/1]
$C_6H_4N_2O_5$	2,3-Dinitrophenol	(303-343)	96.6	(323)		[58/1]
$C_6H_4N_2O_5$	2,4-Dinitrophenol	(293-333)	104.6±4.2	(313)		[58/1] [70/1]
$C_6H_4N_2O_5$	2,5-Dinitrophenol	(278-333)	93.4	(306)		[58/1]
$C_6H_4N_2O_5$	2,6-Dinitrophenol	(293-333)	112.1±4.2	(313)		[58/1] [70/1]
$C_6H_4N_2O_5$	3,4-Dinitrophenol	(328-383)	123.5	(383)		[58/1]
$C_6H_4N_4O_6$	2,4,6-Trinitroaniline	(326-449)	125.3±0.8	(298)	ME	[78/15]
$C_6H_4O_2$	1,4-Benzoquinone		68±5	(262)		[81/4]
			62.8±3.3			[56/5] [77/1]
			68.5±0.6			[53/10]
$(C_6H_4O_2.)$ $(C_6H_6O_2)$	Quinhydrone (quinone-hydroquinone)		88.6±1 U 181.2	(313)		[81/4] [53/10] [60/1]
			NA			[51/6]
$C_6H_4O_5$	Furan-2,5-dicarboxylic acid	(378-402)	121.3	(391)	TE,ME	[83/5]
$C_6H_4S_4$	Tetrathiofulvalene		95.3±1	(345)	TE,ME	[80/22]
		(341-361)	92±6.3	(351)	HSA	[79/7]
$(C_6H_4S_4)$. $(C_{12}H_4N_4)$	(Tetrathiofulvalene) (7,7,8,8-tetracyano-quinodimethane) (TTF-TCNQ)		130±2	(410)	TE,ME	[80/22]
C_6H_5BrO	4-Bromophenol	(260-302)	87.3±0.4	(298)	ME	[71/8]

Molecular formula	Compound	Temp. range (K)	ΔH_S (kJ 1 kcal = 4.184 kJ)	T (K)	Method (see Table 3-4)	Reference
C_6H_5ClO	3-Chlorophenol		53.1			[38/1] [60/1] [70/1]
C_6H_5ClO	4-Chlorophenol		51.9			[38/1] [60/1] [70/1]
$C_6H_5ClO_2$	Chlorohydroquinone	(306-334)	102.9±8.3	(320)		[27/2] [60/1] [70/1]
C_6H_5I	Iodobenzene	(243-255)	43.1			[60/1]
		(248-303)	40.0	(275)	ME	[40/1]
$C_6H_5NO_3$	2-Nitrophenol	(298-310)	73.2±1.3			[35/1] [38/1] [60/1]
$C_6H_5NO_3$	3-Nitrophenol	(325-336)	91.6±1.7			[35/1] [38/1] [60/1]
$C_6H_5NO_3$	4-Nitrophenol	(305-352)	98.8±1	(298)	ME	[71/8]
		(339-351)	91.2±1.7			[35/1] [38/1] [60/1]
$C_6H_5NO_4$	2-Nitro-1,3-dihydroxy-benzene	(253-293)	74.5	(273)		[58/1]
C_6H_6	Benzene	(183-197)	44.4	(298)	TE,ME	[80/24]
			44.1			[60/1]
		(263-270)	46.6	(282)	A	[47/2]
C_6H_6	2,4-Hexadiyne	(282-333)	47±2	(307)	MM	[82/11]
$C_6H_6Cl_6$	α-Hexachlorocyclo-hexane (mp 160 C)	(324-344)	92.9	(334)		[47/1] [60/1]
$C_6H_6Cl_6$	β-Hexachlorocyclo-hexane (mp 314 C)	(368-390)	102.9	(379)		[47/1] [60/1]
$C_6H_6Cl_6$	γ-Hexachlorocyclo-hexane	(293-313)	88.9	(303)	GS	[83/5] [70/8]
	(mp 114°C, Lindane)	(333-365)	115.5			[47/1] [60/1]
$C_6H_6Cl_6$	δ-Hexachlorocyclo-hexane mp 142°C	(328-358)	97.5			[74/1] [60/1]
$C_6H_6F_8O_2$	2,2,3,3,4,4,5,5-Octa-fluoro-1,6-hexanediol		89.2±8.4			[74/18] [77/1]
$C_6H_6N_2O_2$	2-Nitroaniline		82.4±2	(313)		[38/1] [60/1] [35/1]
			90±4.2			[58/1] [70/1]
		(310-319)	79.9±1.7			[34/1]
$C_6H_6N_2O_2$	3-Nitroaniline	(320-384)	93.6±0.7	(351)	ME	[73/5]
		(320-384)	94.6±0.3	(351)	C	[73/5]
			96.5±0.3	(298)	C	[73/5]
		(288-343)	97.6	(316)	ME	[58/1] [70/1]
		(332-341)	88.3±1.7		TE	[34/1]
		(332-341)	88.7±2.5			[38/1] [60/1] [35/1]
$C_6H_6N_2O_2$	4-Nitroaniline		100.9±0.6	(298)	ME	[73/5]
			101.3±0.7	(298)	C	[73/5]
		(303-363)	109.3	(333)	ME	[58/1] [70/1]
			99.3±1.7	(298)	ME	[56/2]

Molecular formula	Compound	Temp. range (K)	ΔH_S (kJ 1 kcal = 4.184 kJ)	T (K)	Method (see Table 3-4)	Reference
		(346–366)	97.5±1.7	(356)	ME	[56/2]
			100.7±2.5	(298)	TE	[38/1]
			98.7±2.5	(361)	TE	[38/1] [60/1]
		(357–367)	103.3±1.7	(362)		[34/1]
$C_6H_6N_4O$	7-Methylhypoxanthine		100.4±13			[78/17]
$C_6H_6N_4O$	9-Methylhypoxanthine	(500–552)	84		HSA	[65/2]
C_6H_6O	Phenol	(263–298)	65.3±3.3	(280)	HSA	[75/3]
		(230–273)	69.7±0.9	(298)	ME	[71/8]
		(282–313)	68.7±0.5		GS	[6C/3] [70/1]
		(278–305)	67.8			[47/1] [60/1]
$C_6H_6O_2$	1,2-Dihydroxybenzene (catechol)		80.8			[38/1] [60/1] [35/1]
$C_6H_6O_2$	1,3-Dihydroxybenzene (resorcinol)	(328–379)	92.3	(353)	GS	[83/3]
		(324–335)	93.3±21.			[68/4]
		(283–323)	93.4	(303)		[58/1]
			95.4±1.7			[38/1] [60/1] [35/1]
$C_6H_6O_2$	1,4-Dihydroxybenzene (hydroquinone)	(341–400)	101.3		GS	[83/3]
			103.9±1	(342)		[81/4]
			103.8			[56/5]
			90.1±0.8			[53/10]
		(326–345)	103.8		GS	[27/2]
$(C_6H_6O_2)\cdot(C_{10}H_8O_2)$	1,4-Hydroquinone. 1,4-naphthoquinone		98.7	(324)	TE,ME	[81/4]
$C_6H_6O_3$	1,2,3-Trihydroxy-benzene	(377–398)	89.1	(387)		[34/2]
C_6H_7N	3-Methylpyridine	(228–254)	62.3		RG,ME	[51/1]
C_6H_7N	4-Methylpyridine	(230–257)	62.8		RG,ME	[51/5]
C_6H_7NO	4-Aminophenol	(403–430)	92.1	(417)	I	[54/8] [60/1]
$C_6H_7NO_3S$	Sulfanilic acid (4-aminobenzene-sulfonic acid)		66.9			[38/1] [60/1]
$C_6H_7N_5$	3-Methyladenine		83.7±9		HSA	[78/17]
$C_6H_7N_5$	9-Methyladenine		92±8		HSA	[78/17]
		(140–185)	121		HSA	[65/2]
$C_6H_8N_2O$	3-Methylcytosine	(487–526)	150.6		HSA	[65/2]
$C_6H_8N_2O_2$	1,3-Dimethyluracil	(313–363)	101.7±2.1	(338)	QR	[80/19]
		(400–454)	46±4.2	(426)	HSA	[78/17]
		(344–370)	92		HSA	[65/2]

Molecular formula	Compound	Temp. range (K)	ΔH_s (kJ 1 kcal = 4.184 kJ)	T (K)	Method (see Table 3-4)	Reference
$C_6H_8N_2O_2$	1-Methylthymine	(378-428)	124.4±1.3	(398)	QR	[80/19]
$C_6H_8O_2$	1,4-Cyclohexanedione		84.4	(289)	TE,ME	[83/7]
			84.2	(298)		[83/7]
$C_6H_8O_4$	Dimethyl fumarate		NA			[72/7]
			84.5±1.7			[34/1]
$C_6H_8O_4$	Dimethyl maleate		44.8			[38/1] [60/1] [35/1]
		(317-341)	41.8±4.2			[34/1]
$C_6H_9N_3O_2$	L-Histidine	(392-492)	142±8	(442)	LE	[77/2]
$C_6H_{10}N_6O_9$	N-(2,2-Dinitropropyl)-2,2-dinitro-N-nitroso-1-propanamine	(323-336)	110.9±8		ME	[73/12] [77/1]
$C_6H_{10}N_6O_{10}$	N-(2,2-Dinitropropyl)-2,2-dinitro-N-nitro-1-propanamine	(398-423)	99.2±0.8		ME	[73/12]
$C_6H_{10}O$	Cyclohexanone		NA			[48/5]
$C_6H_{10}O_4$	Hexanedioic acid	(359-406)	129.3±2.5	(383)	ME	[50/4] [60/4] [70/1]
$C_6H_{10}O_6$	dl-Dimethyl tartrate	(314-339)	112±5.6	(326)	HSA	[81/8]
		(315-358)	113.8	(336)	ME	[54/5] [77/1]
			U 95.0			[38/1] [60/1]
$C_6H_{10}O_6$	d-Dimethyl tartrate	(310-320)	77.4±8	(315)	HSA	[81/8]
		(308-317)	U 113	(312)		[54/5] [77/1]
			88.3			[38/1] [60/1]
$C_6H_{10}O_4$	meso-Dimethyl tartrate		98.3			[38/1] [60/1]
$C_6H_{11}NO$	ε-Caprolactam	(294-314)	83.3±0.8			[53/5] [60/1] [70/1]
$C_6H_{11}NO_2$	1-Amino-1-cyclopentanecarboxylic acid		123.4±4	(455)	ME	[65/1]
$C_6H_{11}N_5O_8$	N-(2,2-Dinitropropyl)-2,2-dinitro-1-propanamine		105.4±4.2			[73/1] [77/1]
C_6H_{12}	Cyclohexane		46.6	(186)	B	[63/6]
		(268-278)	37.2	(273)		[60/1]
		(228-268)	37.7	(248)	A	[47/2]
$C_6H_{12}N_2$	1,4-Diazabicyclo[2.2.2]octane	(324-351)	61.9±3.3	(338)		[60/5] [70/1]
		(353-369)	52.3±3.3	(361)		[60/5] [70/1]
		(323-373)	54.4	(348)		[63/6] [59/]
$C_6H_{12}N_4$	1,3,5,7-Tetraazatricyclo[3.3.1.1^{3,7}]-decane	(302-328)	78.8	(316)	TE,ME	[83/7]
			74.9±2.9	(298)		[60/5] [70/2]
		(281-298)	74.1±0.8	(289)	TE	[60/2]
			75.3			[58/6]

Molecular formula	Compound	Temp. range (K)	ΔH_s (kJ 1 kcal = 4.184 kJ)	T (K)	Method (see Table 3-4)	Reference
$C_6H_{12}O$	Cyclohexanol		NA			[48/5]
$C_6H_{12}O_3$	Paraldehyde	(264-397)	43.14		A	[47/2]
$C_6H_{13}NO$	Hexanamide	(293-303)	98.7±1.7	(298)		[73/19] [77/1]
		(338-368)	95.1±4		GS	[59/3] [70/1]
$C_6H_{13}NO_2$	dl-2-Aminohexanoic acid		120.1±0.4	(455)	ME	[65/1]
$C_6H_{13}NO_2$	2-Amino-3-methyl-pentanoic acid		120.1±0.8	(455)	ME	[65/1]
$C_6H_{13}NO_2$	L-(d)-2-Amino-4-methyl pentanoic acid (L-(d)-leucine)	(323-423)	U 83.7±4	(373)	LE	[77/2]
	D-(l)-Leucine		150.6±0.8	(455)	ME	[65/1] [70/1]
C_6H_{14}	n-Hexane		50.8	(178)	B	[63/6]
$C_6H_{14}N_2O_2$	L-Lysine	(397-497)	U 88±8	(447)	LE	[77/2]
$C_6H_{14}N_4O_2$	L-Arginine	(441-541)	U 134±8	(491)	LE	[77/2]
$C_6H_{14}O_2$	1,6-Hexanediol		83.3±1.7			[72/12] [77/1]
$C_6H_{14}O_2S$	t-Butyl ethyl sulfone		86.6±2.5			[U/3] [70/1]
$C_6H_{18}Si_3O_3$	Hexamethyl-cyclotrisiloxane	(297-335)	55.2±0.4	(316)		[53/11] [60/1]
C_7H_{14}	Perfluoromethylcyclo-hexane		51.6	(234)	B	[63/6] [57/5]
C_7F_{16}	n-Hexadecafluoro-heptane				B	[63/6] [51/7]
$C_7HF_5O_2$	Pentafluorobenzoic acid		91.6±4.2			[69/5] [70/1]
$C_7H_4S_3$	4,5-Benzo-1,2-dithiole-3-thione	(350-361)	102.6±0.4	(355)		[72/16]
			107±0.4	(298)		[72/16]
$C_7H_4S_3$	4,5-Benzo-1,3-dithiole-2-thione		118.8±0.4	(298)		[73/16] [77/1]
$C_7H_5ClO_2$	2-Chlorobenzoic acid		79.5±3.3			[38/1] [60/1] [70/1]
$C_7H_5ClO_2$	3-Chlorobenzoic acid		80.8±3.3			[38/1] [60/1] [70/1]
$C_7H_5ClO_2$	4-Chlorobenzoic acid		87.9±3.3			[38/1] [60/1] [70/1]
$C_7H_5FO_2$	4-Fluorobenzoic acid		91.2±1.3			[69/5] [70/1]
$C_7H_5N_3O_6$	2,4,6-Trinitrotoluene	(301-349)	113.2±1.5	(298)	ME	[79/15]
		(297-330)	99.2±2		GS	[76/3] [77/5]
			104.6±1.7	(298)	ME	[71/4]
		(327-349)	103.3±2.5	(338)		[70/3]
		(323-353)	118.4±4.2		ME	[50/1] [60/1] [70/1]
			NA			[50/2]
$C_7H_5N_3O_7$	2,4,6-Trinitroanisole		133.1±2.1			[50/2] [70/1]

Molecular formula	Compound	Temp. range (K)	ΔH_s (kJ 1 kcal = 4.184 kJ)	T (K)	Method (see Table 3-4)	Reference
$C_7H_5N_3O_7$	3-Hydroxy-2,4,6-trinitrotoluene	(310-365)	111.2±2.1	(298)		[78/15]
		(325-350)	103.3	(337)		[70/3]
			104.6	(298)		[70/3]
$C_7H_5N_5O_8$	2,4,6-N-Tetranitro-N-methylaniline	(335-416)	133.8±1.6	(298)	ME	[78/15]
$C_7H_6N_2O_4$	1,1-Dinitrophenyl-methane	(312-323)	76.1±0.8		ME	[72/6]
$C_7H_6N_2O_4$	2,4-Dinitrotoluene	(332-342)	98.3±2.5	(337)	ME	[70/3] [77/1]
			99.6±2.5	(298)		[70/3] [77/1]
		(277-344)	95.8±1.25	(310)	GS	[76/3] [77/5]
			99.6±1.3		ME	[70/1] [71/4]
$C_7H_6N_2O_4$	2,6-Dinitrotoluene	(277-323)	98.3±0.8	(300)	GS	[76/3] [77/5]
$C_7H_6N_2O_5$	3,5-Dinitro-o-cresol	(290-324)	103.3			[47/1] [60/1]
$C_7H_6O_2$	Benzoic acid	(320-370)	89.1±0.2		C	[82/3]
		(316-391)	89.5±0.05	(353)	DM	[82/12]
		(293-313)	90.6±2		ME	[82/1]
			93.45±1		GS	[81/3]
		(328-398)	U 133.5±4.5		C	[80/20]
		(344-395)	85±2	(369)		[80/25]
		(281-323)	88.3±2.9		LE	[78/17]
			88.5±0.8		C	[76/7]
		(298-343)			ME	[76/11]
		(294-331)	92.5±4		ME	[75/6]
		(293-318)	88.5±1.6		TE	[75/5]
		(273-318)	92.9±0.2		ME	[74/5]
		(293-311)	88.1±0.2		TCM	[73/1]
		(338-383)	89.0±0.4		ME	[73/2]
		(338-383)	89.3±0.4		C	[73/2]
		(290-315)	86.6±1.3		ME,C	[72/9]
		(293-308)	90.±0.3		ME	[72/3]
			89.5±0.2	(298)	C	[72/1] [71/6]
		(299-329)	89.1	(314)		[71/17]
		(324-392)	90.4±0.8	(367)	HSA	[70/9]
			89.7±0.6	(298)	C	[69/2]
		(348-378)	88.9±0.5	(363)	GS	[68/9]
		(291-307)	90.9	(299)	ME	[65/6]
		(243-387)	91.5±0.5	(298)	GS	[54/1] [60/1] [70/1]
			84.2±0.8	(318)	TE	[38/1]

Molecular formula	Compound	Temp. range (K)	ΔH_S (kJ 1 kcal = 4.184 kJ)	T (K)	Method (see Table 3-4)	Reference
		(333-389)	85.8	(383)	T	[34/2]
		(377-394)	84.5±0.5	(364)	I	[27/3]
$C_7H_6O_2$	4-Formylphenol	(303-336)	98.2±1.3	(298)		[71/8]
$C_7H_6O_2$	Tropolone	(273-333)	84.1±0.4		ME	[71/3]
			83.7±0.8	(298)		[51/4] [70/1]
$C_7H_6O_3$	2-Hydroxybenzoic acid	(307-324)	95.7±0.8	(315)	ME	[80/2] [81/11]
		(312-332)	94.9±0.4	(322)	TE	[77/4]
		(312-332)	93.22±0.8	(322)	ME	[77/4]
		(298-328)	99.2±2			[74/5]
		(368-408)	94.8±0.4			[73/2]
		(368-408)	95.1±0.4		GS	[54/1] [60/1] [70/1]
$C_7H_6O_3$	4-Hydroxybenzoic acid	(398-433)	116.3		GS	[54/1] [60/1]
$C_7H_6O_5$	3,4,5-Trihydroxy-benzoic acid	(391-421)	75.1	(406)		[34/2]
C_7H_7NO	2-Aminotropone	(273-333)	71.13±0.4		ME	[71/3]
C_7H_7NO	Benzamide	(323-349)	101.7±1	(298)	C	[82/2]
$C_7H_7NO_2$	2-Aminobenzoic acid (I)		111.6±1.7	(298)		[72/4]
	2-Aminobenzoic acid (II)		100±1	(338)	TE,ME	[79/1]
			99.6±0.5	(378)	C	[74/1]
			104.9±1	(298)	C	[74/1] [77/13]
$C_7H_7NO_2$	3-Aminobenzoic acid		122±1	(374.8)	TE	[79/1]
		(367-389)	122.3±3		C	[74/1]
			128±3.2	(298)	C	[74/1] [77/13]
$C_7H_7NO_2$	4-Aminobenzoic acid		112.3±1	(373)	TE	[79/1]
		(367-389)	114±3.5		C	[74/1]
			116±3.7	(298)	C	[74/1] [77/13]
			U 142			[38/1] [60/1]
$C_7H_7NO_2$	4-Nitrotoluene		79.1±2.5	(298)	ME	[71/4]
		(298-310)	79.1	(298)		[70/3]
C_7H_8	Toluene		43.1	(298)	B	[70/3]
$C_7H_8N_2$	1-Amino-7-imino-1,3,5-cycloheptatriene	(273-333)	49.4±0.4		ME	[71/3]
$C_7H_8N_4O$	9-Ethylhypoxanthine		108.8±13		HSA	[78/17]
			U 83.7		HSA	[65/2]
$C_7H_8N_4O$	1,9-Dimethylhypox-anthine		75.3±13		HSA	[78/17]
C_7H_8O	2-Methylphenol		76.±0.8			[60/3] [70/1]
C_7H_8O	3-Methylphenol	(284-313)	61.7±1.0		GS	[60/3]

Molecular formula	Compound	Temp. range (K)	ΔH_S (kJ 1 kcal = 4.184 kJ)	T (K)	Method (see Table 3-4)	Reference
C_7H_8O	4-Methylphenol		73.9±1.5			[60/3] [70/1]
$C_7H_8O_2S$	Methyl phenyl sulfone		92±2.9			[U/3] [70/1]
$C_7H_8S_3$	4,5-Tetramethylene-1,3-dithiole-2-thione	(340-352)	98.3	(346)		[67/5] [70/1]
			102.1±2.9			[67/5] [70/1]
C_7H_{10}	Norbornene		37.8±0.14	(298)	C	[82/4]
			37.7±0.9	(298)	BG	[78/5]
			38.7±0.5	(298)	C	[76/4]
			33.6±0.08			[73/27]
C_7H_{10}	Nortricyclene		38.7±0.7	(298)	BG	[78/5]
			39.2±1.1	(298)	C	[76/4]
$C_7H_{10}N_2O_2$	1,3-Dimethylthymine	(313-363)	109.2±2.1	(338)	QR	[80/19]
$C_7H_{10}N_2O_2$	1,3,6-Trimethyluracil	(300-340)	106.7±2.5	(320)	QR	[80/19]
$C_7H_{10}O$	Bicyclo[2.2.1]-heptan-2-one	(300-340)	49.0±1.7	(298)	BG	[78/1]
$C_7H_{10}O$	Bicyclo[2.2.1]-heptan-7-one	(300-340)	47.3±2.2	(298)	BG	[78/1]
$C_7H_{10}O_2$	2-Oxabicyclo-[2.2.2]octan-3-one		69.6±2.1			[80/17]
$C_7H_{10}O_3$	2,4,10-Trioxa-adamantane		74.4±0.4	(298)	C	[74/16]
$C_7H_{10}O_3$	Tetramethylsuccinic anhydride		74.1±4.2			[54/3] [70/1]
$C_7H_{10}S_3$	4,5-Tetramethylene-1,2-dithiol-3-thione	(335-350)	101.6	(342)		[72/16]
			105.3	(298)		[72/16]
$C_7H_{10}S_3$	4,5-Tetramethylene-1,3-dithiolan-2-thione	(353-369)	99.0	(360)		[67/5] [70/1]
			103.9±2.9	(298)		[67/5] [70/1]
C_7H_{12}	Bicyclo[2.2.1]heptane		40.26±.32	(298)	C	[82/4]
			40.4±0.8			[78/5]
			39.33±0.13			[73/27]
			40.1±0.8		BG	[71/1] [77/1]
		(284-326)	40.0±0.8	(305)	TSGC	[75/2]
$C_7H_{13}N$	1-Azabicyclooctane		50.8±0.4			[71/10] [77/1]
		(273-362)	50.8±0.2	(298)		[48/3] [60/1] [70/5]
C_7H_{14}	Cycloheptane		53.5	(134)		[63/6]
$C_7H_{15}NO$	Heptanamide (enanthamide)	(345-365)	99.6			[54/4] [60/1]
$C_7H_{15}NO_2$	Hexyl carbamate	(291-314)	96.2±0.8		GS	[59/4]
C_7H_{16}	n-Heptane		57.9	(183)	B	[63/6]

Molecular formula	Compound	Temp. range (K)	ΔH_S (kJ 1 kcal = 4.184 kJ)	T (K)	Method (see Table 3-4)	Reference
$C_8H_4N_2$	1,2-Dicyanobenzene		86.9±1.5	(298)	GS	[80/10]
$C_8H_4N_2$	1,3-Dicyanobenzene		90.1±1.5	(298)	GS	[80/10]
$C_8H_4N_2$	1,4-Dicyanobenzene		88.8±1.5	(298)	GS	[80/10]
$C_8H_4O_3$	Phthalic anhydride	(333–403)	84.4±1.2	(388)	GS	[79/6]
			81±1		C	[71/6]
		(303–333)	88.4±1.2	(318)		[46/2] [60/1] [70/1]
C_8H_5NO	α-Oxo-benzene-acetonitrile		78.7±4.2			[69/9] [77/1]
$C_8H_5NO_2$	Phthalimide		82.8		RG	[56/9]
$C_8H_5N_3$	Pyridinium dicyanomethylide	(403–406)	125.5±1.3		ME	[67/3] [70/1]
$C_8H_6ClNO_3$	Chloromethyl 2-nitrophenyl ketone	(296–327)	103.8			[47/1] [60/1]
$C_8H_6ClNO_3$	Chloromethyl 3-nitrophenyl ketone	(279–343)	109.2	(311)		[47/1] [60/1]
$C_8H_6N_2$	Phthalazine		96.7		ME	[72/10]
$C_8H_6N_2O_2$	3-Aminophthalimide		108.3			[56/9]
$C_8H_6N_2Se$	4-Phenyl-1,2,3-selenadiazole	(275–343)	91.2±1.7		ME	[74/5]
			94.1±0.8	(298)	GS	[73/25]
$C_8H_6N_4$	Monobenzo-1,3a,4,6a tetraapentalene	(323–373)	74.9±2.9	(348)		[67/6]
$C_8H_6N_4$	Monobenzo-1,3a,6,6a tetraapentalene	(323–383)	63.6±2.9	(350)		[67/6]
$C_8H_6O_3$	Piperonal	(293–353)	90.8	(323)		[53/7] [60/1]
$C_8H_6O_4$	Isophthalic acid	(493–563)	106.7±2.2	(523)	GS	[62/2] [70/1]
$C_8H_6O_4$	Terephthalic acid	(523–633)	131.	(573)	GS	[62/2]
		(392–425)	98.24±2.5	(408)		[34/2] [70/1]
C_8H_6S	2,3-Benzothiophene		65.7±0.2	(298)	C	[79/2]
$C_8H_7C_{10}$	Chloromethyl phenyl ketone	(278–323)	90.8	(300)		[47/1] [60/1]
C_8H_7N	Indole	(275–303)	77.8±1.6		ME	[74/5]
			69.9±0.8	(298)		
		(383–328)	74.9	(305)		[54/4] [60/1]
$C_8H_7N_3O_2$	3,6-Diaminophthalimide		98.5			[56/9]
$C_8H_6N_3O_6$	2,2,2-Trinitro-1-phenylethane	(293–308)	84.1±0.4		ME	[72/6] [77/1]
$C_8H_7N_3O_6$	3-Methyl-2,4,6-trinitrotoluene	(319–411)	129.8±1.1	(365)	ME	[78/15]
$C_8H_7N_3O_7$	2,4,6-Trinitro-phenetole		120.5±2.1			[50/5] [70/1]
C_8H_8	Cubane	(239–262)	80.3±1.6	(298)	ME	[66/2] [70/1]
C_8H_8	Cyclooctatetraene		54.4		B	[49/9]
$C_8H_8N_2O_2$	1,2-Benzenedicarboxamide		57.3±4.2		ME	[72/14] [77/1]

Molecular formula	Compound	Temp. range (K)	ΔH_S (kJ 1 kcal = 4.184 kJ)	T (K)	Method (see Table 3-4)	Reference
$C_8H_8N_2O_2$	1,3-Benzenedicarboxamide		54.4±4.2		ME	[71/15] [77/1]
$C_8H_8N_2O_2$	1,4-Benzenedicarboxamide	(373-498)	57.3±4.2			[72/14]
$C_8H_8O_2$	2,5-Dimethyl-1,4-benzoquinone	(273-293)	77.0	(283)		[27/2] [60/1]
$C_8H_8O_3$	2-Methoxybenzoic acid		104.7±0.3	(298)	ME	[78/7]
		(353-368)	90.8±0.4	(360)	GS	[73/2]
		(353-368)	90.9	(360)	GS	[54/1] [60/1]
$C_8H_8O_3$	3-Methoxybenzoic acid		107.5±0.4	(298)	ME	[78/7]
$C_8H_8O_3$	4-Methoxybenzoic acid		109.8±0.6	(298)	ME	[78/7]
$C_8H_8O_2S$	Phenyl vinyl sulfone		82±2.5		B	[69/11] [77/1]
$C_8H_8O_3$	4-Hydroxy-3-methoxy benzaldehyde (vanillin)	(293-353)	88.7	(323)		[53/7] [60/1]
C_8H_9NO	Acetanilide		80.8±0.8			[55/3] [70/1]
$C_8H_9NO_2$	Methyl 2-amino-benzoate (methyl anthranilate)	(383-433)	78.2	(405)	ME	[54/4] [60/1]
$C_8H_{10}NO_5PS$	Methyl parathion (O,O-dimethyl-O-4-nitro-phenylthiophosphate)	(278-288)	108.7	(283)	GS	[83/5] [79/10]
$C_8H_{10}N_2O_2$	N,N-Dimethyl-4-nitroaniline	(344-366)	98.7±1.7	(355)	ME	[56/2]
$C_8H_{10}O$	4-Ethylphenol	(278-317)	80.3±0.5		GS	[63/3] [70/1]
$C_8H_{10}O$	2,3-Dimethylphenol	(283-323)	84.±1.		GS	[60/3] [70/1]
$C_8H_{10}O$	2,4-Dimethylphenol	(282-318)	65.9±0.2		GS	[60/3] [70/1]
$C_8H_{10}O$	2,5-Dimethylphenol	(282-323)	85.0±0.25		GS	[60/3] [70/1]
$C_8H_{10}O$	2,6-Dimethylphenol	(277-313)	75.6±0.17		GS	[60/3] [70/1]
$C_8H_{10}O$	3,4-Dimethylphenol	(282-323)	85.7±0.1		GS	[60/3] [70/1]
$C_8H_{10}O$	3,5-Dimethylphenol	(282-323)	82.8±0.3		GS	[60/3] [70/3]
$C_8H_{10}O_2$	2,5-Dimethylhydro-quinone	(332-361)	1008			[27/2] [60/1]
$C_8H_{10}O_2S$	Methyl-p-tolyl sulfone		100±3.3			[U/3] [70/1]
$C_8H_{10}O_2S$	Methyl benzyl sulfone		99.2±3			[7/3] [70/1]
C_8H_{12}	Bicyclo[2.2.2]octene		43.8±0.4		C	[70/5] [77/1]
C_8H_{12}	3,4-Dimethyl hexadiene		53.1			[56/7] [60/1]
$C_8H_{12}N_2$	Tetramethyl succinonitrile		81.2±1.7			[73/18] [77/1]
$C_8H_{12}O_2$	2,2,4,4-Tetramethyl-1,3-cyclobutanedione		70.3±3.5		HSA	[75/3]
			72.2±0.6			[71/5]
C_8H_{14}	Bicyclo[2.2.2]octane	(323-363)	46.3±0.8		BG	[71/1] [77/1]

Molecular formula	Compound	Temp. range (K)	ΔH_S (kJ 1 kcal = 4.184 kJ)	T (K)	Method (see Table 3-4)	Reference
			47.7±0.8	(298)	BG	[71/1] [77/1]
			48.0±2		C	[70/5] [71/1] [77/1]
$C_8H_{14}O$	3-Oxabicyclo[3.2.2]-nonane		53.1±0.5			[71/10] [77/1]
$C_8H_{14}O_4$	Octanedioic acid	(379-407)	143.1±3.8	(393)	ME	[60/4] [70/1]
$C_8H_{15}N$	3-Azabicyclo[3.2.2]-nonane		57.8±1.3	(298)	C	[70/5]
C_8H_{16}	Cyclooctane		58.7	(166)	B	[63/6]
$C_8H_{17}NO$	Octanamide	(325-374)	110.5±2.9		GS,ME	[59/3] [70/1]
C_8H_{18}	n-Octane		68.1	(216)	B	[63/6]
C_8H_{18}	2,2,3,3-Tetramethyl-butane	(273-338)	43.4±0.2	(298)		[52/1] [70/1]
$C_8H_{18}O$	1-Octanol	(238-251)	105.4±2.5	(244)	ME	[65/6]
			100.4	(298)		[65/6]
$C_8H_{18}O_2S$	Di-n-butyl sulfone		100.4±2.5			[U/3] [70/1]
$C_8H_{18}O_2S$	Di-t-butyl sulfone		94.1±2.9			[U/3] [70/1]
$C_8H_{24}O_4Si_4$	Octamethylcyclotetra-siloxane		64±2		B	[53/11]
$C_9H_5Br_2NO$	5,7-Dibromo-8-hydroxy-quinoline	(363-403)	94.1		ME	[63/5]
C_9H_5ClINO	7-Chloro-5-iodo-8-hydroxyquinoline	(383-413)	131±1.3		ME	[63/5]
$C_9H_5Cl_2NO$	5,7-Dichloro-8-hydroxy-quinoline	(363-393)	92.9		ME	[63/5]
$C_9H_5I_2NO$	5,7-Diiodo-8-hydroxy-quinoline	(403-433)	110.9±8		ME	[63/5]
C_9H_6INO	5-Iodo-8-hydroxy-quinoline	(363-393)	118.8		ME	[63/5]
$C_9H_6O_2$	Coumarin	(293-353)	86.2	(323)	ME	[53/7] [60/1]
$C_9H_6O_6$	1,3,5-Benzenetri-carboxylic acid	(553-593)	159.4	(573)	GS	[62/2]
$C_9H_6S_3$	5-Phenyl-1,2-dithiol-3-thione	(363-373)	117.4±0.4			[72/16]
			123.3±0.4	(298)		[72/16]
C_9H_7NO	8-Hydroxyquinoline	(308-328)	108.8±1.7		ME	[63/5] [70/1]
C_9H_7NO	ω-Cyanoacetophenone		92.5±4.2		ME	[69/9] [77/1]
$C_9H_8N_2O_2$	3-Methylaminophthalimide		104.9		RG	[56/9]
$C_9H_8O_2$	Cinnamic acid	(283-328)	109.6	(305)		[53/7] [60/1]
$C_9H_8O_2S$	Phenyl propadienyl sulfone		105.4±2.5			[69/13] [70/1]
$C_9H_8O_2S$	Phenyl prop-1-ynyl-sulfone		95.4±2.5		B	[69/13] [70/1]
$C_9H_8O_2S$	Phenyl prop-2-ynyl sulfone		105.±2.5		B	[69/13] [70/1]
$C_9H_8O_3$	Endo-5-norbornene-2,3-dicarboxylic anhydride		97±4.2	(298)	MG	[73/17] [77/1]

Molecular formula	Compound	Temp. range (K)	ΔH_S (kJ 1 kcal = 4.184 kJ)	T (K)	Method (see Table 3-4)	Reference
$C_9H_8O_4$	Monomethyl terephthalate	(433-493)	82.8	(473)	GS	[62/2]
$C_9H_9N_3O_6$	2,4,6-Trinitro-mesitylene	(319-397)	103.6±1.2		ME	[78/15]
$C_9H_{10}O$	*trans*-3-Phenyl-2-propen-1-ol		69.5		ME	[54/4]
$C_9H_{10}O_2$	2-Ethylbenzoic acid		101.1±0.4	(298)	ME	[84/1]
			100.7±2.5	(298)	ME	[76/6]
$C_9H_{10}O_2$	3-Ethylbenzoic acid		99.7±0.4	(298)	ME	[84/1]
			99.11±2.5	(298)	ME	[76/6]
$C_9H_{10}O_2$	4-Ethylbenzoic acid		98.9±0.2	(298)	ME	[84/1]
			97.5±2.5	(298)	ME	[76/6]
$C_9H_{10}O_2$	4-Acetylanisole	(276-300)	77.7		V	[59/2]
		(283-333)	93.7	(308)		[54/4] [60/1]
$C_9H_{10}O_2$	3-Methylphenyl acetate	(274-317)	60.7	(295)		[47/1] [60/1]
$C_9H_{10}O_2S$	p-Tolyl vinyl sulfone		82.4±2.5		B	[69/11] [77/1]
$C_9H_{10}O_3$	3-Ethoxy-4-hydroxy-benzaldehyde (vanillal)	(293-353)	101.7	(323)		[53/7] [60/1]
$C_9H_{11}ClN_2O$	3-p-Chlorophenyl-1,1-dimethylurea	(303-379)	114.6±4.9	(341)	ME,C	[72/9]
$C_9H_{11}NO_2$	L-(1)-Phenylalanine	(342-442)	U 90±6.3	(392)	LE	[77/2]
			154±8	(455)	ME	[65/1] [70/1]
$C_9H_{11}NO_3$	L-Tyrosine	(412-512)	101±8	(462)	LE	[77/2]
$C_9H_{12}N_2$	N-Methyl-7-(methylimino)-1,3,5-cyclohepta-trienylamine		49.4±4			[71/3] [77/1]
$C_9H_{12}N_2O_2$	m-Ethoxyphenylurea		75.3±8.3			[54/5] [70/1]
$C_9H_{12}N_2O_2$	p-Ethylphenylurea		83.7±8.3			[54/5] [70/1]
C_9H_{14}	Bicyclo[3.2.2]non-6-ene		48±1.0	(298)	C	[82/4]
C_9H_{14}	Bicyclo[3.3.1]non-2-ene		48.2±0.4	(298)	C	[82/4]
C_9H_{14}	Bicyclo[4.2.1]non-3-ene		49.7±0.8	(298)	C	[82/4]
$C_9H_{14}N_2O_2$	1,3-Diethylthymine		89.8±0.4	(298)	C	[80/7]
		(307-325)	95.0±2.1	(317)	QR	[80/19]
C_9H_{16}	Bicyclo[3.3.1]nonane		50.6±2	(298)	TSGC	[77/9]
$C_9H_{16}NO_2$	2,2,6,6-Tetramethyl-4-oxopiperidine-1-oxyl		83.3±1.7		ME	[65/7] [70/1]
$C_9H_{17}NO$	2,2,6,6-Tetramethyl-4-oxopiperidine		60.8±2.7		ME	[66/4] [70/1]
$C_9H_{17}NO_2$	2,2,6,6-Tetramethyl-1-hydroxy-4-oxopiperidine		80.1±4.6		ME	[65/7] [70/1]
$C_9H_{18}NO_2$	2,2,6,6-Tetramethyl-4-hydroxypiperidine-1-oxyl		101.5±5.2		ME	[66/4] [70/1]

Molecular formula	Compound	Temp. range (K)	ΔH_s (kJ 1 kcal = 4.184 kJ)	T (K)	Method (see Table 3-4)	Reference
$C_9H_{18}N_2O_2S$	3,3-Dimethyl-1-(methylthio)-2-butanone O-(methylamino)carbonyl)-oxime	(298-328)	93.5±6	(308)	ME	[76/11]
$C_9H_{19}NO$	Nonamide	(353-370)	114.6±3.3		ME	[59/3]
$C_9H_{19}NO_2$	2,2,6,6-Tetramethyl-1,4-dihydroxypiperidine		100.4±0.63		ME	[66/4] [70/1]
C_9H_{20}	n-Nonane		74.6	(219)	B	[63/6]
$C_{10}F_8$	Octafluoronaphthalene		79.4±2.5			[74/9]
$C_{10}H_2O_6$	1,2,4,5-Benzenetetracarboxylic dianhydride (pyromellitic dianhydride)		100.4	(559)		[75/11]
$C_{10}H_6O_2$	1,4-Naphthoquinone		90.7±2	(313)	TE,ME	[81/4]
			72.4±3.8			[56/5] [70/1]
$(C_{10}H_6O_2)\cdot(C_{10}H_8O_2)$	(1,4-Naphthoquinone). (1,4-Naphthohydroquinone)		102.3	(342.4)		[81/4]
$2(C_{10}H_6O_2)\cdot(C_{10}H_8O_2)$	2(1,4-Naphthoquinone). (1,4-Naphthohydroquinone)		88.7±3	(328.5)	ME,TE	[81/4]
$C_{10}H_7Br$	2-Bromonaphthalene	(275-378)	64±5	(298)	TE,ME	[81/6]
$C_{10}H_7NO_2$	1-Nitronaphthalene		68.5±1.9			[74/9]
			107.1±2.1			[50/2] [70/1]
$C_{10}H_7NO_2$	1-Nitroso-2-naphthol		86.6±4.2		ME	[68/6] [77/1]
$C_{10}H_7NO_2$	2-Nitroso-1-naphthol		56.5±4.2		ME	[68/6] [77/1]
$C_{10}H_7NO_2$	4-Nitroso-1-naphthol		87.4±4.2		ME	[68/6] [77/1]
$C_{10}H_8$	Azulene	(253-293)	75.8	(273)		[58/1]
			76.8±0.2		C	[72/1]
			95.4±0.4	(298)	ME	[62/1] [70/1]
$C_{10}H_8$	Naphthalene		72.6±0.1	(298)	TE,ME,DM	[83/1] [[81/1]
		(271-285)	72.8±0.3		ME	[82/1]
			72.4±0.7	(298)	C	[82/3]
		(274-353)	72.5±0.1		DM	[81/1]
		(253-273)	72.6±0.6		TE	[80/1]
		(253-273)	74.77±0.4		TE	[77/4]
		(253-273)	73.9±0.2		ME	[77/4]
		(303-329)	74.35±1.7		TSGC	[75/1]
			72.3±0.4		C	[76/1]
		(263-343)	72.5±0.3		DM	[75/4]
		(263-298)	67.8±3.5	(280)	HSA	[75/3]
			72.5	(298)	GS	[74/19]
			72.7±1.7			[74/9]
		(281-290)	64.±5		LE	[73/15]
			72.1±0.25	(298)	C	[72/1]

Molecular formula	Compound	Temp. range (K)	ΔH_s (kJ 1 kcal = 4.184 kJ)	T (K)	Method (see Table 3-4)	Reference
			73.0±0.3	(298)	C	[72/2]
			66.5			[68/1]
		(230-260)	72.7±0.3		KG	[63/1] [70/1]
		(276-283)	66.3		U	[59/2]
		(253-283)	69.2	(268)		[58/1]
		(279-294)	72.4			[53/1] [60/1]
			66.5±1.7	(298)	QF	[38/1]
$C_{10}H_8N_2O_3$	3-Acetamidophthalimide		108.5			[56/9]
$C_{10}H_8O$	1-Naphthol	(296-313)	91.2±0.4		ME	[74/3]
		(279-328)	89.1±1.7		ME	[74/5]
			91.5±3.8	(298)	B	[26/1] [27/1] [70/1]
$C_{10}H_8O$	2-Naphthol	(305-323)	94.2±0.5		ME	[74/3]
		(277-324)	87.4±2.5		ME	[74/5]
			78.7±0.8			[68/1] [77/1]
			83.0±3.8	(298)	B	[26/1] [27/1] [70/1]
$C_{10}H_8O$	1,6-Oxido[10]annulene		80.4±8.4		B	[69/7] [77/1]
$C_{10}H_8O_2$	1,4-Naphthohydroquinone		119±1	(381)	ME,TE	[81/4]
$C_{10}H_8O_2$	2,3-Dihydroxynaphthalene		NA			[79/5]
$C_{10}H_9N$	1-Naphthylamine		90.0±4.2			[47/1] [70/1]
$C_{10}H_9N$	2-Naphthylamine		74.1±1.7			[68/1] [77/1]
			88.3±4.2			[47/2] [70/1]
$C_{10}H_9NO$	B-Cyanopropiophenone		101.7±4.2		ME	[69/9] [77/1]
$C_{10}H_9NO$	2-Methyl-8-hydroxyquinoline	(308-333)	87.9		ME	[63/5]
$C_{10}H_{10}$	Bullvalene		71.8			[81/2]
$C_{10}H_{10}Fe$	Ferrocene		75.6±0.4	(298)	TE,ME, DM	[83/6]
			74.±2	(298)	TE	[81/10]
		(328-398)	70.±2		C	[80/20]
			72.6±1.4	(298)	ME	[80/24]
			74.1±1.7	(298)	TCM	[73/1]
		(385-455)	84.±2		C	[71/6]
			72.7±2	(298)	ME	[69/4]
			76.6±1	(298)	ME	[62/7]
		(323-367)	83.3			[59/6]
		(357-455)	70.5		BG	[52/6]
$C_{10}H_{10}N_2O_2$	3-Dimethylaminophthalimide		90.9		RG	[56/9]

Molecular formula	Compound	Temp. range (K)	ΔH_s (kJ 1 kcal = 4.184 kJ)	T (K)	Method (see Table 3-4)	Reference
$C_{10}H_{10}O_2S$	p-Tolyl propadienyl sulfone		113±2.5		B	[69/13] [70/1]
$C_{10}H_{10}O_2S$	p-Tolyl prop-1-ynyl sulfone		103.3±2.5		B	[69/13] [70/1]
$C_{10}H_{10}O_2S$	p-Tolyl prop-2-ynyl sulfone		107.5±2.5		B	[69/13] [70/1]
$C_{10}H_{10}O_4$	Dimethyl terephthalate	(373-413)	88.3	(393)	GS	[62/2]
$C_{10}H_{11}NO_2$	N-Phenyldiacetamide		90.±0.8	(298)	C	[65/8] [70/1]
$C_{10}H_{11}N_3O_2$	3-Dimethylamino-6-aminophthalimide		108.8			[56/9]
$C_{10}H_{12}O_2S$	p-Tolyl *trans*-prop-1-enyl sulfone		83.7±2.1		B	[69/13] [70/1]
$C_{10}H_{12}O_2S$	p-Tolyl prop-2-enyl sulfone		95.8±2.9		B	[69/13] [70/1]
$C_{10}H_{12}O_2S$	p-Tolyl isopropenyl sulfone		88.7±2.5		B	[69/11] [77/1]
$C_{10}H_{13}NO_2$	p-Phenacetin (p-ethoxyphenyl acetamide)	(312-387)	115.5±2.4		C,ME	[72/9]
$C_{10}H_{14}$	1,2,4,5-Tetramethylbenzene	(318-348)	71.3	(333)	A	[47/4]
$C_{10}H_{14}NO_5PS$	Ethyl parathion (O,O-diethyl-O-p-nitrophenylthiophosphate)	(298-318)	100.6	(308)		[79/10] [83/5]
$C_{10}H_{14}O$	Thymol	(293-323)	89.1±4.5	(303)	HSA	[75/3]
		(229-312)	U 69.0	(270)	TGA	[71/17]
		(273-313)	91.2±4.1			[70/1] [60/1] [47/1]
$C_{10}H_{14}O$	Adamantan-2-one		80.3±2.5	(298)	BG	[78/8]
$C_{10}H_{15}NO$	dl-Carvoxime	(324-343)	101.6±5	(334)	HSA	[81/8]
$C_{10}H_{15}NO$	d-Carvoxime	(324-343)	90.8±4.5	(334)	HSA	[81/8]
$C_{10}H_{16}$	Adamantane (tricyclo-[3.3.1.1^{3,7}]decane)		58.45	(298)	C	[82/4]
		(310-336)	59.7±0.8		TSGC	[75/2]
		(310-336)	59.3±0.2			[71/1]
			60.5±1.3	(298)	BG	[71/1]
		(312-366)	53.6		I	[71/7]
			59.3±0.16			[70/2]
		(313-353)	58.6±0.6		DBM	[67/1]
			62.3±0.6	(298)		[67/1]
$C_{10}H_{16}$	2,2-Dimethyl-3-methylenebicyclo-[2.2.1]heptane (camphene)		46.8		C	[77/12]
$C_{10}H_{16}$	Tricyclo[4.3.1.0^{3,8}] decane	(310-335)	64.9±1.8		TSGC	[75/2]
$C_{10}H_{16}$	Tricyclo[5.2.1.0^{2,6}] decane	(359-443)	52.9±1.3	(298)	BG	[71/1] [77/1]

Molecular formula	Compound	Temp. range (K)		ΔH_S (kJ 1 kcal = 4.184 kJ)	T (K)	Method (see Table 3-4)	Reference
$C_{10}H_{16}NOS$	S-2,3,3-Trichloroallyl N,N-diisopropylthio-carbonate (triallate)	(293-318)		84			[83/5] [78/18]
$C_{10}H_{16}O$	dl-Camphor			51.8±0.8			[77/6]
		(273-293)		51.5±2.6	(283)	HSA	[75/3]
		(273-298)	U	65.8			[60/1] [40/1]
				50.7			[37/2] [60/1]
		(273-453)		53.6	(363)		[10/1] [60/1]
$C_{10}H_{16}O$	*cis*-8-Methyl-2-hydrindanone			60.9±0.2	(298)	C	[70/6] [77/1]
$C_{10}H_{16}O$	Adamantan-1-ol			86.6±2.5	(298)	BG	[78/8]
$C_{10}H_{16}O$	Adamantan-2-ol			88.7±2.5	(298)	BG	[78/8]
$C_{10}H_{16}S_4$	1,3,5,7-Tetramethyl-2,4,6,8-tetrathia-adamantane			117.1±4.1	(298)	TE	[78/3]
$C_{10}H_{18}$	Bicyclo[3.3.2]decane			58.2±2.	(298)		[77/9]
$C_{10}H_{18}$	*cis*-Decahydro-naphthalene			64.8	(230)	B	[63/6]
	trans-Decahydro-napthalene			66.2	(241)	B	[63/6]
$C_{10}H_{18}O$	α-Terpineol	(283-328)		80.3	(305)	ME	[54/4] [60/1]
$C_{10}H_{18}O_4$	Sebacic acid decanedioic acid	(375-403)		160.7±2.5	(289)	ME	[60/4] [70/1]
$C_{10}H_{20}O$	Citronellol	(283-333)		66.1	(308)		[53/7] [60/1]
$C_{10}H_{20}O$	dl-Menthol	(279-299)		78.6±4	(289)	HSA	[81/8]
$C_{10}H_{20}O$	1-Menthol	(279-299)		95.8±4.8	(289)	HSA	[81/8]
$C_{10}H_{20}O_2$	Decanoic acid			118.8±2.2			[68/2] [70/1]
		(290-301)		117.1±1.7	(295)	ME	[61/1]
$C_{10}H_{20}O_3$	Peroxydecanoic acid	(293-303)		117.1±0.8	(298)	ME	[80/23]
$C_{10}H_{21}NO$	Decanamide	(353-370)		125.9±1.3		ME	[59/3]
$C_{10}H_{22}$	Decane			80.3	(298)	B	[80/23]
				84.8	(243)	B	[63/6]
$C_{10}H_{22}O$	Decanol	(264-273)		115.5±6.3	(268)	ME	[65/5]
				112.5±6.3	(298)		[65/5]
$C_{10}H_{24}N_4$	1,4,8,11-Tetraaza-cyclotetradecane	(352-372)		133.9±2.5	(362)	TE	[83/4]
$C_{11}H_6N_4$	Bicyclo[2.2.1.]hept-5-ene-2,2,3,3-tetra-carbonitrile			117.2±5.4	(408)	MG	[72/13] [77/1]
$C_{11}H_8N_4$	1-Methyl-4,4,5,5-tetracyanocyclohex-1-ene			82±2.1	(350)	MG	[71/14] [77/1]
$C_{11}H_8O_2$	1-Naphthoic acid	(340-360)		110.4±0.2	(345)	ME	[74/2] [77/1]
$C_{11}H_8O_2$	2-Naphthoic acid	(347-363)		113.6±0.8	(365)	ME	[74/2] [77/1]
$C_{11}H_{10}$	2-Methylnaphthalene			65.7±0.85		C	[74/10]

Molecular formula	Compound	Temp. range (K)	ΔH_S (kJ 1 kcal = 4.184 kJ)	T (K)	Method (see Table 3-4)	Reference
			61.7±1.7			[68/1] [77/1]
$C_{11}H_{12}N_2O_2$	L-Tryptophane	(340–440)	U 87.9±8	(390)	LE	[77/2]
$C_{11}H_{14}O_2$	2-t-Butylbenzoic acid		99.8±0.4	(315)	ME	[79/4]
$C_{11}H_{14}O_2$	3-t-Butylbenzoic acid		103.±0.5	(327)	ME	[79/4]
$C_{11}H_{14}O_2$	4-t-Butylbenzoic acid		103.8±0.4	(334)	ME	[79/4]
$C_{11}H_{14}O_2$	3,5-Diethylbenzoic acid		104.1±4.2			[74/15] [77/1]
$C_{11}H_{14}O_2S$	p-Tolyl but-1-enyl sulfone		106.3±2.5		B	[69/11] [70/1]
$C_{11}H_{14}O_2S$	p-Tolyl but-2-enyl sulfone		107.5±2.5		B	[69/13] [70/1]
$C_{11}H_{14}O_2S$	p-Tolyl but-3-enyl sulfone		113.4±2.9		B	[69/13] [701]
$C_{11}H_{14}O_2S$	p-Tolyl-isobutenyl sulfone		102.1±2.5		B	[69/11] [77/1]
$C_{11}H_{14}O_2S$	p-Tolyl 2-methyl prop-2-enyl sulfone		106.7±2.9			[69/13] [70/1]
$C_{11}H_{18}$	1-Methyladamantane	(300–342)	67.8±1.3	(298)	BG	[77/11]
		(306–336)	67.6±0.5	(321)		[75/2]
$C_{11}H_{18}$	2-Methyladamantane	(310–330)	67.5±2.1			[75/2]
		(300–340)	68.2±1.3	(298)		[77/11]
$C_{11}H_{20}$	Bicyclo[3.3.3]undecane		63.6±0.8	(298)	C	[75/7] [77/1]
$C_{11}H_{22}O_2$	Undecanoic acid	(303–308)	121.3±1.3	(298)	ME	[68/2] [70/1]
$C_{11}H_{22}O_3$	Peroxyundecanoic acid	(293–303)	125.9±3.4	(298)	ME	[80/23]
$C_{11}H_{23}NO$	N-Methyl decanamide	(303–325)	102.8±0.8		GS	[59/4]
$C_{11}H_{24}$	Undecane		91.5	(236)	B	[63/6]
$C_{12}F_{10}$	Decafluorobiphenyl		85.3±2.3			[74/9]
$C_{12}H_4N_4$	7,7,8,8-Tetracyano-quinodimethane (TCNQ)		126.1±1	(413)	ME,TE	[80/22]
		(433–499)	104.8±10			[80/23]
			105.±9.2	(465)	MG	[63/4] [70/1]
$C_{12}H_5Cl_5$	2,2',4,5,5'-Penta-chlorobiphenyl	(303–313)	92.7	(308)	GS	[81/7]
$C_{12}H_6Cl_4$	2,2',5,5'-Tetrachloro-biphenyl	(303–312)	94.6	(308)	GS	[81/7]
$C_{12}H_7Cl_3$	2',3,4-Trichloro-biphenyl	(303–313)	28.9	(308)	GS	[81/7]
$C_{12}H_8$	Acenaphthylene		73.0±0.4	(298)	C	[72/1]
			71.1±1.3			[65/1] [70/1]
$C_{12}H_8$	Biphenylene		87.3±0.3	(298)	B	[80/27]
			83.8±0.3		C	[72/1]
		(371–381)	U 128.9±2			[55/1] [70/1]
$C_{12}H_8Cl_2$	2,2'-Dichloro-biphenyl	(310–328)	96.2±4.2	(298)	ME	[68/3] [70/1]

Molecular formula	Compound	Temp. range (K)	ΔH_s (kJ 1 kcal = 4.184 kJ)	T (K)	Method (see Table 3-4)	Reference
$C_{12}H_8Cl_2$	4,4'-Dichloro-biphenyl	(303-360)	108.8±4.2	(298)	ME	[68/3] [70/1]
$C_{12}H_8Cl_6O$	1,2,3,4,10,10-Hexa-chloro-6,7-epoxy-1,4,4a,5,6,7,8,8a-octahydro-1,4-endo-exo-5,8-di-methanonaphthalene (dieldrin)	(293-313)	98.7	(303)	GS	[69/4]
$C_{12}H_8F_2$	2,2'-Difluorobiphenyl	(301-318)	95±4.2	(298)	ME	[68/3] [70/1]
$C_{12}H_8F_2$	4,4'-Difluorobiphenyl	(294-318)	91.2±4.2	(298)	ME	[68/3] [70/1]
$C_{12}H_8N_2$	1,10-Phenanthroline		98.3		ME	[72/10]
$C_{12}H_8N_2$	Phenazine		99.9±2.5		ME,GS	[80/6]
		(281-293)	90.4±1.7		LE	[75/1]
			U 81.5			[46/1]
$C_{12}H_8N_2$	Benzo[c]cinnoline	(320-360)	101.7±0.2	(340)	ME	[77/14]
			113		ME	[72/10]
$C_{12}H_8N_2O_4$	4,4'-Dinitrobiphenyl	(441-428)	104.6±1.8	(420)	ME	[53/10] [60/1]
$C_{12}H_8N_4$	Bicyclo[2.2.2]oct-5-ene-2,2,3,3-tetracarbonitrile		111.7±5.4	(433)		[72/13] [77/1]
$C_{12}H_8N_4$	Dibenzo-1,3a,4,6a-tetraazapentalene	(363-433)	70.3±1.7	(400)		[67/6]
$C_{12}H_8N_4$	Dibenzo-1,3a,6,6a-tetraazapentalene	(363-443)	42.3±3.4	(403)		[67/6]
$C_{12}H_8S$	Dibenzothiophene		85.1±0.4	(298)	C	[79/2]
$C_{12}H_8S_2$	Thianthrene	(338-368)	97.5±6.3	(353)	HSA	[79/7]
$C_{12}H_9N$	Carbazole		84.5±0.8			[55/2] [70/1]
$C_{12}H_9N_3O_3$	4-Hydroxy-4'-nitro-azobenzene		146±2.5		TE,ME	[70/4]
$C_{12}H_9N_3O_4$	N-(2,4-Dinitrophenyl)-N-phenylamine		149±3.0		TE,ME	[70/4]
$C_{12}H_9H_3O_5$	N-(2,4-Dinitrophenyl)-N-(4-hydroxyphenyl)amine		155.6±4.2		TE,ME	[70/4]
$C_{12}H_{10}$	Acenaphthene		83.4±1.0	(298)		[75/8] [77/7]
			82.4	(366)	B	[75/8]
			84.7±2.7		ME	[74/9]
		(290-340)	86.2±0.8		ME	[65/3] [70/1]
		(291-310)	82.1±0.4	(300)	V	[59/2]
		(258-308)	81.6			[58/1]
$C_{12}H_{10}$	Biphenyl		77.9±0.3	(298)	C	[79/3]
		(306-332)	80.4±1.6	(319)	TSGC	[75/2]
		(273-313)	76.±4.0		HSA	[75/3]
			83.6±2.5			[74/9]
		(298-318)	75.2		ME	[74/6]
			81.8±0.4	(298)	C	[72/1]

Molecular formula	Compound	Temp. range (K)	ΔH_s (kJ 1 kcal = 4.184 kJ)	T (K)	Method (see Table 3-4)	Reference
			81.6±2			[53/1] [60/1] [70/1]
		(287-307)	75.1±1.7	(297)		[53/10]
		(278-307)	72.8±3	(302)	ME	[51/1]
			68.6±0.8	(292)	QF	[38/1]
$C_{12}H_{10}N_2$	<u>cis</u>-Azobenzene	(298-357)	92.9±0.12	(328)	ME	[77/14]
		(303-333)	U 74.9	(318)	ME	[50/3] [60/1]
$C_{12}H_{10}N_2$	<u>trans</u>-Azobenzene	(299-317)	96.9±0.8	(308)	TE	[77/4]
		(299-317)	94.9±0.8	(308)	ME	[77/4]
		(298-347)	93.8±1.2	(323)	ME	[77/14]
		(303-333)	74.9	(315)		[50/3] [60/1]
$C_{12}H_{10}N_2O_2$	N-(2-Nitrophenyl)-N-phenylamine		101.9±1.7		TE,ME	[70/4]
$C_{12}H_{10}N_2O_2$	N-(4-Nitrophenyl)-N-phenylamine		126.2±1.6		TE,ME	[70/4]
$C_{12}H_{10}N_4$	1,2-Dimethyl-4,4,5,5-tetracyanocyclohexane		107.9±4.2	(378)		[72/13] [77/1]
$C_{12}H_{10}N_4O_4$	4-Amino-4'-nitro-azobenzene		140.4±1.2		TE,ME	[70/4]
$C_{12}H_{10}N_4O_4$	N-(4-Aminophenyl)-N-(2,4-dinitrophenyl)amine		154±2.9		TE,ME	[70/4]
$C_{12}H_{10}O$	2-Acetylnaphthalene	(295-316)	87.9±0.4	(305)	V	[59/2]
$C_{12}H_{10}O_2$	Diphenyl sulfone		106.3±2.9			[U/3] [70/1]
$C_{12}H_{10}O$	Diphenyl ether		82±2.1		E	[58/2] [70/1]
$C_{12}H_{10}S_2$	Diphenyldisulfide		95.±3.0		E	[62/5] [70/1]
$C_{12}H_{11}N$	Diphenylamine		96.7±2.5		TE,ME	[70/4]
		(298-323)	96.7±2.5	(310)	QF	[53/5] [70/1]
$C_{12}H_{12}N_3$	4-Aminoazobenzene	(356-373)	110.9±1.7	(387)	ME	[56/2]
$C_{12}H_{12}$	1,8-Dimethylnaphthalene		79.6	(336)	B	[75/8] [79/5]
			82.7±0.3	(298)	C	[74/14] [77/1]
$C_{12}H_{12}$	2,3-Dimethyl-naphthalene		NA			[79/5]
			81.0		B	[78/8]
		(278-301)	79.9±0.4	(290)	V	[59/2]
$C_{12}H_{12}$	2,6-Dimethyl-naphthalene	(350-383)	84.4±1.9	(366)		[77/7] [75/8]
			82.5	(383)	B	[75/8]
		(279-304)	84.1	(291)	V	[59/2]
$C_{12}H_{12}$	2,7-Dimethyl-naphthalene	(340-369)	83.8±1	(345)		[77/7] [75/8]
			83.2	(369)	B	[75/8]
$C_{12}H_{12}N_2O$	4,4'-Diaminodiphenyl oxide		62.8			[75/11]

Molecular formula	Compound	Temp. range (K)	ΔH_S (kJ 1 kcal = 4.184 kJ)	T (K)	Method (see Table 3-4)	Reference
$C_{12}H_{15}N_3O_2$	1,6-*bis*(dimethylamino) phthalimide		135.3		RG	[58/4]
$C_{12}H_{16}N_2O_5$	1-Methyl-4-t-butyl-3-methoxy-2,6-dinitrobenzene	(293-353)	102.9			[53/7] [60/1]
$C_{12}H_{18}$	Hexamethylbenzene	(303-338)	85.2	(320)	A	[76/2]
			74.7±2		ME	[65/4] [70/1]
			80.8			[57/1] [60/1]
			80.8			[49/8]
$(C_{12}H_{18})\cdot(C_6H_3N_2ClO_4)$	(hexamethylbenzene). (picryl chloride)		93.7			[49/8]
$C_{12}H_{18}$	E,E,E-1,5,9-cyclododecatriene		74.7±0.8			[73/16] [77/1]
$C_{12}H_{18}O$	Exo-4-hydroxy-endo-endo-tetracyclo[6.2.1.1.^{3,5}0^{2,7}] dodecane	(303-343)	79.0±2.5	(298)	TSGC	[80/16]
$C_{12}H_{18}O$	Exo-4-hydroxy-exo-endo-tetracyclo[6.2.1.1.^{3,6}0^{2,7}]dodecane	(323-353)	74.3±1.8		TSGC	[80/16]
			76.3±2.0	(298)		[80/16]
$C_{12}H_{18}O$	Exo-4-hydroxy-exo-exo-tetracyclo[6.2.1.1.^{3,6}0^{2,7}]dodecane	(313-353)	73.9±2		TSGC	[80/16]
			75.9±2.2	(298)		
$C_{12}H_{20}$	2,2-Dimethyladamantane	(300-360)	73.6±1.3	(298)	BG	[77/11]
$C_{12}H_{22}O_4$	Dodecanedioic acid	(375-296)	153.1±2.9	(386)	ME	[60/4] [70/1]
$C_{12}H_{24}O_2$	Dodecanoic acid		132.6±1.7	(298)		[68/2] [70/1]
		(296-314)	140.2±3.3	(304)	ME	[61/1]
$C_{12}H_{24}O_3$	Peroxydodecanoic acid	(293-303)	131.4±1.7	(298)	ME	[80/23]
$C_{12}H_{25}NO$	Dodecanamide	(349-368)	152.7±0.8		ME	[59/3]
$C_{12}H_{26}$	n-Dodecane		100.2	(298)	B	[72/1]
			101.7	(263)	B	[63/6]
$C_{12}H_{26}O$	Dodecanol	(285-294)	130.1±1.2	(290)	ME	[65/6]
			129.3	(298)		[65/6]
$C_{13}H_9N$	Acridine	(281-298)	91.6±2.5	(290)	LE	[75/1]
			78.7		E	[46/1]
$C_{13}H_9N$	3,4-Benzoquinoline		94.6±4	(308)	ME	[75/15]
			107.5		ME	[65/6]
$C_{13}H_9N$	5,6-Benzoquinoline		83.1±3.6	(308)	ME	[75/15]
			106.3		ME	[72/10]
$C_{13}H_9N$	7,8-Benzoquinoline		80.8±2.5	(308)	ME	[75/15]
			100.4		ME	[72/10]
$C_{13}H_{10}$	Fluorene	(350-388)	83.1±1.3			[77/7] [75/8]

Molecular formula	Compound	Temp. range (K)	ΔH_s (kJ 1 kcal = 4.184 kJ)	T (K)	Method (see Table 3-4)	Reference
		(306-322)	82.8			[53/1] [60/1]
			81.8	(388)	B	[75/8]
		(286-300)	80.3±0.8	(293)	TE	[60/2]
$C_{13}H_{10}N_4$	1,5-Diphenyltetrazole	(348-363)	121.5±4.2	(355)	ME	[51/3] [70/1]
$C_{13}H_{10}N_4$	2,5-Diphenyltetrazole	(333-353)	119.7±4.2	(343)	ME	[51/3] [70/1]
$C_{13}H_{10}O$	Benzophenone		94.7±1	(321)	DM	[83/2]
			92±0.83	(298)	C	[74/20] [83/2]
			NA		ME	[80/14]
			84.4±1.13	(298)	C	[78/13]
		(297-317)	93.9±0.5	(307)	TE,ME	[77/4]
		(293-318)	95.0±1.5	(305)	TE	[75/5]
		(294-318)	92.9±0.8	(306)	ME	[75/6]
		(278-311)	77.±2.5	(298)	ME	[74/5]
		(298-318)	89.96	(308)	ME	[74/6]
		(295-304)	94.6±0.8	(298)	TCM	[73/6]
			93.4±0.3	(298)	C	[72/1]
		(293-319)	96.1	(306)		[56/3]
			91.2			[50/2]
		(290-315)	78.2±1.2	(303)		[38/1] [34/1]
			95±2.5	(298)	TE	[32/1] [60/1] [70/1]
		(273-320)	91.2±1.6	(298)	ME	[25/1]
$C_{13}H_{10}O$	Dibenzopyran		112.1±2.1			[58/2] [70/1]
$C_{13}H_{10}O_2$	Phenyl benzoate		99.0±0.4	(298)		[71/2]
			89.5±4.2			[71/11] [77/1]
			96.2±1.7			[47/2] [70/1]
$C_{13}H_{10}O_3$	Diphenyl carbonate		90±8.4	(298)	E	[71/2] [77/1]
$C_{13}H_{10}O_3$	Phenyl salicylate		92±4.2			[47/2] [70/1]
$C_{13}H_{11}NO$	N-Benzoylazepine		113.0			[72/8]
$C_{13}H_{11}NO$	2-Hydroxybenzaldehyde N-phenylimine	(348-408)	129.9	(378)		[58/1]
$C_{13}H_{11}NO$	4-Hydroxybenzaldehyde N-phenylimine	(288-338)	116	(313)		[58/1]
$C_{13}H_{12}$	Diphenylmethane		82.4±8		V	[59/2] [70/1]
		(278-299)	64.0			[51/1] [60/1]
			72.0±0.8	(297)		[38/1]
$C_{13}H_{13}N$	N-Phenyl benzylamine		51.3			[80/8]
$C_{13}H_{22}$	1,3,5-Trimethyl-adamantane	(300-360)	77.8±1.3	(298)	BG	[77/11]
$C_{13}H_{22}O_3$	Dicyclohexyl carbonate	(293-313)	66.5±4.2	(303)	ME	[71/11] [77/1]

Molecular formula	Compound	Temp. range (K)	ΔH_s (kJ 1 kcal = 4.184 kJ)	T (K)	Method (see Table 3-4)	Reference
$C_{13}H_{26}O_2$	Methyl dodecanoate	(262-273)	121.8±2.1	(267)	ME	[65/6]
$C_{13}H_{26}O_2$	Tridecanoic acid		146.4±2.2			[68/2] [70/1]
$C_{13}H_{26}O_3$	Peroxytridecanoic acid	(293-303)	142.7±5		ME	[80/23]
$C_{13}H_{27}NO$	N-Methyl dodecanamide	(323-337)	116.6±0.8	(330)	GS	[59/4]
$C_{13}H_{28}$	n-Tridecane		91.4	(298)	B	[72/1]
$C_{14}H_7NO_4$	1-Nitroanthraquinone		137.9±1.7		TE,ME	[70/4]
			108.9±2.1	(396)	C	[82/3]
$C_{14}H_8O_2$	Anthraquinone		113.±0.8	(298)	C	[73/10]
			107.5±0.8	(434)	ME	[73/10]
		(397-471)	107.9±0.8		ME	[73/10]
		(355-356)	U 105.9		TGA	[71/17]
		(470-590)	127.0±3.0		C	[71/6]
			136.6±3	(298)	C	[71/6]
			116.1±1.7		ME,TE	[70/4]
		(343-403)	126.4	(373)	ME	[58/1]
			112.1	(298)		[56/5] [70/1]
			110.9	(298)		[56/1]
			107.9	(298)		[54/3]
			104.6	(367)	ME	[52/3]
			108.0	(298)	ME	[52/3]
$C_{14}H_8O_2$	9,10-Phenanthraquinone	(353-413)	132	(383)		[58/1]
			91.6±4.6			[56/5] [70/1]
$C_{14}H_8O_3$	1-Hydroxyanthraquinone	(333-383)	120.6	(358)		[58/1]
			101.3±0.4	(407)	HSA	[56/1]
$C_{14}H_8O_4$	1,2-Dihydroxy-anthraquinone		121.9±0.5	(469)	C	[73/4]
			121.5±0.4	(469)	ME	[73/4]
			123.9	(403)	ME	[58/1]
$C_{14}H_8O_4$	1,4-Dihydroxy-anthraquinone		121.9±0.8	(429)	ME	[73/4]
			121.1±4	(429)	C	[73/4]
		(324-351)	U 94.5	(338)	TGA	[71/17]
			123.5	(376)	ME	[58/1]
			103.5±1.3	(409)	HSA	[56/1]
$C_{14}H_8O_4$	1,5-Dihydroxy-anthraquinone		123.2±7		ME	[73/10]
		(363-433)	126.8	(398)	ME	[58/1]
			111.3	(456)		[56/1]
			117.6	(298)		[56/1]

Molecular formula	Compound	Temp. range (K)	ΔH_S (kJ 1 kcal = 4.184 kJ)	T (K)	Method (see Table 3-4)	Reference
$C_{14}H_8O_4$	1,8-Dihydroxy-anthraquinone		116.8		ME	[73/10]
		(333-403)	123	(368)	ME	[58/1]
		(335-356)	U 96.5	(345)	TGA	[71/17]
			105.8±8	(404)		[56/1]
			109.6±8	(298)		[56/1]
$C_{14}H_8O_4$	2,6-Dihydroxy-anthraquinone	(463-533)	173.8	(498)		[58/1]
$C_{14}H_8O_6$	1,4,5,8-Tetrahydroxy-anthraquinone	(403-473)	151.6	(438)		[58/1]
$C_{14}H_9C_{15}$	1,1-Bis(-p-chloro-phenyl)-2,2,2-tri-chloroethane (p,p'-DDT)	(293-313)	117.8	(303)	GS	[72/11]
		(323-463)	117.5	(393)	GS	[56/4] [60/1]
$C_{14}H_9NO_2$	1-Aminoanthraquinone	(361-386)	U 90.9	(374)	TGA	[71/17]
			125.9±2.5		TE,ME	[70/4]
			113±0.4	(463)	HSA	[56/1]
$C_{14}H_9NO_2$	2-Aminoanthraquinone		143.5±2.9		TE,ME	[70/4]
$C_{14}H_9NO_3$	1-Hydroxy-4-amino-anthraquinone		133.5±2.1		TE,ME	[70/4]
$C_{14}H_{10}$	Anthracene	(353-399)	94.3		GS	[83/3]
			97.4±1.1		GS,C	[81/3]
			97.8±0.1		HSA	[80/3]
		(337-361)	104.5±1.5	(298)	TE,ME	[80/24]
		(363-448)	98.8±0.4		HSA	[77/8]
		(328-372)	97.2		ME	[76/8]
		(283-323)	95.8±6		LE	[73/15]
		(353-432)	101.±0.5		ME	[73/2]
			99.7	(393)	C	[73/2]
		(290-358)	84.1		ME,C	[72/9] [71/6]
		(342-359)	98.3±2.1			[64/3] [70/1]
		(327-346)	90±0.13	(337)	TE	[60/2]
			100.8			[58/4]
		(303-373)	103.4±2.9			[58/1] [70/1]
			100.8±4.2			[58/1] [70/1]
		(396-421)	97.5±2		HSA	[53/2]
		(338-353)	102.1±2.1			[53/1] [70/1]
			95.4			[51/6]
		(378-398)	97.3±1.2		RG	[49/3]
			104.6±4.2			[49/6] [70/1]
			95.0			[48/4]

Molecular formula	Compound	Temp. range (K)	ΔH_s (kJ 1 kcal = 4.184 kJ)	T (K)	Method (see Table 3-4)	Reference
			93.3±4.2	(353)		[38/1]
$C_{14}H_{10}$	Diphenylacetylene	(299–321)	90.±4.5	(310)	HSA	[U/1]
		(299–321)	88.7±1.25	(313)		[38/1] [38/2] [60/1]
$C_{14}H_{10}$	Phenanthrene	(315–335)	92.5±2	(298)	TE,ME	[80/24]
			87.2	(372)	B	[75/8]
			90.9±0.4	(298)	C	[72/1] [77/1]
		(279–315)	84.1±2.5	(297)	TE	[60/2]
		(273–333)	95.9	(303)		[58/1] [70/1]
		(310–323)	86.6			[53/1] [70/1] [60/1]
			92.9			[49/6] [70/1]
			84.1±0.8	(313)		[38/1]
$C_{14}H_{10}N_2O_2$	1,4-Diaminoanthraquinone		199.2±2.5		TE,ME	[70/4]
$C_{14}H_{10}O_2$	Benzil	(319–340)	98.4±1.1	(329)		[59/2] [70/1]
			82.8			[38/1] [38/2] [60/1]
$C_{14}H_{10}O_3$	Benzoic anhydride		96.2±4.2	(298)	B	[71/2] [77/1]
			96.7±4.2			[47/2] [70/1]
$C_{14}H_{10}O_4$	Benzoyl peroxide		97.9±2.5	(298)	ME	[75/9]
		(293–313)	89.7±4.2	(303)	ME	[71/11] [77/1]
$C_{14}H_{10}O_4$	Diphenyl oxalate		102.5±8.4		B	[71/2] [77/1]
$C_{14}H_{10}O_5$	O-Phenyl-O,O-benzoylperoxycarbonate		97.9±2.5			[75/9] [77/1]
			133.9±4.2		E	[71/11] [77/1]
$C_{14}H_{12}$	9,10-Dihydroanthracene	(318–379)	92.4±4		ME	[75/12]
			94.2±0.8	(298)	ME	[75/12]
		(319–377)	92.2±0.6		C	[75/12]
			93.9±0.6	(298)	C	[75/12]
		(279–328)	93.3±4	(304)		[58/1] [70/1]
			89.5	(388)		[51/2] [60/1]
$C_{14}H_{12}$	<u>trans</u>-Diphenylethene		100.7±0.4	(298)	SRFG TE,ME,DM	[83/1]
			102.1±0.6		TCM	[73/11]
			99.2±0.4			[72/1]
$C_{14}H_{12}O$	Desoxybenzoin		99.3±4.2			[47/2] [70/1]
$C_{14}H_{12}O_2S$	E-(2-Phenylethenyl)-sulfonyl benzene (phenyl <u>trans</u>-B-styryl sulfone)		105±3.8		B	[69/11] [77/1]
$C_{14}H_{14}$	2,2'-Dimethylbiphenyl	(283–288)	65.7	(285)	ME	[74/6]
$C_{14}H_{14}$	3,3'-Dimethylbiphenyl	(288–308)	71.9	(298)	ME	[74/6]

Molecular formula	Compound	Temp. range (K)	ΔH_S (kJ 1 kcal = 4.184 kJ)	T (K)	Method (see Table 3-4)	Reference
$C_{14}H_{14}$	1,2-Diphenylethane		91.5±0.7	(298)	B	[80/27]
			91.4±0.5	(298)	C	[72/1]
		(286-307)	84.1±0.4		V	[59/2] [70/1]
		(290-317)	72.4±1.3	(304)	ME	[51/1]
			73.2			[38/1] [38/2] [60/1]
$C_{14}H_{14}O_2S$	Dibenzyl sulfone		125.5±2.9			[U/3] [70/1]
$C_{14}H_{14}O_2S$	Di-p-tolyl sulfone		109.6±2.9			[U/3] [70/1]
$C_{14}H_{14}S$	Dibenzyl sulfide		93.3±5		E	[62/5] [70/1]
$C_{14}H_{14}S$	Dibenzyl diselenide	(291-330)	31.2		ME	[74/5]
$C_{14}H_{15}N_3$	N,N-Dimethyl-p-aminoazobenzene		120.9±1.7	(373)	ME	[56/2]
$C_{14}H_{16}$	1,4,5,8-Tetramethyl-naphthalene		99.8±1.4	(298)	C	[74/14] [77/1]
$C_{14}H_{16}O_5$	Benzoyl (3-cyclohexyl-oxy)carbonyl peroxide	(293-313)	96.2±4.2	(303)	ME	[71/11] [77/1]
$C_{14}H_{18}N_2O_5$	2,6-Dimethyl-3,5-dinitro-4-t-butyl-acetophenone	(293-353)	107.9	(323)	ME	[53/7] [60/1]
$C_{14}H_{18}O$	Diamantanone		103.1±.62	(320)	TSGC	[80/4]
$C_{14}H_{20}$	1,8-Cyclotetra-decadiyne	(317-332)	166.±3.2	(325)	ME	[64/1] [70/1]
$C_{14}H_{20}$	1,2,3,4,5,6,7,8-Octahydroanthracene (octhracene)	(438-499)	82.3±1.2	(298)	BG	[71/1] [77/1]
$C_{14}H_{20}$	Diadamantane	(305-333)	96.0±0.8	(319)	TSGC	[75/2]
			117.2±8		B	[71/9]
$C_{14}H_{22}$	1,4-Di-t-butyl-benzene	(285-325)	82.8	(305)	ME	[51/5]
$C_{14}H_{20}O$	Diamantan-1-ol	(319-349)	118.±0.6	(334)		[80/4] [75/2]
$C_{14}H_{20}O$	Diamantan-3-ol	(323=354)	116.1±4.4	(338)		[80/4] [75/2]
$C_{14}H_{20}O$	Diamantan-4-ol	(322-353)	117.8±0.2	(337)		[80/4] [75/2]
$C_{14}H_{22}O_6$	Dicyclohexyl peroxidicarbonate		100.4±4.2			[71/11] [77/11]
		(293-313)	100.4±8.3	(303)	ME	[62/3] [70/1]
$C_{14}H_{24}$	trans-anti-trans Perhydroanthracene	(275-313)	72.7±3.3	(294)	ME	[63/2] [70/1]
$C_{14}H_{24}$	trans-syn-trans Perhydroanthracene	(335-393)	87.4±2.4	(365)	ME	[63/2] [70/1]
$C_{14}H_{24}$	1,3,5,7-Tetramethyl-adamantane	(310-350)	83.7±1.3	(298)	BG	[77/11]
		(295-315)	81.11±0.9	(305)	TSGC	[75/2]
$C_{14}H_{26}O$	Cyclotetradecanone		80.75			[38/1] [60/1]
$C_{14}H_{28}$	Cyclotetradecane	(295-307)	134.8±1.5	(301)	ME	[64/1] [70/1]
$C_{14}H_{28}O$	2-Tetradecanone		130.9±0.45		C	[79/8]

Molecular formula	Compound	Temp. range (K)	ΔH_S (kJ 1 kcal = 4.184 kJ)	T (K)	Method (see Table 3-4)	Reference
$C_{14}H_{28}O_2$	Tetradecanoic acid	(312-325)	139.7±3.8	(318)	ME	[61/1] [70/1]
$C_{14}H_{28}O_3$	Peroxytetradecanoic acid		156.±4.1		ME	[80/23]
$C_{14}H_{29}NO$	Tetradecanamide	(248-375)	167.4±2.5	(352)	ME	[59/3]
$C_{14}H_{30}$	n-Tetradecane		117.6	(298)	B	[72/1]
$C_{14}H_{30}O$	1-Tetradecanol	(293-307)	143.9	(300)	ME	[65/6]
$C_{15}H_{10}O$	Diphenylcyclopropenone	(353-378)	119.7±8	(365)	HSA	[U/2]
		(323-343)	141±4	(333)	ME	[76/9]
$C_{15}H_{10}O_3$	1-Methoxyanthraquinone		106.6	(385)	HSA	[56/1]
$C_{15}H_{10}O_3$	2-Methoxyanthraquinone		118.4±0.4	(419)	HSA	[56/1]
$C_{15}H_{11}NO_2$	1-(N-Methylamino)-9,10-anthraquinone		114.7±3.	(406)	HSA	[56/1]
$C_{15}H_{11}NO_2$	2-(N-Methyl)-9,10-anthraquinone		115.5±0.4	(461)	HSA	[56/1]
$C_{15}H_{12}$	9-Methylanthracene		98.95		RG	[58/4]
$C_{15}H_{14}O$	4,5,6-Trimethyl-benzoxalene		139.7±2.5			[66/3] [70/1]
$C_{15}H_{14}O$	1,3-Diphenyl-2-propanone		89.1±5			[54/2] [70/1] [77/1]
$C_{15}H_{14}O_2S$	(Z)-1-Methyl-4-(2-phenylethenyl) sulfonyl benzene		116.3±3.8		B	[69/11] [77/1]
$C_{15}H_{14}O_2S$	(E)-1-Methyl-4-(2-phenylethenyl) sulfonyl benzene		108.4±2.5		B	[69/11] [77/1]
$C_{15}H_{22}$	1-Methyldiamantane	(310-333)	80.65±0.4	(321)	TSGC	[75/2]
$C_{15}H_{22}$	3-Methyldiamantane	(305-327)	103.1±1.0	(316)	TSGC	[75/2]
$C_{15}H_{22}$	4-Methyldiamantane	(310-333)	79.43±1.25	(321)	TSGC	[75/2]
$C_{15}H_{22}O_2$	3,5-<u>Bis</u>-(2-methyl-propyl)-benzoic acid		108.4±4.2			[74/15] [77/1]
$C_{15}H_{28}O$	Cyclopentadecanone		77.4			[38/1] [60/1] [70/1]
$C_{15}H_{28}O_2$	Pentadecanolide	(283-333)	81.2	(318)	ME	[54/4] [60/1]
$C_{15}H_{30}$	Cyclopentadecane		74.6±0.4			[57/1] [70/1]
$C_{15}H_{30}O$	2-Pentadecanone		139.3±1.6	(298)	C	[79/8]
$C_{15}H_{30}O_2$	Methyl tetradecanoate		137.7±2.1	(281)	ME	[65/6]
$C_{15}H_{31}NO$	N-Methyl tetradecan-amide	(332-347)	130.4±0.8	(340)	GS	[59/4]
$C_{15}H_{32}$	n-Pentadecane		107.8	(298)	B	[72/1]
$C_{16}F_{34}$	n-Perfluorohexadecane	(288-303)	104.6	(295)	ME	[51/9]
$C_{16}H_{10}$	Fluoranthrene		99.2±0.8	(298)	C	[72/1] [77/1]
		(328-353)	102.1±2	(340)	ME	[65/3] [70/1]
		(298-358)	102.6	(328)		[58/1]
$C_{16}H_{10}$	Pyrene	(398-423)	100.2±0.4	(410)	IPM	[80/26]

Molecular formula	Compound	Temp. range (K)	ΔH_S (kJ 1 kcal = 4.184 kJ)	T (K)	Method (see Table 3-4)	Reference
			101.±0.5		C	[74/4]
			100.8±1.5		ME	[74/4]
			95.7		ME	[53/1] [70/1] [77/1]
		(298–363)	100.5	(330)	ME	[58/1]
		(345–358)	94.1	(351)		[52/3] [60/1]
$C_{16}H_{10}O$	2,3,5,6-Dibenzoxalene		129.4±1.3			[66/3] [70/1]
$C_{16}H_{12}S_2$	3,6-Diphenyl-1,2-dithiin		174.5±2.5	(355)		[73/23] [77/1]
			183.1±2.5	(298)		[73/23] [77/1]
$C_{16}H_{13}NO$	9-Acetamidoanthracene		134.8		RG	[58/4]
$C_{16}H_{14}$	9,10-Dimethylanthracene		103.2		RG	[58/4]
$C_{16}H_{14}$	2,7-Dimethyl-phenanthrene		106.7±0.8			[65/5] [70/1]
$C_{16}H_{14}$	4,5-Dimethyl-phenanthrene		104.6±1.3			[65/5] [70/1]
$C_{16}H_{14}$	9,10-Dimethyl-phenanthrene		119.5±1.3			[66/3] [70/1]
$C_{16}H_{14}$	1,4-Diphenylbutadiene		87.0		RG	[58/4]
$C_{16}H_{16}$	[2.2]-Paracyclophane	(353–409)	96.4±1.5		TSGC	[80/15]
			96.3±4.2			[73/13] [77/1]
		(343–383)	92.9±0.84	(363)	ME	[66/11] [70/1]
$C_{16}H_{16}$	[2.2]-Metacyclophane	(308–332)	91.6±1.7	(320)	ME	[69/6] [77/1]
			92.0±2	(298)	ME	[69/6] [77/1]
$C_{16}H_{16}$	[2.2]-Metapara-cyclophane	(311–328)	86.6	(336)	ME	[69/6] [77/1]
			87.5±0.9	(298)	ME	[69/6] [77/1]
$C_{16}H_{16}OS$	2,6-Diphenyl-thiacyclohexan-4-one		136.0	(375)	ME	[72/15]
			144.	(298)	ME	[72/15]
$C_{16}H_{16}O_3$	2-Phenylisopropoxy benzoate	(293–313)	43.1±4.2	(303)	ME	[71/11] [77/1]
$C_{16}H_{18}$	1,3-Diphenylbutane	(288–303)	73.6		ME	[74/6]
$C_{16}H_{28}$	Tricyclo[8.2.2.2] hexadecane	(316–338)	91.6±0.9	(327)	ME	[69/6] [77/1]
$C_{16}H_{30}O$	Cyclohexadecanone		82			[38/1] [60/1]
$C_{16}H_{30}O_4$	Hexadecanedioic acid	(377–398)	151.±3.3	(388)	ME	[60/4]
$C_{16}H_{32}$	Cyclohexadecane		81.8±0.4			[57/1] [70/1]
$C_{16}H_{32}O_2$	Hexadecanoic acid	(320–333)	154.4±4.2	(326)	ME	[61/1] [70/1]
$C_{16}H_{33}NO$	Hexadecanamide	(364–378)	181.6±1.3	(371)	ME	[59/3]
$C_{16}H_{34}$	Hexadecane		135.1	(298)	B	[72/1]
			134.9	(291)	B	[63/6]
		(288–290)	83.4±8		ME	[49/1]

Molecular formula	Compound	Temp. range (K)	ΔH_S (kJ 1 kcal = 4.184 kJ)	T (K)	Method (see Table 3-4)	Reference
$C_{16}H_{34}O$	1-Hexadecanol	(308-320)	167.4±2.1		ME	[65/6]
			169.5±2.1	(298)		[65/6]
$C_{17}H_{10}O$	Benzanthrone		115.5			[52/3] [60/1]
$C_{17}H_{13}N$	N-Methyl-2,3,5,6-dibenzazalene		131.8±1.3			[66/3] [70/1]
$C_{17}H_{16}OS$	Tetrahydro-2,6-diphenyl 4H-thiopyran-4-one		144±3			[72/15] [77/1]
$C_{17}H_{32}O$	Cycloheptadecanone		75.7			[38/1] [60/1] [70/1]
$C_{17}H_{34}$	Cycloheptadecane		66.1±0.6			[57/1] [70/1]
$C_{17}H_{34}O_2$	Methyl hexadecanoate	(291-301)	152.3±2	(296)	ME	[65/6]
$C_{17}H_{35}NO$	N-Methyl hexadecan-amide	(345-355)	144.5±0.8	(350)	GS	[59/4]
$C_{17}H_{36}$	Heptadecane		125.1	(298)	B	[72/1]
		(288-293)	131.3±13	(290)	ME	[49/1] [60/1]
$C_{17}H_{36}O$	1-Heptadecanol		169.5±2.2			[65/6] [70/1]
$C_{18}H_{10}O_2$	1,2-Benzanthra-9,10-quinone		82.8±4.2			[56/5] [70/1]
$C_{18}H_{10}O_2$	5,12-Tetracenequinone		108.8±5.4			[56/5] [70/1]
$C_{18}H_{12}$	1,2-Benzanthracene	(357-454)	120.5	(405)	ME	[67/2]
		(377-403)	104.6±4.2	(390)	ME	[64/3]
		(333-393)	119.7	(363)		[58/1]
$C_{18}H_{12}$	Triphenylene	(381-406)	126.5±4	(298)	TE,ME	[80/24]
		(338-398)	118±4	(368)		[58/1] [70/1]
			107.1			[67/2]
$C_{18}H_{12}$	Naphthacene	(419-446)	143.7±0.5	(298)	TE,ME	[80/24]
			124.7±4			[67/2] [70/1] [77/1]
			117.2		ME	[52/3] [60/1]
			124.3			[51/2] [60/1]
$C_{18}H_{12}$	Tetraphene (1,2-benzanthracene)	(373-396)	123.3±3	(298)		[80/24]
			U 109.2			[51/2] [60/1]
$C_{18}H_{12}$	3,4-Benzphenanthrene		106.3±4			[51/2] [67/2] [70/1]
$C_{18}H_{12}$	Chrysene		131±4	(298)	TE,ME	[80/24]
			117.6±4		ME	[67/2] [70/1]
		(353-418)	121.4	(385)		[58/1]
			117.6			[51/2] [60/1]
$C_{18}H_{14}$	5,12-Dihydrotetracene	(338-398)	115.9±4	(368)		[58/1] [70/1]
			120.5			[51/2] [60/1]
$C_{18}H_{14}$	Diphenylfulvene		104.6±8.3		E	[57/2] [70/1]
$C_{18}H_{14}$	o-Terphenyl		97±1	(298)	B	[71/8]

Molecular formula	Compound	Temp. range (K)	ΔH_S (kJ 1 kcal = 4.184 kJ)	T (K)	Method (see Table 3-4)	Reference
$C_{18}H_{14}$	m-Terphenyl		120±1	(298)	B	[71/8]
		(313-363)	119	(338)		[58/1]
$C_{18}H_{14}$	p-Terphenyl		113±2	(298)	B	[71/8]
			118.4			[67/2]
		(333-393)	120.6	(363)		[58/1]
$C_{18}H_{15}N$	Triphenylamine	(322-373)	87.9±1.3			[78/2]
			92.1±2.5	(298)		[78/2]
$C_{18}H_{15}OP$	Triphenylphosphine oxide		66±6	(298)	B	[78/11]
$C_{18}H_{16}N_2O_2$	9-Diacetylamino-anthracene		106.4		RG	[58/4]
$C_{18}H_{18}$	2,4,5,7-Tetramethyl-phenanthrene		114.2±1.7			[65/5] [70/1]
$C_{18}H_{18}$	3,4,5,6-Tetramethyl-phenanthrene		133.5±3.8			[65/5] [70/1]
$C_{18}H_{20}$	[3.3]-Paracyclophane	(322-343)	103.3±1	(332)	ME	[69/6] [77/1]
$C_{18}H_{30}$	1,3,5-Tri-t-butyl-benzene	(273-315)	79.7±0.4	(294)	ME	[65/6]
$C_{18}H_{30}O$	2,4,6-Tri-t-butyl-phenol	(295-339)	85.6±0.4	(317)		[65/6]
$C_{18}H_{30}O_4$	4-Diacetylbenzene diethyl ketal		112.5			[78/9]
$C_{18}H_{34}O$	cyclo-Octadecanone		77.4			[38/1] [60/1]
$C_{18}H_{36}O_2$	Octadecanoic acid	(331-340)	166.5±4.2	(336)	ME	[61/1] [70/1]
$C_{18}H_{37}NO$	Octadecanamide	(367-379)	195.8±4.2	(373)	ME	[59/3]
$C_{18}H_{38}$	n-Octadecane		152.7	(298)	C	[72/1]
		(288-298)	153.±5	(293)	ME	[49/1] [60/1] [70/1]
$C_{18}H_{38}O$	1-Octadecanol	(318-329)	187.4±1.3	(324)	ME	[65/6]
			191.2±1.3	(298)		[65/6]
$C_{19}H_{16}$	Triphenylmethane	(325-349)	100±0.4	(339)	V	[59/2] [70/1]
$C_{19}H_{36}O$	Cyclononadecanone		82.4			[38/1] [60/1]
$C_{19}H_{38}$	n-Nonadecane		143.6	(298)	C	[72/1]
$C_{19}H_{38}O_2$	Methyl octadecanoate	(299-310)	158.2±2.5	(304)		[65/6]
$C_{19}H_{38}O_2$	Nonadecanoic acid		198.7±5			[68/2] [70/1]
$C_{20}H_{12}$	Perylene	(443-518)	145.2±2.5	(298)	C,ME	[73/7]
			125.5±4.2	(298)		[67/2] [70/1]
		(383-453)	139	(418)		[58/1]
			129.3			[52/3] [60/1]
$C_{20}H_{14}$	9,10-Dihydro-9,10 (1',2') benzo-anthracene (tryptycene)		104.6±12.6			[73/13] [77/1]
$C_{20}H_{14}$	9-Phenylanthracene		115.3			[58/4]

Molecular formula	Compound	Temp. range (K)	ΔH_S (kJ 1 kcal = 4.184 kJ)	T (K)	Method (see Table 3-4)	Reference
$C_{20}H_{16}$	1',9-Dimethyl-1,2-benzanthracene		112.5±3.3			[65/5] [70/1]
$C_{20}H_{16}$	3',6-Dimethyl-1,2-benzanthracene		112.5±3.3			[65/5] [70/1]
$C_{20}H_{16}$	5,6-Dimethylchrysene		130±1.3			[66/3] [70/1]
$C_{20}H_{16}$	9,10-Dimethyl-1,2-benzanthracene		134±1.3			[66/3] [70/1]
		(379–408)	135±2.4	(394)	ME	[64/3]
$C_{20}H_{38}O_4$	Eicosanedioic acid	(380–395)	165.7±3.3	(388)	ME	[60/4]
$C_{20}H_{40}O_2$	Eicosanoic acid	(337–346)	199.6±7.5	(342)	ME	[61/1] [70/1]
$C_{20}H_{42}$	n-Eicosane		170.4	(298)	C	[72/1]
$C_{20}H_{42}O$	1-Eicosanol	(327–341)	218±3.8	(332)	ME	[65/6]
			223±3.8	(298)		[65/6]
$C_{21}H_{16}$	20-Methylcholanthrene	(401–425)	127.2±2.4	(413)	ME	[64/3]
$C_{21}H_{26}$	[1,8]-Paracyclophane	(354–376)	105±1.3	(365)		[69/6] [77/1]
			110.9±2.1	(298)		[69/6] [77/1]
$C_{21}H_{42}O_2$	Methyl eicosanoate	(311–318)	190.8±10	(314)	ME	[65/6]
$C_{22}H_{12}$	Anthranthrene		135.±5.	(479)		[52/3] [60/1]
$C_{22}H_{12}$	1,12-Benzperylene	(454–502)	125.5	(478)	ME	[67/2]
$C_{22}H_{12}O_2$	6,13-Pentacenequinone		116.3±5.9			[56/5] [70/1]
$C_{22}H_{14}$	3,4-Benztetraphene		136.4		ME	[67/2]
$C_{22}H_{14}$	Pentacene	(495–530)	184±10	(298)	TE,WE	[80/24]
		(455–555)	157.7	(505)	ME	[67/2]
$C_{22}H_{14}$	Picene	(425–488)	140.1	(456)	ME	[67/2]
$C_{22}H_{14}$	1,2:3,4-Dibenz-anthracene	(425–452)	159±6	(298)	TE,ME	[80/24]
$C_{22}H_{14}$	1,2:5,6-Dibenz-anthracene	(436–462)	162±6	(298)	TE,ME	[80/24]
		(417–502)	141.8		ME	[67/2]
$C_{22}H_{20}N_2O_4$	N,N'-Bis(2-methoxyphenyl) terephthalamide	(183–197)	197.5±4.2		ME	[73/21] [77/1]
$C_{22}H_{20}N_2O_4$	N,N'-Bis(3-methoxyphenyl) terephthalamide	(183–197)	209.2±8.4		ME	[73/21] [77/1]
$C_{22}H_{20}N_2O_4$	N,N'-Bis(4-methoxyphenyl) terephthalamide		227.6±8.4		ME	[73/22] [77/1]
$C_{22}H_{28}O$	2,4,6-Triisopropyl-benzophenone	(353–364)	116±7	(298)	C	[82/10]
$C_{22}H_{28}O$	3',5'-Diisopropyl-4,4-dimethyl-3-phenyl-1,2-benzocyclobuten-3-ol	(354–364)	117.9	(298)	C	[82/10]
$C_{22}H_{46}O$	1-Docosanol	(335–341)	206.7±10	(330)	ME	[65/6]
			238.5±10	(298)		[65/6]
$C_{24}H_{12}$	Coronene	(476–555)	128.4			[53/2] [60/1]
		(433–513)	147	(473)		[58/1]

Molecular formula	Compound	Temp. range (K)	ΔH_s (kJ 1 kcal = 4.184 kJ)	T (K)	Method (see Table 3-4)	Reference
$C_{24}H_{12}O_2$	3,4:9,10-Dibenzpyrene-5,8-quinone		112.5±5.4			[56/5] [70/1]
$C_{24}H_{14}$	1,2,4,5-Dibenzpyrene	(434–526)	137.6	(480)	ME	[67/2]
$C_{24}H_{14}$	1,2,6,7-Dibenzpyrene	(454–526)	146.9	(490)	ME	[67/2]
$C_{24}H_{18}$	1,3,5-Triphenylbenzene		152±0.3		C,ME	[74/4]
		(384–400)	142.2			[67/2]
			149.7±4.1	(298)	ME	[58/1] [70/1]
$C_{24}H_{20}Pb$	Tetraphenyllead		80.2	(298)	ME	[62/6]
$C_{24}H_{20}Si$	Tetraphenylsilane		149.4±1.7	(298)	ME,TE	[74/12]
		(428–489)	51±0.4			[73/14]
$C_{24}H_{20}Sn$	Tetraphenyltin		U 66.02±1.2	(298)	ME	[62/6]
$C_{24}H_{26}N_2O_2$	1,5-Dipiperidylanthraquinone	(408–458)	173.3	(428)		[58/1]
$C_{24}H_{32}$	[6.6]Paracyclophane	(352–371)	108.8±0.8			[69/6] [77/1]
			115.1±2.1	(298)		[69/6] [77/1]
$C_{25}H_{20}$	Tetraphenylmethane	(396–466)	150.6±4	(298)	TE,ME	[72/5] [77/1]
$C_{26}H_{16}$	Tetrabenzonapthalene	(417–500)	141.8	(458)	ME	[67/2]
$C_{26}H_{18}$	9,10-Diphenylanthracene		116.4			[58/4]
		(393–433)	143.6	(413)		[58/1]
		(208–229)	156.9±4.2		HSA	[53/2] [70/1]
$C_{26}H_{22}$	1,1,2,2-Tetraphenylethane		126.4±1.7	(434)	HSA	[56/1]
$C_{26}H_{26}$	Triple layered [2.2] paracyclophane	(299–412)	119.1±1.5		TSGC	[80/15]
			125.9±2.5	(298)	TSGC	[80/15]
$C_{27}H_{22}O_3$	Di-(p-phenylbenzyl) carbonate		111.7			
$C_{28}H_{14}$	<u>meso</u>-Naphthodianthrene		180.7±5	(605)		[52/3] [60/1]
$C_{28}H_{18}$	9,9'-Bianthryl		128.±4.2			[70/1] [58/1]
		(413–473)	127.9	(443)		[58/1]
			148.1			[51/2] [60/1]
$C_{28}H_{18}$	9,9'-Biphenanthryl		151.5			[51/2] [60/1]
$C_{30}H_{16}$	Pyranthrene		194.5±6.7	(595)		[52/3] [60/1]
$C_{32}H_{14}$	Ovalene		211.7±7.9	(600)		[52/3] [60/1]
$C_{34}H_{18}$	Tetrabenzoperylene	(350–450)	118	(400)	ME	[67/2]
$C_{34}H_{18}$	Isoviolanthrene A (mp 510°C)	(588–724)	218	(590)	ME	[52/3] [60/1]
$C_{34}H_{18}$	Violanthrene		223.8±8.8	(590)		[52/3] [60/1]
$C_{34}H_{18}$	Violanthrene A (mp 478° C)	(562–724)	195.8	(653)	ME	[67/2]
$C_{34}H_{18}$	Violanthrene B (mp 330 ° C)	(555–625)	153.5	(590)	ME	[67/2]

Molecular formula	Compound	Temp. range (K)	ΔH_S (kJ 1 kcal = 4.184 kJ)	T (K)	Method (see Table 3-4)	Reference
$C_{36}H_{30}Si_{12}$	Hexaphenyldisilane		209.2±2.1	(298)	ME,TE	[74/12]
$C_{42}H_{28}$	5,6,11,12-Tetraphenyl-tetracene (rubrene)	(453-523)	160.6±4.2	(488)		[58/1] [70/1]

TABLE 3-6: The Following Compounds Have Either Recently Been Reported or Were Inadvertently Omitted from Table 3-5.

Molecular formula	Compound	Temp. range (K)	ΔH_s (kJ 1 kcal = 4.184 kJ)	T (K)	Method (see Table 3-4)	Reference
CBr_4	Tetrabromomethane monoclinic		54.5±0.7	(298)	C	[A84/2]
$CH_3Cl_3Si.2(C_6H_{15}N_3$	bis-1,3,5-Trimethyl-1,3,5-triazacyclohexane. methyltrichlorosilane		NA			[A84/5]
$C_2H_6N_2O$	N-Methylurea		78.2		E	[A82/1]
$C_4H_{12}B_2Br_4N_2$	Dibromo(dimethylamino) borane dimer		NA			[A83/5]
$C_5H_6N_2$	2-Aminopyridine		78.7±0.8	(298)	C	[A84/6]
$C_5H_6N_2$	3-Aminopyridine		84.0±1.4	(298)	C	[A84/6]
$C_5H_6N_2$	4-Aminopyridine		88.1±1.1	(298)	C	[A84/6]
$C_6H_4N_2$	2-Cyanopyridine		70.7±1.2	(298)	C	[A84/6]
$C_6H_4N_2$	3-Cyanopyridine		72.1±1.8	(298)	C	[A84/6]
$C_6H_4N_2$	4-Cyanopyridine		73.2±0.6	(298)	C	[A84/6]
$C_6H_5NO_2$	3-Pyridine carboxylic acid		123.4±1.2	(298)	C	[84/6]
$C_6H_6O_2$	1,2-Dihydroxybenzene		86.6±1.6	(298)	C	[A84/10]
C_6H_7NS	4-Methylthiopyridine	(347-383)	75.3±3.8	(365)	B	[A74/1]
C_6H_7NS	1-Methyl-4-thiopyridone	(440-465)	188.3±9.2	(452)	B	[A74/1]
$C_6H_{11}NS$	1-Methyl-2-thiopiper-idone	(363-370)	81.2±2.9	(366)	B	[A74/1]
$C_6H_{13}NO$	t-Butylacetamide		NA			[A83/6]
$C_7H_5BrO_2$	2-Bromobenzoic acid		83.3±0.4		DSC	[A83/1]
	2-Chlorobenzoic acid		116.2±0.6		DSC	[A83/1]
	2-Iodobenzoic acid		103.0±0.4		DSC	[A83/1]
$C_7H_6O_2$	Benzoic acid		89.5±0.4		DSC	[A83/1]
C_7H_6ClNO	4-Chlorobenzaldoxime					
	<u>anti</u>		U 60.7±1.7		MS	[A83/2]
$C_7H_6N_2O_3$	2-Nitrobenzaldoxime					
	<u>anti</u>		U 26.4±1.7		MS	[A83/2]
	<u>syn</u>		U 40.2±1.7		MS	[A83/2]
$C_7H_6N_2O_3$	3-Nitrobenzaldoxime					
	<u>anti</u>		U 41.0±1.7		MS	[A83/2]
	<u>syn</u>		U 42.7±1.7		MS	[A83/2]
$C_7H_6N_2O_3$	4-Nitrobenzaldoxime					
	<u>anti</u>		U 56.4±1.7		MS	[A83/2]
$C_7H_7NO_2$	2-Hydroxybenzaldoxime					
	<u>anti</u>		U 51±1.7		MS	[A83/2]

Molecular formula	Compound	Temp. range (K)	ΔH_s (kJ 1 kcal = 4.184 kJ)	T (K)	Method (see Table 3-4)	Reference
	syn		U 65.3±1.7		MS	[A83/2]
$C_7H_7NO_2$	3-Hydroxybenzaldoxime					
	anti		U 52.7±1.7		MS	[A83/2]
	syn		U 57.3±1.7		MS	[A83/2]
$C_7H_7NO_2$	4-Hydroxybenzaldoxime					
	anti		U 54.4±1.7		MS	[A83/2]
$C_7H_7NO_3$	2,4-Dihydroxybenzaldoxime					
	anti		U 76.2±1.7		MS	[A83/2]
	syn		U 93.7±1.7		MS	[A83/2]
$C_7H_8N_{2}0$	2-Aminobenzaldoxime					
	anti		U 33.9±1.7		MS	[A83/2]
	syn		U 63.6±1.7		MS	[A83/2]
$C_7H_8O_2$	3-Methyl-1,2-dihydroxy-benzene		93.2±1.0	(298)	C	[A84/10]
$C_7H_8O_2$	4-Methyl-1,2-dihydroxy-benzene		94.9±1.0	(298)	C	[A84/10]
$C_7H_8O_2S$	4-Methoxy-6-methyl-2-thiopyrone	(402-415)	130.5±5.9	(408)	B	[A74/1]
$C_7H_8O_2S$	6-Methyl-2-methylthio-4-pyrone	(388-433)	87.4±3.8	(410)	B	[A74/1]
$C_8H_8O_2$	2-Methylbenzoic acid		137.7±0.5		DSC	[A83/1]
$C_8H_8O_3$	2-Methoxybenzoic acid		91.6±0.4		DSC	[A83/1]
C_8H_9NO	3-Methylbenzaldoxime					
	anti		U 31±1.7		MS	[A83/2]
C_8H_9NO	4-Methylbenzaldoxime					
	anti		U 36±1.7		MS	[A83/2]
$C_8H_9NO_2$	2-Methoxybenzaldoxime					
	anti		U 20.1±1.7		MS	[A83/2]
	syn		U 32.6±1.7		MS	[A83/2]
$C_8H_9NO_2$	4-Methoxybenzaldoxime					
	anti		U 67.3±1.7		MS	[A83/2]
$C_8H_{12}N_2O_2$	1,3-Dimethyl-5-ethyl-uracil	(300-316)	99.3±0.2	(308)	ME	[A83/7]
		(319)-340)	110±1.2	(330)	QR	[80/19] [A83/7]
$C_9H_{10}O_2$	4-Ethylbenzoic acid	(311-331)	98.9±0.2	(298)	ME	[A84/7]
$C_9H_{10}O_2$	2,3-Dimethylbenzoic acid	(316-337)	102.3±0.4	(326)	ME	[A84/9]
$C_9H_{10}O_2$	2,4-Dimethylbenzoic acid	(312-331)	102.7±0.3	(321)	ME	[A84/9]
$C_9H_{10}O_2$	2,5-Dimethylbenzoic acid	(315-334)	103.6±0.6	(324)	ME	[A84/9]
$C_9H_{10}O_2$	2,6-Dimethylbenzoic acid	(309-324)	98.2±0.2	(317)	ME	[A84/9]
$C_9H_{10}O_2$	3,4-Dimethylbenzoic acid	(325-347)	104.5±0.3	(336)	ME	[A84/9]
$C_9H_{10}O_2$	3,5-Dimethylbenzoic acid	(322-341)	100.8±0.3	(332)	ME	[A84/9]

Molecular formula	Compound	Temp. range (K)	ΔH$_S$ (kJ 1 kcal = 4.184 kJ)	T (K)	Method (see Table 3-4)	Reference
$C_9H_{12}O_2$	3-Isopropyl-1,2-dihydroxybenzene		97.8±1.7	(298)	C	[A/84/10]
C_9H_{14}	Bicyclo[3.3.1]non-2-ene		48.2±0.4	(298)	C	[A82/2]
C_9H_{14}	Bicyclo[4.2.1]non-3-ene		49.7±0.8	(298)	C	[A82/2]
C_9H_{14}	Bicyclo[3.2.2]non-6-ene		48.0±1.0	(298)	C	[A82/2]
$C_{10}D_8$	Naphthalene-d_8	(282-323)	70.6±0.5	(303)	GS	[A83/4]
$C_{10}H_8$	Naphthalene	(283-323)	75.8±1.1	(303)	GS	[A83/4]
			72.6±0.4		DSC	[A83/1]
$C_{10}H_{14}O_2$	4-t-Butyl-1,2-dihydroxybenzene		99.3±1.4	(298)	C	[A84/10]
$C_{10}H_{14}O_2$	6-Methyl-3-isopropyl-1,2-dihydroxybenzene		96.6±0.9	(298)	C	[A84/10]
$C_{10}H_{28}O$	Tri-t-butylmethanol					
	plastic phase	(278-318)	56.5±1.0	(298)	C	[A83/3]
	crystalline phase	(269-300)	63.2±1.2	(298)	C	[A83/3]
$C_{11}H_8O_2$	1-Naphthoic acid		117.6±0.4		DSC	[A83/1]
$C_{11}H_8O_2$	2-Naphthoic acid		119.5±0.6		DSC	[A83/1]
$C_{11}H_{16}N_2O_2$	1,3-Dimethyl-5,6-pentamethyleneuracil	(335-358)	111.9±0.2	(346)	ME	[A83/7]
		(323-338)	108.8±5	(330)	QR	[80/19][A83/7]
		(340-370)	113.4±1.3	(355)	MS	[80/19][A83/7]
$C_{12}H_8$	Acenaphthylene	(238-323)	73.2±0.5	(303)	GS	[A83/4]
$C_{12}H_{10}$	Acenaphthene	(238-323)	86.8±0.9	(303)	GS	[A83/4]
$C_{12}H_{18}$	Hexamethylbenzene		74.9±0.6		DSC	[A84/1]
$2(C_6H_{15}N_3)\cdot SiCl_4$	Bis-1,3,5-trimethyl-1,3,5-triazacyclohexane. tetrachlorosilane		NA			[A84/5]
$C_{13}H_{10}$	Fluorene	(283-323)	88.4±0.6	(303)	GS	[A83/4]
$C_{14}D_{10}$	Phenanthrene-d_{10}	(283-323)	92.2±1.1	(303)	GS	[A83/4]
$C_{14}H_{10}$	Anthracene	(283-323)	91.8±0.9	(303)	GS	[A83/4]
$C_{14}H_{10}$	Phenanthrene	(283-323)	95.±0.6	(303)	GS	[A83/4]
$C_{14}H_{12}$	_trans_-Diphenylethene		U 61.1		MS	[A83/2]
		(293-338)	103.8±2.5	(315)		[A83/9]
$C_{14}H_{14}$	1,2-Diphenylethane	(273-318)	91.2±0.4	(295)		[A83/9]
$C_{14}H_{22}O_2$	3,5-Di-t-butyl-1,2-dihydroxybenzene		100.1±0.6	(298)	C	[A84/10]
$C_{16}H_{10}$	Fluoranthrene	(283-323)	84.6±0.9	(303)	GS	[A83/4]
$C_{16}H_{10}$	Pyrene	(283-323)	91.2±0.5	(303)	GS	[A83/4]
$C_{17}H_{14}FeO$	Benzoylferrocene	(358-382)	116.3±6	(298)	TE,ME	[A83/8]
$C_{18}H_{12}$	Benz[a]anthracene					
	(1,2-benzanthracene)	(283-323)	U 81.3±2.5	(303)	GS	[A83/4]
$C_{18}H_{15}Al$	Triphenylaluminum	(432-477)	1725	(455)	ME,TE	[A84/8]

Molecular formula	Compound	Temp. range (K)	ΔH_S (kJ 1 kcal = 4.184 kJ)	T (K)	Method (see Table 3-4)	Reference
$C_{18}H_{15}B$	Triphenylboron		103.8±2.5	(360)	TE,ME	[A84/3]
$C_{18}H_{22}$	2,3-Dimethyl-2,3-diphenylbutane	(293-348)	96.7±0.8	(320)		[A83/9]
$C_{18}H_{30}$	Hexaethylbenzene		41.3±0.9		DSC	[A84/1]
$C_{22}H_{26}$	1,1'-Diphenyl-1,1'-bicyclopentyl		141.4		E,B	[A83/9]
$C_{24}H_{18}FeO_2$	1,1'Dibenzoylferrocene	(358-381)	109.3±6	(298)	TE,ME	[A83/8]
$C_{24}H_{30}$	1,1-Diphenyl-1,1'-bicyclohexyl		150.2		E,B	[A83/9]
$C_{26}H_{34}$	1,1-Diphenyl-1,1'-bicyclooctyl		174.5		E,B	[A83/9]
$C_{26}H_{38}$	2,3-Dimethyl-2,3-bis-(4-t-butylphenyl)-butane		161.9		E,B	[A83/9]
$C_{30}H_{46}$	3,4-Diethyl-3,4-bis-(4-t-butylphenyl)-hexane		167.8		E,B	[A83/9]
$C_{32}H_{50}$	2,4,5,7-Tetramethyl-4,5-bis-(4-t-butylphenyl)-octane		182.8		E,B	[A83/9]
$C_{32}H_{50}$	4,5-Diethyl-4,5-bis-(4-t-butylphenyl)-octane		182.4		E,B	[A83/9]
$C_{34}H_{54}$	4,5-Dipropyl-4,5-bis-(4-t-butylphenyl)-octane		198.3		E,B	[A83/9]
$C_{38}H_{62}$	5,6-Dibutyl-5,6-bis-(4-t-butylphenyl)-decane		220.9		E,B	[A83/9]

ACKNOWLEDGMENT

I would like to thank Ms. Dietra Lott for typing the manuscript and the numerous additions and corrections this entailed.

REFERENCES

1. Pedley, J.B.; Rylance, J. "N.P.L. Computer-Analysed Thermochemical Data: Organic and Organometallic Compounds"; School of Molecular Sciences, University of Sussex: Brighton, UK, 1977.
2. Cox, J.D.; Pilcher, G. "Thermochemistry of Organic and Organometallic Compounds". Academic Press: New York, 1970.
3. Stull, D.R.; Westrum, E.F., Jr.; Sinke, G.C. "The Chemical Thermodynamics of Organic Compounds". Wiley: New York, 1969.
4. Sandman, D.J.; Epstein, A.J.; Chickos, J.S.; Ketchum, J.; Fu, J.S.; Scheraga, H.A. *J. Chem. Phys.* **1979,** *70,* 305.
5. Chickos, J.S., Garin, D.L.; Hitt, M.; Schilling, G. *Tetrahedron,* **1981,** *37,* 2255.
6. Domalski, E.S.; *Environ. Sci. Technol.* **1971,** *5,* 443.
7. (a) Mirsky, K. *Acta Crystallogr. Sect. A* **1976,** *32,* 199. (b) Govers, H.A.J. ibid, **1975,** *31,* 380. (c) Fyfe, C.A.; Harold-Smith, D. *Can. J. Chem.* **1976,** *54,* 769.

8. Spencer, W.F.; Cliath, M.M. *Residue Rev.* **1983,** *85,* 57.
9. Melia, T.P.; Merrifield, *R. J. Inorg. Nucl. Chem.* **1970,** *32, 1489.*
10. Chickos, J.S. *J. Chem. Educ.* **1975,** *52,* 134.
11. Dygdala, R.S.; Stefanski, K. *Chem. Phys.* **1980,** *53,* 51. Dygdala, R.S.; Stefanski, K.; Wonikowski, J. *Bull. Acad. Pol. Sci. Ser. Sci. Math. Astron. Phys.* **1977,** *25,* 439.
12. Clark, L.B.; Peschel, G.G.; Tinoco, I., Jr. *J. Phys. Chem.* **1965,** *69,* 3615.
13. Nowak, M.J.; Szczepaniak, K.; Barski, A.; Shugar, D. *Z. Naturforsch. C: Biosci.* **1978,** *33,* 876.
14. Dygdala, R.S.; Stefanski, K.; Wonikowski, J. *Bull. Acad. Pol. Sci. Ser. Sci. Math. Astron. Phys.* **1977,** *25,* 439.
15. Dygdala, R.S.; Stefanski, K. *Chem. Phys.* **1980,** *53,* 51.
16. Nowak, M.J.; Szczepaniak, K.; Barski, A.; Shugar, D. *Z. Naturforsch. C: Biosci.* **1978,** *33,* 876.
17. Kwart, H.; Hoster, D.P. *J. Org. Chem.* **1967,** *32,* 1867.
18. Pracejus, G. *Justus Liebigs Ann. Chem.* **1959,** *622,* 10.
19. Zahorsky, U.I.; Musso, H. *Chem. Ber.* **1973,** *106,* 3608.
20. Fales, H.M.; Wright, G.J. *J. Am. Chem. Soc.* **1977,** *99,* 2339.
21. Garin, D.L.; Greco, D.J.C.; Kelley, L. *J. Org. Chem.* **1977,** *42,* 1249.
22. Jacques, J.; Collet, A.; Wilen, S.H. "Enantiomers, Racemates and Resolutions". Wiley: New York, 1981.
23. Coppock, J.B.M.; Goss, F.R. *J. Chem. Soc.* **1939,** 1789.
24. Guette, J.P.; Boucherot, D.; Horeau, A. *Tetrahedron Lett.* **1973,** 465.
25. Horeau, A.; Guette, J.P. *Tetrahedron* **1974,** *30,* 1923.
26. Ambrose, D.; Sprake, C.H.S. *J. Chem. Soc. A* **1971,** 1261.
27. Atlk, Z.; Ewing, M.B.; McGlashan, M.L. *J. Phys. Chem.* **1981,** *85,* 3300.
28. Takagi, S.; Fujiskiro, R.; Amaya, R. *Chem. Commun.* **1968,** 480.
29. Herndon, W.C.; Vincenti, S.P. *J. Am. Chem. Soc.* **1984,** *105,* 6174.
30. Walden, W.; Zimmerman, C.; Kolbe, A. Pracejus, H. *Tetrahedron,* **1977,** *33,* 419.
31. Chickos, J.S. Unpublished results.
32. Wolf, K.L.; Weghofer, H. *Z. Phys. Chem. B* **1938,** *39,* 194.
33. Crowell, T.I.; Jones, G.L., Jr. *J. Phys. Chem.* **1954,** *58,* 666.
34. Meyerhoffer, W. *Berichte* **1904,** *37,* 2604.
35. Van Ekeren, P.J.; Jacobs, M.H.G.; Offringa, J.C.A.; DeKruif, C.G. *J. Chem. Thermodyn.* **1983,** *15,* 409.
36. Murata, S.; Sakiyama, M.; Seki, S. *J. Chem. Thermodyn.* **1982** *14,* 707, 723.
37. Colomina, M.; Jiménez, P.; Turrión, C. *J. Chem. Thermodyn.* **1982,** *14,* 779.
38. DeKruif, C.G.; Blok, J.G. *J. Chem. Thermodyn.* **1982,** *14,* 201.
39. Cox, J.D. *Pure Appl. Chem.* **1974,** *40,* 399, 424; **1977,** *49,* 1437.
40. Benson, S.W. "Thermochemical Kinetics", 2nd ed. Wiley: New York, 1976, Chapter 2.
41. DeKruif, C.G. *J. Chem. Thermodyn.* **1980,** *12,* 243.
42. Clarke, E.C.W.; Glew, D.N. *Trans. Faraday Soc.* **1966,** *62,* 539.
43. Dulong, P.L.; Petit, A.T. *Ann. Chim. Phys.* **1819,** *10,* 395; Boltzmann, L. *Sitzb. K. Akad. Wiss. Wien* **1871,** *63:2,* 679; Lewis, G.N.; Randall, M. "Thermodynamics", 2nd ed., revised by Pitzer, K.S.; and Brewer, L.; McGraw-Hill: New York, 1961.
44. Rae, A.I.M.; Mason, R. *Proc. R. Soc. London* **1968,** *304,* 487.
45. Walden, P. *Z. Elektrochem.* **1908,** *14,* 713.
46. Trouton, F. *Phil. Mag.* **1884,** *18,* 54.
47. Klages, F. *Chem. Ber.* **1949,** *82,* 358.
48. Albert, A.; Willis, J.B. *Nature* **1946,** *157,* 341; Willis, J.B. *Trans. Faraday Soc.* **1947,** *43,* 97.
49. McEachern, D.M.; Sandoval, O.; Iniguez, J.C. *J. Chem. Thermodyn.* **1975,** *7,* 299.
50. Arshadi, M.R. *J. Chem. Thermodyn.* **1980,** *12,* 903.
51. Mackle, H.; Mayrick, R.G. *Trans. Faraday Soc.* **1962** *58,* 238.
52. Laidler, K.J. *Can. J. Chem.* **1956,** *34,* 626; **1960,** *38,* 2367.
53. Cottrell, T.L. "The Strength of Chemical Bonds", 2nd ed., Butterworths: London, 1958, Section 7.4.
54. Mackle, H.; Mayrick, R.G. *Tran. Faraday Soc.* **1962,** *58,* 33.
55. Wenner, R.R. "Thermochemical Calculations". McGraw-Hill: New York, 1941, pp. 14–16.
56. Reid, R.C.; Sherwood, T.K. "The Properties of Gases and Liquids", 2nd ed.; McGraw-Hill:

New York, 1966. Janz, G.J. "Thermodynamic Properties of Organic Compounds". Academic Press: New York, 1967.
57. Watson, K.M. *Ind. Eng. Chem.* **1943,** *35,* 398; **1931,** *23,* 360.
58. Thomson, G.W. In "Physical Methods of Organic Chemistry", 2nd ed., Pt. 1, Weissberger, A., Ed.; Wiley-Interscience: New York, 1949, pp. 235–244.
59. Sturtevant, J.M. In "Physical Methods of Organic Chemistry", 2nd ed., Pt. 1, Weissberger, A., Ed.; Wiley-Interscience: New York, 1949, Chapter 14.
60. Green, J.H.S. *Q. Rev. Chem. Soc.* **1961** *15,* 125.
61. Morawetz, E. *J. Chem. Thermodyn.* **1972,** *4,* 455, 461.
62. Morawetz, E. *J. Chem. Thermodyn.* **1972,** *4,* 139, 145.
63. Gardner, P.J.; Hussain, K.S. *J. Chem. Thermodyn.* **1972,** *4,* 819.
64. Stridh, G. *J. Chem. Thermodyn.* **1976,** *8,* 895.
65. Stridh, G.; Sunner, S.; Svensson, C. *J. Chem. Thermodyn.* **1977,** *9,* 1005.
66. Månnson, M.; Sellers, P.; Stridh, G.; Sunner, S. *J. Chem. Thermodyn.* **1977,** *9,* 91.
67. Reid, R.C.; Prausnitz, J.M.; Sherwood, T.R. "Properties of Gases and Liquids", 3rd ed.; McGraw-Hill: New York, 1978, Chapter 6.
68. Cremer, R.D., III. *J. Am. Chem. Soc.* **1980,** *102,* 1837, 1849.
69. DuCros, M.; Greison, J.F.; Sannier, H. *Thermochim. Acta* **1980,** *36,* 39.
70. Chickos, J.S. Hyman, A.S.; Ladon, L.H.; Liebman, J.F. *J. Org. Chem.* **1981,** *46,* 4294.
71. Gambill, W.R. *Chem. Eng.* **1958,** *65,* 147.
72. Bondi, A. *J. Chem. Eng. Data* **1963,** *8,* 371.
73. Inokuchi, H.; Shiba, S.; Hauda, T.; Akamatu, H. *Bull. Chem. Soc. Japan.* **1952,** *25,* 299.
74. Aihara, A. *Bull. Chem. Soc. Japan* **1959,** *32,* 1242.
75. Davies, M. *J. Chem. Educ.* **1971,** *48,* 591.
76. Mnyukh, Y.V. *J. Phys. Chem. Solids* **1963,** *24,* 631.
77. Madsen, H.E.L.; Boistelle, R. *Acta Crystallogr. Sect. A* **1976,** *32,* 828.
78. Broadhurst, M.G. *J. Res. Natl. Bur. Stand.* **1962,** *66A,* 241.
79. Leclercq, M.; Collet, A.; Jacques, J. *Tetrahedron* **1976,** *32,* 821.
80. Mirsky, K.; Cohen, M.D. *Chem. Phys.* **1978,** *28,* 193.
81. Ashcroft, S.J. *Thermochim. Acta* **1971,** *2,* 512.
82. Jones, A.H. *J. Chem. Eng. Data* **1960,** *5,* 196.
83. Thomson, G.W.; Douslin, D.R. In Physical Methods of Chemistry, Part V, "Determination of Thermodynamic and Surface Properties", Weissberger, A.; and Rossiter, B.W., Eds.; Wiley-Interscience: New York, 1971, Chapter 2.
84. Steele, W.V. *J. Chem. Thermodyn.* **1977,** *9,* 311.
85. Holborn, L.; Bauman, A. *Ann. Phys.* **1910,** *31,* 945.
86. de Wilde, J.H. *Z. Anorg. Chem.* **1937,** *233,* 411.
87. Sabbah, R.C.R. *Hebd. Seances Acad. Sci. Ser. C* **1979,** *289,* 153; *Chem. Abstr.* **1980,** *92,* 58049 p.
88. Clark, T.; Knox, T.; Mackle, H.; McKervey, M.A.; Rooney, J.J. *J. Chem. Soc. Faraday Trans. 1* **1975,** *71,* 2107.
89. Radchenko, L.G.; Kitaigorodskii, A.I. *Zh. Fiz. Khim.* **1974,** *48, 2702; Engl. Ed. 1595.*
90. Parsons, G.H.; Rochester, C.H.; Wood, C.E.C. *J. Chem. Soc. B* **1971,** 533.
91. Andon, R.J.L.; Biddiscombe, D.P.; Cox, J.D.; Handley, R.; Harrop, D.; Herington, E.F.G.; Martin, J.F. *J. Chem. Soc.* **1960,** 5246.
92. Balson, E.W. *Trans. Faraday Soc.* **1947,** *43,* 54.
93. Wu, P.-J.; Hsu, L.; Dows, D.A. *J. Chem. Phys.* **1971** *54,* 2714.
94. DeKruif, C.G.; Govers, H.A.J. *J. Chem. Phys.* **1980** *73,* 553.
95. Matsumoto, M.; Amaya, K. *Bull. Chem. Soc. Japan* **1980,** *53,* 3510.
96. "Handbook of Chemistry and Physics", 63rd ed., Weast, R.C., Ed.; CRC Press: Boca Raton, Fla., 1983, pp. C727–C731.

REFERENCES TO TABLE 5

1910
10/1 Holborn, L.; Bauman, A. *Ann. Phys.* **1910,** *31,* 945.
1920
20/1 Baxter, G.P.; Bezzenberger, F.K.; Wilson, C.H. *J. Am. Chem. Soc.* **1920,** *42,* 1386.

1925
25/1 Volmer, M.; Kirchhoff, P.Z. *Z. Phys. Chem.* **1925,** *115,* 223.
25/2 Perry, J.H.; Bardwell, D.C. *J. Am. Chem. Soc.* **1925,** *47,* 2629.
1926
26/1 Andrews, D.H.; Gynn, G.; Johnston, J. *J. Am. Chem. Soc.* **1926,** *48,* 1274.
1927
27/1 May, O.E.; Berliner, J.F.T.; Lynch, D.F.J. *J. Am. Chem. Soc.* **1927,** *49,* 1012.
27/2 Coolidge, A.S.; Coolidge, M.S. *J. Am. Chem. Soc.* **1927** *49,* 100.
27/3 Klosky, S.; Woo, L.P.L.; Flanigan, R.J. *J. Am. Chem. Soc.* **1927,** *49,* 1280.
1928
28/1 Steiner, L.E.; Johnston, J. *J. Phys. Chem.* **1928,** *32,* 912.
1930
30/1 Coolidge, A.S. *J. Am. Chem. Soc.* **1930,** *52,* 1874.
1932
32/1 Neumann, K.; Volker, E. *Z. Phys. Chem. Abt. A5* **1932,** *161,* 33.
1933
33/1 Broadway, L.F.; Fraser, R.G.J. *J. Chem. Soc.* **1933,** 429.
33/2 Yost, D.M.; Stone, W.E. *J. Am. Chem. Soc.* **1933,** *55,* 1889.
33/3 Tanneberger, H. *Chem. Ber.* **1933,** *66B,* 484.
1934
34/1 Wolf, K.L.; Trieschmann, H.G. *Z. Phys. Chem. B* **1934,** *27,* 376.
34/2 Hirsbrunner, H. *Helv. Chim. Acta* **1934,** *17,* 477.
1935
35/1 Trieschmann, H.G. Dissertation, Inst. Phys. Chem. Elektrochem, Universität Kiel, Germany, 1935.
1936
36/1 Aston, J.G.; Messerly, G.H. *J. Am. Chem. Soc.* **1936** *58,* 2354.
1937
37/2 de Wilde, J.H. *Z. Anorg. Chem.* **1937,** *233,* 411.
1938
38/1 Wolf, K.L.; Weghofer, H. *Z. Phys. Chem. B* **1938,** *39,* 194.
38/2 Weghofer, H. Dissertation, Phys. Chem. Inst. Universität Halle, Germany, 1938.
1940
40/1 Zibberman-Granovskaya, A.A. *J. Phys. Chem. USSR* **1940,** *14,* 759, 1004; *Chem. Abstr.* **1941,** *35,* 3867.
40/2 Darkis, F.R.; Vermillion, H.E.; Gross, P.M. *Ind. Eng. Chem.* **1940,** *32,* 946.
40/3 Messerly, G.H.; Aston, J.G. *J. Am. Chem. Soc.* **1940,** *62,* 886.
1941
41/1 Nitta, I.; Seki, S. *J. Chem. Soc. Japan* **1941,** *62,* 581; *Chem. Abstr.* **1942,** *36,* 154.
1943
43/1 Nitta, I.; Seki, S. *J. Chem. Soc. Japan,* **1943,** *64,* 475; *Chem. Abstr.* **1947,** *41* 3359a.
43/2 Ketelaar, J.A.A.; Kruyer, S. *Rec. Trav. Chim. Pays-Bas* **1943,** *62,* 550; *Chem. Abstr.* **1944,** *38,* 1926.
1946
46/1 Albert, A.; Willis, J.B. *Nature* **1946,** *157,* 341.
Willis, J.B. *Trans. Faraday Soc.* **1947,** *43,* 97.
46/2 Crooks, D.A.; Feetham, F.M. *J. Chem. Soc.* **1946,** 899.
1947
47/1 Balson, E.W. *Trans. Faraday Soc.* **1947,** *43,* 54.
47/2 Stull, D.R. *Ind. Eng. Chem.* **1947,** *39,* 517.
47/3 Ivin, K.J.; Dainton, F.S. *Trans. Faraday Soc.* **1947,** *43,* 32.
47/4 Balson, E.W.; Denbeigh, K.G.; Adam, N.K. *Trans. Faraday Soc.* **1947,** *43,* 42.
1948
48/1 Nitta, I.; Seki, S.; Momotani, M.; Sato, K. *J. Chem. Soc. Japan* **1948,** *69,* 85; *Chem. Abstr.* **1950,** *44,* 9204e.
48/2 See [43/1]
48/3 Brown, H.C.; Sujishi, S. *J. Am. Chem. Soc.* **1948,** *70,* 2878.
48/4 Nitta, I.; Seki, S.; Momotani, M.; Sato, K. *J. Chem. Soc. Japan* **1948,** *69,* 430; *Chem. Abstr.* **1951,** *45,* 6448i.

48/5 Nitta, I.; Seki, S. *J. Chem. Soc. Japan; Pure Chem. Sect.* **1948,** *69,* 141, 143; *Chem. Abstr.* **1952,** *46,* 1320i.
48/6 Pace, E.L.; Aston, J.G. *J. Am. Chem. Soc.* **1948,** *70,* 566.

1949

49/1 Bradley, R.S.; Shellard, A.D. *Proc. R. Soc. London Ser. A* **1949,** *198,* 239.
49/2 Winstrom, L.O.; Kulp, L. *Ind. Eng. Chem.* **1949,** *41,* 2584.
49/3 Sears, G.W.; Hopke, E.R. *J. Am. Chem. Soc.* **1949,** *71,* 1632.
49/4 Dreisbach, R.R.; Martin, R.A. *Ind. Eng. Chem.* **1949,** *41,* 2875.
49/5 Parks, G.S.; Moore, G.E. *J. Chem. Phys.* **1949,** *17,* 1151.
49/6 Klages, F. *Chem. Ber.* **1949,** *82,* 358.
49/7 Sears, G.W.; Hopke, E.R. *J. Am. Chem. Soc.* **1949,** *71,* 2575.
49/8 Nitta, I.; Seki, S.; Chihara, H. *J. Chem. Soc. Japan, Pure Chem. Sect.* **1949,** *70,* 387; *Chem. Abstr.* **1951,** *45,* 2737d.
49/9 Scott, D.W.; Gross, M.E.; Oliver, G.D.; Huffman, H.M. *J. Am. Chem. Soc.* **1949,** *71,* 1634.

1950

50/1 Edwards, G.T. *Trans. Faraday Soc.* **1950,** *46,* 423.
50/2 Nitta, I.; Seki, S.; Momotani, M.; Sato, K. *J. Chem. Soc. Japan* **1950,** *71,* 378; *Chem. Abstr.* **1951,** *45,* 6448g.
50/3 Bright, N.F.H.; Carson, T.; Dyson, T.A. *Research* **1950,** *3,* 185.
50/4 Nitta, I.; Seki, S.; Momotani, M.; Suzuki, K.; Nakagawa, S. *Proc. Jap. Acad.* **1950,** *26,* 11; *Chem. Abstr.* 1951, 45, 4545b.
50/5 Nitta, I.; Seki, S.; Momotani, M. *J. Chem. Soc. Japan, Pure Chem. Sect.* **1950,** *71,* 450; *Chem. Abstr.* **1951,** *45,* 6448i.

1951

51/1 Bright, N.F.H. *J. Chem. Soc.* **1951,** 624.
51/2 Magnus, A.; Hartman, H.; Becker, F. *Erdol Kohle* **1951,** *4,* 115; *Z. Phys. Chem.* **1951** *197,* 75.
51/3 Nitta, I.; Seki, S.; Suzuki, K. *Bull. Chem. Soc. Japan* **1951,** *24,* 63.
51/4 Nicholson, G.R. Quoted by Cook, J.W.; Gibb, A.R.; Raphael, R.A.; Somervile, A.R. *J. Chem. Soc.* **1951,** 503.
51/5 Hopke, E.R.; Sears, G.W. *J. Chem. Phys.* **1951,** *19,* 1345.
51/6 Nitta, I.; Seki, S.; Chihara, H.; Suzuki, K. *Sci. Pap. Osaka Univ.* No 29, 1951; *Chem. Abstr.* **1952,** *46,* 3820.
51/7 Oliver, G.D.; Grisard, J.W. *J. Am. Chem. Soc.* **1951,** *73,* 1688.
51/8 Burger, L.L.; Cady, G.H. *J. Am. Chem. Soc.* **1951,** *73,* 4243.
51/9 Bradley, R.S.; Waghorn, G.C.S. *Proc. R. Soc. London Ser. A* **1951,** *206,* 65.

1952

52/1 Scott, D.W.; Douslin, D.R.; Gross, M.E.; Oliver, G.D.; Huffman, H.M. *J. Am. Chem. Soc.* **1952,** *74,* 883.
52/2 Bradley, R.S.; Cotson, S.; Cox, E.G. *J. Chem. Soc.* **1952,** 740.
52/3 Inokuchi, H.; Shiba, S.; Handa, T.; Akamatu, H. *Bull. Chem. Soc. Japan* **1952,** *25,* 299.
52/4 Aihara, A. *J. Chem. Soc. Japan Pure Chem. Sect.* **1952,** *73,* 855; *Chem. Abstr.* **1953,** *47,* 3634 d.
52/5 Bywater, S. *J. Polym. Sci.* **1952,** *9,* 417.
52/6 Kaplan, L.; Kester, W.L.; Katz, J.J. *J. Am. Chem. Soc.* **1952,** *74,* 5531.

1953

53/1 Bradley, R.S.; Cleasby, T.G. *J. Chem. Soc. London* **1953,** *2,* 1681.
53/2 Stevens, B. *J. Chem. Soc.* **1953,** 2973.
53/3 Jenkins, A.D.; Style, D.W.G. *J. Chem. Soc.* **1953,** 2337.
53/4 Bradley, R.S.; Cotson, S. *J. Chem. Soc.* **1953,** 1684.
53/5 Aihara, A. *J. Chem. Soc. Japan, Pure Chem. Sect.* **1953,** *74,* 437, 631, 634; *Chem. Abstr.* **1954,** *48,* 1088h.
53/6 Edwards, G. *Trans. Faraday Soc.* **1953,** *49,* 152.
53/7 Serpinskii, V.V.; Voitkevich, S.A.; Lyuboshits, N.Y. *Zh. Fiz. Khim.* **1953,** *27,* 1032; *Chem. Abstr.* **1955,** *49,* 3594e.
53/8 Ewald, A.H. *Trans. Faraday Soc.* **1953** *49,* 1401.
53/10 Seki, S.; Suzuki, K. *Bull. Chem. Soc. Japan* **1953,** *26,* 209, 372.

53/11 Osthoff, R.C.; Grubb, W.T.; Burkhard; C.A. *J. Am. Chem. Soc.* **1953,** *75,* 2227.
53/12 Bradley, R.S.; Care, A.D. *J. Chem. Soc.* **1953,** *1688.*

1954

54/1 Davies, M.; Jones, J.I. *Trans. Faraday Soc.* **1954,** *50,* 1042.
54/2 Springall, H.D.; White, T.R. *J. Chem. Soc.* **1954,** 2764.
54/3 Jordan, T.E. "Vapor Pressure of Organic Compounds". Wiley: New York, 1954.
54/4 Serpinskii, V.V.; Voitkevich, S.A.; Lyuboshits, N.Y. *Zh. Fiz. Khim.* **1954** *28,* 810, 1969; *Chem. Abstr.* **1955,** *49,* 6677; **1956,** *50,* 4573h.
54/5 Crowell, T.I.; Jones, G.L. *J. Phys. Chem.* **1954,** *58,* 666.
54/6 Abrams, A.; Davis, T.W. *J. Am. Chem. Soc.* **1954,** *76,* 5993; Mooney, R.B.; Ludlam, E.B. *Proc. R. Soc. Edinburgh* **1929,** *49,* 160.
54/7 Lord, G.; Woolf, A.A. *J. Chem. Soc.* **1954,** 2546.
54/8 Dunn, S.A. *J. Am. Chem. Soc.* **1954,** *76,* 6191.

1955

55/1 Cass, R.C.; Springall, H.D.; Quincey, P.G. *J. Chem. Soc.* **1955,** 1188.
55/2 Armstrong, G.T.; Brickwedde, F.G.; Scott, R.B. *J. Res. Natl. Bur. Stand.* **1955,** *55,* 39.
55/3 Aihara, A. *J. Chem. Soc. Japan Pure Chem. Sect.* **1955,** *76,* 492; *Chem. Abstr.* **1955,** *49,* 15323c.
55/4 Neale, E.; Williams, L.T.D. *J. Chem. Soc.* **1955,** 2485.

1956

56/1 Beynon, J.H.; Nicholson, G.R. *J. Sci. Instrum.* **1956,** *33,* 376.
56/2 Majury, T.G. *Chem. Ind. (London)* **1956,** 349.
56/3 Serpinskii, V.V. *Fiz. Chem.* **1956,** *30,* 177.
56/4 Dickinson, W. *Trans. Faraday Soc.* **1956,** *52,* 31.
56/5 Magnus, A. *Z. Phys. Chem. (Frankfurt)* **1956,** *9,* 141.
56/6 Suzuki, K.; Onishi, S.; Koide, T.; Seki, S. *Bull. Chem. Soc. Japan* **1956,** *29,* 127.
56/7 Seki, S.; Suzuki, K.; *J. Chem. Soc. Japan* **1956,** *77,* 346; *Chem. Abstr.* **1956,** *50,* 5343i.
56/8 Milazzo, G. *Ann. Chim. (Rome)* **1956,** *46,* 1105.
56/9 Barber, E.J.; Cady, G.H. *J. Phys. Chem.* **1956,** *60,* 504.

1957

57/1 VanKamp, A. Dissertation, Free University of Amsterdam, 1957.
57/2 Day, J.H.; Oestruch, C. *J. Org. Chem.* **1957,** *22,* 214.
57/3 McErvan, W.S.; Rigg, M.W. *J. Am. Chem. Soc.* **1951,** *73,* 4725.
57/4 Saggiomo, A.J. *J. Org. Chem.* **1957,** *22,* 1171.
57/5 Rowlinson, J.S.; Thacker, R. *Trans. Faraday Soc.* **1957,** *53,* 1.

1958

58/1 Hoyer, H.; Peperle, W. *Z. Elektrochem.* **1958,** *62,* 61.
58/2 Cass, R.C.; Fletcher, S.E.; Mortimer, C.T.; Springall, H.D.; White, T.R. *J. Chem. Soc.* **1958,** 1406.
58/3 Peppel, W.J. *Ind. Eng. Chem.* **1958,** *50,* 767.
58/4 Klochkov, V.P. *Zh. Fiz. Khim.* **1958,** *32,* 1177; *Chem. Abstr.* **1958,** *52,* 19394f.
58/5 Ungnade, H.; McBee, E.T. *Chem. Rev.* **1958,** *58,* 249.
58/6 Klipping, V.G.; Stranski, I.N. *Z. Anorg. Allg. Chem.* **1958,** *297,* 23.

1959

59/1 Tagaki, S.; Chihara, H.; Seki, S. *Bull Chem. Soc. Japan* **1959,** *32,* 84.
59/2 Aihara, A. *Bull Chem. Soc Japan* **1959,** *32,* 1242.
59/3 Davies, M.; Jones, A.H.; Thomas, G.H. *Trans. Faraday Soc.* **1959,** *55,* 1100.
59/4 Davies, M.; Jones, A.H. *Trans. Faraday Soc.* **1959,** *55,* 1329.
59/5 Bradley, R.S.; Drury, T. *Trans. Faraday Soc.* **1959,** *55,* 1844.
59/6 Cordes, J.F.; Schreiner, S. *Z. Anorg. Allg. Chem.* **1959,** *299,* 87.
59/7 Klipping, I.; Stranski, I. *Z. Anorg. Allg. Chem.* **1959,** *297,* 23.
59/8 Stern, J.H.; Gregory, N.W. *J. Phys. Chem.* **1959,** *63,* 556.
59/9 Farkas, A.; Mills, G.A.; Ernef, W.E.; Maerker, J.B. *J. Chem. Eng. Data* **1959,** *4,* 334.

1960

60/1 Jones, A.H. *J. Chem. Eng. Data* **1960,** *5,* 196.
60/2 Budurov, S. *Izv. Khim. Inst. Bulg. Akad. Nauk* **1960,** *7,* 281.
60/3 Andon, R.J.L.; Biddiscombe, D.P.; Cox, J.D.; Handley, R.; Harrop, D.; Herington, E.F.G.; Martin, J.F. *J. Chem. Soc.* **1960,** 5246.
60/4 Davies, M.; Thomas, G.H. *Trans. Faraday Soc.* **1960,** *56,* 185.

60/5 Wada, T.; Kishida, E.; Tomiie, Y.; Suga, H.; Seki, S.; Nitta, I. *Bull. Chem. Soc. Japan* **1960,** *33,* 1317.
60/6 Hirt, R.C.; Steger, J.E.; Simard, G.L. *J. Polym. Sci.* **1960,** *43,* 319; *Chem. Abstr.* **1960,** *54,* 16968b.
60/7 Woodman, A.L.; Murbach, W.L.; Kaufman, M.H. *J. Phys. Chem.* **1960,** *64,* 658.

1961

61/1 Davies, M.; Malpass, V.E. *J. Chem. Soc.* **1961,** 1048.
61/2 Walsh, P.N.; Smith, N.O. *J. Chem. Eng. Data* **1961,** *6,* 33.
61/3 Zimmermann, H. *Z. Elektrochem.* **1961,** *65,* 368.

1962

62/1 Bauder, A.; Gunthard, H.H. *Helv. Chim. Acta* **1962,** 1698.
62/2 Kraus, M.; Beranek, L.; Kochloeff, K.; Bazant, V. *Chem. Prumsyl.* **1962,** *12,* 649; *Chem. Abstr.* **1963,** *58,* 10746c.
62/3 Rabinovich, I.B.; Tel'noi, V.I.; Terman, L.M.; Kirillova, A.S.; Razuvaev, G.A. *Dokl. Akad. Nauk SSSR* **1962,** *143,* 133; *Engl. Ed.* 171; *Chem. Abst.* **1962,** *57,* 4557e.
62/4 Tjebbes, J. *Acta Chem. Scand.* **1962,** *16,* 916.
62/5 Mackle, H.; Mayrick, R.G. *Trans. Faraday Soc.* **1962,** *58,* 238.
62/6 Carson, A.; Cooper, R.; Stranks, D.R. *Trans. Faraday Soc.* **1962,** *58,* 2125.
62/7 Edwards, J.W.; Kingston, G.L. *Trans. Faraday Soc.* **1962,** 1323.

1963

63/1 Miller, G.A.; *J. Chem. Eng. Data* **1963,** *8,* 69.
63/2 Margrave, J.L.; Frisch, M.A.; Bautista, R.G.; Clarke, R.L.; Johnson, W.S. *J. Am. Chem. Soc.* **1963,** *85,* 546.
63/3 Biddiscombe, D.P.; Handley, R.; Harrop, D.; Head, A.J.; Lewis, G.B.; Martin, J.F.; Sprake, C.H.S. *J. Chem. Soc.* **1963,** 5764.
63/4 Boyd, R.H. *J. Chem. Phys.* **1963,** *38,* 2524.
63/5 Horton, G.R.; Wendlandt, W.W. *J. Inorg. Nucl. Chem.* **1963,** *25,* 241.
63/6 Bondi, A. *J. Chem. Eng. Data,* **1963,** *8,* 374.

1964

64/1 Frisch, M.A.; Bautista, R.G.; Margrave, J.L.; Parsons, C.G.; Wotez, J.H. *J. Am. Chem. Soc.* **1964,** *86,* 335.
64/3 Kelley, J.D.; Rice, F.O. *J. Phys. Chem.* **1964,** *68,* 3794.

1965

65/1 Svec, H.; Clyde, D.D. *J. Chem. Eng. Data* **1965,** *10,* 151.
65/2 Clark, L.B.; Peschel, G.G.; Tinoco, I. *J. Phys. Chem.* **1965,** *69,* 3615.
65/3 Boyd, R.H.; Christensen, R.L.; Pua, R. *J. Am. Chem. Soc.* **1965,** *87,* 3554.
65/4 Frankosky, M.; Aston, J.G. *J. Phys. Chem.* **1965,** *69,* 3126.
65/5 Karnes, H.A.; Kybett, B.D.; Wilson, M.H.; Margrave, J.L.; Newman, M.S. *J. Am. Chem. Soc.* **1965,** *87,* 5554.
65/6 Davies, M.; Kybett, B. *Trans. Faraday Soc.* **1965,** *61,* 1608, 1893.
65/7 Kalashnikova, L.A.; Rozantsev, E.G.; Chaikin, A.M. *Izv. Akad. Nauk SSSR* **1965,** 2525; *Chem. Abstr.* **1965,** *63,* 4972c.
65/8 Wadsö, I. *Acta Chem. Scand.* **1965,** *19,* 1079.

1966

66/1 Boyd, R.H. *Tetrahedron* **1966,** *22,* 119.
66/2 Kybett, B.D.; Carroll, S.; Natalis, P.; Bonnell, D.W.; Margrave, J.L.; Franklin, J.L. *J. Am. Chem. Soc.* **1966,** *88,* 626.
66/3 Geiseler, G.; Quitzsch, K.; Rauh, H.-J.; Schaffernicht, H.; Walther, H.J. *Ber. Phys. Chem.* **1966,** *70,* 551.
66/4 Lebedev, Y.A.; Rosantsev, E.G.; Kalashnikova, L.A.; Lebedev, V.P.; Nieman, M.B.; Apin, Y.A. *Dokl. Acad. Nauk SSSR* **1966,** *168,* 104, *Engl. Ed.* 460.
66/5 Månsson, M.; Sunner, S. *Acta Chem. Scand.* **1966,** *20, 845.*

1967

67/1 Bratton, W.K.; Szilard, I.; Cupas, C.A. *J. Org. Chem.* **1967,** *32,* 2019.
67/2 Wakayama, N.; Inokuchi, H. *Bull. Chem. Soc. Japan* **1967,** *40,* 2267.
67/3 Boyd, R.H.; Guha, K.R.; Wuthrich, R. *J. Phys. Chem.* **1967,** *71,* 2187.
67/4 Miroshnichenko, E.A.; Lebendev, Y.A.; Shevelev, S.A.; Gulevskaya, V.I.; Fainzi-l'berg, A.A.; Apin, A.Y. *Zh. Fiz. Khim.* **1967,** *41,* 1477; *Engl. ed.* 783.
67/5 Geiseler, G.; Schaffernicht, H.; Walther, H.J. *Ber. Phys. Chem.* **1967,** *71,* 196.
67/6 Chia, Y.T.; Simmons, H.E. *J. Am. Chem. Soc.* **1967,** *89,* 2638.

1968
68/1 Karyakin, N.V.; Rabinovich, I.B. Pakhomov, L.G. *Russ. J. Phys. Chem.* **1968,** *42,* 954.
68/2 Baccanari, D.P.; Novinski, J.A.; Pan, Y.-C.; Yevitz, M.M.; Swain, H.A. *Trans. Faraday Soc.* **1968,** *64,* 1201.
68/3 Smith, N.K.; Gorin, G.; Good, W.D.; McCullough, J.P. *J. Phys. Chem.* **1964,** *68,* 940.
68/4 Desai, P.D.; Wilhoit, R.C.; Zwolinski, B.J. *J. Chem. Eng. Data* **1968,** *13,* 334.
68/5 Pepekin, V.I.; Miroshnichenko, E.A.; Levedev, Y.A. Apin, A.Y. *Russ. J. Phys. Chem.* **1968,** *42,* 1583.
68/6 Hamilton, J.V.; Fagley, T.F. *J. Chem. Eng. Data* **1968,** *13,* 523.
68/7 Baroody, E.E.; Carpenter, G.A.; Robb, R.A.; Zimmer, M.F. *J. Chem. Eng. Data* **1968,** *13,* 215.
68/8 Mason, R.; Rae, A.I.M. *Proc. R. Soc. 1968 London, Ser. A* **304,** 501.
68/9 Mertl, I. *Chem. Listy* **1968,** *62,* 584.
1969
69/1 Stull, D.R.; Westrum, E.F., Jr; Sinke, G.C. "The Chemical Thermodynamics of Organic Compounds". Wiley: New York, 1969.
69/2 Chastel, R.; Steckel, Tachoire, H. "Proceedings of the First International Conference on Calorimetry, and Thermodynamics", Warszana. Polish Scientific Publishers, 5/2, 1969; as cited in [82/1].
69/3 Crimmons, F.T. *U.S. At. Energy Comm.* **1969,** UCRL-50704. *Chem. Abstr.* **1970,** *72,* 70876a.
69/4 Spencer, W.F.; Cliath, M.M. *Environ. Sci. Technol.* **1969,** *3,* 670.
69/5 Cox, J.D.; Gundry, H.A.; Harrop, D.; Head, A.J. *J. Chem. Thermodyn.* **1969,** *1,* 77.
69/6 Shieh, C.-F.; McNally, D.; Boyd, R.H. *Tetrahedron,* **1969,** *25,* 3653.
69/7 Bremser, W.; Hagen, R.; Heilbronner, E.; Vogel, E. *Helv. Chim. Acta* **1969,** *52,* 418.
69/8 Månsson, M.; Morawetz, E.; Nakase, Y.; Sunner, S. *Acta Chem. Scand.* **1969,** *23,* 56.
69/9 Lebedeva, N.D.; Dneprovskii, A.S.; Katin, Y.A. *Russ. J. Phys. Chem.* **1969,** *43,* 770.
69/10 Miroshnichenko, E.A.; Lebedev, Y.A. *Khim. Geterotsikl. Soedin.* **1969,** *5,* 963.
69/11 Mackle, H.; McNally, D.V.; Steele, W.V. *Trans. Faraday Soc.* **1969,** *65,* 2060, 2069.
69/12 Rosen, J.M.; Dickinson, C. *J. Chem. Eng. Data* **1969,** *14,* (1), 120.
69/13 Mackle, H.; Steele, W.V. *Trans. Faraday Soc.* **1969,** *65,* 2069, 2073.
69/14 Andrews, J.T.S.; Westrum, E.F., Jr; *J. Organomet. Chem.* **1969,** *17,* 349.
1970
70/1 Cox, J.D.; Pilcher, G. "Thermochemistry of Organic and Organometallic Compounds". Academic Press: New York, 1970.
70/2 Månsson, M.; Rapport, N.; Westrum, E.F., Jr. *J. Am. Chem. Soc.* **1970,** *92,* 7296.
70/3 Lenchitz, C.; Velicky, R. *J. Chem. Eng. Data* **1970,** *15,* 401.
70/4 Kojima, H. *Sen-i Gakkaishi (Japan)* **1970,** *26,* 530.
70/5 Westrum, E.F., Jr.; Wong, W.K.; Morawetz, E. *J. Phys. Chem.* **1970,** *74,* 2542.
70/6 Sellers, P. *Acta Chem. Scand.* **1970,** *24,* 2453.
70/8 Spencer, W.F.; Cliath, M.M. *J. Agric. Food Chem.* **1970,** *18,* 529.
70/9 Melia, T.P.; Merrifield, R. *J. Inorg. Nucl. Chem.* **1970,** *32,* 1489.
1971
71/1 Boyd, R.H.; Sanwal, S.N.; Shary-Tehrany, S.; McNally, D. *J. Phys. Chem.* **1971,** *75,* 1264.
71/2 Carson, A.S.; Fine, D.H.; Gray, P.; Laye, P.G. *J. Chem. Soc. B* **1971,** 1611.
71/3 Jackson,, W.; Hung, T.S.; Hopkins, H.P., Jr. *J. Chem. Thermodyn.* **1971,** *3,* 347.
71/4 Lenchitz, C.; Velicky, R.W.; Silvestro, G.; Schlosberg, L.P. *J. Chem. Thermodyn.* **1971,** *3,* 689.
71/5 Sellers, P. *Acta Chem. Scand.* **1971,** *25,* 2194.
71/6 Beech, G.; Lintonbon, R.M. *Thermochim Acta* **1971,** *2,* 86.
71/7 Wu, P.-J.; Hsu, L.; Dows, D.A. *J. Chem. Phys.* **1971,** *54,* 2714.
71/8 Parson, G.H.; Rochester, C.H.; Wood, C.E.C. *J. Chem. Soc. B* **1971,** 533.
71/9 Carson, A.S.; Laye, P.G.; Steele, W.V.; Johnston, D.E.; McKervey, M.A. *J. Chem. Thermodyn.* **1971,** *3,* 915.
71/10 Wong, S.-W.S.; Westrum, E.F., Jr. *J. Chem. Thermodyn.* **1971,** *3,* 105.
71/11 Kiparisova, E.G.; Rabinovitch, I.B.; *Dok. Akad. Nauk SSSR* **1971,** 199, 1075; *Engl. ed.* 675.
71/12 Choi, J.K.; Joncich. M.J. *J. Chem. Eng. Data* **1971,** *16,* 87.

71/13 Rapport, N.J.; Westrum, E.F., Jr.; Andrews, J.T.S. *J. Am. Chem. Soc.* **1971,** *93,* 4363.
71/14 Rogers, F.E. *J. Phys. Chem.* **1971,** *75,* 1734.
71/15 Hamilton, W.S.; Witt, L.C. *J. Chem. Eng. Data* **1971,** *16,* 234.
71/16 Morawetz, E. Dissertation, Univeristy of Lund, Sweden 1971; see [83/7].
71/17 Ashcroft, S.J. *Thermochim. Acta* **1971,** *2,* 512.

1972

72/1 Morawetz, E.; *J. Chem. Thermodyn.* **1972,** 4, 139, 455, 461.
72/2 Irving, R.J.; *J. Chem. Thermodyn.* **1972,** 4, 793.
72/3 Colomina, M.; Monzon, C.; Turrión, C.; Laynez, J. Paper presented at the the Fifth Experimental Thermodynamics Conference, Lancaster; as cited in [82/1].
72/4 Arnold, P.R.; Jones, F. *Mol. Cryst. Liq. Cryst.* **1972,** *19,* 133.
72/5 Kana'an, A.S. *J. Chem. Thermodyn.* **1972,** *4,* 893.
72/6 Pepekin, V.I.; Matyushin, Y.N.; Rozantsev, G.G.; Shevelev, S.A.; Apin, A.Y. *Izv. Akad. Nauk SSSR, Ser. Khim* **1972,** *21,* 2703; *Engl. ed.* 2634.
72/7 Lebedeva, N.D.; Katin, Y.A. *Zh. Fiz. Khim.* **1972,** 46, 1061; *Engl. ed.* 623.
72/9 Wiedeman, H.G. *Thermochim. Acta* **1972,** *3,* 355.
72/10 Mills, G.D. *Diss. Abstr. Int. B* 33, **1972** 1485; *Chem. Abstr.* **1973,** *78,* 42653f.
72/11 Spencer, W.F.; Cliath, M.M. *J. Agric. Food Chem.* **1972,** *20,* 645.
72/12 Gardner, P.J.; Hussain, K.S. *J. Chem. Thermodyn.* **1972,** *4,* 819.
72/13 Rogers, F.E. *J. Phys. Chem.* **1972,** 76, 106.
72/14 Hamilton, W.S.; Witt, L.C. *J. Chem. Eng. Data* **1972,** *17,* 138.
72/15 Geiseler, G.; Sawistowsky, J. *Z. Phys. Chem (Leipzig)* **1972,** *250,* 43.
72/16 Geiseler, Von G.; Rauh, H.J. *Z. Phys. Chem. (Leipzig)* **1972,** *249,* 376.

1973

73/1 DeKruif, C.G.; Oonk, H.A. *J. Chem. Ing. Technol.* **1973,** 45, 455; *Chem. Abstr.* **1973,** 78, 152198.
73/2 Malaspina, L.; Gigli, R.; Bardi, G. *J. Chem. Phys.* **1973,** *59,* 387.
73/3 Barnes, D.S.; Mortimer, C.T.; Mayer, E. *J. Chem. Thermodyn.* **1973,** *5,* 481.
73/4 Malaspina, L.; Bardli, G.; Gigli, R. *J. Chem. Thermodyn.* **1973,** 5, 699.
73/5 Malaspina, L.; Gigli, R.; Bardi, G.; DeMaria, G. *J. Chem. Thermodyn.,* **1973,** *5,* 699.
73/6 See reference 73/1.
73/7 Gigli, R.; Malaspina, L.; Bardi, G. *Ann. Chim. (Rome)* **1973,** 63, 627 *Chem. Abstr.* **1975,** *82,* 77775g.
73/8 Lebedeva, N.D.; Katin, Y.A. *Zh. Fiz. Khim.* **1973,** *47,* 1620; *Engl. ed.* 922.
73/9 Behrendt, W.; Gattow, G. *Z. Anorg. Allg. Chem.* **1973,** *398,* 198.
73/10 Bardi, G.; Malaspina, L.; Piacenti, V. *J. Chem. Eng. Data* **1973,** *18,* 126.
73/11 See Reference [73/1].
73/12 Pepekin, V.I.; Gafurov, R.G.; Lebedev, Y.A.; Eremenko, L.T.; Sogomonyan, E.M.; Apin, A.Y. *Izv. Akad. Nauk SSSR, Ser. Khim* **1973,** *22,* 318; *Engl. ed.* 304.
73/13 Rodgers, D.L.; Westrum, E.E., Jr.; Andrews, J.T.S. *J. Chem. Thermodyn.* **1973,** *5,* 733.
73/14 McCauley, J.A.; Smith, N.O. *J. Chem. Thermodyn.* **1973,** *5,* 31.
73/15 McEachern, D.M.; Sandoval, O. *J. Phys. E* **1973,** *6,* 155.
73/16 Rauh, H.-J.; Geyer, W.; Schmidt, H.; Geiseler, G. *Z. Phys. Chem. (Leipzig)* **1973,** *252,* 395.
73/17 Rogers, F.E.; Quan, S.W. *J. Phy. Chem.* **1973,** *77,* 828.
73/18 See Reference [73/8].
73/19 Lebedeva, N.D.; Katin, Y.A. *J. Appl. Chem. USSR* **1973,** *46,* 2131.
73/20 Hamilton, J.V.; Mitchell, G.M.; Ayers, D.A. *J. Chem. Eng. Data* **1973,** *18,* 366.
73/21 Hamilton, J.V.; Mitchell, G.M. *J. Chem. Eng. Data* **1973,** *18,* 38, 364.
73/22 See Reference [73/21].
73/23 Geiseler, G.; Sawistowsky, J. *Z. Phys. Chem. (Leipzig)* **1973,** *253,* 333.
73/24 Rauh, H.-J.; Geiseler, G. *Z. Phys. Chem. (Leipzig)* **1973,** *252,* 395.
73/25 Arshadi, M.R.; Shabrang, M. *J. Chem. Soc. Perkin Trans. 2* **1973,** 1732.
73/26 See reference 73/25.
73/27 Hall, H.K.; Smith, C.D.; Baldt, J.H. *J. Am. Chem. Soc.* **1973,** *95,* 3197.

1974

74/1 Sabbah, R.; Chastel, R.; Laffitte, M. *Can. J. Chem.* **1974,** *52,* 2201.
74/2 Colomina, M.; Roux, M.V.; Turrión, C. *J. Chem. Thermodyn.* **1974,** *6,* 149.
74/3 Colomina, M.; Roux, M.V.; Turrión, C. *J. Chem. Thermodyn.* **1974,** *6,* 571.

74/4 Malaspina, L.; Bardi, G.; Gigli, R. *J. Chem. Thermodyn.* **1974,** *6,* 1053.
74/5 Arshadi, M.R. *J. Chem. Soc. Faraday, Trans. 1* **1974,** *70,* 1569.
74/6 Pribilova, J.; Poulchy, J. *Collect. Czech. Chem. Commun.* **1974,** *39,* 3233.
74/7 Teplitskii, A.B.; Sukhodub, L.F.; Yanson, I.K. *Fiz. Kondens, Sostoyanuja* **1974,** *32,* 68; *Chem. Abstr.* **1976,** *84,* 170535e.
74/8 Yanson, I.K.; Verkin, B.I.; Shlyarevskii, O.I.; Teplitskii, A.B. *Stud. Biophys.* **1974,** *46,* 29. *Chem. Abstr.* **1975,** *83,* 8884a.
74/9 Radchenko, L.G.; Kztaigoradskii, *A.I. Zh. Fiz. Khim.* **1974,** *48,* 2702; *Engl. ed.* 1595.
74/10 Sabbah, R.; Chastel, R.; Laffitte, M. *Thermochim. Acta* **1974,** *10,* 353.
74/11 Pepekin, V.I.; Matyushin, Y.N.; Lebedev, Y.A. *Izv. Akad. Nauk SSSR, Ser. Khim.* **1974,** *23,* 1786; *Engl. ed.* 1707.
74/12 Calle, L.M.; Kanaan, A.S. *J. Chem. Thermodyn.* **1974,** *6,* 935.
74/13 See Reference [74/5].
74/14 Månsson, M. *Acta Chem. Scand. Ser. B* **1974,** *28,* 677.
74/15 Roux, M.V.; Turrión, C.; Colomina, M.; Pérez-Ossario, R. *An. Quim.* **1974,** *70,* 201; *Chem. Abstr.* **1974,** *81,* 119718c.
74/16 Månsson, M. *Acta Chem. Scand. Ser. B* **1974** *28,* 895.
74/17 Engel, P.S.; Wood, J.L.; Sweet, J.A.; Margrave, J.L. *J. Am. Chem. Soc.* **1974,** *96,* 2381.
74/18 Cox, C.D. *Pure Appl. Chem.* **1974,** *40,* 399.
74/19 Sinke, G.C. *J. Chem. Thermodyn.* **1974,** *6,* 311.
74/20 Sabbah, R. "Energetique de Liaison Inter- et Intramoleculaires". C.N.R.S.: Marseille, 1974.

1975

75/1 McEachern, D.M.; Sandoval, O.; Iniguez, J.C. *J. Chem. Thermodyn.* **1975,** *7,* 299.
75/2 Clark, T.; Knox, T.; Mackle, H.; McKervey, M.A.; Rooney, J.J. *J. Chem. Soc. Faraday Trans. 1* **1975,** *71,* 2107.
75/3 Chickos, J.S. *J. Chem. Educ.* **1975,** *52,* 134.
75/4 Ambrose, D.; Lawrenson, I.J.; Sprake, C.H.S. *J. Chem. Thermodyn.* **1975,** *7,* 1173.
75/5 DeKruif, C.G.; van Ginkel, C.H.D.; Voogd, J. In "Comptes Rendus de la IVme Conférence Internationale sur la Thermodynamique et la Chimie, 1975", Rouquerol, J., and Sabbah, R., Eds.; Centre de Recherche sur la Microcalorimetrie Thermochimique, CNRS: Marseilles, France, as cited in Reference [82/1].
75/6 van Ginkel, C.H.D.; DeKruif, G.C.; DeWaal, F.E.B. *J. Phy. E Sci. Instrum.* **1975,** *8,* 490.
75/7 Parker, W.; Steele, W.V.; Stirling, W.; Watt, I. *J. Chem. Thermodyn.* **1975,** *7,* 795.
75/8 Osborn, A.G.; Douslin, D.R. *J. Chem. Eng. Data* **1975,** *20,* 229.
75/9 Carson, A.S.; Laye, P.G.; Morris, H. *J. Chem. Thermodyn.* **1975,** *7,* 993.
75/10 Grovers, H.A.J. *Acta Crystallogr. Sect. A* **1975,** *31,* 380.
75/11 Bagrov, G.V.; Andreev, S.N.; Smirnova, M.F.; Vol'f, L.A. *Deposited Doc.* **1975,** VINITI 1196; *Chem. Abstr.* **1977,** *87,* 59208y.
75/12 Malaspina, L.; Gigli, R.; Bardi, G.; DeMaria, G. In "Comptes Rendus de la IVme Conférence Internationale sur la Thermodynamique et la Chimie, 1975" by Rouquerol, J.; and Sabbah, R., Eds.; Centre de Rechérche", sur la Microcalorimétrie Thermochimique, CNRS: Marseille, France *Chem. Abstr.* **1976,** *84,* 179387r.
75/13 Anthoney, M.E.; Carson, A.S.; Laye, P.G. Conférence Internationale sur la Thermodynamique et la Chimie, 1975", Rouquerol, J.; and Sabbah, R., Eds., Centre de Recherche sur la Microcalorimétrie Thermochimique; CNRS: Marseille, France; *Chem. Abstr.* **1976,** *84,* 163961n.
75/14 Teplitskii, A.B.; Yanson, I.K. *Biofizika* **1975,** *20,* 189; *Chem. Abstr.* **1975,** *83,* 16712h.
75/15 McEachern, D.M.; Iniquez, J.C.; Ornelas, H.C. *J. Chem. Eng. Data* **1975,** *20,* 226.
75/16 Yanson, I.K.; Teplitskii, A.B. *Zh. Fiz. Khim.* **1975,** *49,* 736; *Engl. Ed.* 428.
75/17 Bogdanova, K.A.; Berlin, A.A.; Kompanets, V.Z.; Rakova, G.V.; Miroshnichenko, E.A.; Lebedev, Y.A.; Enikolopyan, N.S. *Vysokomol. Soedin. Ser. A* **1975,** *17,* 658; *Chem. Abstr.* **1975,** *83,* 43841f.
75/18 Barnes, D.S.; Pilcher, G. *J. Chem. Thermodyn.* **1975,** *7,* 377.

1976

76/1 Ferro, D.; Piacente, V.; Gigli, R.; D'Ascenzo, G. *J. Chem. Thermodyn.* **1976,** *8,* 1137.
76/2 Ambrose, D.; Lawrenson, I.J.; Sprake, C.H.S. *J. Chem. Thermodyn.* **1976,** *8,* 503.
76/3 Pella, P.A. *Anal. Chem.* **1976,** *48,* 1634.

76/4 Kozina, M.P.; Bychikhina, L.V.; Gal'chenko, G.V.; Milwitskaya, E.M.; Ordubade, M.; Plate, A.F. *Dokl. Akad. Nauk SSSR* **1976,** *226,* 1105; *Engl. ed.* 143.

76/5 Anthoney, M.E.; Carson, A.S.; Laye, P.G.; Yürekle, M. *J. Chem. Thermodyn.* **1976,** *8,* 1009.

76/6 Colomina, M.; Jiménez, P.; Pérez-Ossorio, R.; Turrión, C. *J. Chem. Thermodyn.* **1976,** *8,* 439.

76/7 Miroshnichenko, E.A.; Lebedev, Y.A. In "Mezhnol. Vzaimodeistvie Konform. Mol. Tezisy Dokl. Vses. Simp. 3rd" **1976,** *40; Chem. Abstr.* **1979,** *91,* 56104h.

76/8 Taylor, J.W.; Crookes, R.J. *J. Chem. Soc. Faraday Trans. 1* **1976,** *72,* 723.

76/9 Hopkins, H.P., Jr.; Bostwick, D.; Alexander, C.J. *J. Am. Chem. Soc.* **1976,** *98,* 1355.

76/10 Barnard, M. *Thermochim. Acta* **1976,** *16,* 267.

76/11 Depablo, R.S. *J. Chem. Eng. Data* **1976,** *21,* 141.

1977

77/1 Pedley, J.B.; Rylance, J. "N.P.L. Computer Analysed Thermochemical Data: Organic and Organometallic Compounds"; School of Molecular Sciences, University of Sussex: Brighton, UK, 1977.

77/2 Gaffney, J.S.; Pierce, R.C.; Friedman, L., *J. Am. Chem. Soc.* **1977,** *99,* 4293.

77/3 Ngauv, S.N.; Sabbah, R.; Laffite, M. *Thermochim. Acta* **1977,** *20,* 371.

77/4 DeKruif, C.G., van Ginkel, C.H.D. *J. Chem. Thermodyn.* **1977,** *9,* 725.

77/5 Pella, P.A. *J. Chem. Thermodyn.* **1977,** *9,* 301.

77/6 Steele, W.V. *J. Chem. Thermodyn.* **1977,** *9,* 311.

77/7 Finke, H.L.; Messerly, J.F.; Lee, S.H.; Osborn, A.G.; Douslin, D.R. *J. Chem. Thermodyn.* **1977,** *9,* 937.

77/8 Dygdala, R.S.; Stefanski, K.; Wonikowski, J. *Bull. Acad. Pol. Sci. Ser. Sci., Math. Astron. Phys.* **1977,** *25,* 439.

77/9 Parker, W.; Steele, W.V.; Watt, I. *J. Chem. Thermodyn.* **1977,** *9,* 307.

77/10 Steinwandter, H. *Chemosphere,* **1977,** *6,* 59; *Chem. Abstr.* **1977,** *87,* 17285r.

77/11 Steele, W.V.; Watt, I. *J. Chem. Thermodyn.* **1977,** *9,* 843.

77/12 Kozina, M.P.; Bychikhina, L.V.; Gal'chenko, G.L. *Zh. Fiz. Khim.* **1977,** *51,* 2142; *Engl. ed.* 1258.

77/13 Nabavian, M.; Sabbah, R.; Chastel, R.; Laffitte, M. *J. Chim. Phys. Phys-Chim. Biol.* **1977,** *74,* 115.

77/14 Schulze, F.-W.; Petrick, H.-J.; Canmenga, H.K.; Klinge, H. *Z. Phys. Chem. (Wiesboden)* **1977,** *107,* 1.

77/15 Sabbah, R.; Nabavian, M.; Laffitte, M. *C.R. Hebd. Seances Acad. Sci. Ser. C* **1977,** *284* (23), 953; *Chem. Abstr.* **1977,** *87,* 151519y.

77/16 Cox, C.D. *Pure Appl. Chem.* **1977,** 49, 1437.

1978

78/1 Steele, W.V. *J. Chem. Thermodyn.* **1978,** *10,* 585.

78/2 Steele, W.V. *J. Chem. Thermodyn.* **1978,** *10,* 441.

78/3 Heath, G.A.; Hefter, G.T. Steele, W.V. *J. Chem. Thermodyn.* **1978,** *10,* 395.

78/4 Sabbah, R.; Laffitte, M. *Bull. Soc. Chem. Fr.* **1978,** *1,* 50.

78/5 Steele, W.V. *J. Chem. Thermodyn.* **1978,** *10,* 919.

78/6 Berney, C.V.; Spickerman, D. *J. Chem. Thermodyn.* **1978,** *10,* 637.

78/7 Colomina, M.; Jiménez, P.; Roux, M.V.; Turrión, C. *J. Chem. Thermodyn.* **1978,** *10,* 661.

78/8 Arora, M.; Steele, W.V. *J. Chem. Thermodyn.* **1978,** *10,* 403.

78/9 Karyakin, N.V.; Kamelova, G.P.; Sapozhnikov, V.N. *Termodin. Org. Soedin.* **1978,** *7,* 52. *Chem. Abstr.* **1980,** *92,* 75658d.

78/10 Vilcu, R.; Perisanu, S.; Ciocazanu, I. *Bul. Inst. Politek "Gheorghe Gheorghiu-Dej" Bucuresti Ser. Chim.-Metal* **1978,** *40,* 9–12; *Chem. Abstr.* **1979,** *91, 28258t.*

78/11 Jorge, R.A.; Airolde, C.; Chagas, A.P. *J. Chem. Soc. Dalton Trans.* **1978,** 1102.

78/13 Sabbah, R.; Laffitte, M. *Thermochim. Acta* **1978,** *23,* 196.

78/14 Chickos, J.S.; Sherwood, D.E. *J. Org. Chem.* **1978,** *43,* 1146.

78/15 Cundall, R.B.; Palmer, T.F.; Wood, C.E.C. *J. Chem. Soc. Faraday Trans. 1* **1978,** *74,* 1339.

78/16 Calis-Van Ginkel, C.H.D.; Calis, G.H.M.; Timmermans, C.W.M.; DeKruif, C.G.; Oonk, H.A.J. *J. Chem. Thermodyn.* **1978,** *10,* 1083.

78/17 Nowak, M.J.; Szczepaniak, K.; Barski, A.; Shugar, D. *Z. Naturforsch. C: Biosci.* **1978,** *33,* 876.

78/18 Grover, R.; Spencer, W.F.; Farmer, W.J.; Shoup, T.D. *Weed Sci.* **1978,** *26,* 505.
78/19 Trimble, L.E.; Voorhoeve, R.J.H. *Analyst* **1978,** *103,* 759.

1979

79/1 DeKruif, C.G.; Voogd, J.; Offringa, J.C.A. *J. Chem. Thermodyn.* **1979,** *11,* 651.
79/2 Sabbah, R. *Bull. Soc. Chem. Fr.* **1979** (9–10, Pt. 1), 434.
79/3 Sabbah, R.C.R. *Hebd. Seances Acad. Sci. Ser. C* **1979,** *289,* 153; *Chem. Abstr.* **1980,** *92,* 58049p.
79/4 Colomina, M.; Jiménez, P.; Pérez-Ossorio, R.; Turrión, C. *J. Chem. Thermodyn.* **1979,** *11,* 1179.
79/5 Colomina, M.; Jiménez, P.; Roux, M.V.; Turrión, C. *An. Quim.* **1979,** *75,* 620; *Chem. Abstr.* **1980,** *92,* 75685k.
79/6 Das, D.; Dharwadkar, S.R.; Chandrasekharaiah, M.S. *Thermochim. Acta* **1979,** *30,* 371.
79/7 Sandman, D.J.; Epstein, A.J.; Chickos, J.S.; Ketchum, J.; Fu, J.S.; Scheraga, H.A. *J. Chem. Phys.* **1979,** *70,* 305; ibid, **1981,** *74*(1), erratum.
79/8 Sunner, S.; Svensson, C.H.; Zelepuga, A.S. *J. Chem. Thermodyn.* **1979,** *11,* 491.
79/9 Kimura, T.; Takagi, S. *J. Chem. Thermodyn.* **1979,** *11,* 47.
79/10 Spencer, W.F.; Shoup, T.D.; Cliath, M.M.; Farmer, W.J.; Hague, R. *J. Agric. Food. Chem.* **1979,** *27,* 273.
79/11 Daamen, H. *Thermochim. Acta* **1979,** *34,* 69.

1980

80/1 DeKruif, C.G.; *J. Chem. Thermodyn.* **1980,** *12,* 243.
80/2 Colomina, M.; Jiménez, P.; Roux, M.V.; Turrión, C. *J. Calorim. Anal. Therm.* **1980,** *11,* 3-19-1/3-19-16 (Span.); *Chem. Abstr.* **1981,** *95,* 121878n; see Reference [81/11].
80/3 Dygdala, R.S.; Stefanski, K. *Chem. Phys.* **1980,** *53,* 51.
80/4 Clark, T.; Knox, T.M.; McKervey, M.A.; Mackle, H. *J. Chem. Soc. Perkin Trans, 2* **1980,** 1686.
80/5 Sabbah, R. *Thermochim. Acta* **1980,** *41,* 33.
80/6 Arshadi, M.R. *J. Chem. Thermodyn.* **1980,** *12,* 903.
80/7 Sabbah, R.; Komorowski, S. *Thermochim. Acta* **1980,** *41,* 379.
80/8 Aihara, A. *Denki Tsushin Daigaku Gakuho* **1980,** *31,* 65; *Chem. Abstr.* **1981,** *94,* 174045z.
80/9 Ferro, D.; Bencivenni, L.; Teghil, R.; Mastromarino, R. *Thermochim. Acta* **1980,** *42,* 75.
80/10 Sato-Toshima, T.; Sakiyama, M.; Saki, S. *Bull. Chem. Soc. Japan* **1980,** *53,* 2762.
80/11 Lebedeva, N.D.; Kalin, Y.A.; Raydnenko, V.L.; Stapanov, V.B.; Kozlova, N.M. *Zh. Prikl. Khim. (Leningrad)* **1980,** *53,* 2588; *Chem. Abstr.* **1981,** *94,* 156270q.
80/12 Rodionova, G.N.; Krutovskaya, I.V.; Rodinov, A.N.; Tuchin, Y.G.; Karpov, V.V. *Zh. Prikl. Spektrosk.* **1980,** *32,* 638.
80/13 Ferro, D.; Bencivenni, L.; Teghil, R.; Pelino, M. *J. Indian Chem. Soc.* **1980,** *57,* 629.
80/14 Colomina, M.; Jiménez, P.; Turrión, C.; Fernandez, J.A.; Monzén, C. *An. Quim. Ser. A* **1980,** *76,* 245.
80/15 Nishiyama, K.; Sakiyama, N.; Seki, S.; Horita, H.; Otsubo, T.; Misumi, S. *Bull. Chem. Soc. Japan* **1980,** *53,* 869.
80/16 Steele, W.V. *J. Chem. Thermodyn.* **1980,** *12,* 187.
80/17 Andruzzi, F.; Pilcher, G.; Hacking, J.M.; Cavell, S. *Makromol. Chem.* **1980,** *181,* 923; *Chem. Abstr.* **1980,** *93,* 8632y.
80/18 Bardi, G.; Bencivenni, L.; Ferro, D.; Martini, B.; Nunziante, C.S.; Teghil, R. *Thermochim. Acta* **1980,** *40,* 275.
80/19 Teplitskii, A.B.; Yansen, I.K.; Glukhova, O.T.; Zielenkiewicz, A.; Zielenkiewicz, W.; Wierzchowski, K.L. *Biophys. Chem.* **1980,** *11,* 17.
80/20 Murray, J.P.; Cavell, K.J.; Hill, J.O. *Thermochim. Acta* **1980,** *36,* 97.
80/21 Sabbah, R. *Thermochim. Acta* **1980,** *35,* 73.
80/22 DeKruif, C.G.; Govers, H.A.J. *J. Chem. Phys.* **1980,** *73,* 553.
80/23 Swain, H.A., Jr.; Kwan, Y.C.-Y.; Sung, H.-N. *J. Phys. Chem.* **1980,** 1347.
80/24 Calado, J.C.R.; Dias, A.R.; Mina de Piedade, M.E.; Martinho Simoes, J.A. *Rev. Port. Quim.* **1980,** *22,* 53; see Reference [83/6].
80/25 Sachinides, J.; Hill, J.O. *Thermochim. Acta* **1980,** *35,* 99.
80/26 Smith, N.K.; Stewart, R.C., Jr.; Osborn, A.G.; Scott, D.W. *J. Chem. Thermodyn.* **1980,** *12,* 919.
80/27 Osborn, A.G.; Scott, D.W. *J. Chem. Thermodyn.* **1980,** *12,* 429.

1981

81/1 DeKruif, C.G.; Kuipers, T.; van Miltenburg, J.C.; Schaake, R.C.F.; Stevens, G. *J. Chem. Thermodyn.* **1981,** *13,* 1081.

81/2 Månsson, M.; Sunner, S. *J. Chem. Thermodyn.* **1981,** *13,* 671.

81/3 Brostow, W.; McEachern, D.M.; Valdez, J.A. *Mater. Chem.* **1981,** *6,* 187.

81/4 DeKruif, C.G. Smit, E.J.; Gover, H.A. *J. Chem. Phys.* **1981,** *74,* 5838.

81/5 Sabbah, R.; Minadakis, C. *Thermochim. Acta* **1981,** *43,* 269.

81/6 Ferro, D.; Pelino, M. *Rev. Roum. Chim.* **1981,** *26,* 9.

81/7 Westcott, J.W.; Simon, C.G.; Bidleman, T.F. *Environ. Sci. Technol.* **1981,** *15,* 1375.

81/8 Chickos, J.S., Garin, D.L.; Hitt, M.; Schilling, G. *Tetrahedron* **1981,** *37,* 2255.

81/9 Pirogov, A.I.; Azarova, G.I.; Klopov, V.I. *Izv. Vyssh. Uchebn. Zaved., Khim. Khim. Tekhnol.* **1981,** **24** (7), 827–829; *Chem. Abstr.* **1981,** *95,* 138901n.

81/10 Pelino, M.; Tomassetti, M.; Piacente, V.; D'Ascenzo, G. *Thermochim. Acta* **1981,** *44,* 89.

81/11 Colomina, M.; Jiménez, P.; Roux, M.V.; Turrión, C. *An. Quim.* **1981,** *77,* 114.

1982

82/1 Colomina, M.; Jiménez, P.; Turrión, C. *J. Chem. Thermodyn.* **1982,** *14,* 779.

82/2 Gomez, T.; Alfonso, L.; Sabbah, R. *Thermochim. Acta* **1982,** *58,* 311.

82/3 Murata, S.; Sakiyama, M.; Seki, S. *J. Chem. Thermodyn.* **1982,** *14,* 707, 723.

82/4 Jochems, R.; Dekker, H.; Mosselman, C.; Somsen, G. *J. Chem. Thermodyn.* **1982,** *14,* 395, 799.

82/5 Menadakis, C.; Sabbah, R. *Thermochim. Acta* **1982,** *55,* 147.

82/6 Wren, D.J.; Vikis, A.C. *J. Chem. Thermodyn.* **1982,** *14,* 435.

82/7 Inagaki, S.; Murato, S.; Sakiyama, M. *Bull. Chem. Soc. Japan* **1982,** *55,* 2808.

82/8 Gomez, T.; Alfonso, L.; Sabbah, R. *Thermochim. Acta* **1982,** *57,* 67.

82/9 Iniguez, J.C.; Lopez, M.E.; McEachern, D.M. *Rev. Soc. Quim. Mex.* **1982,** *26,* 122; *Chem. Abstr.* **1982,** *97,* 144252j.

82/10 Imagaki, S.; Murata, S.; Sakiyama, M.; Ito, Y.; Umehara, Y.; Hijiya, T.; Matsuura, T. *Bull. Chem. Soc. Japan* **1982,** *55,* 2803.

82/11 Batley, M.; Mraw, S.; Staveley, L.A.K.; Overs, A.H.; Owen, M.C.; Thomas, R.K.; White, J.W. *Mol. Phys.* **1982,** *45,* 1015.

82/12 DeKruif, C.G.; Blok, J.G. *J. Chem. Thermodyn.* **1982,** *14,* 201.

1983

83/1 Van Ekeren, P.J.; Jacobs, M.H.G.; Offringa, J.C.A.; DeKruif, C.G. *J. Chem. Thermodyn.* **1983,** *15,* 409.

83/2 DeKruif, C.G.; Van Miltenburg, J.C.; Blok, J.G. *J. Chem. Thermodyn.* **1983,** *15,* 129.

83/3 Bender, R.: Bieling, V.; Mauer, G. *J. Chem. Thermodyn.* **1983,** *15,* 585.

83/4 Clay, R.M.; Corr, S.; Keenan, G.; Steele, W.V. *J. Am. Chem. Soc.* **1983,** *105,* 2070.

83/5 Spencer, W.F.; Cliath, M.M. *Residue Rev.* **1983,** *85,* 57.

83/6 Jacobs, M.H.G.; van Ekeren, P.J.; DeKruif, C.G. *J. Chem. Thermodyn.* **1983,** *15,* 619.

83/7 DeWit, H.G.M.; Van Miltenburg, J.C.; DeKruif, C.G. *J. Chem. Thermodyn.* **1983,** *15,* 651.

1984

84/1 Colomina, M.; Jiménez, P.; Roux, M.V.; Turrión, C. *J. Chem. Thermodyn.* **1984,** *16,* 379.

U **(Unpublished Results)**

U/1 Chickos, J.S.; Liebman, J.F.

U/2 Chickos, J.S.; Liebman, J.F.; Greenberg, A.

U/3 Mackle, H.; Steele, W.V.; McNally, D.V.

U/4 Stull, D.

U/5 Chickos, J.S.

REFERENCES TO TABLE 6

1974

A 74/1 Beak, P.; Mueller, D.S.; Lee, J. *J. Amer. Chem. Soc.* **1974,** *96,* 3867.

1982

A 82/1 Airoldi, C.; Chagas, A.P.; Assuncao, F.P. *Can. J. Chem.* **1982,** *60,* 2132.

A 82/2 Jochems, R.; Dekker, H.; Mosselman, C.; Somsen, G. *J. Chem. Thermodyn.* **1982,** *14,* 799.

1983

A 83/1 Holdiness, M.R. *Thermochimica Acta* **1983,** *68,* 375.

A 83/2 Majer, J.R.; Azzouz, A.S.P. *J. Chem. Soc. Faraday Trans. I* **1983,** *79,* 675.

A 83/3 Maskill, H.; Steele, W.V. *J. Chem. Thermodyn.* **1983,** *15,* 481.

A 83/4 Sonnefeld, W.J.; Zoller, W.H.; May, *Anal. Chem.* **1983,** *55,* 275.

A 83/5 Spitsyn, V.I.; Kolli, I.D.; Balabanova, E.A.; Smirnov, E.V.; Murav'eva I.A. *Zh. Neorg. Khim.* **1983,** *28,* 2698.

A 83/6 Zielenkiewicz, A.; Zielenkiewicz, W.; Colomina, M.; Jimenez, P.; Cesari, E.; Sanahiya, A.; Torra, V. *An. Quim. Ser. A* **1983,** 79 (3, Suppl: 2) 677. *Chem. Abst.* **1984,** *101,* 22920f.

A 83/7 Colomina, M.; Jiménez, P. Turrión, C.; Kaminski, M.; Zielenkiewicz, W., *Thermochimica Acta* **1983,** *68,* 371.

A 83/8 Pelino, M.; Gigli, R.; Tomassette, M. *Thermochimica Acta* **1983,** *61,* 301.

A 83/9 Kratt, G.; Bechhaus, H.-D.; Bernlohr, W.; Ruchardt, C. *Thermochimica Acta,* **1983,** *62,* 279.

1984

A 84/1 Holdiness, M.R. *Thermochimica Acta* **1984,** *78,* 435.

A 84/2 Bickerton, J.; Minas Da Piedade, M.E.; Pilcher, G. *J. Chem. Thermodyn.* **1984,** *16,* 661.

A 84/3 Govorchin, S.W.; Kana'an, A.S.; Kanameuller, J.M. *J. Chem. Thermodyn.* **1984,** *16,* 703.

A 84/5 Gol'din, G.S.; Levites, E.I.; L'vov, A.I.; Sinko, N.L. *Koord. Khim.* **1984,** *20,* 563.

A 84/6 Bickerton, J.; Pilcher, G.; Al-Takhin, G. *J. Chem. Thermodyn.* **1984,** *16,* 373.

A 84/7 Colomina, M.; Jiménez, P.; Roux, M.V.; Turrión, C. *J. Chem. Thermodyn.* **1984,** *16,* 379.

A 84/8 Govorchin, S.W.; Kana'an, A.S. *J. Chem. Thermodyn.* **1984,** *16,* 437.

A 84/9 Colomina, M.; Jiménez, P.; Roux, M.V.; Turrión, C. *J. Chem. Thermodyn.* **1984,** *16,* 1121.

A 84/10 Ribeiro Da Silva, M.D.M.C.; Ribeiro Da Silva, M.A.V.; Pilcher, G. *J. Chem. Thermodyn.* **1984,** *16,* 1149.

CHAPTER 4

Experimental Thermochemistry

Kenneth B. Wiberg

Yale University, New Haven, Connecticut

CONTENTS

1. INTRODUCTION

The energy of a compound, along with its structure, represents the most important quantities by which it may be characterized. A knowledge of energies is essential for any study of equilibria. The equilibrium constant is, of course, related to the free energy change, but whereas entropies may frequently be satisfactorily calculated via statistical mechanics,[1] energies must be obtained experimentally.[2] Thermochemical data also are necessary for any detailed examination of kinetics. The changes in activation energy for any series of related compounds may be due to changes in relative ground state energies between reactants and products, and/or to specific effects that appear only in the activated complexes. Thermochemical data are required

© 1987 VCH Publishers, Inc.
MOLECULAR STRUCTURE AND ENERGETICS, Vol. 2

to resolve this question. The design of industrial processes that produce chemical substances also depends on the availability of thermochemical data for all the substances involved in the process.

Energies of molecules may be derived in several ways. In the case of organic compounds, the most general method is oxygen bomb calorimetry.[3] Here the compound is burned in oxygen to yield carbon dioxide and water, and the energy released is measured. Another experimental procedure is reaction calorimetry,[4] in which the energy change during a reaction is determined. In this case, if the heats of formation of all but one of the compounds are known, that for the remaining compound can be determined from the heat of reaction.

It is clearly not possible to determine the energies of all known organic compounds experimentally. In many cases, particularly with hydrocarbons, the heats of formation may be estimated. The simplest and least accurate method makes use of an empirical correlation between structures and energies, such as Franklin's group equivalents[5] or Allen's bond energies.[6] Molecular mechanics offer a more realistic modeling of intramolecular potentials.[7] It can provide rather satisfactory estimates of both structures and energies of most hydrocarbons, and with somewhat less accuracy, it can provide corresponding data for substituted compounds. Here, a major problem is the relative lack of the data that are needed in estimating the intramolecular potentials for heteroatoms. Semiempirical molecular orbital calculations[8] provide a useful alternative to molecular mechanics calculations and are able to handle a larger variety of structural features. However, since some of the parameters are determined empirically, the calculations frequently lead to large errors when applied to compounds different from those used in developing the parameters.[9] Ab initio molecular orbital calculations[10] can provide very good estimates of both structures and energies for any type of organic compound, but they are expensive with regard to computer time and are practical only for smaller compounds (C_2–C_6) or larger compounds having some symmetry, a property that decreases the amount of calculation needed.

This chapter is largely concerned with experimental aspects of determining the energies of organic compounds, but also briefly considers the methods for estimating energies. Molecular mechanics is considered in more detail by Ōsawa and Konematsu in Chapter 7 of Volume 3 in this series.

2. COMBUSTION CALORIMETRY

In a combustion experiment, a weighed pellet of a solid compound or a weighed sample of a liquid sealed into a thin glass or Mylar container is placed in a crucible. The crucible is inserted into a calorimetric bomb (Figure 4-1) along with a fuse, which may be ignited electrically, and 1 ml of water, which is used to dilute any nitric acid that may be formed. The bomb

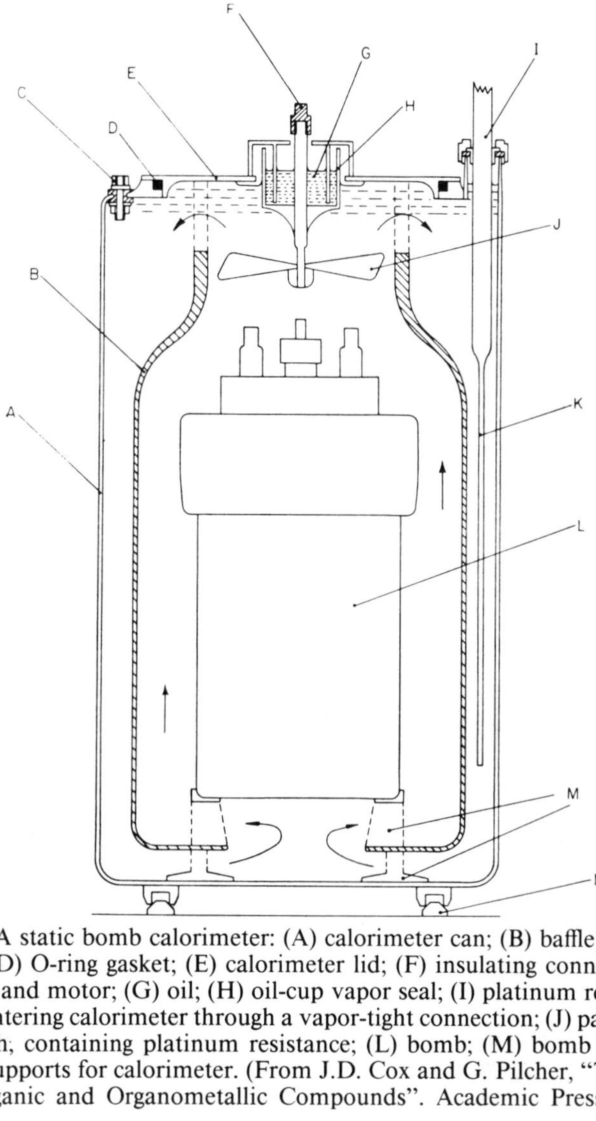

Figure 4-1 A static bomb calorimeter: (A) calorimeter can; (B) baffle; (C) electrical terminals; (D) O-ring gasket; (E) calorimeter lid; (F) insulating connector between stirrer shaft and motor; (G) oil; (H) oil-cup vapor seal; (I) platinum resistance thermometer, entering calorimeter through a vapor-tight connection; (J) paddle; (K) thin metal sheath, containing platinum resistance; (L) bomb; (M) bomb supports; (N) insulating supports for calorimeter. (From J.D. Cox and G. Pilcher, "Thermochemistry of Organic and Organometallic Compounds". Academic Press: New York, 1970.)

is sealed and is pressurized to 300 psi with pure oxygen. The bomb is placed in a double-walled cannister containing an accurately known amount of water ($\simeq 1$ kg), which absorbs the energy liberated in the combustion process. The system is allowed to come to thermal equilibrium in a thermostat controlled to $\pm 0.0005°$, and then ignition of the sample is effected by passing electric current through the fuse. The temperature rise due to combustion is measured, and knowing the calibration constant for the calorimeter, the heat liberated is calculated and converted to a per-mol basis. During the experiment, the temperature is measured with an uncertainty of about 0.0001° using a platinum resistance thermometer or a quartz thermometer. The calibration constant for the calorimeter is usually obtained by burning a compound such as benzoic acid for which the heat of combustion is known with good accuracy.

This may sound like a rather simple and routine procedure, but in fact combustion calorimetry is one of the most difficult of experimental procedures. Consider cyclohexane, which is an average-sized molecule. Combustion in oxygen gives a ΔH of reaction of -944.79 kcal/mol (note that the negative sign indicates that heat is released in the process). We should like to have an uncertainty in the measurement of ± 0.2 kcal/mol or less, which requires that the precision of the experiment be $\pm 0.02\%$. This includes errors in weighing the sample and the water in the calorimeter jacket, errors in temperature measurement, errors in corrections to the standard state, and errors involved in the combustion process (incomplete combustion forming either carbon monoxide or soot). It also includes errors due to sample impurities, and of these, one of the most difficult is water, since many compounds are hygroscopic. Thus, errors in any part of the experiment must be kept under $\pm 0.01\%$. Good quality quantitative chemical analysis usually strives for $\pm 0.1\%$ uncertainty. The combustion experiment must be 10 times more precise.

In a typical experiment, 500 mg of the compound is used and must be weighed with an uncertainty of less than 50 μg. The combustion process leads to a temperature rise of about 3°, which must be determined with an uncertainty of less than 0.3 millidegree. To ensure complete combustion, the contents of the bomb are swept through a carbon dioxide absorbent, which is weighed. The ratio of the amount of carbon dioxide found to that expected should be between 1.0001 and 0.9999. All the other experimental variables must be determined equally well. Finally, the data must be reduced to standard conditions (25°C and 1 atm pressure) making use of the Washburn corrections.[11] It is a tribute to the skill of calorimetrists that they have been able to surmount the problems in most cases, and it is not surprising that some of the measurements fall short of this standard.

The procedure above is useful for compounds containing carbon, hydrogen, oxygen, and nitrogen, which can be obtained in very pure form and in relatively large amount (3–5 g). Microbomb calorimeters have been developed for studying smaller samples,[12] but they are not applicable to all types of compounds and generally do not permit the analysis of the combustion products. Other possible errors also increase when the scale is markedly reduced. In the case of compounds containing sulfur or halogens, which lead to sulfuric acid or hydrogen halides as reaction products, one must ensure that the contents of the bomb are completely mixed. This requires the use of a rotating bomb calorimeter,[13] which is considerably more complex than a standard oxygen bomb calorimeter. As a result, far fewer data are available for these compounds.

It must be remembered that the important quantity is accuracy, not precision. The reported precision of a set of measurements refers only to random errors. When such high precision is achieved, systematic errors may well be considerably larger than the random errors, leading to an inaccuracy that is much larger than the uncertainty interval.

Once the heat of combustion has been obtained, the more useful quantity is the heat of formation from the elements. If we again consider cyclohexane, we may write the following thermodynamic cycle:

$$
\begin{array}{ll}
C_6H_{12}(l) + 9O_2(g) \rightarrow 6CO_2(g) + 6H_2O(l) & \Delta H = -944.79 \\
\underline{6C(s) + 6H_2(g) + 9O_2(g) \rightarrow 6CO_2(g) + 6H_2O(l)} & \underline{\Delta H = -974.22} \\
6C(s) + 6H_2(g) \rightarrow C_6H_{12}(l) & \Delta H = -29.43 \text{ kcal/mol}
\end{array}
$$

The heat of combustion of graphite (the standard form for carbon) is -94.05 kcal/mol and the heat of combustion of hydrogen gas (the standard form for hydrogen) is -68.32 kcal/mol. The expected heat of combustion of 6 gram-atoms of graphite and 6 mols of hydrogen would then be -974.22 kcal/mol. The difference between this and the observed heat of combustion of cyclohexane, -29.43 kcal/mol, is the heat of formation of cyclohexane from the elements at 298 K (25°C). It may be noted that the quantity of interest is obtained as the small difference between two large numbers.

How may we determine the accuracy of the heats of formation? One way is to carry out the measurement in several different laboratories. We would be unlikely to find systematic errors that were the same from laboratory to laboratory. Therefore, if the values agree within their uncertainty intervals, the inaccuracy and the uncertainty interval may be equated. Unfortunately, such replicate experiments are not available for most compounds, and there is no way to assess the accuracy from just one set of combustion data. This includes many key compounds such as cyclopropane and cyclobutane. Other methods for assessing the accuracy of heats of formation are discussed below.

Most of the available data derived from oxygen bomb calorimetry have been summarized, along with estimates of the "best" values when more than one determination has been carried out.[14]

3. REACTION CALORIMETRY

Reaction calorimetry (other than combustion)[4] is not generally capable of determining heats of formation directly. However, the heats of formation may frequently be combined with other data to provide heats of formation. The precision requirements for heats of reaction are considerably less than for heats of combustion, since one measures a small quantity directly rather than obtaining it as the small difference between two large quantities.

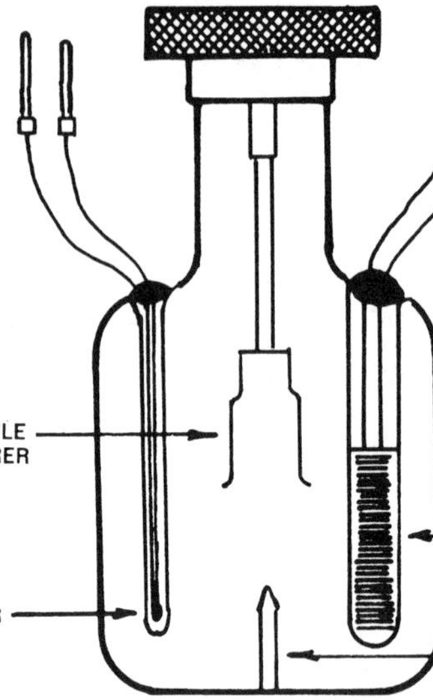

Figure 4-2. A reaction cell for the LKB reaction calorimeter. (After a design of Sunner and Wadsö.[15])

Reaction calorimetry may be carried out in a number of different fashions. The simplest is constant environment calorimetry, sometimes known as isoperibol calorimetry. Here, one of the reactants is sealed into a small ampoule, and the other is dissolved in a suitable solvent. The solution is placed in the calorimetric vessel, and the ampoule is placed in a holder (frequently the stirrer) (Figure 4-2).[15] The vessel is enclosed in a container, which is placed in a thermostat that will regulate the temperature to $\pm 0.0005°$. The system is allowed to come to thermal equilibrium, and the ampoule is broken to initiate the reaction.

Throughout the experiment the temperature is measured as a function of time, giving a time–temperature curve such as in Figure 4-3. The curve is divided into three portions, the foreperiod (up to the initiation of the reaction), the reaction period, and the afterperiod, during which thermal equilibrium is again achieved. One of the important quantities is the change in temperature during the reaction period. The foreperiod data are fit to an equation such as:

$$T_i = at_i^2 + bt_i + c$$

and the temperature (T) at the time (t) corresponding to the end of the foreperiod is obtained from the equation. The temperature at the beginning of the afterperiod is obtained in the same fashion. The observed temperature rise is due to several sources: the reaction, heat transfer from the external bath, and the stirring energy. To correct for the latter two quantities, one uses Newton's law:

$$\frac{\Delta T}{\Delta t} \sim \frac{\partial T}{\partial t} = k_1(T - T_b) + k_2$$

where T_b is the bath temperature (the thermostat), t is the time, k_1 is the heat transfer constant to or from the thermostat, and k_2 represents the heat input from stirring. Data for both foreperiod and afterperiod may be combined and used to determine k_1 and k_2 for the experiment. Then, the temperature correction may be determined from:

$$\Delta T_{\text{corr}} = \int_{t_1}^{t_2} [k_1(T - T_b) + k_2]\, dt$$

where t_1 is the time at the beginning of the reaction period and t_2 is the time at its end. The correction is subtracted from the observed temperature rise to give that due to the reaction.

To convert the temperature increase into the heat of reaction, the heat capacity of the system must be known. It may be determined by measuring the temperature rise for a process having a known heat of reaction, but a more precise determination is made via an electrical calibration in which electrical power is supplied to a heater in the reaction vessel for a known amount of time. A system has been developed for carrying out the entire experiment automatically under computer control.[16]

Other types of reaction calorimeters also have been used. An adiabatic calorimeter, in which the jacket is maintained at the same temperature as the reaction vessel, has the advantage that no corrections need to be applied to the observed temperature rise. Two designs have been employed. In the

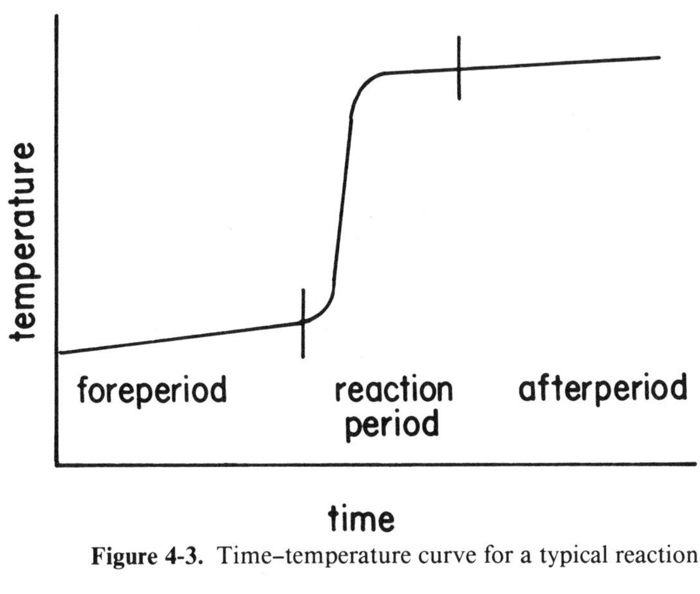

Figure 4-3. Time–temperature curve for a typical reaction.

first, the difference in temperature between the vessel and the jacket is monitored, and when a difference is found, the jacket is heated accordingly.[17] In the other, the cell is cooled via the Pellitier effect to keep it at the temperature of the jacket.[18] Both methods have been used successfully, but both are more complex to implement than constant-environment calorimetry.

Finally, some reactions that occur very slowly at room temperature may be conveniently studied using a temperature-programmed calorimeter. In one design, the temperature is increased at a steady rate and the heat of reaction observed as a difference in temperature between a cell containing the reacting substances in an appropriate solvent and a cell containing just the solvent.[19] The commercially available differential thermal analysis instruments could in principle carry out this type of experiment, but generally these are limited by heat transfer problems and the difficulty of handling completely sealed samples.

Hydrogenation of alkenes is the reaction that has been most widely studied calorimetrically. The first classic experiments, carried out by Kistiakowsky and his co-workers[20] in the 1930s and 1940s, still represent the best work on this reaction. One of the primary difficulties with catalytic hydrogenation is the question of whether the catalyst is unchanged during the course of the reaction. Any thermal effects related to the catalyst will appear as an error in the measured heat of hydrogenation.

Kistiakowsky measured the heats of reaction in the gas phase by passing a mixture of the alkene and hydrogen into a chamber containing the catalyst and allowing the reaction to occur with a constant flow of reactants until a steady temperature rise was found. Then the temperature was recorded and the exit stream was diverted to a combustion chamber, wherein the product was converted to carbon dioxide, which was determined analytically. When sufficient conversion had been effected, the flow to the combustion chamber was stopped and the final temperature noted. The amount of carbon dioxide collected and the temperature rise provide the information necessary in calculating the heat of hydrogenation with all substances in the gas phase. Since equilibrium was achieved before data were collected, it is fairly certain that no change in catalyst occurred during the course of the experiment.

All subsequent experiments have been carried out in solution, which is more convenient although less accurate. Most of these data come from the work of Turner and his co-workers.[21] Here, a known amount of alkene was dissolved in acetic acid and placed in the calorimetric vessel. Platinum oxide was placed in an ampoule, which was inserted into the vessel. The calorimeter was allowed to come to equilibrium in an atmosphere of hydrogen, and the reaction was initiated by breaking the ampoule. The temperature rise corresponded to the heat of reduction of the catalyst to colloidal platinum plus the heat of hydrogenation of the alkene in solution to the product, also in solution. The heat of reduction of the catalyst was determined independently and was subtracted from the observed temperature rise, giving just the heat of hydrogenation.

Besides the problem associated with the possible change in the catalyst, this experiment gives data for all compounds in solution. The values for the pure liquid or gas state would be much more valuable, since they could then be combined with other thermochemical data. Most of the other investigators have tried to avoid this problem. In one procedure, an ampoule containing the pure alkene is broken into acetic acid containing prereduced catalyst.[22] These data, combined with the heat of solution of the product, would give the heat of hydrogenation in the pure liquid phase. The other approach is to use cyclohexane as the solvent and to assume that the heats of solution of alkene and alkane in the solvent would be the same within experimental error.[18,23] The former would be the preferred mode of procedure, although the latter appears to be successful. Most of the available data for alkene hydrogenation have been summarized.[24] They have been used to calculate resonance energies, as well as to examine the stabilization of double bonds by alkyl groups, the difference in energy between exo and endo double bonds, and a wide variety of related questions. The hydrogenolysis of alkyl halides[25] and of carbon–carbon single bonds in strained hydrocarbons[18] also has been examined calorimetrically.

One of the disadvantages of hydrogenation in studying organic compounds is that it removes the functional group and replaces it with just a hydrocarbon. Organic chemists are usually more interested in compounds containing functional groups. In this context, one desirable reaction is the hydration of an alkene. This reaction usually occurs too slowly to be useful for calorimetry. However, the same information may be obtained via trifluoroacetolysis:[26]

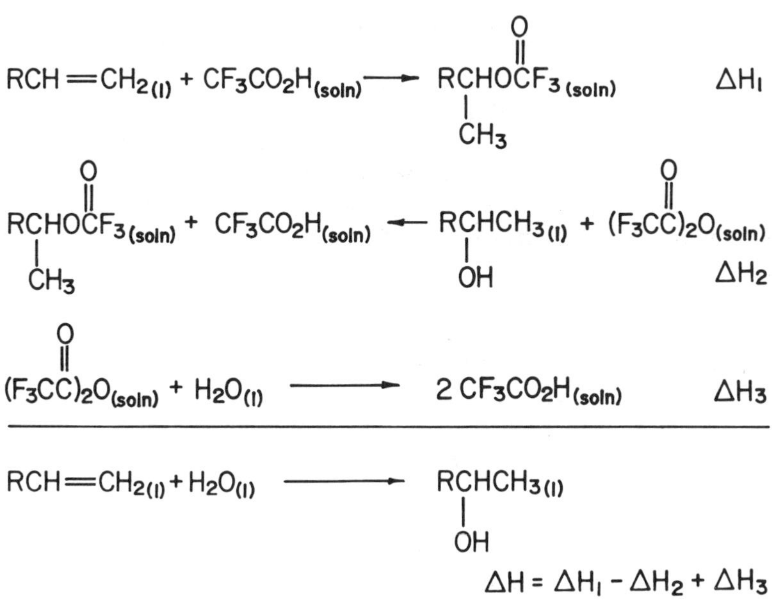

The heats of each of these reactions may be determined calorimetrically with high precision. The combination of these data allows the calculation of the heat of hydration in the pure liquid phase.

This type of experiment allows one to obtain several pieces of information. Consider the reaction of the five n-heptenes with trifluoroacetic acid:[27]

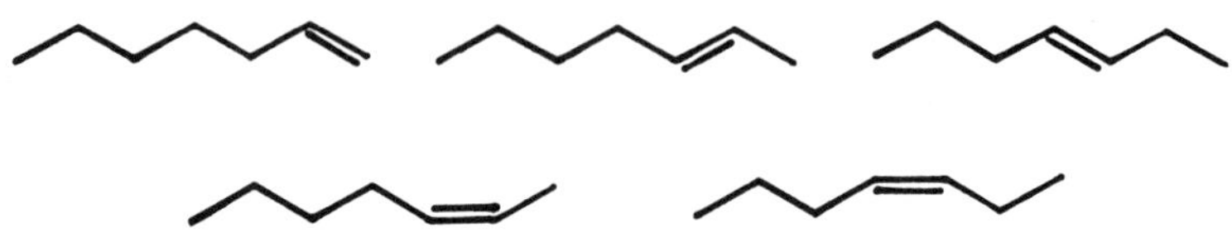

The reactions lead to mixtures of 2-, 3-, and 4-heptyl trifluoroacetates, but the differences in energy between the esters may readily be determined via equilibration experiments. The heats of reaction with trifluoroacetic acid to give the equilibrium mixture of trifluoroacetates could then be determined. The precision of the experiments was ± 25 cal/mol, which is significantly better than has been achieved in hydrogenation studies.

The heats of combustion of all five n-heptenes have been determined in a set of experiments that had rather small uncertainties.[28] It was then possible to combine the data from trifluoroacetolysis and combustion calorimetry as follows[27]: the error in the heat of formation for compound i derived from the combustion measurement may be represented by δ_i, and that for the heat of reaction of compound i with trifluoroacetic acid may be represented by γ_i. Then

$$\delta_i = A_i' - A_i$$
$$\gamma_i = A_i + B_i' - X$$

where A_i' is the observed heat of formation, A_i is the true value of ΔH_f, B_i' is the observed heat of reaction with trifluoroacetic acid, and X is the constant sum of the true values of A_i and B_i. We now minimize F, which is the sum of $\delta_i^2 + \gamma_i^2$ subject to appropriate weighting factors (a_i and β_i). The latter may be taken as 1000/stated error (cal/mol). Thus, we minimize F with respect to the A_i's and X:

$$F = \Sigma(a_i\delta_i^2 + \beta_i\gamma_i^2)$$

which gives n equations of the form:

$$(a_i + \beta_i)A_i - \beta_i X = a_i A_i' - \beta_i B_i'$$

where the index runs from 1 to n, and:

$$\Sigma\beta_i(A_i - X) = \Sigma - \beta_i B_{i'}$$

This set of $n + 1$ normal equations may be solved by matrix inversion, giving the values of A_i. This is a generally useful approach in combining the results of different types of calorimetric measurement.

Reactions such as alkene hydration may be used in relating the energies of a set of substituted compounds. Examples of possible reactions are:

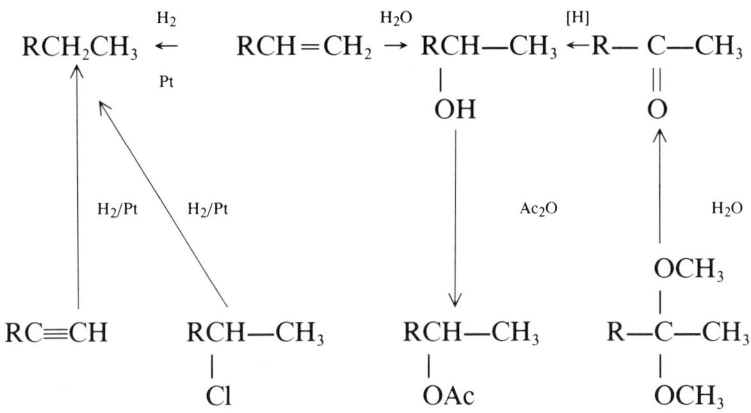

Sets of reactions such as those above would allow one to relate the heats of formation of a variety of compounds to that of the corresponding saturated hydrocarbon. Since there is a large body of experimental data for the latter, this would allow the determination of the heats of formation of a variety of compounds. Besides hydrogenation of alkenes and alkynes, some of these reactions, such as the reduction of alkyl halides to alkanes[25] and the hydrolysis of ketals,[29] have been studied.

Many other reactions have been studied calorimetrically. They include the hydrolysis of esters,[30] thiolesters,[31] anhydrides,[32] amides,[33] acid chlorides,[34] esters of inorganic compounds,[35] and organometallic compounds.[36] The heats of ionization of phenols,[37] carboxylic acids,[38] alcohols,[39] and cyanocarbon acids[40] have been determined as well as the heats of protonation of amines.[41] The heats of formation of carbocations have been determined by measuring the heats of reaction of alkenes with strongly acidic solutions.[42] Some of these reactions have also been studied in the gas phase, and a comparison of ΔH_r in the gas phase and in solution provides valuable information concerning the solvation of ions.[43a]

4. TEMPERATURE DEPENDENCE OF HEATS OF FORMATION

Heats of formation are usually determined at 25°C for the liquid or solid phase, whereas reactions are usually carried out at other temperatures, and possibly in other phases. Furthermore, most theoretical studies refer to the gas phase. The conversion of the heat of formation in the liquid or solid phase to that for the gas phase requires a correction for the heat of vaporization or of sublimation. The measurement of these quantities has been discussed elsewhere.[43b] If we have values for the substances in the gas phase at 25°C, how can we adjust these data to correspond to a different temperature? If there were a way to obtain the heat capacities of all the substances, the correction would be simply the difference in heat capacity between reactants and products integrated over the temperature range.

Although heat capacities can be measured, it is more common to estimate them for the gas phase using the methods of statistical mechanics.[44] The quantities that contribute to the heat capacity in the gas phase are the translational energy, the rotational energy, and the vibrational energy. The translational component depends only on the mass of the substance in question, the rotational component depends only on the moments of inertia (which can easily be calculated if the structure is known), and the vibrational component depends only on the vibrational frequencies. In many cases, the frequencies may be determined from an analysis of the infrared and Raman spectra, and in other cases, it is often possible to estimate them.

The heat capacity also determines the entropy, and so a combination of a calorimetric measurement and the statistical treatment above can yield the heat of formation, the heat capacity, the entropy, and the free energy of formation at any desired temperature. The temperature effect on the thermochemical quantities are given by:[44]

$$\frac{H^\circ - H^\circ_0}{T} = 7.9488 + 1.9872 \sum \left(\frac{x_i}{e^{x_i} - 1} \right)$$

$$\frac{G^\circ - H^\circ_0}{T} = 6.8635 \log M + 18.3027 \log T + 2.2878 \log (I_x I_y I_z \times 10^{117})$$

$$- 4.5757 \log \sigma - 1.9872 \sum [\ln (1 - e^{-x_i})] - 10.2960$$

$$S^\circ = \frac{H^\circ - H^\circ_0}{T} - \frac{G^\circ - H^\circ_0}{T}$$

$$C_p = 7.9488 + 1.9872 \sum \left[\frac{x_i^2 e^{x_i}}{(e^{x_i} - 1)^2} \right]$$

Here, M is the mass, T is the temperature, $I_x I_y I_z$ is the product of the three moments of inertia in grams centimeter squared on a per molecule basis, and σ is the symmetry number. The quantity x_i is related to the observed vibrational frequencies ν_i, in reciprocal centimeters by:

$$x_i = \frac{hc\nu_i}{kT} = \frac{1.4387\nu_i}{T}$$

In addition, H°_0 is the value of the enthalpy at 0 K and is equal to G°_0. At this temperature, $S^\circ = 0$. To convert a heat of formation at one temperature to that at another temperature, calculate the change in $H^\circ - H^\circ_0$ between the two temperatures and subtract the corresponding quantities for each of the elements in its normal state.[45] The result is the change in heat of formation on going from one temperature to the other. Since the entropy is obtained at the same time, the free energy may also be calculated at any temperature. The foregoing applies to rigid molecules; when internal rotors are present, they must be treated in a slightly different fashion.[6,44]

5. STRUCTURE–ENERGY CORRELATIONS

It is important to be able to verify the accuracy of heats of formation, regardless of how they are derived. As mentioned above, one way is to obtain the value via two separate, preferably different experiments. Thus, for example, if the heats of formation of an alkene and the corresponding alkane have been determined by combustion calorimetry, they may be related via measurement of the heat of hydrogenation. If all three data agree within their uncertainty intervals, one may reasonably conclude that the accuracy corresponds the uncertainty intervals. Another example was provided above in which the heats of trifluoroacetolysis of alkenes could be used to check the heats of formation.

Another method of verifying experimental heats of formation makes use of structure–energy correlations. It is known that the increment per methylene group in an acyclic system is constant.[46] Table 4-1 shows some data for straight-chain alkanes and some of their derivatives. It can be seen that the methylene increment is essentially the same for all series.[47] Furthermore, it is clear that the heat of formation of *n*-heptanol must be significantly in error. Similar correlations may be derived for branched-chain compounds. Besides providing a check on experimental data, such correlations allow one to make rather good estimates of the heats of formation for compounds in a series that has not been studied experimentally.

Another way in which data may be examined is via calculations of energies for converting one group, such as methyl, to another. Group conversion is more satisfactory than replacement of a hydrogen by a functional group because to a considerable extent nonbonded interactions will cancel in the former case, but not in the latter.

Figure 4-4 shows the experimental data for replacing a methyl group by hydroxy.[48] It can be seen that an essentially constant change in energy is found for groups of related compounds, that is, primary alcohols as one group, secondary as a second, and tertiary alcohols as a third. Similar relationships should apply for replacement of a methyl by halogen, sulfur, or nitrogen.

In many cases, we wish to estimate the heats of formation of compounds for which experimental data are not available and for which the replacement schemes above are not appropriate. The simplest approach makes use of an empirical relationship between structure and energy. This was pioneered by Franklin with his group equivalents,[5] and similar schemes using bond energies have been described.[6,14] Much new and more precise experimental data have become available since the initial derivations of the group equivalents, however, and the values should be recalculated using all the currently available data.

The main limitation for such estimates of energy is that they cannot adequately take into account structural deformations (eg, bond angle bending) or steric effects other than the simplest kinds. Here, a far better approach is

TABLE 4-1. Enthalpies of Formation of Liquid Alcohols and Alkanes

	ΔH_f (kcal/mol)		
Compound	Observed[a]	Calculated	Difference
	1-Hydroxyalkanes		
Ethanol	−66.42 ± 0.08	−66.43	0.01
1-Propanol	−72.51 ± 0.30	−72.40	−0.11
1-Butanol	−78.29 ± 0.13	−78.37	0.08
1-Pentanol	−84.27 ± 0.17	−84.34	0.07
1-Hexanol	−90.65 ± 0.24	−90.31	−0.34
1-Heptanol	−95.29 ± 0.20	−96.28	1.01
1-Octanol	−102.30 ± 0.26	−102.26	−0.04
		Methylene increment = 5.97 kcal/mol	
	***n*-Alkanes**		
n-Butane	−35.38 ± 0.16	−35.36	−0.02
n-Pentane	−41.49 ± 0.17	−41.45	−0.04
n-Hexane	−47.46 ± 0.18	−47.54	0.08
n-Heptane	−53.59 ± 0.22	−53.63	0.04
n-Octane	−59.78 ± 0.25	−59.72	−0.06
		Methylene increment = 6.09 kcal/mol	
	2-Hydroxyalkanes		
2-Propanol	−76.02 ± 0.12	−75.98	−0.04
2-Butanol	−81.87 ± 0.23	−81.89	0.02
2-Pentanol	−87.74 ± 0.18	−87.79	0.05
2-Hexanol	−93.68 ± 0.21	−93.70	0.02
2-Heptanol	−99.65 ± 0.16	−99.61	−0.04
		Methylene increment = 5.91 kcal/mol	
	2-Methylalkanes		
2-Methylpropane	−37.02 ± 0.13	−36.96	−0.06
2-Methylbutane	−42.88 ± 0.17	−42.93	0.06
2-Methylpentane	−48.91 ± 0.24	−48.92	0.01
2-Methylhexane	−54.85 ± 0.27	54.91	0.06
2-Methylheptane	−60.95 ± 0.36	−60.89	−0.06
		Methylene increment = 5.98 kcal/mol	

[a]Liquid phase data (References 14, 47). The methylene increments include that for the heats of vaporization.

molecular mechanics, which attempts to model all intramolecular interactions (bond length distortion, bond angle bending, torsional interactions, and both attractive and repulsive nonbonded interactions). In the case of hydrocarbons that do not have very unusual structures leading to rehybridization, molecular mechanics has been shown to give energies with an average uncertainty of ±1 kcal/mol, along with rather good structures.[7]

Molecular mechanics remains a largely empirical method, in which the parameters (particularly the nonbonded interactions) are developed by fitting a body of experimental data. The success with hydrocarbons results from the large body of data that may be used in developing the parameters.

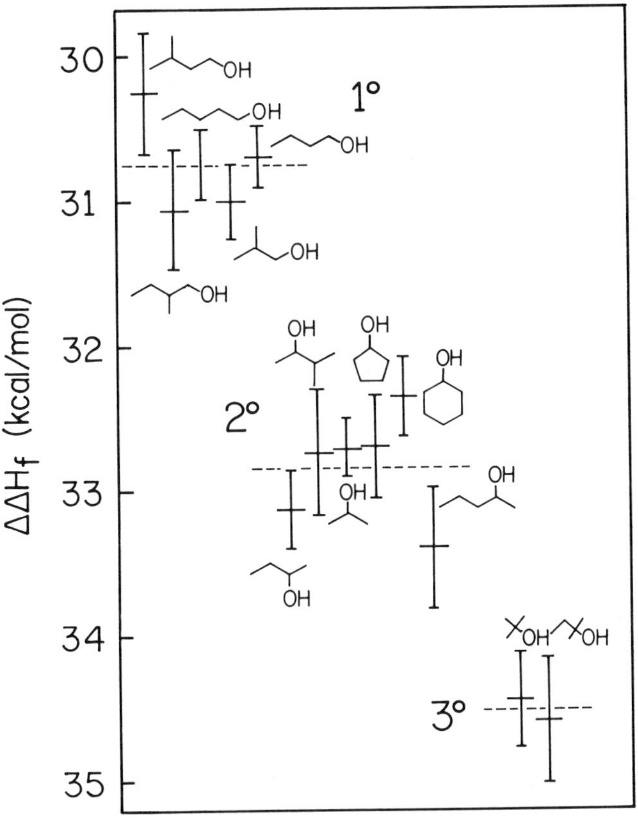

Figure 4-4. Differences in heats of formation between alcohols and the corresponding hydrocarbons in which the hydroxy has been replaced by methyl.

Far fewer data are available for compounds having substituents such as hydroxy, carbonyl, and halo. As a result, the parameters for such compounds are far less satisfactory.

The nonbonded interaction frequently manifest themselves as steric effects. They remain as one of the less well understood "effects" in chemistry. Although their existence is well documented, and their effects on the rates of a group of related reactions may be correlated by Taft's E_s values,[49] their origin cannot be said to be well understood.

There are two problems associated with obtaining data on steric effects from kinetic experiments. First, the structure of the activated complex is not well defined and may change in response to steric interactions. Second, the quantity that may be determined accurately is the free energy of activation. It would be helpful to be able to separate the free energy changes into energy and entropy changes. Unfortunately, it is difficult to obtain $\Delta H^{\ddagger}$ with an uncertainty of better than 0.5 kcal/mol, and many of the effects are not much larger than this.

More quantitative information may be obtained by studying the energy change involved in converting one substance into another, for here the structures of the reaction components either are known or could be determined. Direct calorimetric measurements are able to give energy differences with high precision (and hopefully, accuracy). Steric effects are commonly found in the interconversion of trigonal and tetrahedral carbons. In the former case, the bond angles are on the order of 120°, and this decreases to around 110° in the latter. As a result, the groups come closer together, and nonbonded repulsion will increase. Some examples of such conversions are the hydration of alkenes discussed above and the following:

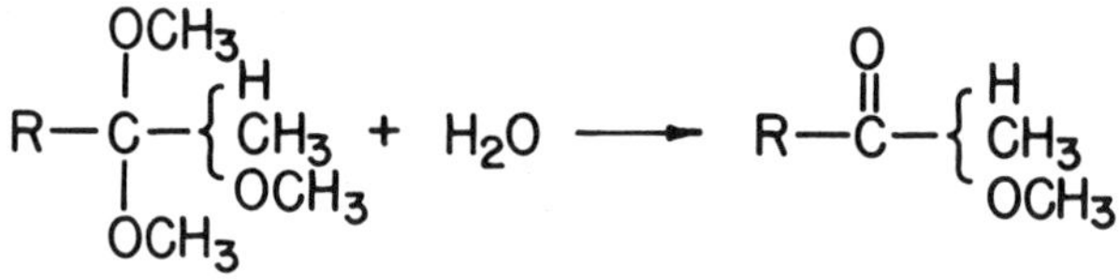

Considerable experimental data are now available for the hydrolysis of acetals, ketals, and orthoesters.[50] The study of these reactions has shown that the steric effect has a large entropy component.

The reactions above are not well suited for determining better oxygen nonbonded interactions because the range of effects is not large and the compounds exist in a mixture of conformers. Here, it would be more satisfactory to examine rigid compounds that can have only one principal conformation. The norbornyl ketals provide a good example. The ring system may be prepared with methyl groups in a variety of positions,

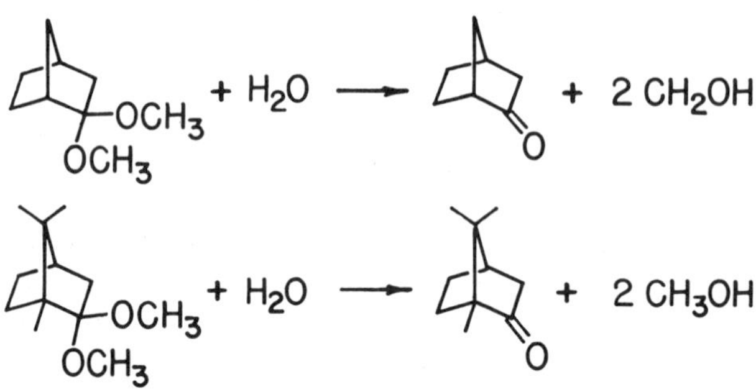

and the enthalpies of hydrolysis of many of them have been measured. The change in heat of reaction on going from norbornyl ketal to camphor ketal is 5 kcal/mol.[51]

Many of the most interesting compounds cannot be isolated in pure form because of their high reactivity or cannot be obtained in large enough amount for calorimetric measurements because of the difficulty in preparation. They often are not good candidates for energy estimation via molecular mechanics because of their unusual hybridization. Here, we may resort to theoretical calculations. The more practical of these is the semiempirical molecular orbital calculations using methods such as those involving modified intermediate neglect of diatomic overlap (MINDO/3 or MNDO), and they frequently are successful. However, they have not been overly successful with some groups of compounds, such as small-ring alkanes or branched hydrocarbons.[9]

It is likely that the major problem with these calculations is not that they estimate some of the terms in a semiempirical fashion instead of calculating them directly, but rather that they provide an insufficiently flexible basis set. The current programs use a minimal basis set (ie, a fixed size for an atomic orbital on a given type of atom), and it is known from ab initio calculations that such a basis set will not correctly reproduce the energy difference between alkenes and alkanes.[52] The problem is simply that the carbon orbitals in alkanes and alkenes have different optimal sizes, and this discrepancy should be accommodated in the basis set. This is the reason for the use of a split valence basis sets in ab initio calculations (3–21G, 4–31G, etc). Furthermore, for some severely deformed compounds, it is necessary to include d orbitals at carbon to improve the flexibility of the basis set.[53] There is probably no reason for the semiempirical calculations not to use split valence orbitals, and even to include d orbitals. However, so far this has not been done.

The ab initio molecular orbital calculations, although costly and still limited to rather small molecules, provide the best means for estimating both the energies and the structures of all sets of molecules. It has been found that the structures calculated using split valence basis sets generally give bond angles as good as most of those that have been measured, and bond lengths generally about 0.01 Å too short.[54] The difference between calculated and observed bond lengths results from at least two factors. First, the calculated lengths correspond to the bottom of the potential well, whereas the observed length is the average value at the zero-point vibrational level. Since the stretching of most bonds is anharmonic, the average value will be greater than the value at the bottom of the well. Second, most calculations neglect electron correlation, leading to a slightly more concentrated charge distribution between a given pair of nuclei, and leading to a bond that is too short.

The energies calculated by the standard Hartree–Fock approach cannot be used directly to calculate heats of formation because they are not corrected for electron correlation. It has been found that the correlation energy is essentially constant for any type of group, and therefore it is possible to develop a set of group equivalents for converting the energies into heats of formation. This should give ΔH_f at 0 K. However, since group equivalent schemes are also able to account for zero-point energies and the change in energy on going from 0 to 298 K, all may be combined into single group equivalents.[55] They allow the estimation of the heats of formation for a vari-

TABLE 4-2. Results of Group Equivalent Analysis (6–31G*)

Compound	Energy, H	ΔH_f (kcal/mol) Observed	Calculated[a]	Difference
Compounds Used to Determine Group Equivalents				
Ethane	−79.22876	−20.04	−20.08	−0.04
Propane	−118.26365	−25.02	−25.11	0.09
Butane	−157.29840	−30.03	−30.10	−0.07
Isobutane	−157.29896	−32.07	−31.99	−0.08
Pentane	−196.33302	−35.10	−35.01	−0.09
Cyclopropane	−117.05887	12.73	13.50	−0.77
Cyclobutane	−156.09703	6.78	6.37	0.41
Cyclopentane	−195.16124	−18.44	−17.11	−1.33
Cyclohexane	−234.20796	−29.50	−29.61	0.11
Bicyclo[2.1.0]pentane	−193.92697	37.70	36.95	0.75
Bicyclo[2.2.0]hexane	−232.96520	29.90	29.77	0.13
Bicyclo[2.2.1]heptane	−272.06116	−12.40	−13.65	1.25
Bicyclo[2.2.2]octane	−311.10358	−23.67	−23.41	−0.26
Ethylene	−78.03172	12.50	11.99	0.51
Propene	−117.07147	4.88	3.50	1.38
1-Butene	−156.10487	−0.20	−0.08	0.12
1,4-Pentadiene	−193.94093	25.30	27.05	−1.75
Cyclopentene	−193.97717	8.23	6.04	2.19
trans-2-Butene	−156.11041	−2.99	−4.49	1.50
Cyclobutene	−154.89962	37.45	37.89	−0.44
Bicyclo[2.1.0]pent-2-ene	−192.71022	79.70	80.61	−0.91
Compounds Used to Test Group Equivalents				
Cyclopropene	−115.82305	66.2	68.7	−2.5
Bicyclo[1.1.0]but-1(3)-ene	−153.58015		146.7	
Tetrahedrane	−153.59788		134.1	
Bicyclo[1.1.0]butane	−154.87177	51.9	54.9	−3.0
1,3-Butadiene	−154.91965	26.1	24.9	1.2
Cyclopentadiene	−192.79172	31.9	29.3	2.6
Bicyclo[2.1.0]pent-1(4)-ene	−192.62148		137.5	
Bicyclo[2.1.0]pent-1(2)-ene	−192.63036		131.2	
Bicyclo[2.1.0]pent-1(5)-ene	−192.63312		129.5	
[1.1.1]Propellane	−192.69106		88.5	
Bicyclo[1.1.1]pentane	−193.90569		50.4	
Bicyclo[2.2.0]hex-1(4)-ene	−231.71870		93.3	
Bicyclo[3.1.0]hex-1(5)-ene	−231.72055		92.1	
[2.1.1]Propellane	−231.72188		86.0	
Bicyclo[2.2.0]hex-2(3)-ene	−231.76788	62.5	61.0	1.5
Bicyclo[2.1.1]hex-2-ene	−231.77417		57.0	
1,5-Hexadiene	−232.97615	20.2	23.0	−2.8
Bicyclo[2.1.1]hexane	−232.98929		14.7	
Norbornadiene	−269.65244	57.4	57.5	−0.1
[2.2.1]Propellane	−270.75955		79.1	
[2.2.2]Propellane	−309.80908		64.8	
Windowpane	−347.56427		144.5	
Cubane	−307.39361	148.7	144.1	4.6

[a] $\Delta H_f = 627.5[E_{calc} + 39.59841 \times n(CH_3) + 39.02671 \times n(CH_2) + 38.45291 \times n(CH) + 37.87600 \times n(C_{sat}) + 37.88026 \times n(C_{unsat})]$; 1 hartree = 627.5 kcal/mol.

ety of compounds with an uncertainty on the order of ± 2 kcal/mol (Table 4-2).

As methods for correcting for electron correlation continue to improve, and computers become faster, it should be possible to calculate the energies of a variety of *small* molecules with an accuracy that rivals the best calorimetric values. However, since the time required for the calculations increases as the fourth or fifth power of the number of orbitals involved, it is clear that experimental thermochemical data will continue to be needed for many years.

6. CONCLUSIONS

Thermochemical data are important for a variety of purposes, ranging from the detailed analysis of chemical reactions to the design of facilities for the commercial preparation of chemical substances. There is a need for more thermochemical data, particularly for substituted organic compounds. It will never be possible to have direct calorimetric measurements for most organic compounds, since they can be prepared far faster than it is possible to carry out thermochemical measurements. What is needed is sufficient experimental data for a wide variety of types of compound so that improved methods for estimating thermochemical quantities can be developed. Unfortunately, there is a trend toward fewer laboratories making the needed measurements.

ACKNOWLEDGMENT

These thermochemical studies were supported by the Office of Basic Energy Sciences, U.S. Department of Energy.

REFERENCES

1. Janz, G.J. "Estimation of Thermodynamic Properties of Organic Compounds". Academic Press: New York, 1967.
2. It is important to recognize that whereas statistical mechanics may provide information on the change in the heat of formation with temperature, it cannot by itself establish an absolute value at any temperature. In contrast, the absolute value of the entropy may be calculated via statistical mechanics because the entropy of a pure crystalline substance is zero at 0 K.
3. Prosen, E.J. In "Experimental Thermochemistry", Vol. 1, Rossini, F.D., Ed.; Wiley-Interscience: New York, 1956, p. 126.
4. Skinner, H.A. In "Experimental Thermochemistry", Vol. II, Skinner, H.A., Ed.; Wiley-Interscience: New York, 1962, p. 147. Skinner, H.A.; Sturtevant, J.M.; Sunner, S., In ibid, p. 157.
5. Franklin, J.L. *Ind. Eng. Chem.* **1949,** *41,* 1070.

6. Allen, T.L. *J. Chem. Phys.* **1959,** *31,* 1039. Benson, S.W. "Thermochemical Kinetics". Wiley: New York, 1976.
7. Burkert, U.; Allinger, N.L. "Molecular Mechanics". American Chemical Society Monograph 177. ACS: Washington, D.C., 1982.
8. Dewar, M.J.S. "The Molecular Orbital Theory of Organic Chemistry". McGraw-Hill: New York, 1969. Pople, J.A.; Beveridge, D.L. "Approximate Molecular Orbital Theory". McGraw-Hill: New York, 1970. Murrell, J.N.; Harget, A.J. "Semi-empirical Self-consistent-field Molecular-orbital Theory of Molecules". Wiley: New York, 1972.
9. Cf Schulman, J.M.; Disch, R.L. *J. Am. Chem. Soc.* **1984,** *106,* 1202. Dewar, M.J.S.; Thiel, W. ibid, **1977,** *99,* 4907.
10. Pople, J.A. *Mod. Theor. Chem.* **1977,** *4,* 1.
11. Hubbard, W.N.; Scott, D.W.; Waddington, G. In "Experimental Thermochemistry". Rossini, F.D., Ed.; Wiley-Interscience: New York, 1956, p. 75.
12. Månsson, M. *J. Chem. Thermodyn.* **1973,** *5,* 721; ibid, **1974,** *6,* 1018.
13. Good, W.D.; Scott, D.W. In "Experimental Thermochemistry", Vol. II, Skinner, H.A., Ed.; Wiley-Interscience: New York, 1962, p. 15.
14. Cox, J.D.; Pilcher, G. "Thermochemistry of Organic and Organometallic Compounds". Academic Press: London, 1970. Pedley, J.B.; Rylance, J. "Sussex-N.P.L. Computer-Analyzed Thermochemical Data, Organic and Organometallic Compounds". University of Sussex, Brighton, UK, 1977.
15. Sunner, S., Wadsö, I. *Acta Chem. Scand.* **1959,** *13,* 97.
16. Wiberg, K.B.; Squires, R.R. *J. Chem. Thermodyn.* **1979,** *11,* 773.
17. Skinner, H.A.; Sturtevant, J.M.; Sunner, S. In "Experimental Thermochemistry," Vol. II, Skinner, H.A., Ed.; Wiley-Interscience: New York, 1962, pp. 216–217.
18. Roth, W.R.; Lennartz, H.-W. *Chem. Ber.* **1980,** *113,* 1806. Roth, W.R.; Klarner, F.-G.; Lennartz, H.-W. ibid, **1980,** *113,* 1818. Roth, W.R.; Kirmse, W.; Hoffman, R.; Lennartz, H.-W. ibid, **1982,** *115,* 2508. Roth, W.R.; Scholz, B.P.; Breuckmann, R.; Jelich, K.; Lennartz, H.-W. ibid, **1982,** *115,* 1934.
19. Oth, J.F.M. *Rec. Trav. Chim. Pays-Bas* **1968,** *87,* 1185. Kabakoff, D.S.; Bünzli, J.-C.G.; Oth, J.F.M.; Hammond, W.B.; Berson, J.A. *J. Am. Chem. Soc.* **1975,** *97,* 1510.
20. Kistiakowsky, G.B.; Romeyn, H., Jr.; Ruhoff, J.R.; Smith, H.A.; Vaughan, W.E. *J. Am. Chem. Soc.* **1935,** *57,* 65. Kistiakowsky, G.B.; Ruhoff, J.R.; Smith, H.A.; Vaughan, W.E. ibid, **1935,** *57,* 876; **1936,** *58,* 146, 237. Doliver, M.A.; Gresham, T.L.; Kistiakowsky, G.B.; Vaughan, W.E. ibid, **1937,** *59,* 831. Doliver, M.A.; Gresham, T.L.; Kistiakowsky, G.B.; Smith, E.A.; Vaughan, W.E. ibid, **1938,** *60,* 440. Conn, J.B.; Kistiakowsky, G.B.; Smith, E.A. ibid, **1939,** *61,* 1868. Conant, J.B.; Kistiakowsky, G.B. *Chem. Rev.* **1937,** *20,* 181.
21. Turner, R.B.; Meador, W.E.; Winkler, R.E. *J. Am. Chem. Soc.* **1957,** *79,* 4116, 4122. Turner, R.B.; Meador, W.R.; Doering, W.v.E.; Knox, L.H.; Mayer, J.R.; Wiley, D.W. ibid, **1957,** *79,* 4127. Turner, R.B.; Meador, W.R. ibid, **1957,** *79,* 4133. Turner, R.B.; Garner, R.H. ibid, **1958,** *80,* 1424. Turner, R.B.; Nettleton, D.E., Jr.; Perelman, M. ibid, **1958,** *80,* 1430. Turner, R.B.; Prinzbach, H.; Doering, W.v.E. ibid, **1960,** *82,* 3451. Roth, W.R.; Bang, W.B.; Goebel, P.; Sass, R.L.; Turner, R.B.; Yue, A.P. ibid, **1964,** *86,* 3178. Turner, R.B. ibid, **1964,** *86,* 3586. Turner, R.B.; Goebel, P.; Mallon, J.; Doering, W.v.E.; Coburn, J.F., Jr.; Pomerantz, M. ibid, **1968,** *90,* 4315.
22. Flitcroft, T.; Skinner, H.A.; Whiting, M.C. *Trans. Faraday Soc.* **1957,** *53,* 784. Flitcroft, T.; Skinner, H.A. ibid, **1958,** *54,* 47. Skinner, H.A.; Snelson, A. ibid, **1959,** *55,* 404. Williams, R.B. *J. Am. Chem. Soc.* **1942,** *64,* 1395.
23. Rogers, D.W.; Skanupong, S. *J. Phys. Chem.* **1974,** *78,* 2569. Rogers, D.W.; Hoyte, O.P.A.; Ho, R.K.C. *J. Chem. Soc. Faraday Trans.* **1978,** *74,* 46. Rogers, D.W.; Choudhoury, D.N. ibid, **1978,** *74,* 2868. Rogers, D.W.; von Voithenberg, H.; Allinger, N.L. *J. Org. Chem.* **1978,** *43,* 380. Rogers, D.W.; Dagdagan, O.A.; Allinger, N.L. *J. Am. Chem. Soc.* **1979,** *101,* 671. Rogers, D.W.; Choi, L.S.; Girellini, R.S.; Holmes, T.J.; Allinger, N.L. *J. Phys. Chem.* **1980,** *84,* 1810. Allinger, N.L.; Glasser, J.A.; Davis, H.E.; Rogers, D.W. *J. Org. Chem.* **1981,** *46,* 658. Allinger, N.L.; Dodziuk, H.; Rogers, D.W.; Naik, S.N. *Tetrahedron* **1982,** *38,* 1593. Rogers, D.W.; Crooks, E.L. *J. Chem. Thermodyn.* **1983,** *15,* 1087.
24. Jensen, J.L. *Prog. Phys. Org. Chem.* **1976,** *12,* 189. Fuchs, R. Peacock, L.A. *J. Phys. Chem.* **1979,** *83,* 1975.
25. Wiberg, K.B.; Wasserman, D.J. *J. Am. Chem. Soc.* **1981,** *103,* 6563.
26. Wiberg, K.B.; Wasserman, D.J.; Martin, E. *J. Phys. Chem.* In press.
27. Good, W.D.; Smith, N.K. *J. Chem. Thermodyn.* **1979,** *11,* 111.

28. Lacher, J.R.; Kianpour, A.; Oetting, F.; Park, J.D. *Trans. Faraday Soc.* **1956,** *52,* 1500. Lacher, J.R.; Kianpour, A.; Park, J.D. *J. Phys. Chem.* **1956,** *60,* 1454. Lacher, J.R.; Emery, E.; Bohmfalk, E. ibid, **1956,** *60,* 692 (1956). Lacher, J.R.; Kianpour, A.; Park, J.D. ibid, **1957,** *61,* 1125. Davies, J.; Lacher, J.R.; Park, J.D. *Trans. Faraday Soc.* **1965,** *61,* 2413. Fowell, P.; Lacher, J.R.; Park, J.D. ibid, **1965,** *61,* 1324.
29. Wiberg, K.B.; Squires, R.R. *J. Am. Chem. Soc.* **1979,** *101,* 2204; ibid, **1981,** *103,* 4473.
30. Wadsö, I. *Acta Chem. Scand.* **1958,** *12,* 630; ibid, **1960,** *14,* 561.
31. Wadsö, I. *Acta Chem. Scand.* **1957,** *11,* 1745; ibid, **1960,** *14,* 561; ibid, **1962,** *16,* 487.
32. Conn, J.B.; Kistiakowsky, G.B.; Roberts, R.M.; Smith, E.A. *J. Am. Chem. Soc.* **1942,** *64,* 1747. Wadsö, I. *Acta Chem. Scand.* **1962,** *16,* 471.
33. Wadsö, I. *Acta Chem. Scand.* **1960,** *14,* 903; ibid, **1962,** *16,* 471, 479. Hill, J.O.; Wadsö, I. ibid, **1968,** *22,* 1950.
34. Carson, A.S.; Skinner, H.A. *J. Chem. Soc. London* **1950,** 272. Pritchard, H.O.; Skinner, H.A. ibid, **1950,** 272. Carson, A.S.; Pritchard, H.O.; Skinner, H.A. ibid, **1950,** 565. Quitzsch, K.; Koehler, H.; Schneider, L.; Geiseler, G. *Z. Chem.* **1962,** *2,* 378. Finch, A.; Gardner, P.J.; Radcliffe, K. *J. Chem. Eng. Data* **1968,** *13,* 176.
35. Cox, J.R., Jr.; Wall, R.E.; Westheimer, F.H. *Chem. Ind. (London)* **1959,** 565. Davis, R.E. *Proc. Indiana Acad. Sci.* **1960,** *70,* 106. Pagdin, N.; Pine, A.K.; Tillett, J.G.; van Woerden, H.F. *J. Chem. Soc.* **1962,** 3835.
36. Skinner, H.A.; Bennett, J.E.; Pedley, J.B. *Pure Appl. Chem.* **1961,** *2,* 17. Fowler, P.; Mortimer, C.T. *J. Chem. Soc.* **1961,** 3793. Organometallic thermochemistry is reviewed by H.A. Skinner and J.A. Connor in Chapter 6 of this volume.
37. O'Hara, W.F.; Hepler, L.G. *J. Phys. Chem.* **1961,** *65,* 2107. Millero, F.J.; Ahluwaha, J.C.; Hepler, L.G. *J. Chem. Eng. Data* **1964,** *9,* 192, 319.
38. Breslauer, K.J. *J. Chem. Thermodyn.* **1979,** *11,* 527.
39. Briere, G.; Crochon, B.; Felici, N. *C. R. Acad. Sci.* **1962,** *254,* 4458. Briere, G.; Felici, N.; Piot, E. ibid, **1962,** *255,* 107. Kundu, K.K.; Das, M.N. *J. Chem. Eng. Data* **1964,** *9,* 82. Arnett, E.M.; Small, L.E. *J. Am. Chem. Soc.* **1977,** *99,* 808.
40. Boyd, R.H.; Wang, C.-H. *J. Am. Chem. Soc.* **1965,** *87,* 430.
41. Arnett, E.M.; Burke, J.J. *J. Am. Chem. Soc.* **1966,** *88,* 2340, 4308.
42. Arnett, E.M.; Larsen, J.W. *J. Am. Chem. Soc.* **1968,** *90,* 791, 792; ibid, **1969,** *91,* 1438. Arnett, E.M.; Petro, C.; Schleyer, P.v.R. ibid, **1979,** *101,* 522. Arnett, E.M.; Pienta, N.J. ibid, **1980,** *102,* 3329.
43. (a) Arnett, E.M.; Chawla, B.; Bell, L.; Taagepera, M.; Hehre, W.J.; Taft, R.W. *J. Am. Chem. Soc.* **1977,** *99,* 5729. (b) Reid, R.C.; Prausnitz, J.M.; Sherwood, T.K. "The Properties of Gases and Liquids". McGraw-Hill: New York, 1977.
44. Stull, D.R.; Westrum, E.F., Jr.; Sinke, G.C. "The Chemical Thermodynamics of Organic Compounds". Wiley: New York, 1969.
45. Stull, D.R.; Prophet, H. "JANAF Thermochemical Tables". Dow Chemical Co. (National Bureau of Standards), 1971.
46. Prosen, E.J.; Johnson, W.H.; Rossini, F.D. *J. Res. Natl. Bur. Stand.* **1946,** *37,* 51.
47. Wiberg, K.B.; Wasserman, D.J.; Martin, E. *J. Phys. Chem.* **1984,** *88,* 3684.
48. Wiberg, K.B.; Wasserman, D.J. To be published.
49. Taft, R.W., Jr. In "Steric Effects in Organic Chemistry", Newman, M., Ed., Wiley: New York, 1956, Chapter 13.
50. Wiberg, K.B.; Squires, R.R. *J. Am. Chem. Soc.* **1979,** *101,* 2204; ibid, **1981,** *103,* 4473.
51. Wiberg, K.B.; Cunningham, W. To be published.
52. Latham, W.A.; Hehre, W.J.; Pople, J.A. *J. Am. Chem. Soc.* **1971,** *93,* 808.
53. Wiberg, K.B. *J. Am. Chem. Soc.* **1983,** *105,* 1227.
54. DeFrees, D.J.; Raghavachari, K.; Schlegel, H.B.; Pople, J.A. *J. Am. Chem. Soc.* **1982,** *104,* 5576.
55. Wiberg, K.B. *J. Comput. Chem.* **1984,** *5,* 197.

CHAPTER 5

Thermochemical Methods in the Structural Analysis of Organic Compounds

Kalevi Pihlaja

University of Turku, Turku, Finland

CONTENTS

1. INTRODUCTION

This chapter discusses various instrumental methods for collecting thermochemical data for structural and stereochemical analysis. Special empha-

© 1987 VCH Publishers, Inc.
MOLECULAR STRUCTURE AND ENERGETICS, Vol. 2

sis is laid on methods like NMR spectroscopy and mass spectrometry, which are often utilized to determine energy parameters even though their usefulness in the context of thermochemical studies is not generally recognized. The goal, however, is to concentrate on describing applications and on considering the available data to support further conclusions and calculations rather than to illustrate the methods themselves. Hence throughout, the emphasis is on the chemical knowledge obtained by applications of the different thermochemical methods rather than on the underlying principles. The reader interested in experimental procedures is provided with suitable citations in the text.

Sections 2–4 consider new enthalpy of formation data and the use of an Allen-type group increment scheme for the evaluation of enthalpies of formation, nonbonded interactions, and conformational energies. Section 5 describes briefly the use of heats of hydrogenation in acquiring thermochemical knowledge, and Section 6 concentrates on the determination of different energy parameters with the aid of chemical equilibration. Sections 7 and 8 illustrate the use of mass spectrometry and NMR spectroscopy, respectively, in collecting thermochemical information, and the last section describes some applications that do not fall under the other headings.

2. STANDARD ENTHALPIES OF FORMATION

A. The Need for Data

Combustion calorimetry[1–4] is the only direct means of determining enthalpies of formation of organic compounds. Although the experimental details and rules of standardization can be found in many outstanding texts[1–4] and the necessary equipment is not too expensive, there is far too little experimental work done in this field today. Of course, it is understandable that one does not wish to destroy 5–10 g of extremely pure ($\gtrsim$99.99% with a known water content) compound to acquire finally a single experimental result (ie, the enthalpy of combustion of a given compound and from it the enthalpy of formation[5]), since a sample 100 times smaller lets us collect a vast amount of spectroscopic information without destroying the material at all. This topic has also been discussed by Wiberg in Chapter 4 of this volume.

However, there are many reasons for not giving up. First, reliable enthalpy of formation data for a number of key compounds are still desperately needed for the safe estimation and tabulation of data. Second, any semiempirical procedure must have its roots in good experiments, and the success of truly ab initio calculations can be judged only on the basis of precise experimental results. Perhaps the development of micro methods[1] together with an increasing demand of new, highly accurate combustion calorimetric results will reactivate combustion calorimetry.

B. New Experimental Values

Since the comprehensive compilations by Cox and Pilcher[2] and Pedley and Rylance,[6] enthalpies of formation of a number of organic molecules have become available (Table 5-1). Some data for heterocyclic compounds have been reviewed also by Pihlaja and Taskinen.[7]

McEachern and co-workers[8] measured the enthalpies of combustion and sublimation of 3,4-, 5,6-, and 7,8-benzoquinolines **(1–3)** and calculated their enthalpies of formation. Taking into account the systematic error[7] in the enthalpy of sublimation of **1** and the combined error limits the enthalpies of formation of gaseous **1–3**, these values do not differ significantly from each other (Table 5-1).

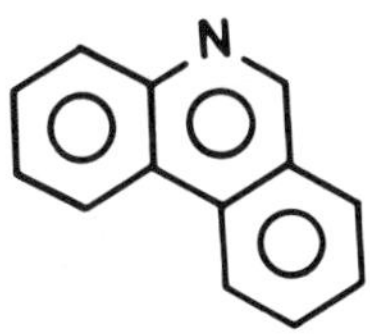

1

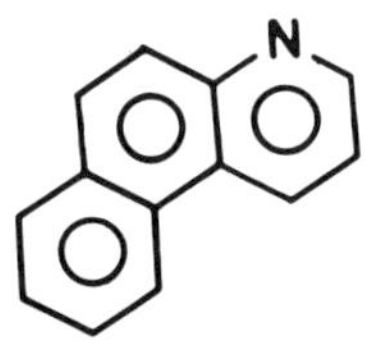

2

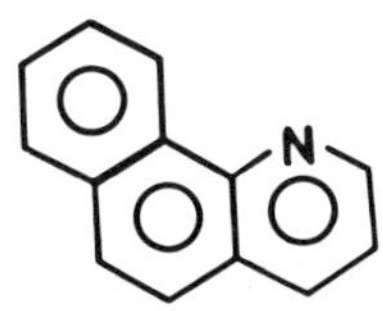

3

Engel and associates[9,10] determined enthalpies of formation of 11 diazenes (**4–14**; Table 5-1) and Montgomery and co-workers[11] those of di-*t*-butyldiazene (**8**) and 3,3,4,4-tetramethyl-Δ^1-1,2-diazetine **(15).** All the acyclic derivatives **(4–8)** were trans forms.

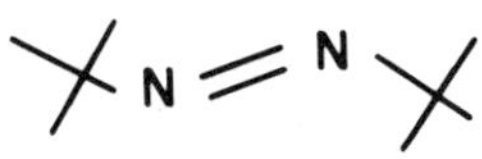

8

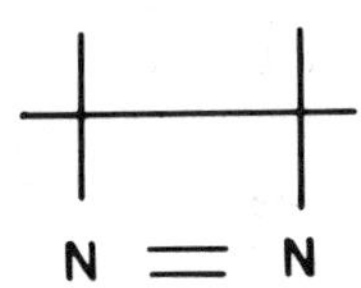

15

The values above can then be used to recalculate enthalpies of formation of the gaseous dimethyl- **(16),** diethyl- **(17),** and methylethyldiazenes **(18),**[12] which are also important synthetic reagents. The value for **16** is based on the postulation[12] that:

$$\Delta_f H^\circ_m\{\text{dibutyldiazene } (\mathbf{7}), \text{g}\} - \Delta_f H^\circ_m\{\text{methylbutyldiazene } (\mathbf{4}), \text{g}\}$$
$$= \Delta_f H^\circ_m\{\text{methylbutyldiazene } (\mathbf{4}), \text{g}\} - \Delta_f H^\circ_m\{\text{dimethyldiazene } (\mathbf{16}), \text{g}\} \quad (5\text{-}1)$$

from which $\Delta_f H^\circ_m(\mathbf{16}, \text{g})$ is 148.6 kJ/mol (Table 5-1). The enthalpies of formation of **17** and **18** can be obtained from those of **7** and **4** by subtracting

TABLE 5-1. Experimentally Determined Standard Enthalpies of Formation

	Compound	State	$-\Delta_f H_m^\circ$(cr or l) (kJ/mol)	$-\Delta_f H_m^\circ$(g) (kJ/mol)	Ref.
1	3,4-Benzoquinoline	cr	−148.9 ± 3.7	−243.5 ± 5.5	8
2	5,6-Benzoquinoline	cr	−150.6 ± 6.2	−233.7 ± 7.4	8
3	7,8-Benzoquinoline	cr	−149.7 ± 4.0	−230.5 ± 5.0	8
4	Methylbutyldiazene	l	−42.5 ± 2.2	−78.9 ± 2.2	9
5	Di-*n*-propyldiazene	l	−11.5 ± 3.5	−51.3 ± 3.5	10
6	Diisopropyldiazene	l	0.3 ± 3.5	−35.6 ± 3.6	10
7	Di-*n*-butyldiazene	l	40.1 ± 3.8	−9.2 ± 3.8	9
8	Di-*t*-butyldiazene	l	75.5 ± 2.7	36.4 ± 2.75	10,11
9	*t*-Butyl(1,1,3,3-Me_4-butyl)diazene	l	172.9 ± 5.6	119.3 ± 5.6	10
10	Di(1,1,3,3-Me_4-butyl)diazene	l	263.3 ± 9.3	196.8 ± 9.4	10
11	1,1,3,3-Tetramethyltrimethylenediazene	cr	22.3 ± 3.5	−39.3 ± 3.6	10
12	1,1,4,4-Tetramethyltetramethylenediazene	l	8.1 ± 4.6	−42.0 ± 4.6	10
13	2,3-Diazabicyclo[2.2.1]heptene-2	cr	−152.1 ± 2.6	−207.4 ± 2.7	10
14	1,4-Dimethyl-2,3-diazabicyclo[2.2.2]octene-2	cr	−20.4 ± 4.4	−92.4 ± 4.4	10
15	3,3,4,4-Tetramethyl-Δ^1-1,2-diazetine	cr	−88.0 ± 2.7	−150.3 ± 2.9	11
16	Dimethyldiazene	—	—	−148.6 ± 5.1	12
17	Diethyldiazene	—	—	−92.2 ± 3.8	12
18	Methylethyldiazene	—	—	−120.4 ± 2.3	12
19	2-Nonanone	l	397.2 ± 1.7	340.8 ± 1.7	14
20	2-Dodecanone	l	476.1 ± 2.4	404.3 ± 2.5	14
21	Norbornane	cr	102.0 ± 3.3	61.6 ± 3.4	15
22	Norbornene	cr	−53.5 ± 3.2	−91.2 ± 3.3	15
23	Norbornadiene	cr	−213.8 ± 3.0	−247.6 ± 3.1	15
24	Nortricyclene	l	−43.4 ± 2.0	−82.1 ± 2.1	15
25	Quadricyclane	l	−302.1 ± 2.2	−339.1 ± 2.3	15
26	Bicyclo[3.3.1]nonane	cr	178.2 ± 0.8	127.6 ± 2.5	22
27	Bicyclo[3.3.2]decane	cr	164.0 ± 7.1	105.9 ± 7.5	22
28	1,7,7-Me_3-Bicyclo[2.2.1]heptan-2-one	cr	319.2 ± 2.5	267.4 ± 2.9	24
29	Bicyclo[2.2.1]heptan-2-one	cr	216.6 ± 2.7	167.8 ± 3.3	25
30	Bicyclo[2.2.1]heptan-7-one	cr	181.3 ± 2.3	133.9 ± 2.9	25
31	1-Me-Adamantane	cr	242.7 ± 1.7	174.9 ± 2.1	28
32	2-Me-Adamantane	cr	224.7 ± 1.7	156.5 ± 2.1	28
33	2,2-Me_2-Adamantane	cr	256.5 ± 2.5	182.8 ± 2.9	28
34	1,3-Me_2-Adamantane	cr	286.6 ± 2.5	218.8 ± 2.9	28
35	1,3,5-Me_3-Adamantane	cr	332.6 ± 4.2	254.8 ± 4.2	28
36	1,3,5,7-Me_4-Adamantane	cr	378.7 ± 3.8	295.0 ± 3.8	28
37	Adamantan-2-one	cr	310.9 ± 3.8	230.5 ± 4.6	30
38	Adamantan-1-ol	cr	397.5 ± 2.5	310.9 ± 3.3	30
39	Adamantan-2-ol	cr	387.9 ± 3.8	299.2 ± 4.6	30
40	1,3,5,7-Me_4-2,4,6,8-Tetrathiaadamantane	cr	154.0 ± 2.9	36.8 ± 5.0	32
41	Octanenitrile	l	107.4 ± 1.5	50.6 ± 1.5	36
42	Decanenitrile	l	158.4 ± 1.8	91.6 ± 1.8	36
43	Undecanenitrile	l	184.5 ± 2.0	113.4 ± 2.0	36
44	Tetradecanenitrile	l	260.1 ± 2.6	174.8 ± 2.7	36
45	Acetonitrile	l	40.59 ± 0.30	74.04 ± 0.37	37
46	1-Dodecene	l	226.2 ± 2.1	165.4 ± 2.1	38
47	1-Hexadecene	l	330.5 ± 3.0	250.2 ± 3.0	38
48	1-Bromododecane	l	344.7 ± 2.6	269.9 ± 2.6	39
49	1-Bromohexadecane	l	444.5 ± 2.8	350.1 ± 3.2	39
50	Di-*n*-propyldiazene-*N*-oxide	l	82.7 ± 1.4	31.0 ± 1.4	40
51	Di-*t*-butyldiazene-*N*-oxide	l	153.5 ± 2.1	107.6 ± 2.1	40
52	1,3,5-Triazine	cr	−171.69 ± 0.77	−225.87 ± 0.89	41
53	Oxazole	l	48.03 ± 0.54	15.52 ± 0.54	42
54	Isoxazole	l	−42.13 ± 0.54	−78.58 ± 0.54	42

TABLE 5-1. Experimentally Determined Standard Enthalpies of Formation, contd.

	Compound	State	$-\Delta_f H_m^\circ$(cr or l) (kJ/mol)	$-\Delta_f H_m^\circ$(g) (kJ/mol)	Ref.
55	3-Me-Isoxazole	l	4.14 ± 0.63	−35.65 ± 0.67	43
56	5-Me-Isoxazole	l	5.61 ± 0.75	−34.06 ± 0.75	43
57	2-OH-Pyridine	cr	166.3 ± 0.8	79.7 ± 1.5	44
58	3-OH-Pyridine	cr	132.0 ± 1.1	43.7 ± 1.7	44
59	4-OH-Pyridine	cr	144.6 ± 1.3	40.8 ± 2.1	44
60	2-Me-3-OH-Pyridine	cr	173.6 ± 1.2	84.5 ± 1.8	44
61	2-Me-4-OH-Pyridine	cr	184.7 ± 1.1	71.7 ± 1.7	44
62	2-Me-5-OH-Pyridine	cr	166.0 ± 1.5	69.8 ± 2.6	44
63	2-Me-6-OH-Pyridine	cr	212.3 ± 2.1	120.3 ± 2.5	44
64	2-CN-Pyridine	cr	−210.0 ± 0.9	−280.7 ± 1.5	45
65	3-CN-Pyridine	cr	−205.8 ± 0.9	−277.9 ± 2.0	45
66	4-CN-Pyridine	cr	−210.3 ± 0.9	−283.5 ± 1.1	45
67	2-NH_2-Pyridine	cr	−39.4 ± 0.8	−118.1 ± 1.1	45
68	3-NH_2-Pyridine	cr	−60.2 ± 0.8	−144.2 ± 1.6	45
69	4-NH_2-Pyridine	cr	−41.8 ± 0.8	−129.9 ± 1.4	45
70	7-*n*-Hexadecylspiro[4.5]decane	l	590.4 ± 2.4	—	48
71	Spiro[4.5]decane	l	200.0 ± 2.2	145.1 ± 2.2	48,49
72	Spiro[5.5]undecane	l	243.6 ± 3.1	187.5 ± 3.1	48,49
73	2,2,4,6,6-Me_5-Heptane	l	348.7 ± 1.4	265.2 ± 1.5	50
74	1-Pentene	l	46.97 ± 0.64	—	51
75	*cis*-2-Pentene	l	53.49 ± 0.62	—	51
76	*trans*-2-Pentene	l	57.98 ± 0.76	—	51
77	2-Me-1-Butene	l	60.96 ± 0.84	—	51
78	3-Me-1-Butene	l	51.60 ± 0.62	—	51
79	2-Me-2-Butene	l	68.07 ± 1.30	—	51
80	Bullvalene	cr	−262.3 ± 3.3	−334.1 ± 3.3	52
81	Pyrene	cr	−125.5 ± 1.2	−225.7 ± 1.3	53
82	1,2′-Binaphthylmethane	cr	−162.0 ± 1.9	—	54
83	1′,2′,3′,4′-Tetrahydro-1,2′-binaphthylmethane	cr	−44.1 ± 1.8	—	54
84	Triphenylene	cr	−151.8 ± 1.5	—	54
85	Decahydrotriphenylene	cr	193.9 ± 1.65	—	54
86	Biphenylene	cr	333.4 ± 1.1	—	54
87	1,2,3,4-Tetrahydronaphthalene	l	28.6 ± 1.0	—	55
88	2,5,8-Me_3-Tetrahydronaphthalene	l	135.6 ± 1.2	—	55
89	2-*t*-Bu-Tetrahydronaphthalene	l	26.3 ± 1.5	—	55
90	4-Me-Diphenylmethane	l	−61.5 ± 1.2	—	55
91	2,5-Me_2-Diphenylmethane	l	−24.6 ± 1.4	—	55
92	2,2′,4,4′,5,5′-Me_6-Diphenylmethane	cr	154.5 ± 1.9	—	55
93	Phenylcyclohexane	l	76.6 ± 1.1	16.7 ± 1.1	55
			79.2 ± 5.6	19.2 ± 5.6	56
94	Bicyclohexyl	l	273.3 ± 1.2	215.3 ± 1.2	55
			276.4 ± 3.1	218.4 ± 3.1	56
95	Bicyclopentyl	l	179.3 ± 1.0	—	55
96	Biphenyl	cr	−96.75 ± 2.6	−113.4 ± 2.65	56
97	1,4-Dihydro-1,4-epoxynaphthalene	cr	−62.1 ± 5.8	—	57
98	1,1,4,4-Me_4-Cyclodecane	l	338.1 ± 1.2	—	58
99	1,1,5,5-Me_4-Cyclodecane	l	328.6 ± 1.7	—	58
100	2-MeO-Naphthalene	cr	97.9 ± 1.8	—	59
101	2-EtO-Naphthalene	cr	134.0 ± 2.0	—	59
102	2,3-Dihydroxynaphthalene	cr	316.4 ± 1.5	—	59
103	2,7-Dihydroxynaphthalene	cr	326.1 ± 1.7	—	59
104	*o*-Ethylbenzoic acid	cr	441.3 ± 1.8	340.2 ± 1.8	60,63
105	*m*-Ethylbenzoic acid	cr	446.0 ± 1.7	346.3 ± 1.7	60,63
106	*p*-Ethylbenzoic acid	cr	454.5 ± 1.9	355.6 ± 1.9	60,63

Table 5-1. Experimentally Determined Standard Enthalpies of Formation, contd.

	Compound	State	$-\Delta_f H^\circ_m$(cr or l) (kJ/mol)	$-\Delta_f H^\circ_m$(g) (kJ/mol)	Ref.
107	*o*-Methoxybenzoic acid	cr	538.5 ± 1.2	433.8 ± 1.2	61
108	*m*-Methoxybenzoic acid	cr	553.5 ± 1.2	446.0 ± 1.3	61
109	*p*-Methoxybenzoic acid	cr	561.7 ± 1.2	451.9 ± 1.4	61
110	*o*-*t*-Butylmethoxybenzoic acid	cr	476.2 ± 1.9	376.1 ± 2.1	62
111	*m*-*t*-Butylmethoxybenzoic acid	cr	504.3 ± 1.6	400.8 ± 1.9	62
112	*p*-*t*-Butylmethoxybenzoic acid	cr	502.9 ± 1.7	398.5 ± 2.0	62
113	Phenazine	cr	−243.7 ± 2.3	−343.6 ± 2.8	64
114	2,1,3-Benzoxadiazole	cr	−235.8 ± 0.9	−300.7 ± 2.1	64
115	Cyclohexylamine	l	147.7 ± 1.3	104.9 ± 1.3	65
	Cyclohexylamine·HCl	cr	408.2 ± 1.5	—	65
116	Methyl-*t*-butylamine	l	141.2 ± 3.8	108.3 ± 4.3	66
117	Isopropyl-*t*-butylamine	l	203.1 ± 3.1	165.0 ± 3.7	66
118	Di-*t*-butylamine	l	212.3 ± 2.5	171.9 ± 3.2	66
119	2,2,6,6-Me_4-Piperidine	l	206.9 ± 2.0	159.9 ± 2.8	66
120	2-Me-6-*t*-Bu-Aniline	l	108.6 ± 2.5	44.8 ± 3.5	66
121	2,4-Pentanedione	l	425.5 ± 1.0	—	67
	Keto	l	416.3 ± 1.1	374.4 ± 1.3	67
	Enol	l	427.6 ± 1.1	384.4 ± 1.3	67
122	2,2-Me_2-Heptane-3,5-dione	l	527.1 ± 2.2	—	68
	Enol	l	527.8 ± 2.3	470.9 ± 2.3	68
123	2,6-Me_2-Heptane-3,5-dione	l	526.8 ± 2.0	—	68
	Enol	l	527.3 ± 2.1	470.4 ± 2.1	68
124	2,2,6-Me_3-Heptane-3,5-dione	l	568.1 ± 2.0	—	68
	Enol	l	568.4 ± 2.1	510.7 ± 2.1	68
125	2,2,6,6-Me_4-Heptane-3,5-dione	l	587.7 ± 3.8	—	68
	Enol	l	587.9 ± 3.9	528.4 ± 3.9	68
126	Benzoylacetone (enol)	l	335.1 ± 2.8	251.3 ± 2.9	68
127	2-Ethylhexanoic acid	l	635.1 ± 2.0	559.5 ± 2.1	69
128	Dimethyl oxalate	cr	756.3 ± 0.8	709.3 ± 1.0	70
129	L-Proline	cr	515.18 ± 0.52	—	71
130	Cytosine	cr	221.3 ± 2.3	—	72
131	L-Histidine	cr	466.7 ± 2.8	—	72
132	Uracil	cr	424.4 ± 2.5	—	72
133	1,1′-Dimethyl-4,4′-bipyridinium cation (paraquat)	aq	−183.2 ± 3.7	—	73
134	$(CF_2)_3(COOCH_3)_2$	l	1959.8 ± 4.2	1907.4 ± 4.7[a]	74
135	$(C_5F_{11})_2O$	l	4783 ± 15	4734 ± 16	74
136	1,1,2-Trifluoro-1,2,2-trichloroethane	l	745.0 ± 4.3	716.8 ± 4.3	75
137	1,1,1-Trifluoro-2-chloro-2-bromoethane	l	720.0 ± 4.9	690.4 ± 4.9	76
138	1,1,2-Trifluoro-2-chloro-1-bromoethane	l	675.3 ± 6.4	644.8 ± 6.7[a]	76
139	Trifluorochloroethene	l	522.7 ± 4.7	505.5 ± 5.1[a]	77
140	2,2,4,4,6-Me_5-1,3-Dioxane	l	567.1 ± 3.6	520.9 ± 3.8	78
141	4,4,6,6-Me_4-1,3-Dioxane	l	518.0 ± 2.6	470.6 ± 3.3	78
142	2,4,4,6,6-Me_5-1,3-Dioxane	l	559.6 ± 3.1	512.8 ± 3.7	78
143	Tetrahydro-1,3-oxazine	l	198.9 ± 1.9	157.2 ± 2.7	79
144	2-Me-Tetrahydro-1,3-oxazine	l	246.2 ± 1.6	206.9 ± 2.5	79
145	3-Me-Tetrahydro-1,3-oxazine	l	204.8 ± 3.1	164.0 ± 3.7	79
146	3,6-Me_2-Tetrahydro-1,3-oxazine	l	240.3 ± 3.5	196.4 ± 4.0	79
147	Tetrahydro-1,4-oxazine	l	185.7 ± 2.7	142.8 ± 3.3	79
148	4-Me-Tetrahydro-1,4-oxazine	l	196.1 ± 2.9	155.5 ± 3.6	79
149	2-Methyl-2-oxazoline	l	169.5 ± 0.9	130.5 ± 0.9	80
150	2-Ethyl-2-oxazoline	l	193.1 ± 1.1	148.9 ± 1.2	80

[a]Estimated enthalpies of vaporization. The assigned uncertainty is ±2 kJ/mol in all cases.

TABLE 5-2. Values of A, B (The CH_2 Increment), and $2s_B$ (Equation 5-3) for a 14-Homologue Series.[a]

Compounds	$-A$ (kJ/mol)	$-B$ (kJ/mol)	n
1. *n*-Alkanes	42.65	20.72 ± 0.035	≥2
2. 2-Methylalkanes	111.62	20.97 ± 0.20	≥2
3. 1-Cyclohexylalkanes	130.44	20.74 ± 0.15	≥2
4. 1-Alkenes	−41.30	20.71 ± 0.07	≥2
5. 1-Alkanols	194.64	20.18 ± 0.04	≥2
6. 2-Alkanones	197.54	20.52 ± 0.12	≥2
7. 1-Methoxycarbonylalkanes	389.72	20.56 ± 0.09	≥4
8. 1-Cyanoalkanes	−95.39	20.83 ± 0.10	≥3
9. 1-Mercaptoalkanes	5.20	20.73 ± 0.09	≥2
10. 1-Chloroalkanes	69.58	21.01 ± 0.11	≥2
11. 1-Bromoalkanes	26.83	20.24 ± 0.12	≥4
12. 1-Aminoalkanes[b]	3.27	22.20	≥2
13. 1-Methoxyalkanes[c]	175.73	20.56	≥2
14. 1-Thiomethylalkanes[d]	18.98	20.72	≥2

[a]From Reference 13.
[b]Average deviation for the 3 compounds, ±0.25 kJ/mol; see the text, p. 189.
[c]Average deviation for the 4 compounds, ±0.23 kJ/mol; see the text, p. 189.
[d]Average deviation for the 4 compounds, ±0.70 kJ/mol; see the text, p. 189.

the energy associated with four and two CH_2 increments, respectively[13] (cf Table 5-2).

Sellers[14] determined the enthalpies of formation of 2-nonanone **(19)** and 2-dodecanone **(20)** to provide more data for the examination of the CH_2 increment in different homologous series.[13]

Steele[15] measured the standard enthalpies of formation of five C_7 bridged-ring hydrocarbons: norbornane **(21)**, norbornene **(22)**, norbornadiene **(23)**, nortricyclene **(24)**, and quadricyclane **(25)** for which inconsistent values had been presented earlier. (The interrelations between **21–25** are shown in the scheme.) All Steele's results are reasonable and find support, for example:

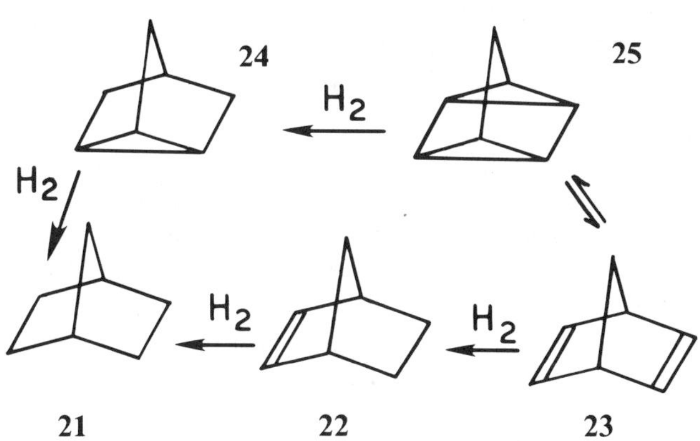

ple: in the equilibrium measurement of Walsh and Wells[16,17] for **22** and **23** and in the isomerization study of **23** and **25** by Kabakoff and co-workers.[18] Furthermore Traeger and McLoughlin[19] estimate from the ionization energy of cyclopentadiene (8.566 ± 0.01 eV), from the standard enthalpies of formation of cyclopentadiene (133.2 ± 1.3 kJ/mol) and acetylene (226.7 ± 0.8 kJ/mol) and from the appearance potential (9.76 ± 0.03 eV) of cyclopentadienyl cation from norbornadiene **(23)** that the standard enthalpy of the latter should be equal to or greater than 244.8 ± 5.9 kJ/mol, in good agreement with the value 247.6 ± 3.1 kJ/mol reported by Steele.[15] Jochems and associates[20] also report enthalpies of sublimation for **21** and **22** in excellent agreement with those obtained by Steele.[15] The reader is reminded of Chapter 3 in this volume, in which Chickos includes a full compendium of enthalpy of sublimation data. At any rate there is no doubt that the results of Hall and co-workers[21] (included in Reference 6, unfortunately!) for **21–25** must be *disregarded,* as suggested by Steele.[15]

Parker and co-workers[22] determined the standard enthalpies of formation of bicyclo[3.3.1]nonane **(26)** and bicyclo[3.3.2]decane **(27)** using a semimicro static combustion bomb calorimeter both in the crystalline state and, after measuring the enthalpies of sublimation, in the gaseous state. The experimental values agree well with the ones calculated using the Schleyer force field.[23]

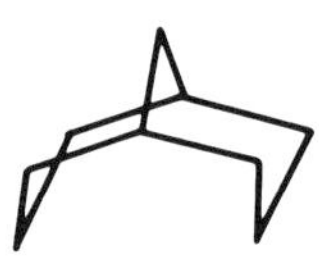

26

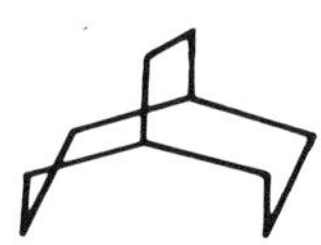

27

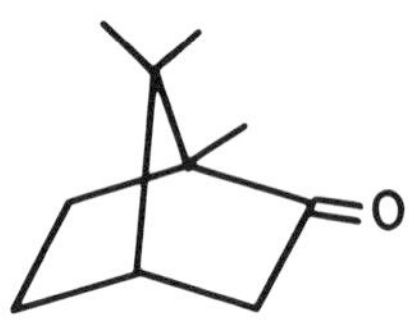

28

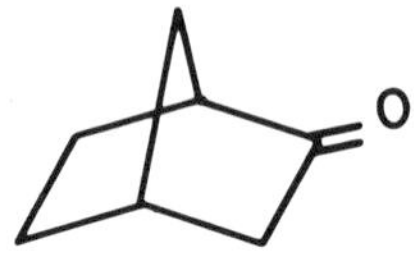

29

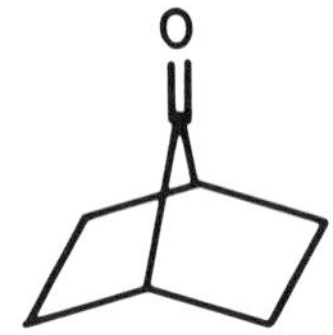

30

Steele has also measured the standard enthalpies of formation of three bicyclic ketones. The combustion of 1,7,7-trimethylbicyclo[2.2.1]heptan-2-one **(28)** was undertaken to compare results obtained from a "small" bomb calorimeter with those obtained in a conventional bomb calorimeter.[24] The enthalpies of formation of bicyclo[2.2.1]heptan-2- **(29)** and -7-ones **(30)**[25] have been determined to test the empirical force field calculations.[26] The $\Delta_f H°(g)$ values for **29** and **30** show that the latter is more strained by 33.9 kJ/mol. On the other hand, when going from norbornane **(21)** to **29,** the strain increases by only about 9 kJ/mol, as can be seen from Equation 5-2:

norbornane (**21**) → 2-norbornanone (**30**)
$-\Delta H = 106.2$ kJ/mol
cyclopentane[6] → cyclopentanone[6] (5-2)
$-\Delta H = 115.3$ kJ/mol
propane[27] → acetone[6]
$-\Delta H = 112.5$ kJ/mol

Steele and Watt[28] have also determined the standard enthalpies of six methyladamantanes **(31–36)** in the crystalline and gaseous states. Their results confirmed the postulation reached by the force field[23] that increasing methyl substitution at the bridgeheads decreases the total strain within the system. Actually the Schleyer force field[23] gives values in excellent agreement with the experimental findings,[28] whereas the Allinger force field[29] reproduces the observed values slightly less accurately. Furthermore Steele[28] has questioned the experimental energies of combustion of methyladamantanes reported by Clark, McKervey, and associates.[30]

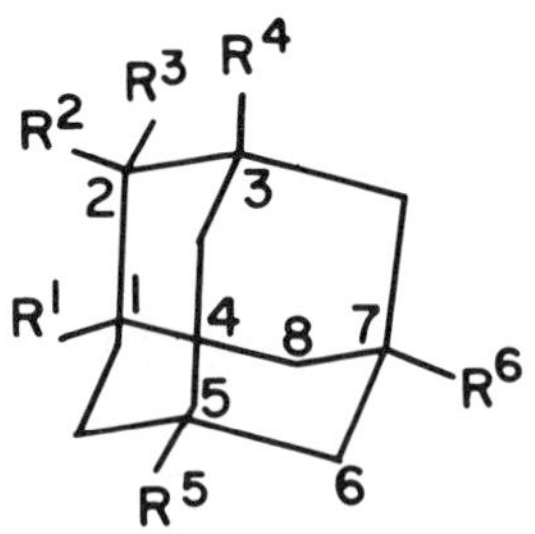

	R^1	R^2	R^3	R^4	R^5	R^6
31	CH_3	H	H	H	H	H
32	H	H	CH_3	H	H	H
33	H	CH_3	CH_3	H	H	H
34	CH_3	H	H	CH_3	H	H
35	CH_3	H	H	CH_3	CH_3	H
36	CH_3	H	H	CH_3	CH_3	CH_3
37	H	=O		H	H	H
38	OH	H	H	H	H	H
39	H	H	OH	H	H	H

Arora and Steele[31] also measured the standard enthalpies of combustion of adamantan-2-one **(37)** and adamantan-1- **(38)** and -2-ols **(39)** and derived from them their standard enthalpies of formation in the crystalline and gaseous states. Compound **39** is relatively 11.7 ± 5.9 kJ/mol more stable than **38,** in keeping with the situation in methyladamantanes.[28] The Allinger 1977 force field reproduced the observed enthalpies of formation for **37–39** fairly well.[30]

Heath, Hefter, and Steele[32] determined the enthalpy of formation of 1,3,5,7-tetramethyl-2,4,6,8-tetrathiaadamantane **(40)** in the crystalline and gaseous states. Using group increment calculations, it can be shown that **40,** like **36,** is practically strain-free.[32]

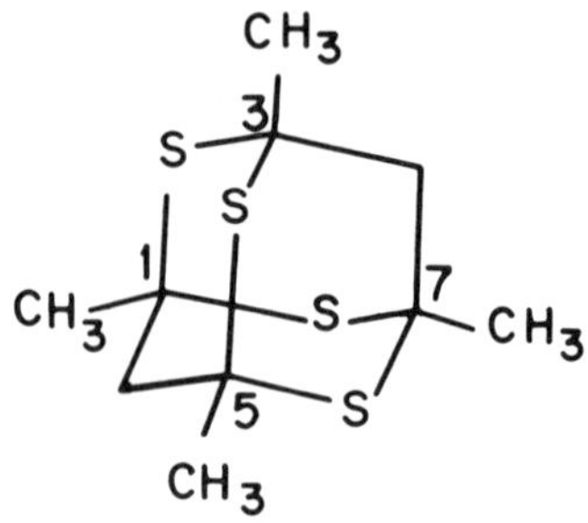

40

The Lund group has determined enthalpies of vaporization and formation for many compounds to examine the constancy of the CH_2 increment in homologous series (Section 3).[13,33] Månsson and co-workers[34] determined enthalpies of vaporization for four *n*-alkanes, seven 1-alkanols, mercaptodecane, three chloroalkanes, and eleven methyl-*n*-alkanoates for this purpose.[13] Svensson[35] remeasured the enthalpies of vaporization of 1-decanol and 1-dodecanol with results that are in excellent agreement with those of Månsson.[34]

Stridh and associates[36] derived the enthalpies of formation of octane- **(41)**, decane- **(42)**, undecane- **(43)**, and tetradecanenitriles **(44)** from the energies of combustion and enthalpies of vaporization at 298.15 K. It was, however, only recently that Xu-Wu and Månsson measured the enthalpies of combustion and formation of acetonitrile **(45)**, the parent member of the 1-nitrile family.[37]

Stridh also measured the enthalpies of formation of 1-dodecene **(46)**,[38] 1-hexadecene **(47)**,[38] 1-bromododecane **(48)**,[39] and 1-bromohexadecane **(49)**[39] to inspect the CH_2 increments in 1-alkenes and 1-bromoalkanes.

Byström has derived enthalpies of formation of di-*n*-propyldiazene *N*-oxide **(50)**[40] and di-*t*-butyldiazene *N*-oxide **(51)**[40] in the liquid and gaseous states and those of 1,3,5-triazine **(52)** in the solid and gaseous states.[41] Her results will be discussed in some detail later in Subsection D, below.

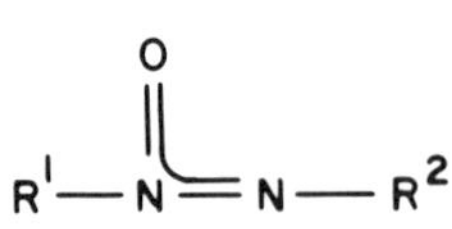

$R^1 = R^2$

50 n$-C_3H_7$

51 t$-$Bu

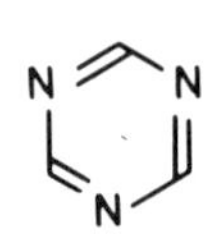

52

Hamilton and associates have measured enthalpies of combustion of oxazole **(53)**,[42] isoxazole **(54)**,[42] 3-methylisoxazole **(55)**,[43] and 5-methylisoxazole **(56)**.[43]

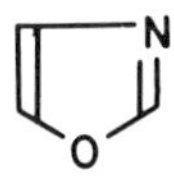

53

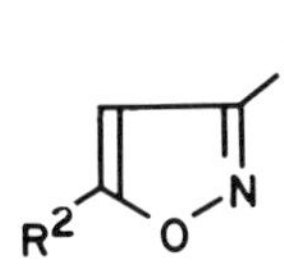

	R^1	R^2
54	H	H
55	CH_3	H
56	H	CH_3

Pilcher and his co-workers[44] measured enthalpies of combustion of 2- **(57)**, 3- **(58)**, and 4-hydroxypyridines **(59)** and 2-methyl-3- **(60)**, 2-methyl-4- **(61)**, 2-methyl-5- **(62)**, and 2-methyl-6-hydroxypyridines **(63)** and derived from them their molar enthalpies of formation in the crystalline and gaseous states.

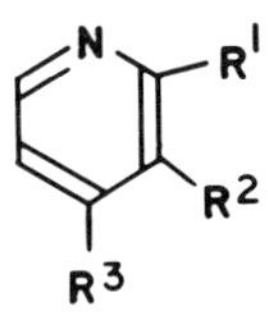

	R^1	R^2	R^3
57	OH	H	H
58	H	OH	H
59	H	H	OH
64	CN	H	H
65	H	CN	H
66	H	H	CN
67	NH_2	H	H
68	H	NH_2	H
69	H	H	NH_2

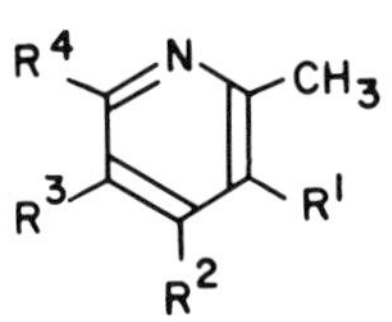

	R^1	R^2	R^3	R^4
60	OH	H	H	H
61	H	OH	H	H
62	H	H	OH	H
63	H	H	H	OH

Bickerton and associates[45] reported the standard enthalpies of formation of 2- **(64)**, 3- **(65)**, and 4-cyanopyridines **(66)** and the corresponding aminopyridines (**67–69**, respectively) in crystalline and gaseous states and compared differences between the derived enthalpies of formation of some of the gaseous compounds with theoretically predicted values.[44,46,47] The latter correctly predict that 2-*R*-pyridines are more stable than 4-*R*-pyridines.

Subach and Zwolinski[48] measured the enthalpy of combustion of 7-*n*-hexadecylspiro[4.5]decane **(70)** and derived its enthalpy of formation at 298.15 K in the liquid state. They also measured the enthalpies of vaporization of spiro[4.5]decane **(71)** and spiro[5.5]undecane **(72)** to derive their enthalpies of formation in the gaseous state to reach ultimately a better understanding of the energetics of spiroalkanes.[48,49]

70 R = n − $C_{16}H_{33}$

71 R = H

72

Melaugh and co-workers[50] determined the enthalpies of vaporization and the relative enthalpies of combustion of *n*-dodecane and 2,2,4,6,6-pentamethylheptane **(73)** to determine the energy of isomerization of the former into **73:** −12.2 ± 1.4 kJ/mol for the liquid and −24.5 ± 1.4 kJ/mol for the gaseous state. Since the enthalpies of formation of liquid and gaseous *n*-dodecane are known accurately, the data above let us derive the respective values for **73** (Table 5-1).

Good and Smith[51a] redetermined the enthalpies of formation of all isomeric pentenes **(74–79)** in the liquid state. These results are discussed in some detail later (Section 5).

Recently, Wiberg and Wasserman[51b] applied successfully the enthalpies of trifluoroacetolysis of the five *n*-hexenes as determined by an automated reaction calorimetry system[51c] to determine the enthalpies of formation of the isomeric 2- and 3-hexenes in relation to that of 1-hexene. The approach allowed them to adjust the enthalpies of formation of *cis*- and *trans*-2-hexenes at −80.1 and −83.2 ± 0.8 kJ/mol and those of *cis*- and *trans*-3-hexenes at −78.1 and −82.1 ± 0.8 kJ/mol, respectively. It could be concluded that the reported enthalpies of formation[6] for the *trans*-3- and *cis*-2-hexenes are almost surely incorrect (see also p. 199).[51b]

Månsson and Sunner[52] measured the enthalpies of combustion and sublimation of bullvalene **(80)** and derived its enthalpies of formation in the crystalline and gaseous states. The apparent strain energy in **80** amounts to 146 kJ/mol.

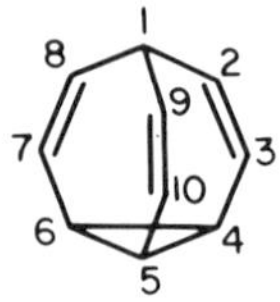

80

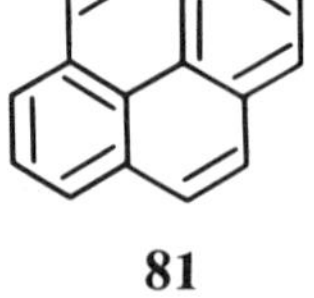

81

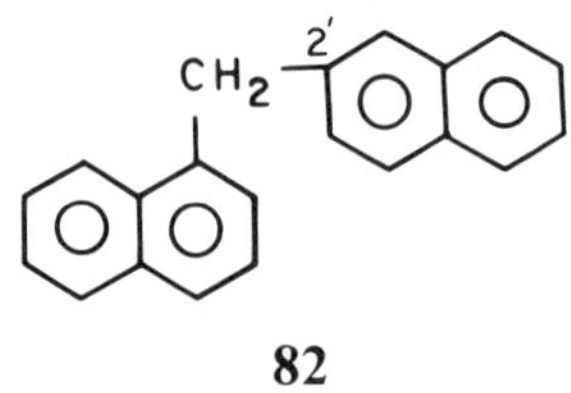

82

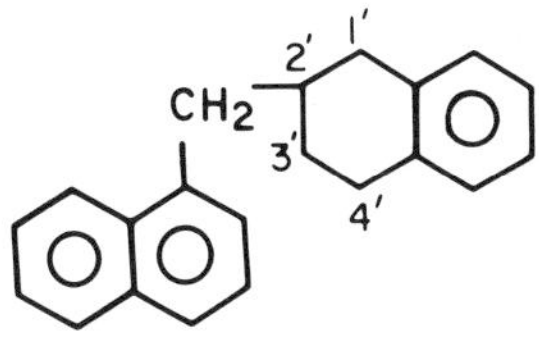

83

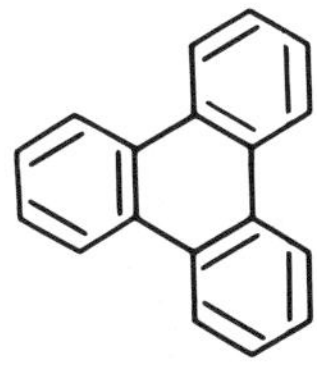

84

85

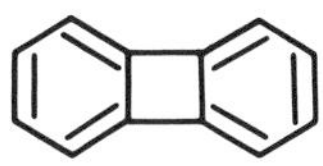

86

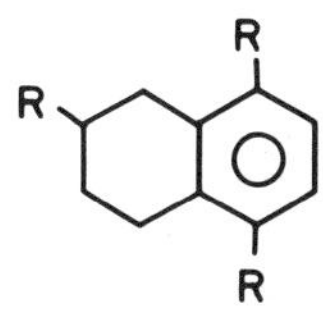

87 R = H
88 R = CH_3

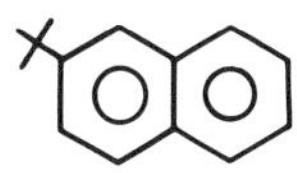

89

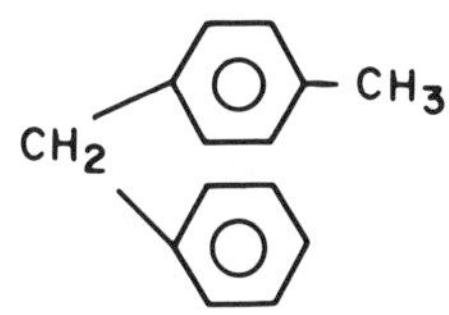

90

Pyrene (**81**; crystal I and the gaseous state) has been studied by Smith and co-workers.[53] Good and Lee have studied derivatives of polynuclear aromatic hydrocarbons[54] **(82–86)** and naphthalenes **(87–89)**,[55] diphenylmethanes **(90–92)**,[55] and bicyclic hydrocarbons **(93–95)**[55] in the condensed state to develop fuels having a high enthalpy of combustion per unit mass.

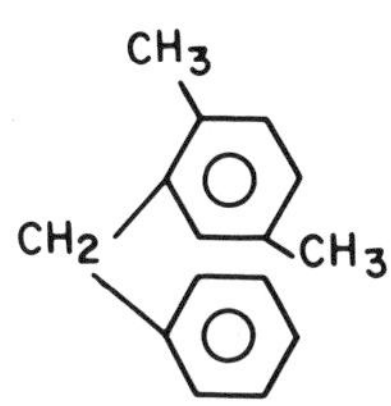

91

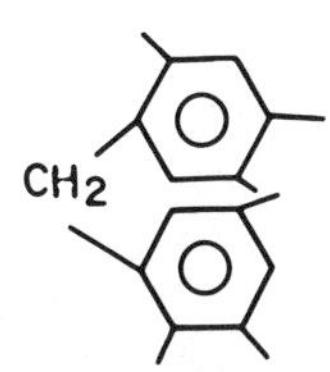

92

93

94

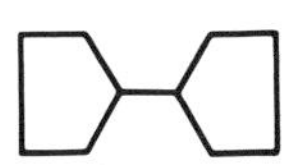

95

Montgomery and Rossini[56] also measured the enthalpies of combustion and vaporization or sublimation of **93** and **94** as well as those of biphenyl **(96)**. Their standard enthalpies of formation for **93** and **94** seem to be less accurate than those of Good and Lee[55] but are, however, in substantial accord with them within the limits of uncertainty (Table 5-1). Enthalpies of combustion and formation of crystalline 1,4-dihydro-1,4-epoxynaphthalene **(97),** an oxidizing agent with unusual properties, have been determined by Havel and associates.[57]

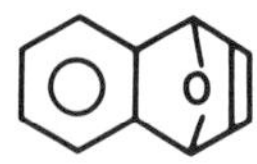

97

Dekker and Mosselman[58] derived the standard enthalpies of formation of liquid 1,1,4,4- **(98)** and 1,1,5,5-tetramethylcyclodecane **(99)** from their enthalpies of combustion at 298.15 K and found that the former is 9.5 ± 1.4 kJ/mol more stable.

Colomina and co-workers[59–63] have continued their investigations of the influence of polar and steric effects of substituents on the thermochemical stability of various aromatic compounds and have determined the enthalpies of combustion and formation for 2-methoxy- **(100),**[59] 2-ethoxy- **(101),**[59] 2,3-dihydroxy- **(102),**[59] and 2,7-dihydroxynaphthalenes **(103),**[59] and *o*-, *m*-, and *p*-ethyl- **(104–106)**, [60,63] *o*-, *m*-, and *p*-methoxy- **(107–109)**[61], and *o*-, *m*-, and *p*-*t*-butylbenzoic acids **(110–112).**[62]

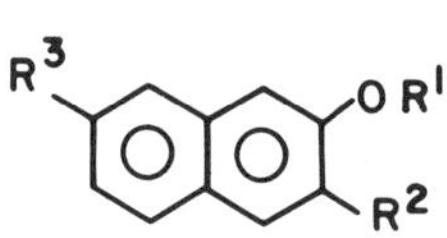

	R^1	R^2	R^3
100	CH_3	H	H
101	C_2H_5	H	H
102	H	OH	H
103	H	H	OH

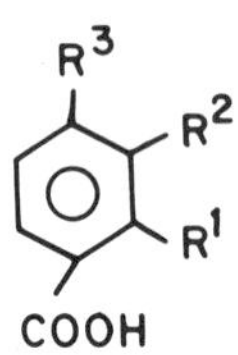

	R^1	R^2	R^3
104	C_2H_5	H	H
105	H	C_2H_5	H
106	H	H	C_2H_5
107	OCH_3	H	H
108	H	OCH_3	H
109	H	H	OCH_3
110	t-Bu	H	H
111	H	t-Bu	H
112	H	H	t-Bu

Arshadi[64] has determined the enthalpies of combustion, sublimation, and formation of phenazine **(113)** and 2,1,3-benzoxadiazole **(114)** and has calculated the C—N and O—N bond energy terms in condensed aromatic rings to be 439 and 284.5 kJ/mol, respectively.

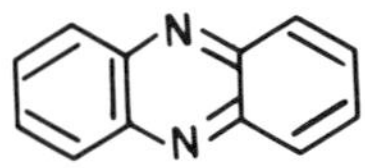

113

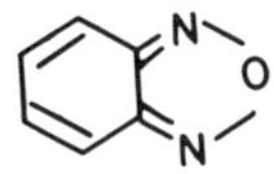

114

Steele[65] has measured the enthalpies of formation of cyclohexylamine **(115)** and its hydrochloride, and five sterically hindered amines (**116–120;** Table 5-1) have been studied by Suradi and co-workers.[66] These results will be discussed in some detail later (pp. 194, 195).

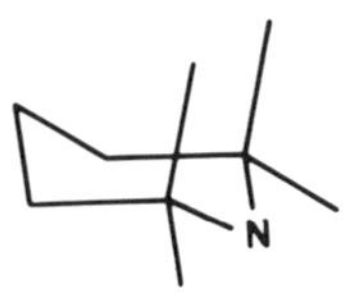

119

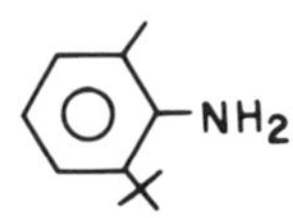

120

Pilcher and colleagues have determined enthalpies of combustion of 2,4-pentanedione **(121),**[67a] 2,2-dimethyl- **(122),**[68] 2,6-dimethyl- **(123),**[68] 2,2,6-trimethyl- **(124),**[68] and 2,2,6,6-tetramethylheptane-3,5-diones **(125),**[68] and benzoylacetone **(126)**[68] and have derived the standard enthalpies of formation for the keto–enol equilibrium mixtures in the condensed state and for the liquid and gaseous enol forms (in the case of **121** for the keto form, also). The data for the enols were extracted by utilizing the known compositions of the keto–enol mixtures, the known enthalpies of vaporization for the enols,[68b,c] and the enthalpies of enolization of **121.**[67,68] The enthalpies of formation of β-diketones are necessary in evaluating the enthalpies of formation of metal β-diketonate coordination complexes from reaction calorimetric studies.

Stridh[69] has studied 2-ethylhexanoic acid **(127)** and Anthoney and co-workers[70] have investigated dimethyloxalate **(128).**

The enthalpies of formation of crystalline L-proline **(129),**[71] cytosine **(130),**[72] L-histidine **(131),**[73] and uracil **(132)**[72] have been determined.

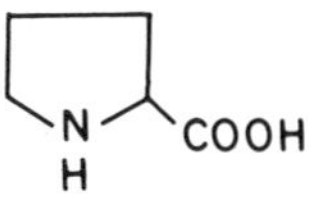

129

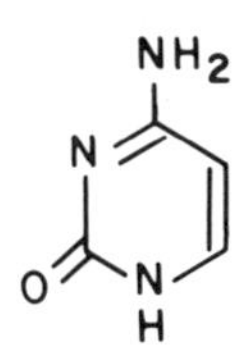

130

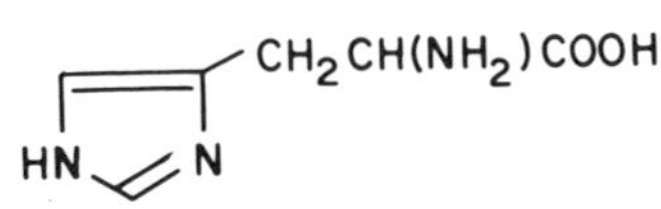

131

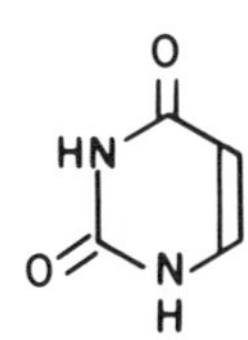

132

Gundry and associates[73] have measured the enthalpies of combustion (-6608.3 ± 3.4 kJ/mol) and solution (10.9 ± 0.1 kJ/mol) of paraquat dichloride (**133**) and have derived the standard enthalpy of formation of the 1,1′-dimethyl-4,4′-bipyridinium cation.

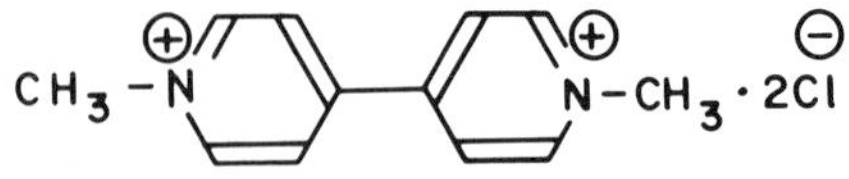

133

Kolesov and co-workers[74–77] have determined enthalpies of formation of many organic halogen compounds (**134–139,** Table 5-1) and have discussed their use in structural correlations.

Pihlaja and colleagues have measured enthalpies of combustion of some methyl-substituted 1,3-dioxanes (**140–142**)[78] and tetrahydro-1,3- (**143–146**)[79] and -1,4-oxazines (**147** and **148**).[79] These results will be discussed in detail later (pp. 196–198, 193 and 194).

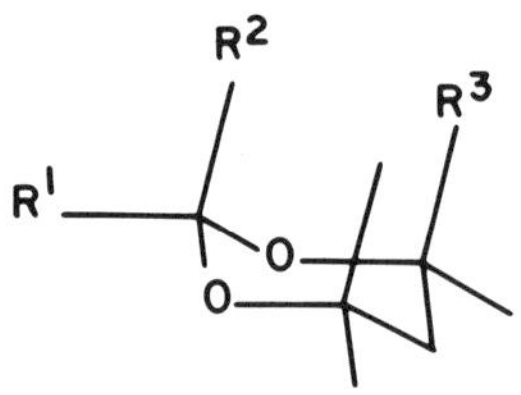

	R^1	R^2	R^3
140	CH_3	CH_3	H
141	H	H	CH_3
142	CH_3	H	CH_3

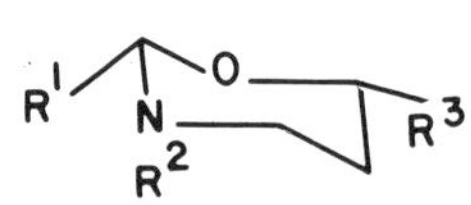

	R^1	R^2	R^3
143	H	H	H
144	CH_3	H	H
145	H	CH_3	H
146	H	CH_3	CH_3

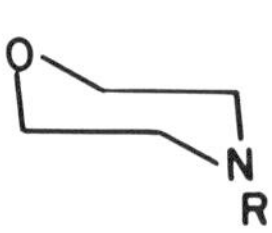

147 R = H
148 R = CH_3

The enthalpies of combustion of 2-methyl- (**149**) and 2-ethyl-2-oxazolines (**150**) and their standard molar enthalpies of formation have been reported,[80] also.

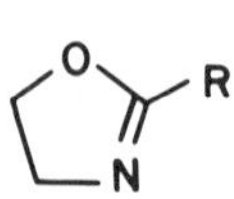

149 R = CH_3
150 R = C_2H_5

Finally there are three reports on the enthalpies of formation of 25 different explosives[81–83] and on their detonation properties.[83]

3. IS THERE A UNIVERSAL METHYLENE CH_2 INCREMENT?

The concept of the possible existence of a universal methylene increment attracted especially the late Professor Sunner, and the Lund group carried out many extremely careful measurements of enthalpies of combustion and vaporization/sublimation to derive accurate standard enthalpies of formation for *n*-alkanes,[13,34] 2-methylalkanes,[13] 1-cyclohexylalkanes,[13] 1-alkenes,[13,34,38] 1-alkanols,[13,34,35] 2-alkanones,[13,14,34] 1-cyanoalkanes,[13,36,37] 1-mercaptoalkanes,[13,34] 1-chloroalkanes,[13,34] 1-bromoalkanes,[13,34,39] and methyl-alkanoates.[13,34]

Sunner and associates[13,33,35,36,38,39] have used the equation:[84]

$$\Delta_f H_m^\circ[\mathrm{Y} - (\mathrm{CH_2})_n - H, \mathrm{g}]/(\mathrm{kJ/mol}) = A + Bn + \delta \qquad (5\text{-}3)$$

where A is a constant peculiar to the end group, Y; B is a constant for all normal alkyl series (the CH_2 increment), independent of Y; and δ has a small finite value at least for $n = 1$.[13,85] Table 5-2 lists the values of A, B, and $2s_B$ for the homologous series above.[13] We have added to the list 1-methoxyalkanes (ethane,[6,86] $-\Delta_f H_m^\circ(\mathrm{g}) = 216.4 \pm 0.7$; propane,[6,86,87] 237.7 ± 0.8; butane,[6,86,87] 258.2 ± 1.3; and decane,[86,87] 381.2 ± 2.1 kJ/mol), 1-methylthioalkanes (ethane,[6,86] $-\Delta_f H_m^\circ = 59.6 \pm 1.1$; propane,[6,86] 82.2 ± 0.9; butane,[6] 102.2 ± 0.7; and pentane,[6,9,86] 122.0 ± 2.5 kJ/mol), and 1-aminoalkanes (ethane,[6,86] $-\Delta_f H_m^\circ(\mathrm{g}) = 47.5 \pm 0.7$; propane,[6,86] 70.2 ± 0.4; and butane,[2,6,86] 91.9 ± 0.9 kJ/mol).

The average CH_2 increment (B) from Table 5-2 (excluding items 5 and 11–14) is 20.75 ± 0.15 kJ/mol, which is practically equal to the respective value 20.8 kJ/mol from the Allen-type correlations (cf Subsection B). The 1-bromoalkanes and especially 1-alkanols exhibit significantly different B values. Although the CH_2 increments for 1-aminoalkanes may also differ significantly from the mean value above, more standard molar enthalpies for them would be needed to prove or disprove the case. It is interesting that in the case of alkanols both Γ_{CCC} and Γ_{CCO} group increments again differ significantly from those for other compounds, lending support to the special character of the OH group.

Later Sunner and Wulff[33] tried to fit their data[13] with supplemental data for $\Delta_f H_m^\circ(\mathrm{Y} - H, \mathrm{g})$ from the compilation of Pedley and Rylance[6] into:[37]

$$\Delta_f H_m^\circ[\mathrm{Y} - (\mathrm{CH_2})_n - \mathrm{H,g}] = \Delta_f H_m^\circ(\mathrm{Y} - \mathrm{H, g}) + \beta n \exp\left(\frac{\alpha}{n}\right) \qquad n \geq 1 \qquad (5\text{-}4)$$

and found $\beta = 20.63 \pm 0.09$ kJ/mol, in addition to α_γ values varying from 0 to 0.5. However, at the present we can still conclude that for practical purposes Equation 5-3 is satisfactory, and especially that Allen-type correlations yield reliable calculated values for all the *n*-alkanes in Table 5-2, although in some cases (cf 1-alkanols) the Γ_{CCC} and Γ_{CCY} increments may have slightly deviating values[86] (Table 5-4, below).

4. ALLEN-TYPE GROUP INCREMENTS FOR ESTIMATING ENTHALPIES OF FORMATION OF ALKANES AND THEIR OXYGEN, SULFUR, AND NITROGEN ANALOGUES AND ALIPHATIC KETONES

A. General Treatment

The background to the Allen-type group increment compilation of thermochemical data can best be obtained from Cox and Pilcher[2] or from the pioneer paper of Skinner.[88] Several reports[89a–c] have described the bonded and nonbonded interactions in saturated hydrocarbons with the aid of numerous buttressing and branching effects to allow accurate prediction of their enthalpies of formation. Kao and Chung-Phillips[89d] treated similarly some alcohols, thiols, thioethers, and amines. Although all these approaches have their own merits, they are fairly tedious; hence an easy-to-apply method that allows accurate prediction of enthalpies of formation for a wide range of compounds would be preferable for practical purposes. An Allen-type correlation fulfills this requirement, although for some highly branched alkanes it is not as accurate as the more elaborate schemes.[89a–c]

Our general approach was first to adjust the coefficients in the equation:[2,86,90,91]

$$\Delta_a H^\circ(\mathrm{g}) - \Sigma E_b(\mathrm{g}) = n_1\Gamma_{CCC} + n_2\Gamma_{CCO} + n_3\Gamma_{COC} + n_4\Gamma_{OCO} + n_5\Delta_{CCC} + n_6\Delta_{CCO} + n_7\Delta_{OCO} + n_8 S_{15}^{HH}(\mathrm{C}) + n_9 S_{15}^{HH}(\mathrm{O}) + n_{10} S_{16}^{HH}(\mathrm{C}) + n_{11} S_L \quad (5\text{-}5)$$

where $\Delta_a H^\circ(\mathrm{g})$ is the enthalpy of atomization of the compound in question and $\Sigma E_b(\mathrm{g})$ the sum of the respective bond energies in the gaseous state (Table 5-3). The structural parameters (n_1, n_2, etc.) correspond to the numbers of the different interactions in the molecule. The general term Γ_{ABC} illustrates the effect of the interaction A–B–C, and Δ_{ABC} means the trigonal interaction because of the three nonbonded atoms A, B, and C attached to the same carbon atom. The contributions of the gauche-*n*-butane type of nonbonded interactions between fifth-neighbor hydrogen atoms over a C or O atom are represented by S_{15}^{HH}(C or O); S_{16}^{HH} (C) is the contribution of the nonbonded interactions between sixth-neighbor hydrogen atoms; finally, S_L is the so-called locking parameter in alkanes.[86,92] The enthalpies of formation

TABLE 5-3. Enthalpies of Formation of the Gaseous Atoms and Bond Energies in the Gaseous State[86] Used in the Evaluation of the Group Increments in Table 5-4

Atom	$-\Delta_f H°(g)$ (kJ/mol)
C	716.7
H	218.0
O	249.2
N	472.7
S	277.0
Bond	**$E_b(g)$ (kJ/mol)**
C−C	330.1
C−H	415.85
C−O	327.95
C−N	273.05
C−S	274.15
O−H	463.5
N−H	390.9
S−H	366.8
N−N	158.6

of 71 alkanes, 21 alcohols, 17 ethers, and 8 acetals were used to estimate the values of the different group increments in Equation 5-5 (Table 5-4).

In this context it is interesting to note that the alkanols fitted into the scheme much better after their Γ_{CCC} and Γ_{CCO} increments were given values separate from those of alkanes and the other compounds (Table 5-4). In the other words, our Allen-type correlation supports Sunner's postulation[13] that the CH_2 increment for 1-alkanols differs significantly from its universal value (20.18 vs 20.75 kJ/mol). Our correlation leads to the values 20.25 and 20.8 kJ/mol, respectively, in excellent agreement with the values above.

The group increments, Γ_{CCC}, Δ_{CCC}, $S_{15}^{HH}(C)$, and S_L calculated for alkanes and their oxygen analogues are applicable to their sulfur and nitrogen analogues as well as to aliphatic ketones. The additional group increments can then easily be solved, and the notations (Table 5-4) are the same as in Equation 5-5.[86] The notation ${}^{N}\Delta_{ABC}$ stands for the interaction caused by the atom trio ABC attached to a nitrogen atom. In the case of ketones C_d stands for a carbonyl carbon and $S_{15}^{H\pi}$ corresponds to a gauche-1,4 interaction between a π orbital of the C=O bond and a hydrogen atom.[86] The latter calculations were carried out using data for 45 sulfur and 21 nitrogen compounds and 18 ketones. The average differences between the calculated and estimated enthalpies of formation were ± 1.1 kJ/mol for alkanes and their oxygen analogues (118 compounds), ± 0.8 kJ/mol for the sulfur and nitrogen compounds, and ± 1.0 kJ/mol for the aliphatic ketones. The standard enthalpies of formation of the gaseous atoms and the values of bond energies in the gaseous state used in the compilations[86] are given in Table 5-3.

TABLE 5-4. Optimum Values of the Allen-Type Group Increments Calculated for Saturated Alkanes and Their Oxygen, Nitrogen, and Sulfur Analogues and Aliphatic Ketones[a]

Increment	Contribution to $-\Delta_f H°(g)$	(kJ/mol)
Γ_{CCC}	11.71 ± 0.03	(10.8)
Γ_{CCC} (alcohols)	11.16 ± 0.09	
Γ_{CCO}	24.01 ± 0.45	(23.7)
Γ_{CCO} (alcohols)	24.49 ± 0.49	
Γ_{COC}	23.72 ± 0.77	(24.7)
Γ_{OCO}	54.75 ± 0.85	(54.8)
Γ_{CCN}	17.13 ± 0.35	(17.2)
Γ_{CNC}	19.85 ± 0.78	(18.0)
Γ_{CNN}	24.40 ± 0.55	(24.3)
Γ_{NCN}	44.9[b]	
Γ_{CCS}	14.62 ± 0.18	(13.8)
Γ_{CSC}	11.18 ± 0.37	(12.4)
Γ_{SCS}	20.3[b]	
Γ_{CSS}	16.29 ± 0.28	(18.5)
$\Gamma_{CC(=O)C}$	89.50 ± 0.88	
Γ_{CCC_d}	12.70 ± 0.55	
Γ_{OCN}	31.7 ± 2.4[b]	
Δ_{CCC}	−3.27 ± 0.14	(−2.3)
Δ_{CCO}	−6.52 ± 0.56	(−6.0)
Δ_{OCO}	−14.30 ± 0.84	(−15.4)
Δ_{OOO}	−31.4 ± 1.4[b]	
Δ_{CCN}	−3.79 ± 0.43	(−4.2)
${}^{N}\Delta_{CCC}$	−12.86 ± 2.64	(−6.8)
${}^{N}\Delta_{CCN}$	−11.25 ± 1.82	(−9.8)
Δ_{CCS}	−5.49 ± 0.22	
Δ_{CCC_d}	−7.50 ± 0.54	
$S_{15}{}^{HH}(C)$	−1.94 ± 0.16	(−2.5)
$S_{15}{}^{HH}(O)$	−2.79 ± 0.95	
$S_{16}{}^{HH}(C)$	−10.58 ± 0.50	
S_L	−2.43 ± 0.18	
$S_{15}{}^{HN:}(N)$	1.67 ± 1.04	
$S_{15}{}^{HS:}(C)$	−1.52 ± 0.28	
$S_{15}{}^{H\pi}(C)$	−1.71 ± 0.33	
$S_{15}{}^{HH}(N)$	−4.1 ± 1.1[b]	

[a]Values in parentheses are from Reference 2.
[b]Increments derived in the text.

B. Nonbonded Interaction Energies Between Next-nearest-neighbor and Next-next-nearest-neighbor Heteroatoms

a. Next-nearest-neighbor Interactions

The only next-nearest-neighbor interaction that has been solved with the multilinear regression program[86] together with the other group increments

(Table 5-4) is that between a pair of next-nearest-neighbor oxygen atoms, Γ_{OCO} = 54.75 kJ/mol.

Månsson[93] measured the enthalpies of combustion and vaporization for 3,5-dithiaheptane **(151)** and 3,6-dithiaoctane **(152)** and derived their standard molar enthalpies in the gaseous state: -65.4 ± 1.5 and -83.3 ± 1.5 kJ/mol, respectively. The enthalpy of formation of **151** leads to a value of 20.3 kJ/mol for the Γ_{SCS} increment with the parameters from Tables 5-3 and 5-4. From the enthalpies of formation[79] of **143, 145,** and **146,** we can estimate Γ_{OCN} + CRSE (conventional ring-strain energy) in these compounds at 31.6, 34.5, and 28.9 kJ/mol. On the average, we can postulate that $\Gamma_{OCN} = 31.7 \pm 2.4$ kJ/mol, since the CRSE in tetrahydro-1,3-oxazines is obviously very small as in cyclohexane, piperidine, thiane, and 1,3-dioxane (1.4, 1.6, -1.3, and 1.5 kJ/mol, respectively) from which actually all but thiane and 1,3-dioxane (see also pp. 222–223) are included in the group increment correlations.

$(CH_3CH_2S)_2CH_2$

151

$CH_3CH_2SCH_2CH_2SCH_2CH_3$

152

Similarly the Γ_{NCN} increment can be estimated from the enthalpy of formation (-17.6 ± 1.4 kJ/mol) of gaseous *N,N,N′,N′*-tetramethylmethylenediamine[6] at 44.9 kJ/mol, an intermediate value between those of Γ_{OCO} and Γ_{SCS} (Table 5-4), as could be expected.

b. Next-next-nearest-neighbor Interactions

It has been known for quite some time[94] that a pair of next-next-nearest-neighbor oxygen atoms in a straight-chain compound gives rise to a destabilization as large as -7.8 kJ/mol, as can be seen from the difference between the observed[94] (-408.2 kJ/mol) and calculated (Tables 5-3 and 5-4) standard molar enthalpies of 3,6-dioxaoctane. The corresponding S$-$C$-$C$-$S moiety has no effect, since **152** fits into our correlation with a deviation of only 0.1 kJ/mol.

Obviously the O$-$C$-$C$-$O interaction plays a role in cyclic compounds as well, since by taking into account two such destabilizing interactions in 1,4-dioxane[6] ($\Delta_f H_m^\circ$(g) = -316.0 ± 0.7 kJ/mol), its CRSE reduces to 1.3 kJ/mol, in fair agreement with the strain in other six-membered cyclanes (see above). The N$-$C$-$C$-$N interaction cannot be estimated because it is counterbalanced by the stabilizing $S_{15}^{HN:}$(N) increment due to an attraction between the *N*-hydrogen atom and the lone pair orbital of the other nitrogen atom in hydrazines.[86] Obviously this problem could be overcome by determining the enthalpy of sublimation of piperazine to derive its standard molar enthalpy of formation in the gaseous state.[6]

The interaction between the next-next-nearest-neighbor oxygen and nitrogen atoms can be estimated from the enthalpies of formation (Table 5-1) of tetrahydro-1,4-oxazine **(147)** and its 4-methyl derivative **(148),** both of which include two such interactions. On the average we obtain -3.9 ± 2.2 kJ/mol for the O—C—C—N interaction. The ring strain (CRSE) in **147** and **148** has been assumed to be negligible, as in tetrahydro-1,3-oxazines.

C. Some Applications of Group Increments to Structural and Stereochemical Analysis

In principle many applications of the group increments have been discussed in some depth in earlier reviews.[2,5,7] This is why it is fairly easy on the basis of earlier illustrations to recalculate, for instance, the CRSE,[7] with the revised values of the group increments. It is, however, advantageous to give some examples about the use of group increments for structural and stereochemical analysis featuring models where the treatment gives us new useful thermochemical information as well.

a. The Enthalpy of Formation of Cyclohexylamine

Since the standard molar enthalpy of cyclohexylamine **(115)** has recently become available,[65] it is interesting to prove that the experimental value does not deviate significantly from the one calculated by:

$$-\Delta_f H°(\mathbf{115}, \text{g}) = \Sigma E_b(\text{g}) + 6\Gamma_{CCC} + 2\Gamma_{CCN} + \Delta_{CCN} - \Sigma\Delta_f H°(\text{g, atoms}) \quad (5\text{-}6)$$

from which $-\Delta_f H°$ (**115**, g) is equal to 103.6 kJ/mol (Tables 5-3 and 5-4). (Values for the appropriate E_b(g) and Δ_fH (atoms) are taken from Table 5-3 while the values for Γ_{CCC}, Γ_{CCN} and Δ_{CCN} are from Table 5.4) The experimental value (Table 5-1) is -104.9 ± 1.3 kJ/mol, in excellent agreement with the calculated one.

b. Strain Energies in Some Acyclic Compounds

Since only the S_{16}^{HH}(C) increment was included in the compilation of the Allen increments, it is interesting to compare the corresponding interactions over O, N, and S atoms with it.[66,94] Standard molar enthalpies of formation

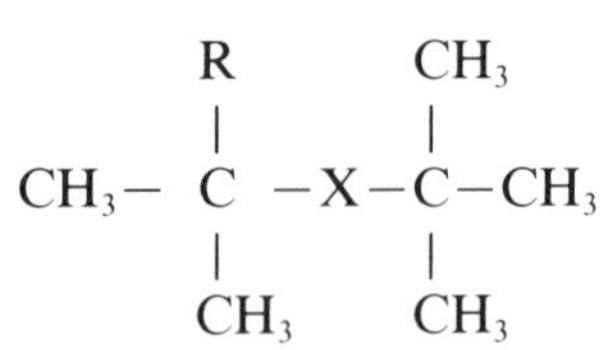

153

are available for **153** when R = H (except X = S) and when R = CH_3 (Table 5-5). In the latter case there are two S_{16}^{HH}(X)-type interactions, which furthermore are both locked. Since at the present it is not possible to estimate the locking parameters for O, N, and S compounds separately,[86] the

TABLE 5-5. S_{16}-Type Interactions in **153** When R = H or CH_3 and X = CH_2, O, NH, and S

X	CH_2	O	NH	S	Interaction
R = H	10.6	7.6[a]	10.3[b]	—[c]	$S_{16}{}^{HH}(X)$
R = CH_3	26.0	44.0[d]	46.3[e]	21.0[f]	Two locked $S_{16}{}^{HH}(X)$

[a]Based on $-\Delta_f H°(g)$ = 357.6 ± 5.0 kJ/mol from References 2 and 87.
[b]Based on the data for **117** (Table 5-1).
[c]Not available.
[d]Based on $-\Delta_f H°(g)$ = 363.4 ± 1.6 kJ/mol, which is a combination of the results in References 87 and 95.
[e]Based on the data for **118** (Table 5-1).
[f]Based on the enthalpy of formation, −188.9 ± 0.7 kJ/mol, taken from Reference 6.

strain energy in **153** when R = CH_3 and X = CH_2 has been taken equal to $2\{S_{16}^{HH}(C) + S_L\}$ (Table 5-4).

In this context it is worth mentioning that $\Delta_f H_m^\circ(g) = -292.6 \pm 5.4$ kJ/mol for methyl-*t*-butyl ether, obtained by combining its enthalpy of combustion by Smutny and Bondi[95] and its enthalpy of vaporization by Fenwick and co-workers,[87] shows that this compound is not strained except for the two $S_{15}^{HH}(O)$ interactions (Table 5-4). However, its standard molar enthalpy of formation[6,87] by the latter authors seems to be somewhat in error.

Since **153** (R = H) includes in addition to the $S_{16}^{HH}(X)$ interaction one $S_{15}^{HH}(X)$-type interaction that has not been estimated for X = NH, the enthalpies of formation of triethylamine (−92.9 ± 0.8 kJ/mol) and methyl *t*-butylamine[66] (**116,** Table 5-1) have been used to estimate the $S_{15}^{HH}(N)$ increment with the aid of the parameters given in Tables 5-3 and 5-4. Triethylamine has three and **116** two of the interactions above; hence on the average $S_{15}^{HH}(N) = -4.1 \pm 1.1$ kJ/mol.

Table 5-5 lists the strain energies in **153** when R = H and X = CH_2, O, and NH and the sum of two locked $S_{16}^{HH}(X)$ when R = CH_3 and X = CH_2, O, NH, and S. It is interesting that the normal $S_{16}^{HH}(X)$ interactions are smaller for X = O and NH than for X = CH_2, whereas the locked interactions follow more closely the expected order, S < CH_2 < 0 ≃ NH.

There are three orthooesters, **154–156,** the enthalpies of formation of which are known in the gaseous state.[6,96,97] The value quoted for **155** has been recalculated from the enthalpies of hydrolysis reported by Hine and Klueppel[97a] and Wiberg and Squires.[97b]

		R^1	R^2	$-\Delta_f H_m^\circ$/(kJ/mol)
$R^1C(OR^2)_3$	**154**[96]	H	CH_3	537.3 ± 2.2
	155[96]	H	C_2H_5	635.4 ± 3.0
	156[6,97]	CH_3	CH_3	571.4 ± 3.0

They lead to $\Delta_{OOO} = -31.4 \pm 1.4$ kJ/mol, which is in line with Δ_{CCO} and Δ_{OCO} interactions (Table 5-4), although it includes also the possible contribution of the anomeric effects.

D. Bond–Bond Interactions in Dialkyldiazenes

When discussing the enthalpies of formation of **50** and **51,** Byström[40] also derived some Allen-type group increments for dialkyldiazenes. Since her approach includes some obvious mistakes, we repeat the procedure here with the aid of standard molar enthalpies of gaseous methylbutyldiazene (**4**), di-*n*-propyldiazene (**5**), di-*n*-butyldiazene (**7**), and di-*t*-butyldiazene (**8**) (78.9, 51.3, 9.2, and −36.4 kJ/mol, respectively; cf Table 5-1). The unknown group increments for these diazenes can be evaluated from:

$$\Sigma(\mathrm{N{=}N}) + n_1\Sigma_{\mathrm{CCN}_d} + n_2\Delta_{\mathrm{CCN}_d} = \Delta_a H°(\mathrm{g}) - [n_3E_{\mathrm{CC}} + n_4E_{\mathrm{CH}} + n_5\Gamma_{\mathrm{CCC}} + n_6\Delta_{\mathrm{CCC}}] \qquad (5\text{-}7)$$

where $\Sigma(\mathrm{N{=}N}) = E_{\mathrm{N=N}} + 2E_{\mathrm{CN}} + 2\Gamma_{\mathrm{CN=N}}$ is common to all diazenes. The results are $\Sigma(\mathrm{N{=}N})$ = 1043.35 kJ/mol, Γ_{CCN_d} = 18.7 kJ/mol, and Δ_{CCN_d} = −7.7 kJ/mol. The latter two increments are very reasonable if we compare them with the Γ_{CCN}, $\Gamma_{\mathrm{CCC}_d}/\Gamma_{\mathrm{CCC}}$, and $\Delta_{\mathrm{CCC}}/\Delta_{\mathrm{CCC}_d}$ increments in Table 5-4, but they differ fundamentally from the values suggested by Byström.[40] Using the foregoing increments and Equation 5-7, we estimate $\Delta_f H°(\mathrm{g})$ for **16–18** (148.4, 92.6, and 120.5 kJ/mol, respectively), in excellent agreement with the estimates[12] shown in Table 5-1. For diisopropyldiazene (**6**) the calculated value (28.9 kJ/mol) deviates, however, substantially from the experimental finding of 35.6 ± 3.6 kJ/mol. In this context it should be mentioned that Engel and co-workers also have discussed the isomerization of some *cis*- and *trans*-diazenes[98] and the ring strain of cyclic 1,2-dialkyldiazenes.[99]

E. Estimation of Conformational Interactions with the Aid of Enthalpies of Formation

Enthalpies of formation can be used to estimate conformational energies in, for example, (hetero)cyclohexanes. The best known example is, of course, the determination of conformational energies in methylcyclohexanes.[100,101] Pihlaja and co-workers have discussed some other applications in various reports.[5,7,102–105]

In this context we will reestimate the chair–twist enthalpy difference and the values of 2,4- and 4,6- syn–axial interactions in 1,3-dioxanes using experimental enthalpies of formation and the values derived for the hypothetical strainfree molecules using the Allen group increments (Tables 5-3 and 5-4).

The standard molar enthalpy of formation of *trans*-2,2,4,6-tetramethyl-1,3-dioxane **(157)** is −482.7 ± 3.3 kJ/mol,[102,103] and the hypothetical value for a strain-free molecule with the same substitution pattern from the Allen increments is −512.4 kJ/mol. It has been shown that **157** prefers greatly a 2,5-twist from;[107] hence the difference 512.4 − 482.7 = 29.7 kJ/mol represents the chair–(2,5-twist) enthalpy difference, in good agreement with other recent estimates.[104,106] The implicit assumptions about **157** are (a) that it has the same conformation in the liquid and gaseous states and (b) that it exhibits no residual nonbonded interactions other than those due to the twist form.

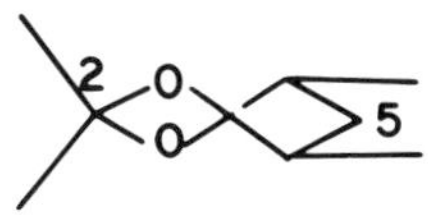

157

2,2,4,4,6-Pentamethyl-1,3-dioxane **(140)** has been shown to exist in a chair form[104,108] like the other 2,2,4,4, tetramethyl derivatives. Total conformational interaction (ΔH_s) in **140,** which is due to syn–axial 2,4-Me,Me, 2,6-Me,H-, 4,6-Me,H interactions, is equal to $\Delta_f H°$ (**140,** g, obs)[78] − $\Delta_f H°$ (**140,** g, calc). Using the group increments (Tables 5-3 and 5-4), the standard molar enthalpy of formation can be estimated for the hypothetical strain-free **140:**

$$
\begin{aligned}
&-\Delta_f H^\circ\ (\mathbf{140},\ \text{g, calc}) = \Sigma E_b\,(\text{g}) - \Sigma\Delta_f H^\circ\ (\text{g, atoms})\\
&+ 6\Gamma_{\text{CCC}} + 9\Gamma_{\text{CCO}} + 2\Gamma_{\text{COC}} + \Gamma_{\text{OCO}} + \Delta_{\text{CCC}} + 6\Delta_{\text{CCO}} + 2\Delta_{\text{OCO}} \qquad (5\text{-}8)\\
&= 552.6\ \text{kJ/mol}
\end{aligned}
$$

Thus ΔH_s amounts to 552.6 − 520.9 = 31.7 kJ/mol, from which the value 19.5 kJ/mol can be obtained for the syn–axial 2,4-Me,Me interaction by subtracting the values of the syn–axial 2,6-Me,H (8.4 kJ/mol) and 4,6-Me,H (3.8 kJ/mol) interactions.[102,105,107] The result shows that the 1,3-dioxane ring must be appreciably deformed to minimize the syn–axial 2,4 interaction.[105] Actually the value above is surprisingly close to the syn–axial Me,Me interaction in cyclohexanes,[100] namely, 15.5 kJ/mol.

Similarly, the total strain (ΔH_s) in 4,4,6,6-Me_4 **(141)** and 2,4,4,6,6-Me_5 **(142)** would be 38.0 and 38.6 kJ/mol, respectively, (for values of $\Delta_f H°$ (g), cf Table 5-1). On the average, the syn–axial 4,6-Me,Me interaction would then be:[102]

$$
\begin{aligned}
\Delta H_s - 2\Delta H_{4a-\text{Me},2a-\text{H}} &= (38.3 - 16.8)\ \text{kJ/mol} \qquad (5\text{-}9)\\
&= 21.5\ \text{kJ/mol}
\end{aligned}
$$

a value that is close to the estimate above for the syn–axial 2,4-Me,Me interaction but even slightly higher. It must be emphasized that the syn–axial interactions calculated above demonstrate only the increase in total strain in comparison with compounds without syn–axial Me,Me interactions. They do not necessarily tell the exact values.

Recently the enthalpy of formation (−159.9 kJ/mol; Table 5-1) of 2,2,6,6-tetramethylpiperidine **(119)** became also available.[66] The group increments (Table 5-3 and 5-4) allow the evaluation of $\Delta_f H°$ (**119,** g) for the hypothetical strain-free molecule from:

$$-\Delta_f H^\circ\,(\mathbf{119}, \text{g}) = \Sigma E_b(\text{g}) - \Sigma\Delta_f H^\circ(\text{g, atoms}) + 9\Gamma_{CCC} + 6\Gamma_{CCN} + \Gamma_{CNC} + 2\Delta_{CCC} + 6\Delta_{CCN} = 196.8 \text{ kJ/mol} \quad (5\text{-}10)$$

hence $\Delta H_s = 196.8 - 159.9 = 36.9$ kJ/mol. The 2,4-diaxial Me,H interactions[109] are both about 4.0 kJ/mol; hence the contribution of the syn–axial 2,6-Me,Me interaction would be 28.9 kJ/mol, which is surprisingly high in comparison with the estimates above for 1,3-dioxanes. This type of situation corresponds to an $S_{16}^{HH}(X)$ interaction, that is locked, and thus we can compare the syn–axial Me,Me interactions in (hetero)cyclohexanes with those in compounds **153** when R = CH_3:

	$S_{16}^{HH}(X)$ (kJ/mol)	
X	(Hetero)cyclohexane	**153** R=CH_3
CH_2	15.5[a]/21.5[b]	13.0
O	19.5	22.0
NH	28.9	23.2

[a]Cyclohexane. [b] 1,3-Dioxane.

Even in the light of the foregoing comparison, the value of the $S_{16}^{HH}(X)$ interaction for piperidine seems to be somewhat too large.

5. ENTHALPIES OF HYDROGENATION AND SOME RELATED DATA

Since the enthalpy of hydrogenation is a measure of the difference between the enthalpies of formation of the saturated and unsaturated compounds, it offers—if the enthalpy of formation of the saturated compound is well established—a useful method for the derivation of the enthalpy of formation of the unsaturated compound.

On the other hand, there are some weaknesses in the hydrogenation route to, for example, enthalpies of formation of the alkenes, as pointed out by Good and Smith.[51a] If, however, the necessary precautions[110,111] are taken, the enthalpy changes accompanying the hydrogenation of olefins can be fundamental to structural analysis.[112]

Considering the importance of hydrogen calorimetry, relatively few research groups have entered the field[112] and very little new data have become available since the last reports by Turner and co-workers.[113,114] Most of the available data have been reviewed earlier[86,112,115]; hence the following discussion is very brief.

In 1973 Turner reported enthalpies of hydrogenation for 7 acetylenic[114] and 16 miscellaneous alkenes in acetic acid solution at 298 K. The results show that cyclooctyne and cyclononyne are strained by about 42.3 and 12.1 kJ/mol, whereas in 10- and 12-membered rings the triple-bond strain is negligible:

			$-\Delta_f H°$ (kJ/mol)
octyne-4	$\xrightarrow{H_2}$	*cis*-4-octene	148.1
cyclooctyne	$\xrightarrow{H_2}$	*cis*-cyclooctene	190.3
cyclononyne	$\xrightarrow{H_2}$	*cis*-cyclononene	160.2
cyclodecyne	$\xrightarrow{H_2}$	*cis*-cyclodecene	149.8
cyclododecyne	$\xrightarrow{H_2}$	*cis*-cyclododecene	148.1

They also demonstrated the steric interactions in *cis*-alkenes along with the increasing size of the alkyl substituents:

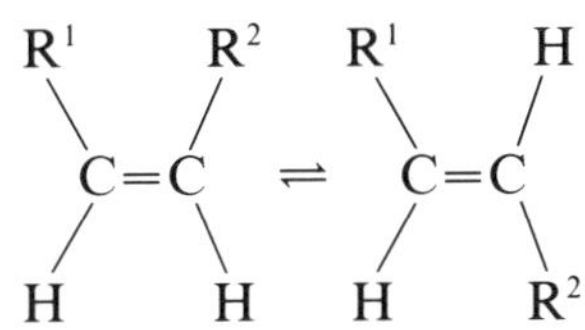

cis *trans*

R^1	R^2	$-\Delta H°_{isom}$ (kJ/mol)
CH_3	CH_3	4.2
CH_3	$CH(CH_3)_2$	3.8
$(CH_3)_2CH$	$CH(CH_3)_2$	8.0
CH_3	$C(CH_3)_3$	18.0
$(CH_3)_3C$	$C(CH_3)_3$	38.9
$(CH_3)_3CCH_2$	$CH_2C(CH_3)_3$	3.8

Even in *cis*-2,2,6,6-tetramethyl-4-octene, the substituents can avoid steric repulsions by rotation and angle widening, as can be seen from the enthalpies of isomerization, which are lower than expected. Recently Kapteijn and co-workers[116] carried out the gas phase cis–trans isomerization of both 2-butene and 2-pentene over a wide temperature range with $-\Delta H°_{isom} = 4.34 \pm 0.04$ and 4.90 ± 0.08 kJ/mol, in fair agreement with the foregoing values based on hydrogenation. From the standard molar enthalpies of formation[51] of *cis*- and *trans*-2-pentenes (Table 5-1), one can obtain $-\Delta H°_{isom} = 4.49 \pm 0.59$ kJ/mol, also in excellent accord with the values quoted above. Correspondingly,[117] for 2- and 3-heptenes $-\Delta H°_{isom} = 4.39 \pm 0.88$ and 5.06 ± 0.91 kJ/mol, respectively, giving further support to Turner's postulations.[113]

Enthalpies of hydrogenation[113] also point out that there is a strain of about 30.5 kJ/mol between the ethanobridges of **160,** since the enthalpy of hydrogenation of **158** to **159** is in the normal range (cf norbornene,[118] $\Delta_h H° = -138.5$ kJ/mol).

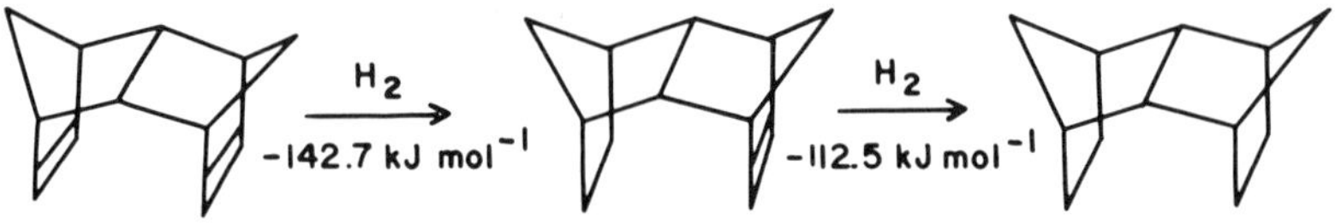

158 **159** **160**

In another paper 1973 Turner and associates[114] reported enthalpies of hydrogenation of various cyclic dienes and trienes, 1,5-hexadiene and *cis*- and *trans*-1,3,5-hexatrienes in acetic acid solution at 298 K and discussed different nonbonded interactions in them.

In 1971 Rogers and McLafferty[112a] constructed a calorimeter for the measurement of the enthalpies of hydrogenation of liquid alkenes on palladium black catalyst in acetic acid or cyclohexane solution, which had the advantages of repetitive sample injection, fairly small sample requirement, and very short reaction time. The calorimeter was applied to measure enthalpies of hydrogenation of 16 allyl- and vinyl-substituted and other alkenes. Comparison with some enthalpies of hydrogenation known from previous work showed that the data are as reliable as usual in this field. Special emphasis has been laid on the structure and strain energies of hexamethyl Dewar benzene **(161)** and its hydrogenation product **(162a).**[112]

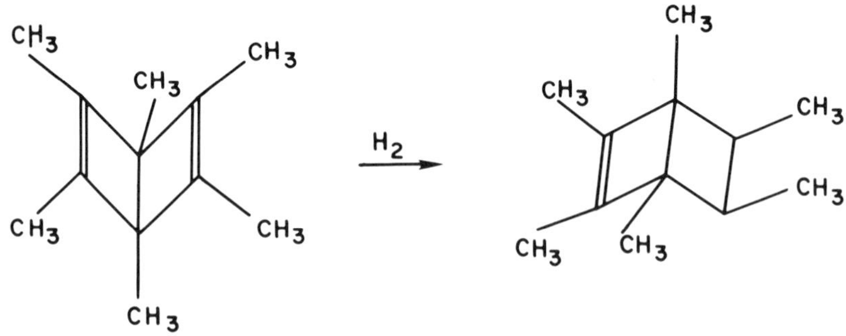

161 **162a**

More than a decade later, Roth and Lennartz[112b] described a method based on an isothermal titration calorimetry for the measurement of enthalpies of hydrogenation. In the test runs they were able to reproduce the known enthalpies of hydrogenation of 1-heptene, cyclopentene, cyclohexene, cycloheptene, and *cis*-cyclooctene, reasonably well.[112b] Later on Roth and colleagues[112c] applied their method to determine the enthalpies of hydrogenation of bicyclo[2.1.0]pent-2-ene (**162b;** 175.3 kJ/mol) and its 5,5-dimethyl derivative (**162c;** 175.7 kJ/mol). As the differences ($\Delta H_1 - \Delta H_2$) of the enthalpies of the following reactions:

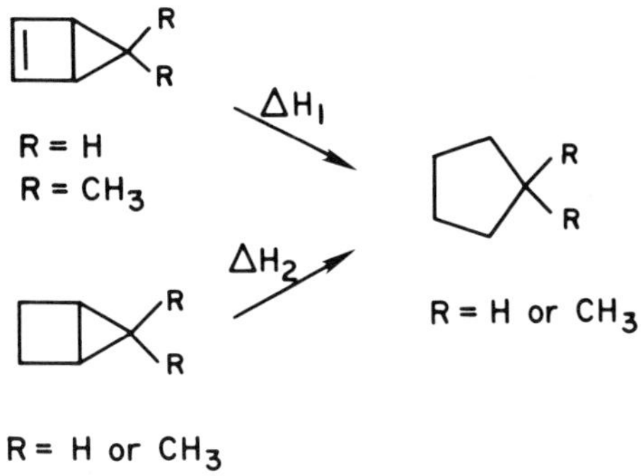

162b R = H
162c R = CH_3

Furthermore Roth[112d] utilized enthalpies of hydrogenation when studying the influence of fluoro substituents on the isomerization of 1,1-difluoro-2-vinyl- and *cis*-1,1-difluoro-2-methyl-3-vinylcyclopropane. In their most recent communication, Roth, Vogel and associates[112e] discuss the resonance energies (E_r) of the two bridged [10]annulenes shown below and derive their enthalpies of formation from their enthalpies of hydrogenation and from the enthalpies of formation of the products of reduction taken from force field calculations.

E_R/(kJ mol^{-1})

72.0	27.2

6. CHEMICAL EQUILIBRATION

Chemical equilibration can be regarded as one of the most accurate and versatile tools for measuring conformational and other energy differences between stereoisomers or other structural isomers. Methods depend on the problem to be solved,[5,7,100] although the analysis of the equilibrium mixtures occurs most often by gas chromatography and sometimes by (^{1}H and ^{13}C) NMR spectroscopy.

Chemical equilibration is especially advantageous in the case of epimerizable compounds,[5,7,100,105,119] although it can be applied to various other systems as well. Since its use has been frequently reviewed,[5,7,100,105,119] only some typical illustrative examples are given in this context.

Taskinen and co-workers[120–136] have been carrying out thorough studies on the thermodynamics of vinylethers by equilibrating them at different temperatures in cyclohexane solution with iodine as catalyst or in neat solution with $Hg(OAc)_2$ as catalyst. In some cases ether,[120a] CCl_4,[136] and 1,4-dioxane[120a,131a,120d] have been used as solvent and *p*-toluene sulfonic acid as catalyst.[120a,d]

For instance, it has been shown by means of some isomerization reactions that the thermodynamically more stable geometrical isomer of 3-methoxy-4-methyl-2-pentene **(163)** has the *E* configuration,[120b] which has been confirmed by [^{13}C]NMR shift data.[137] The same configurational assignment is applicable to other related isomer pairs unless the group R is too bulky (eg, *t*-Bu), whereby the relative stabilities are reversed.[120c,133] By means of [^{13}C] NMR shift data, it was confirmed that the diminished thermodynamic stability of the *Z* form arises from the inability of the CH_3O group to assume a planar (s-trans and/or s-cis) structure about the O−C (sp^2) bond. Consequently the conjugative interaction in the −O−C=C moiety of a *Z* configuration is considerably weaker than that in the corresponding *E* isomer.[137] The findings above, together with the relatively low thermodynamic stability of some heterocyclic six-membered vinyl ethers such as 2-methyleneoxane[138] and 4-methylene-1,3-dioxane[123c] are explainable only if the simplest vinylether, methylvinylether, can attain the planar s-cis and s-trans conformations only but not the nonplanar gauche form,[139] as proposed by several authors. That is, which of these are energy minima and which are maxima?

163
R = iPr

E Z

Taskinen and associates also showed that the thermodynamic stability of an −O−C=C−Ẍ system, where Ẍ is an electron donor through its conjugative interaction with the C=C bond, is weaker than could be expected on the basis of the thermodynamic stabilities of the −O−C=C and C=C−Ẍ systems taken separately. This effect is most enhanced when Ẍ is a strong electron donor in the conjugative sense, such as −O− or F.[120d,125b,129]

On the other hand, if X is an electron acceptor such as a carbonyl group, the thermodynamic stability of the −O−C=C−X system is greatly enhanced.[136] Obviously the X (C=O) moiety strengthens the ability of the −O− atom to conjugate with the adjacent C=C bond, whereas in the former systems the conjugative interactions of the −O− and −X moieties with the C=C bond oppose each other, hence weaken the total stabilization.

Furthermore Taskinen and colleagues have studied the effects of the size and nature of the alkoxy group on the thermodynamics of various isomerization reactions like 5-11:[121,125a,127]

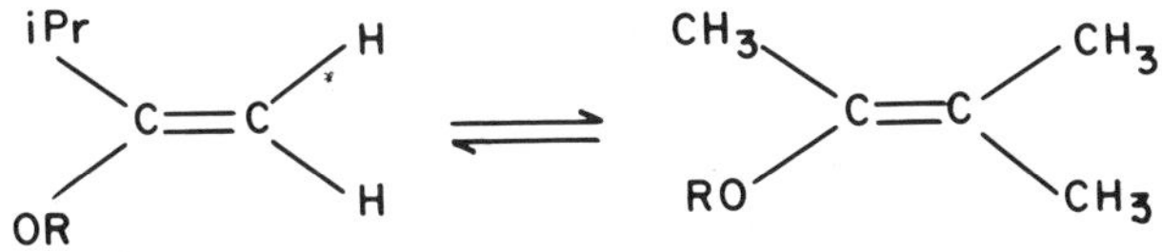

(5-11)

Other topics include the effects of α- and β-substituents,[122,124,128,130,131b,134a,135] the relative ring-strain energies of 3- to 8-membered carbocyclic rings,[123b] 1,2-dialkoxyethylenes,[125b] unsaturated acetals,[131a] alkyl-substituted divinyl ethers,[132] evaluation of some cis interaction energies,[133] and relative stabilities of the isomeric methyl enol ethers of acetonylacetone[134b] and geometrical isomers of 8- and 15-membered 1-methoxycycloalkenes.[126c] The references to other relevant papers on the thermodynamics of vinylethers can be found in the reports of Taskinen and others.[120–139]

Virtanen[140] studied structure–stability relationships in vinyl sulfides by equilibrating their geometrical isomers in tetrahydrofuran in the presence of a catalytic amount of *p*-toluene sulfonic acid. In isomerizations where the double bond migrated from the (α,β) position to the (β,γ) position *t*-BuOK appeared to be a preferable catalyst and DMSO a suitable solvent.

The thermodynamic data allow the estimation of various stabilization energies; for example, the stabilization caused by an ethylthio group at the β carbon of an α-methyl-substituted vinyl sulfide is about 16 kJ/mol when cis to the methyl group **(164)** but only about 5 kJ/mol if there is a methyl group also at the β carbon **(165).** The 3p-2p$_\pi$ conjugation energy in methyl vinyl sulfides has been estimated at about 8 kJ/mol and the 2p-2p$_\pi$ conjugation energy is methyl vinyl ethers at about 27 kJ/mol. The results speak, in general, for a strong participation of the 3d orbitals of the sulfur atom in the conjugation.[140]

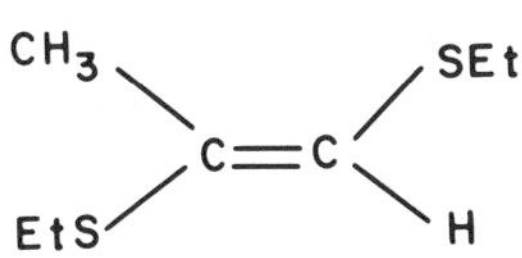

164

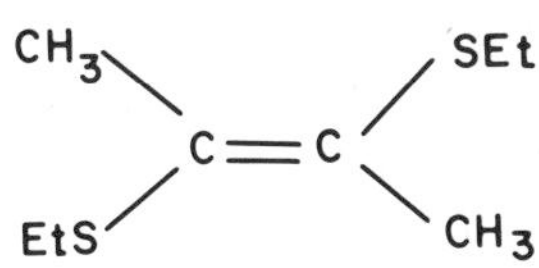

165

Gardner and others have studied the influence of steric and polar factors[141] as well as solvent[142] on the equilibrium position of several cis–trans or *E–Z* isomeric alkenes. The equilibrations were carried out (a) by bromine and sunlight in neat solution at 303 K, (b) in sealed Pyrex tubes in a furnace at 483 K, or (c) in methanol with a trace of NaOMe. When studying the solvent effects, the first method of equilibration was used.

As mentioned before, chemical equilibration is especially useful in the case of epimerizable compounds[5,7,100,105,119] and complements often, for example, the information given by NMR spectroscopy.[143–147] For instance, the equilibration of the cyclic sulfites *r*-2-oxo-*cis*-4,*cis*-6- **(166)** and *r*-2-oxo-*trans*-4,*trans*-6-dimethyl-1,3,2-dioxathianes **(167)** in CCl_4 and CH_3OH using

CF_3COOH as catalyst showed that the axial S=O group is stabilized in respect with the equatorial one by 14.8 and 9.5 kJ/mol at 352 K, respectively.[143] From the *cis*- and *trans*-2-oxo-4,4,6-trimethyl-1,3,2-dioxathianes, the latter was still 6.3 kJ/mol more stable than the former in CCl_4 at 352 K, despite the syn-axial arrangement of the S=O and one of the methyl groups.[143]

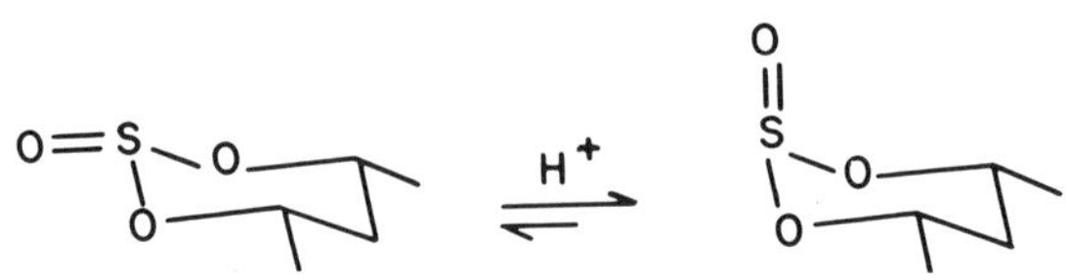

166 **167**

Chemical equilibration of some cis–trans isomeric 2-alkyl-4-methyl-5-oxo-1,3-oxathiolanes in CCl_4 using CF_3COOH as catalyst allowed the determination of the conformational energies of the axial 2-Me, 2-*t*-Bu, and 4-Me groups at 7.6, 9.8, and 1.2 kJ/mol, respectively[144] and the collected data were shown to be in agreement with a ring conformation in which the sulfur atom is situated at the tip of an envelope form[144]:

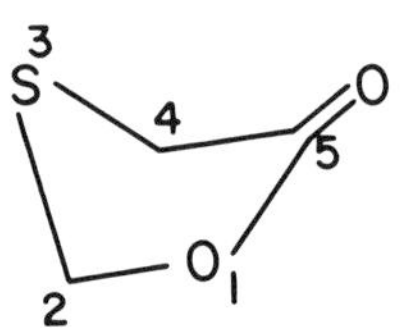

Recently, the equilibration of *cis*- **(168)** and *trans*-2,4,4,6-tetramethyl-1,3-dioxane **(169)** was carried out successfully in Et_2O using equimolar amounts of the substrate (0.1 M) and the catalyst (CF_3COOH) at 298, 313, and 333 K.[148] The gas chromatographic analysis showed that the $-\Delta G^\circ$ value remained practically constant (in other words, $\Delta S^\circ \simeq 0$), 12.9 ± 0.5 kJ/mol, within the limits of experimental accuracy. Taking into account that $-\Delta H^\circ(4a\text{-}CH_3)$ in the cis isomer[102] is equal to 12.2 kJ/mol, we obtain the value 25.1 kJ/mol for the chair-1,4–twist enthalpy difference, $\Delta_{CT}H(1,4)$ in close agreement with another recent estimate.[104] On the other hand it shows that a microcalorimetric determination[149] of $\Delta_{CT}H(1,4)$ is in error by more than 10 kJ/mol.

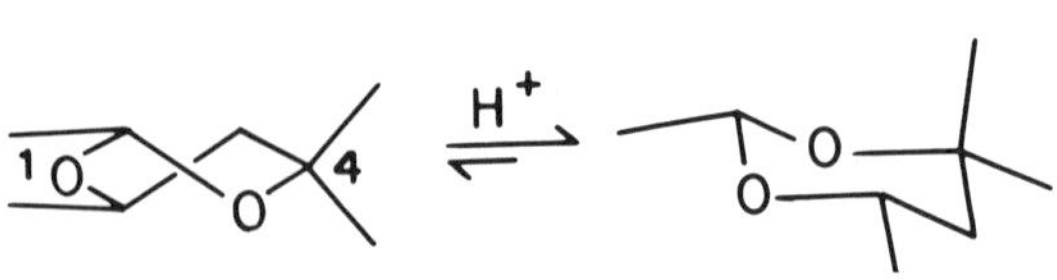

169 **168**

In this context it should be mentioned that in principle microcalorimetry offers a good method for chemical equilibration,[5] since the reactions shown below lead directly to all thermodynamic functions ($\Delta H°$, K, $\Delta G°$, hence $\Delta S°$).

$$\text{isomer I} \xrightarrow[298\text{ K}]{\Delta H_1} \text{equilibrium mixture} \qquad (5\text{-}12)$$

$$\text{isomer II} \xrightarrow[298\text{ K}]{\Delta H_2} \text{equilibrium mixture}$$

where $\Delta H° = \Delta H_1 + \Delta H_2$[146]; $K = \Delta H_2/\Delta H_1$.

In practice, however, the results have not been very promising[146] (see also above), or a rather complicated approach is necessary to avoid various sources of experimental errors.[150] The reader is reminded of Chapter 4 in this volume by Wiberg, which includes more detailed description of this method.

Sometimes chemical equilibration can be used to complement kinetic studies as shown by Lajunen and co-workers with *exo*- and *endo*-5-acetyl-2-norbornenes[151] and 5-methyl-2-norbornenols and 5-methylene-2-norborneols.[152] These equilibrations were carried out in NaOMe/MeOH and in aqueous perchloric acids (0.1 and 1.0 mol/dm^3), respectively.

Sometimes several conformers take part in the state of equilibrium. For instance, *r*-2-methyl-2-*t*-butyl, *trans*-4-methyl-1,3-dithiane **(170)** is a mixture of a chair form (C_{2a}) and a 2,5-twist form (T_2), whereas the *r*-2-methyl-2-*t*-butyl, *cis*-4-methyl derivative **(171)** attains exclusively the 2,5-twist form (T_1). Both twist forms are practically equally stable. Hence

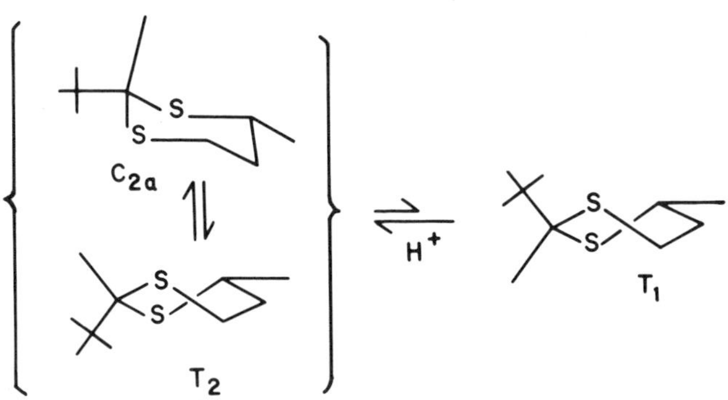

170 **171**

$$K = \frac{T_2 + C_{2a}}{T_1} = 1 + \frac{C_{2a}}{T_1} \qquad (5\text{-}14)$$

If we take into account that $C_{2a}/T_1 = K_T = K - 1$, the results of equilibration [K 7.35 (338.15 K), 5.88 (358.15 K), and 4.83 (378.15 K)] let us estimate $\Delta H° = -13.5 \pm 0.2$ kJ/mol and $\Delta S° = -24.4 \pm 0.4$ J/mol K for the equilibrium between C_{2a} and T_1.[153] Since[154] $\Delta H°(2a - CH_3) \simeq \Delta G°(2a - CH_3) = 8.0$ kJ/mol, we obtain further that $\Delta_{CT}H°(2,5) = 21.5$ kJ/mol and

$\Delta_{CT}S^{\circ}$ = 24.4 J/mol K, in reasonable agreement with previous estimates.[154,155] In general, isomeric alkyl-substituted 1,3-dithianes offer several examples of equilibria that must be treated with the aid of more than one conformation on one or both sides of the epimer equilibrium.[153]

Saturated hydrocarbons can be equilibrated over palladium in the liquid phase at elevated temperatures as shown by Allinger and Pamphilis with *cis*- and *trans*-1,2-dimethylcycloheptanes.[156] They equilibrated also some six- and seven-membered cyclic alkenes in boiling pentane in the presence of a catalytic amount of concentrated sulfuric acid.[156] Pasto and Rao[157] studied the equilibria between four stereoisomeric 2,5-di-*t*-butylcyclohexanols over Raney nickel in cyclohexane at temperatures between 351 and 531 K.

Beak and co-workers[158] used the equilibration of alkyltropic isomers by the catalytic function of the common alkylated derivative to obtain quantitative information about the imidate–amide (*A*), pyridine–pyridone (*A*), and thiopyridine–thiopyridone (*B*) equilibria:

$$\mathrm{CH_3{-}X{-}Y{=}Z} \xrightleftharpoons{[\mathrm{CH_3{-}X \cdots Y \cdots Z{-}CH_3}]^+} \mathrm{X{=}Y{-}Z{-}CH_3}$$

A: X = O, Y = C, Z = NR
B: X = S, Y = C, Z = NR (5-15)

Blackburne and colleagues[159] determined the cis–trans equilibria of some 4-phospha analogues of 2,6-dimethylcyclohexanone at 310 K in 1,4-dioxane in which D_2O including 6% of Na was added twice (at the beginning and after 6h). The analysis of the equilibrium mixtures was carried out by [^{1}H] NMR spectroscopy and showed that in many cases there were more than 1 + 1 chair forms present.

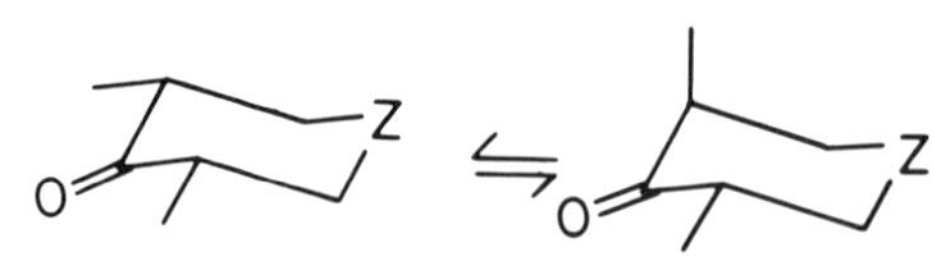

Z = PPh, P(O)Ph, P(S)Ph, P(Se)Ph (5-16)

Stoddart and his group have equilibrated diacetals[160] by acid and γ-butyrolactones[161] by base to determine the relative stabilities of 1,3:2,4- and 1,4:2,3-diacetals and configurational equilibria of 2,4-disubstituted γ-butyrolactones, respectively.

7. MASS SPECTROMETRY

Ionization and appearance energies (also called ionization and appearance potentials) can give thermochemical information about the interactions present in the molecular ground states when comparing their experimental values for positional or stereoisomers with each other. The isomeric compounds (M and M_1) must then form a common or at least an energetically similar ion in order to make the difference in the enthalpies of formation equal to the difference in the appearance energies, assuming further that the excess energy terms for the two isomers are equal:

$$\Delta_f H°(\mathrm{M}) - \Delta_f H°(\mathrm{M_1}) = AE(\mathrm{M_1}) - AE(\mathrm{M}) \quad (5\text{-}17)$$

Both the applicability and the limitations of the appearance energy approach and Equation 5-17 in solving stereochemical and other structural problems have been frequently and thoroughly reviewed,[162–164] hence only one example is presented here.

The appearance energy of the primary $[M-SO_2]^+$ ion is higher for **173** than for **172,** in agreement with its lower enthalpy of formation, and the difference of 0.125 ± 0.022 eV (= 12.1 ± 2.1 kJ/mol) between the appearance energies[165] leads to a good estimate of the gas phase stabilization of the axial S=O group, since **172**a and **172**b are practically equal in stability.[143] The result above is close to the one obtained in CCl_4 by chemical equilibration (p. 204).[143,166] Similarly, the appearance energies of the primary $[M-HSO_3]^+$ fragment ions in the case of *cis*- and *trans*-2-oxo-4,4,6-trimethyl-1,3,2-dioxathianes differ by 6.9 ± 2.0 kJ/mol (the latter is stabilized), in agreement with the result of chemical equilibration.[143]

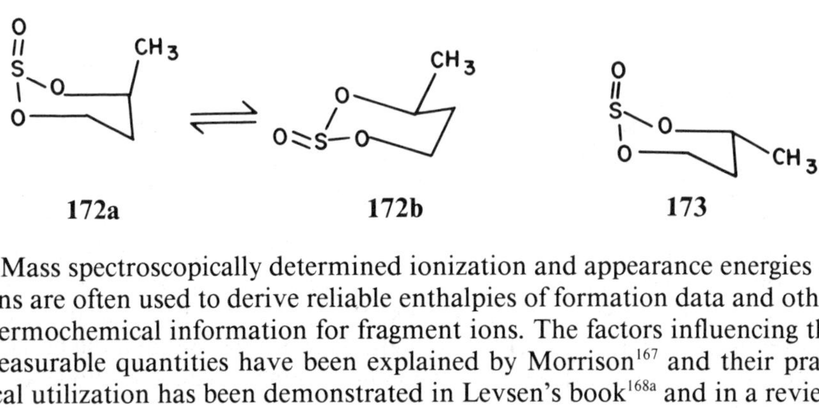

172a **172b** **173**

Mass spectroscopically determined ionization and appearance energies of ions are often used to derive reliable enthalpies of formation data and other thermochemical information for fragment ions. The factors influencing the measurable quantities have been explained by Morrison[167] and their practical utilization has been demonstrated in Levsen's book[168a] and in a review by Maccoll.[168b] Hence the discussion can be limited again to some illustrative applications.

Franklin and Sen Sharma[169] used the Warren method[170] to determine appearance energies by electron impact for the reaction:

$$CH_3R + e^- \rightarrow CH_3^+ + R\cdot + 2e^- \quad (5\text{-}18)$$

to derive enthalpies of formation of several free radicals. The approach included a correction for the excess energy (E^*), which was calculated from the translational energy in the center of mass, $\bar{\epsilon}_t$, by means of the equation:

$$E^* = \alpha N \bar{\epsilon}_t \quad (5\text{-}19)$$

where N is the number of vibrational degrees of freedom and α an empirical constant ($\simeq$0.44). When employing a time-of-flight instrument, this method has given satisfactory values for the enthalpies of formation of several ions and free radicals. The use of a modified sector-field mass spectrometer[169] led to acceptable values for most of the hydrocarbon radicals studied, whereas the value for $\Delta_f H(NH_2)$ seemed to be erroneously low. Similarly, the enthalpies of formation of ethyl, diethyl, and triethyl amines were not very satisfactory. Obviously the correction based on Equation 5-19 could not be made very accurately in addition to other sources of error in the appearance energies.[162–164,168] The Warren method is, however, clearly less accurate[171b] than, for example, the semilog plot method described by Lossing and associates[171c] or some other newer approaches.[168]

In this context we emphasize that in their recent paper[171d] McMillen and Golden discussed and reviewed thoroughly hydrocarbon bond dissociation energies for the determination of which the enthalpies of formation of free radicals are also crucial.

Conrath and co-workers[172] studied the structures of ions (F^+) generated from acyclic, 1,3-diketones (M) by loss of CO, CH_2CO, or alkenes (N) by comparing the enthalpies of formation of these ions [$\Delta_f H(F^+)$] with data for ions of known structure generated from other precursors:

$$M + e^- \rightarrow F^{+\cdot} + N + 2e^- \qquad (5\text{-}20)$$
$$\Delta_f H[F^{+\cdot}] = \Delta_f H(M) + AE(F^{+\cdot}) - \Delta_f H(N) - \epsilon$$

The ejected neutrals were assumed to be formed in their electronic ground state [$\Delta_f H(N)$] and the excess energy term (ϵ) was approximated by the simplified rate expression.[171,173] The appearance (AE) and ionization (IE) energies were determined by the Warren method,[170] and the enthalpies of formation [$\Delta_f H(M)$] of 1,3-diketones were found via combustion calorimetry. The inspection of the thermochemistry connected with Equation 5-20 showed that the loss of CO from the molecular ion of acyclic 1,3-diketones is accompanied by migration of an alkyl group to the C_2 atom, thus yielding keto-type ions **(174a)** or enol-type ions **(174b)** as stated by Conrath.[172] The elimination of a ketene moiety also produces enol-type ions and the possible expulsion of an alkene 1,3-diketo ion **(174c)** which can also be enolized.

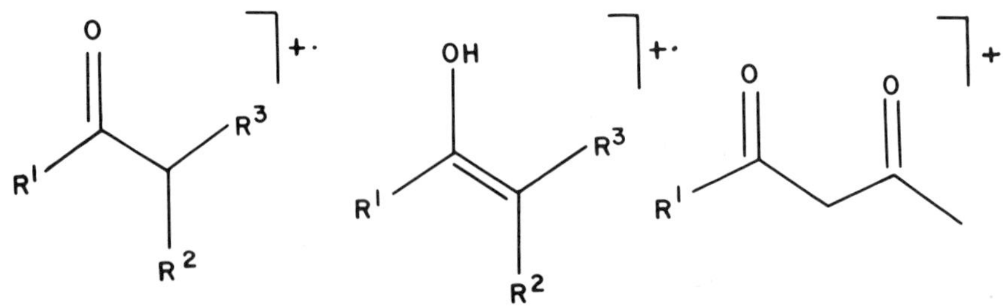

174a **174b** **174c**

The appearance and ionization energy measurements let us also calculate the enthalpies of formation of stable ions by correlating them with thermochemical data.[168,174] In this way it can be pointed out that, for example, **175** and **176** as well as **178** and **179** form common ions **177** and **180,** the enthalpies of which are about 1315 and 1350 kJ/mol, respectively.[174] In other words, electron-impact-induced fragmentation of aromatic and isomeric acetylenic compounds gives open-chain as well as cyclic ion structures.

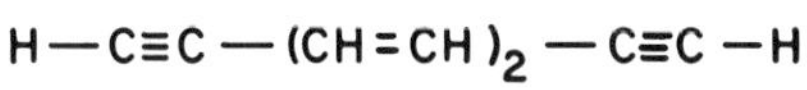

175

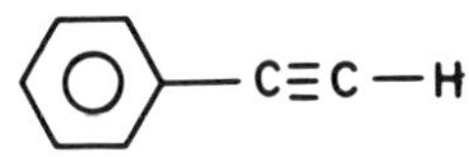

176

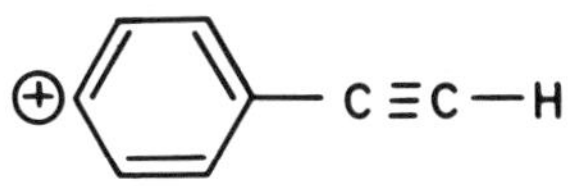

177

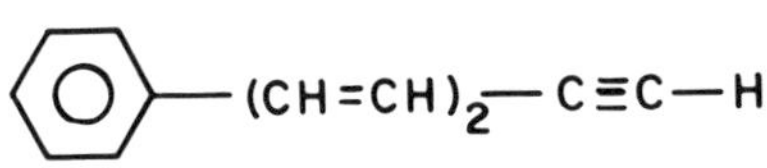

178

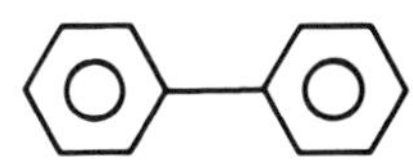

179

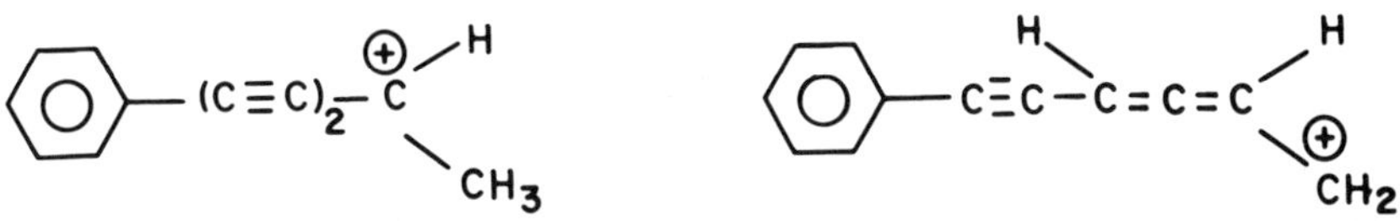

180

One of the pioneers in the use of ionization and appearance energies in thermochemical and structural studies is F.P. Lossing, who has developed a method for measuring these energies based on impact of an energy-resolved electron beam from an electrostatic electron monochromator.[175] For instance, some time ago Lossing and others[175b,176] determined the ionic enthalpies of formation of C_1–C_7 alkyl radicals by measuring the ionization potentials for the free radicals with a suggested uncertainty of ± 12 kJ/mol. The hydride ion affinities $D(R^+ - H^-)$ were found to be almost constant for each of the primary, secondary, and tertiary ions up from and at about C_4 ($\simeq$ 1100, 1035, and 970 kJ/mol, respectively).[176]

Holmes and Lossing[177] reported the gas phase enthalpies of formation of 13 enol-positive ions (Equation 5-21) of aliphatic aldehydes, ketones, acids, and esters and compared them with those of the corresponding keto ions. The former are more stable[177–180] by 58–130 kJ/mol, which is in contrast to the neutral tautomers, where the keto forms are in general the more stable.[180,181] The results of Holmes and Lossing[178] prove also that the changes in ionic enthalpies of formation resulting from methyl substitution on C and O atoms illustrate the charge distribution on these atoms, as was proposed earlier.[182]

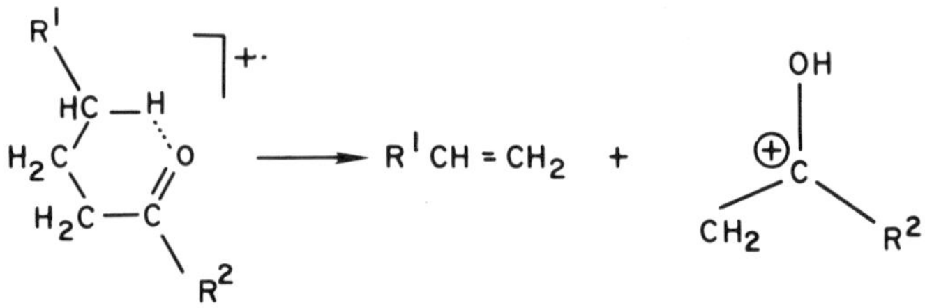

(5-21)

Several other reports[183–186] consider the mechanisms and energetics of the keto–enol tautomerism in the gas phase. Especially interesting thermochemically are the enthalpies of formation of the ionic and neutral enols of acetaldehyde and acetone (757 vs −125 and 661 vs 159 kJ/mol, respectively).[187]

The energetics and ion chemistry of isomeric $[C_2H_3O]^+$ ions,[188–189] $[C_2H_4O]^+$ ions,[190–195] $[C_4H_4O]^+$ ions,[191] $C_2H_6O^+$ ions,[192] and $[C_7H_7O]^+$ ions[193] have been discussed both from experimental and theoretical points of view. Furthermore, the studies on alkane elimination from ionized alkanols[194] and fragmentation of stereoisomeric methylcyclohexanols[195] have been shown to benefit from a thermochemical approach.

Herman and Harrison[196] used chemical ionization (CI) mass spectrometry very elegantly together with thermochemical data to show that a series of formate (methyl through butyl) and acetate (methyl through pentyl) esters protonated in the gas phase by the Brønsted acids H_3^+, N_2H^+, CO_2H^+, N_2OH^+, and HCO^+ fragment via the following decomposition modes:

$$R'-\overset{\overset{\displaystyle OH}{|}}{C}-\overset{+}{O}R \rightarrow R'-C{\equiv}O^+ + HOR \qquad (5\text{-}22a)$$

$$\rightarrow R'-\overset{\overset{\displaystyle OH}{|}}{C}-\overset{+}{O}H + RH \qquad (5\text{-}22b)$$

$$\rightarrow R'-\overset{\overset{\displaystyle O}{\|}}{C}-OH + R^+ \qquad (5\text{-}22c)$$

The route shown in reaction 5-22a is unimportant for formates.

In addition to the reports cited earlier, there are several mass spectrometric investigations dealing mainly or partially with thermochemistry of isomeric $C_3H_6^{+\cdot}$ ions,[197] $C_6H_{10}^{+\cdot}$ ions,[198] alkane molecular ions,[199] gaseous alkyne ions,[200] $C_6H_{12}^{+\cdot}$ ions,[201] and different butyl ions.[202,203] Furthermore there are some reports on the enthalpies of formation and thermochemistry of α-aminoalkyl radicals,[204,205] stable $C_2H_5X^+$ (X = Cl, Br) radical cations of structure

[$CH_3CHXH^{+\cdot}$],[206] stable α-arylvinyl cations,[207] and organometallic compounds,[208] although it is not possible to illustrate them all in detail here.

Finally a paper of Büchler and Vogt[209] reports the electron impact mass spectra of cyanoacetylenes X−C≡C−C≡N (X = H, F, Cl, Br, I), together with ionization and appearance energies from which the enthalpies of formation and the C−X and C≡C bond energies are derived and compared with reference data.

8. NUCLEAR MAGNETIC RESONANCE SPECTROSCOPY

NMR spectroscopy offers another versatile tool for collecting thermochemical information, although it is not a proper thermochemical method. In addition to dynamic NMR,[210−213] vicinal coupling constants, chemical shifts, and signal intensities can be used for solving stereochemical and structural problems in the context of thermochemical studies, or vice versa.

A. Vicinal Coupling Constants

Vicinal coupling constants are especially suited for conformational analysis of systems in which the limiting structures and at the same time model values for the coupling constants can be reliably defined.[5,143,145,146,212,214−220] The model values are usually taken from anancomeric model compounds or observed directly at low temperatures, where the distinct conformers may freeze out. For example, the conformational equilibrium constant for 5-methyl-1,3-oxathiane **(181)** can be estimated from:

$$K = \frac{\Sigma J_{\text{obs}} - \Sigma J\,(\mathbf{181}\text{A})}{\Sigma J\,(\mathbf{181}\text{E}) - \Sigma J_{\text{obs}}} \tag{5-23}$$

where $\Sigma J\,(\mathbf{181}\text{E}) = 30.0$ Hz, $\Sigma J\,(\mathbf{181}\text{A}) = 10.7$ Hz, and $\Sigma J_{\text{obs}} = 26.45$ Hz.[214] Hence $K = 4.4$ and $-\Delta G° = 3.7$ kJ/mol at 298 K. If the denominator in Equation 5-23 becomes very small, a more accurate value for a conformational equilibrium constant can be obtained by using the values of the vicinal coupling constants (eg, J_{aa} and J_{ee} in the case of six-membered rings) separately as model values.[143,145,214,220,221] The values of the vicinal coupling constants for a conformer mixture can also be solved at several temperatures and then used for calculating $K = f(T)$, hence also $\Delta H°$ and $\Delta S°$.[222,223]

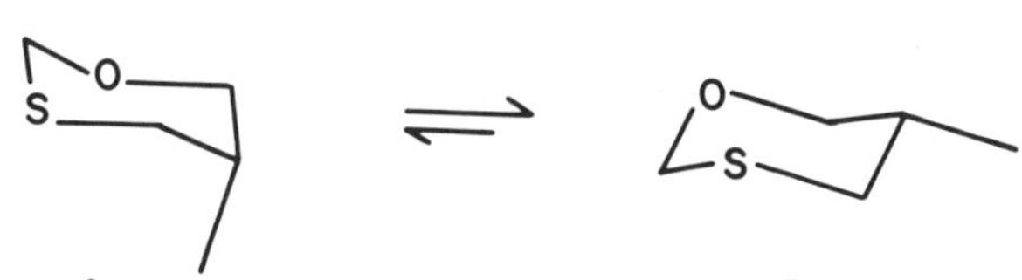

A E

181

Another way to utilize vicinal coupling constants in structural analysis is the application of Karplus equation,[215] often even in its simplest form ($J = A \cos^2\theta$) to solve rotamer populations, hence also stability differences.[216,217,224–226]

Booth and Lemieux[227] showed that the N—H axial conformation is the dominant or sole conformation for tetrahydro-1,3-oxazines (**182** and **183**) and 1-methyltetrahydro-1,3-diazine (**184**) by measuring the CH—NH coupling constants at low temperatures.

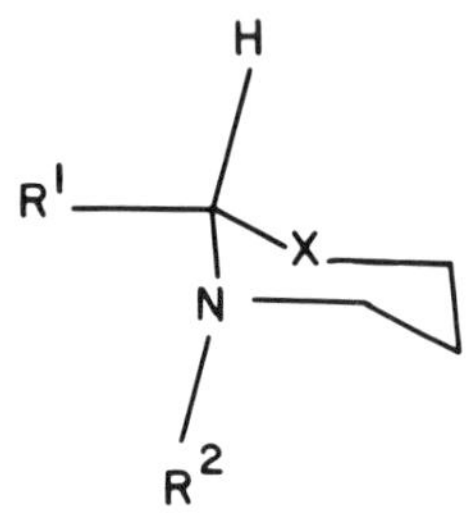

	R	R	X
182	H	H	O
183	CH_3	H	O
184	H	CH_3	NH

In addition to the $^1H-^1H$ vicinal coupling constants, the vicinal $^1H-^{19}F$,[228] $^{13}C-^{19}F$,[228] $^1H-^{13}C$,[229] and $^1H-^{31}P$[220] couplings can be applied to solve conformer populations, hence molecular energetics. Sometimes even the signal width, which is a function of the different coupling constants in question, can be used for these purposes.[230]

B. Chemical Shifts

Chemical shifts measured by ^{13}C NMR spectroscopy can be especially useful in determining conformer populations for different compounds using the approach originally presented by Dalling and Grant[231] as extended later by Pihlaja and others.[232–238]

The general approach by the foregoing and many other authors[239–243] includes the calculation of substituent effects by a multilinear regression analysis from the equation:

$$\delta C(x) = \delta_p C(x) + \Sigma n_x^y SE(x) \qquad (5\text{-}24)$$

where $\delta C(x)$ is the chemical shift of a given carbon atom in a substituted derivative, $\delta_p C(x)$ is that of the same carbon atom in the parent compound, and n_x^y are the numbers of parameters $SE(x)$ caused by the substitution y and acting at the carbon x.

When calculating substituent effects of cycloalkanes and heterocyclanes by Equation 5-24, the basic principle is that all derivatives included in the calculations have the same ring conformation (eg, the chair form). The conformational heterogeneity of the latter can be taken into account by entering two sets of independent parameters, each weighed by the pertinent mole fractions.[232,234,236–238,244]

The method has been proved to suit very well for solving conformational interactions within a certain family of compounds using ^{13}C NMR chemical shifts determined at 298 K only. Obviously the substituent effects do not depend on temperature, although the total chemical shifts may vary with it.[245]

As an example, the ^{13}C chemical shift parametrization of methyl-substituted tetrahydro-1,3-oxazines allows a complete conformational analysis and leads to the following conformational energies $\Delta G°$ (kJ/mol): 2*a*-CH_3 16.0, 4*a*-CH_3 10.9, 5*a*-CH_3 5.4, and 6*a*-CH_3 12.3.[246]

Although their smaller range and the lack of accurate model values makes them less useful than ^{13}C shifts, ^{1}H NMR chemical shifts can also be used.[143,147,212,219]

C. Signal Intensities

Signal intensities determined by integration—or sometimes, if the signals are very sharp, simply by peak heights—can often be used for calculating different equilibrium ratios between structural isomers. If these measurements are carried out at several temperatures, they do allow the determination of $\Delta H°$ and $\Delta S°$ in addition to $\Delta G°$ for the equilibrium in question. Since this approach has been applied very often, only a few descriptive examples are given.

Pihlaja and Aaljoki[247] studied the ring–chain tautomerism of alkyl-substituted 1,3-oxazolidines (Equation 5-25) and determined the values of the equilibrium constants at different temperatures with the aid of the intensities of suitable ^{1}H NMR signals.

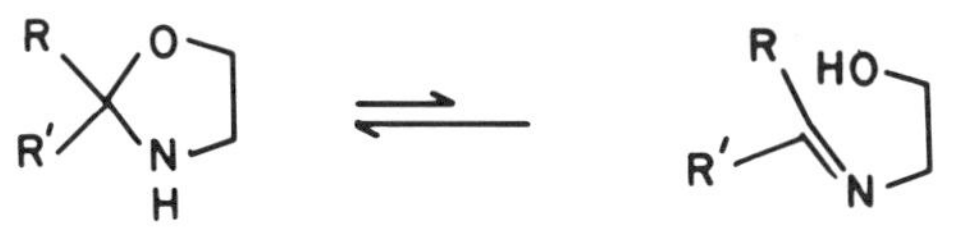

(5-25)

Clark and Parker[248] carried out a thermodynamic study of the tautomerization (Equation 5-26) of 2-(*N*-cyclohexylimino)-1,3-diphenylpropane in DMSO-d_6 also by a variable-temperature ^{1}H NMR study over a temperature range of 310–411 K:

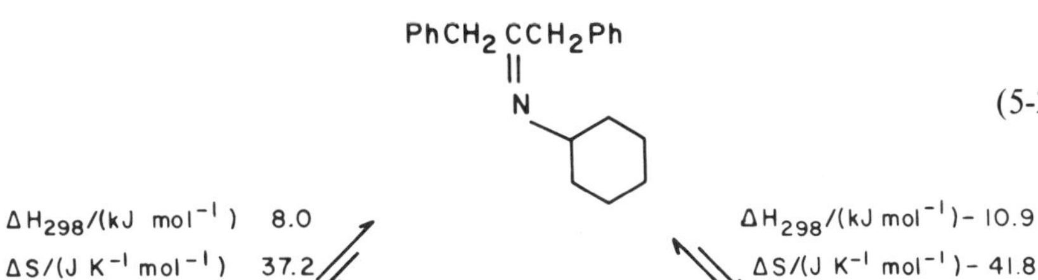

(5-26)

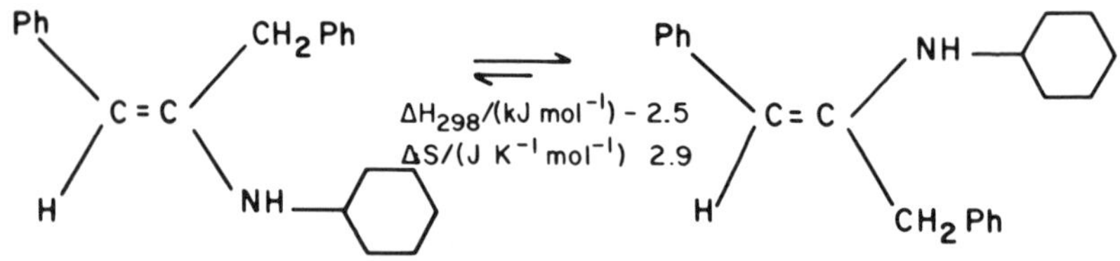

Several authors[241,249,250] have used low-temperature ^{13}C NMR spectra to compute conformational equilibria in cyclohexanes and saturated heterocycles by integration or by measuring the heights of the well-separated signals due to corresponding carbon atoms in the two conformations. It should be reemphasized that the ^{13}C NMR chemical shifts offer a beneficial opportunity to solve conformational equilibria at 298 K without corrections for temperature effects that should be taken into account if one intends to extrapolate the low-temperature data to 298 K.[234,250]

Although tetrahydro-1,3-oxazines usually exist almost exclusively in the ring form, they can exhibit a ring–chain tautomerism similar to that of 1,3-oxazolidines (Equation 5-25). Even though both forms—if present in reasonable amounts—can be seen separately in an NMR experiment,[245] the equilibrium is established fairly fast. Since we know that the amount of the open-chain form remains almost negligible[246,251] in the case of alkyl-substituted tetrahydro-1,3-oxazines, we concluded that ^{13}C NMR signal intensities can be used to determine the epimer equilibria.

In another study,[246] a multilinear correlation of ^{13}C chemical shifts with the substitution patterns was applied to carry out a conformational analysis of methyl tetrahydro-1,3-oxazines at 298 K. The tabulation that follows compares three $\Delta G°$ values of the epimeric equilibria (Equation 5-27)[251] with

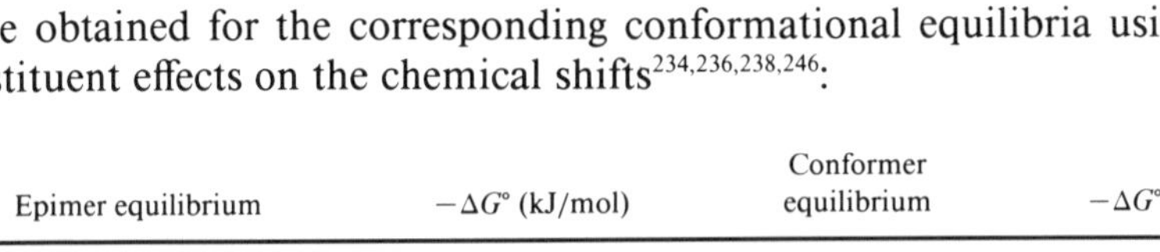

(5-27)

those obtained for the corresponding conformational equilibria using the substituent effects on the chemical shifts[234,236,238,246]:

Epimer equilibrium	$-\Delta G°$ (kJ/mol)	Conformer equilibrium	$-\Delta G°$ (kJ/mol)
cis-2,5-Me_2 ⇌ *trans*-2,5-Me_2	5.2 ± 0.2	5a-Me ⇌ 5e-Me	5.4 ± 0.1
r-2,*c*-4,-*t*-6-Me_3 ⇌ *r*-2,*t*-4,*c*-6-Me_3	1.2 ± 0.1	4e6a-Me_2 ⇌ 4a6e-Me_2	1.4 ± 0.4
r-2,*t*-4,*t*-5-Me_3 ⇌ *r*-2,*c*-4,*c*-5-Me_3	5.4 ± 0.1	4a5e-Me_2 ⇌ 4a5a-Me_2	5.0 ± 0.2

It is easy to see that the results are in excellent agreement with each other. In the case of epimer equilibria (Equation 5-27), the experiments are very easy to repeat at various temperatures to determine ΔH° and ΔS° as well. However in the case of methyl tetrahydro-1,3-oxazines we did not, in general, find palpable entropy differences.[251]

D. Dynamic Nuclear Magnetic Resonance Spectroscopy

Since the scope of dynamic NMR is very wide,[210–213] we can present here only few recent examples about its use in structural analysis and thermochemistry. The method seems to be used most often to obtain information about the energetics of rotational isomerism,[252–269] ring reversal (and nitrogen inversion) in six-membered rings,[270–275] and conformational motion in medium-sized rings.[276–280] Dynamic NMR can, however, be useful also when studying organic solids[281] and inorganic complexes.[282]

a. Rotational Isomerism

T. Schaefer and co-workers[252–255] have determined barriers to the internal rotation of the dichloromethyl group in several polyhalosubstituted (at least at the 2 and 6 positions) dichloromethylbenzene and related compounds.

For example, for **185** and **186** a complete line shape analysis of the ^{1}H spectra gave the following values for ΔH^*, ΔS^*, and ΔG^{*298} of the internal rotation in toluene-d_8[252]:

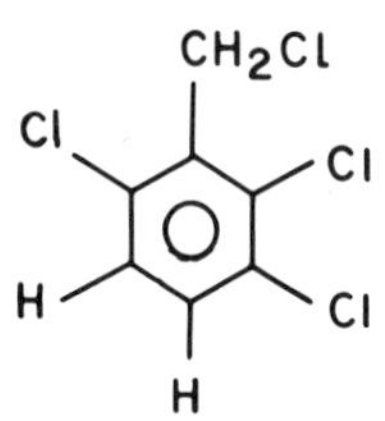

185

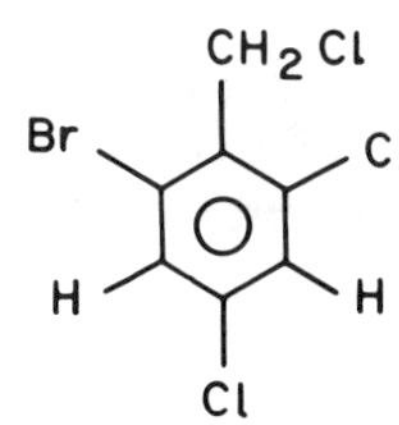

186

	185	186
$\Delta H^*/(\text{kJ mol}^{-1})$	55.6 ± 1.3	61.1 ± 1.3
$\Delta S^*/(\text{JK}^{-1}\text{ mol}^{-1})$	− 29.3 ± 2.9	−4.6 ± 2.9
$\Delta G^*/(\text{kJ mol}^{-1})$	64.4 ± 0.4	62.5 ± 0.4

Olah and colleagues[258] found that several 9,9′-bifluorenyls **(187)** exhibit hindered motion around the central C—C bond and determined the energy barriers for the interconversion of the equivalent gauche conformers in which the molecules freeze out at low temperatures by complete line shape analysis of the ^{1}H NMR spectra at various temperatures with the results shown in Table 5-6. In the case of **187f,** restricted rotation about the axial bond of the *t*-butyl group with a ΔG^* of 60.2 kJ/mol was also observed. It should be noted that the barrier for **187b** is exceptionally low and the authors were not able to give a unequivocal reason for it.[258]

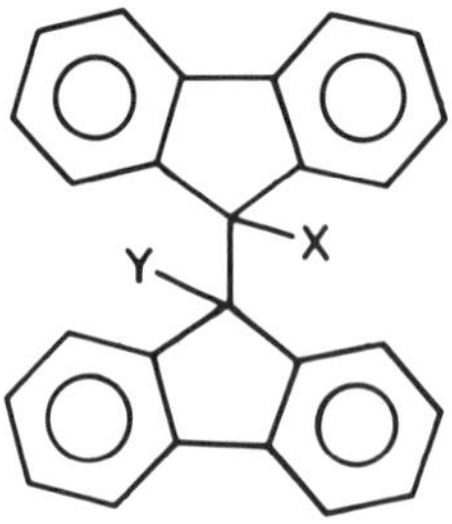

A $X = Y = H$
B $X = Y = CH_3$
C $X = Y = Cl$
D $X = Y = Br$
E $X = t - Bu, Y = H$

187

TABLE 5-6. Energetics for the Interconversion of Bifluorenyls[a]

Compound	ΔH^* (kJ/mol)	ΔS^* (J/mol K)	ΔG^*_{240} (kJ/mol)
187a	31.4 ± 2.1	-45.2 ± 4.2	42.2
187b	26.4 ± 2.9	-64.0 ± 8.4	41.8
187c	51.0 ± 2.1	-20.1 ± 8.4	55.8
187d	51.0 ± 4.2	-20.9 ± 8.4	56.0
187e	39.3 ± 2.1	-32.2 ± 2.1	47.0

[a]From Reference 258.

Evans and Miller[260] used ^1H, ^{13}C, and ^{15}N NMR spectroscopy at low temperatures to study conformations and dynamics of carcinogenic *N*-substituted 2-aminofluorene compounds. They found that **188** and **189** consist of 83 and 23% of the cis form, respectively, and determined barriers to internal rotation about the amide bond at 60.2 and 57.3 kJ/mol, respectively.

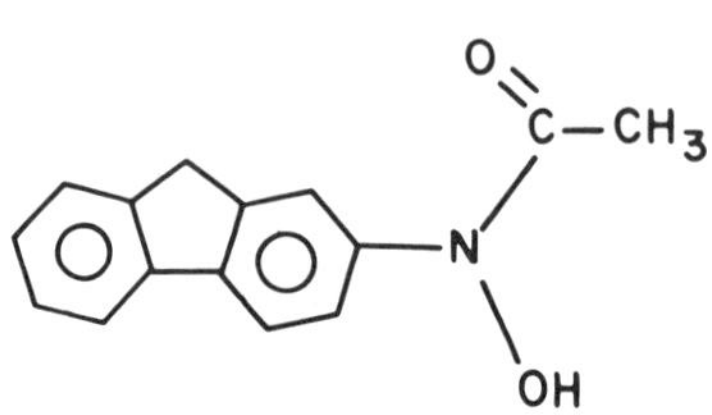

188

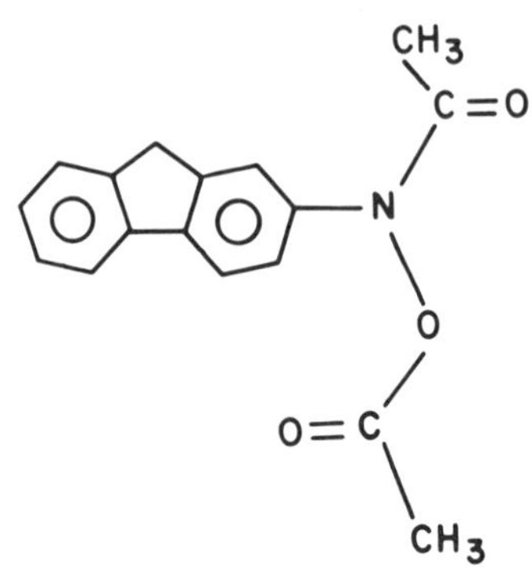

189

Lunazzi and associates[261] pointed out that in a crowded amine such as 3-pentyl diisopropylamine, the barrier to rotation becomes higher than that to *N*-inversion ($\Delta G^* < 25.1$ kJ/mol for the *N*-inversion but $= 38.5 \pm 0.4$ kJ/mol for the rotation).

TABLE 5-7. Equilibrium Rotamer Populations and Gibbs Energies of Activation[a] to Hindered Rotation of the N–N Bonds in **190–195**

		Gibbs energies (kcal/mol)		
Compound	E(%)/Z(%)	ΔG^{*}_{ZE}	ΔG^{*}_{EZ}	ΔG^{*}_{av}
190	50/50	93.3	93.3	93.3
191	50/50	84.9	84.9	84.9
192	62/38	77.4	79.1	78.2
193	81/19	94.6	100.0	97.1
194	80/20	82.8	87.9	85.4
195	92/8	76.5	84.1	80.3

[a]Calculated from the coalescence temperatures using equations developed by Shanin-Atidi and Bar-Eli.[267b]

Dürst,[262] Boyd,[263] and their co-workers studied the dynamic stereochemistry of *N,N*-dialkylbenzamides and imines and their derivatives, respectively.

Several reports apply dynamic ^{1}H, ^{13}C, and ^{19}F NMR together with force field calculations to study dynamic gearing in ethanes with many alkyl substituents,[264] barriers to internal rotation in neopentylbenzenes,[265,266] and intramolecular rotations in *syn*- and *anti-o*-tolyl-di-*t*-butyl carbinols.[267]

Haky and co-workers[267a] determined by ^{13}C NMR the isomer populations and N–N rotational barriers in five-membered cyclic nitrosamines **190–195** (Table 5-7).

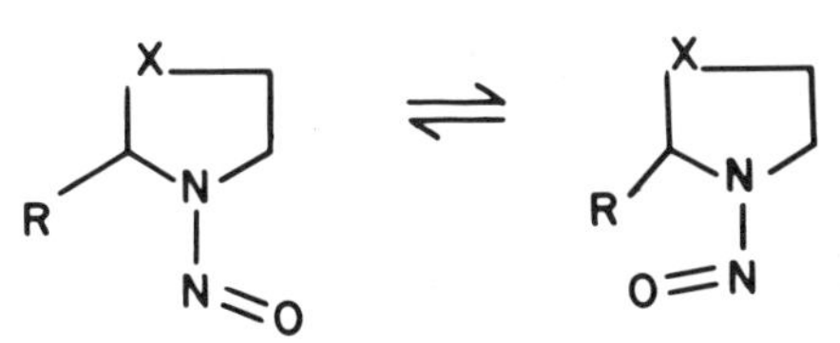

E Z

	X	R
190	CH_2	H
191	S	H
192	O	H
193	CH_2	CH_3
194	S	CH_3
195	O	CH_3

Finally, Yamamoto and Ōki[268] have revealed by an ^{1}H dynamic NMR study that 9-(2-alkylphenoxy)-1,4-dimethyltriptycenes (alkyl = methyl or isopropyl) show two conformers (ap and $\pm$sc) at room temperature and that the energy barrier for gear rotation between them is about 76 kJ/mol.

b. Ring Reversal and Nitrogen Inversion

Doddrell and colleagues studied the chair–chair interconversion of 1,1,3,3-tetramethylcyclohexane **(196)** and its 2,2- **(197)** and 5,5-difluoro **(198)** derivatives and found by the comparison of the activation parameters deter-

TABLE 5-8. Thermodynamic Functions of Activation (kJ/mol) for the Cyclohexanes **196–198**

Compound	ΔH^*	ΔS^*	ΔG^* at T_c
196	26.8 ± 0.8	−55.2 ± 4.2	37.5 ± 0.4 (193 K)
197	41.0 ± 0.4	−4.2 ± 2.1	42.0 ± 0.4 (235 K)
198	24.3 ± 0.8	−37.7 ± 4.2	30.8 ± 0.4 (173 K)

mined from the variable-temperature ^{13}C NMR spectra of **196** in CS_2 with those determined from a variable-temperature ^{19}F NMR study of **197** and **198** in $(C_2H_5)_2O$ that the introduction of the *gem*-difluoro group causes significant steric interactions (cf Table 5-8).

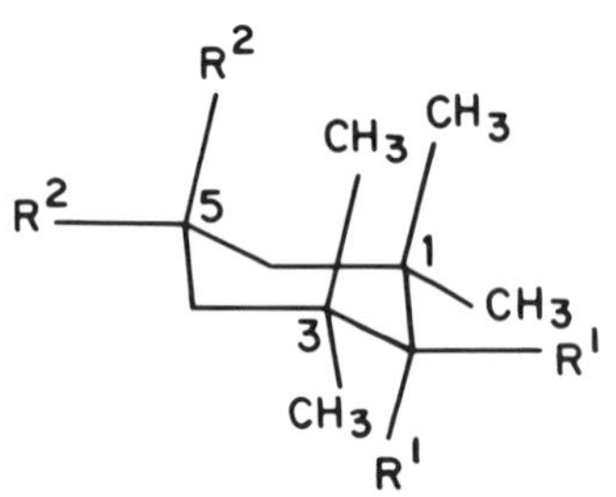

196 $R^1 = R^2 = H$
197 $R^1 = F, R^2 = H$
198 $R^1 = H, R^2 = F$

A very interesting report[270a] is that of Poupko and Luz, who studied ring inversion of cyclohexane-d_{12} in liquid crystalline solvents. The kinetic parameters were derived by line shape theory for two interacting nuclei of spin $I = 1$ including dipolar and quadrupolar interactions. Nevertheless the values of the parameters obtained for the ring inversion ($\Delta H^* = 43.1$ kJ/mol and $\Delta S^* = 2.1$ J/mol K) are close to those obtained in solvent CS_2.[270b]

Anderson and associates[271] found that barriers to the ring inversion of CH_2-X-CH_2-bridged perinaphthalenes increase along the series O < S < Se < Te (<26.4, 31.0, 31.8, and 33.5 kJ/mol, respectively), which is in marked contrast to the situation in heterocyclohexanes (O > S > Se > Te: 43.1, 39.3, 34.3, and 30.5 kJ/mol, respectively).

An elegant paper of Ogawa and colleagues[273] discusses the barriers to rotation and inversion in *meso*-1,1′-bi(2-methylpiperidine)s **(199–201)** in the light of ^{13}C dynamic NMR and molecular mechanics calculations of the corresponding hydrocarbons (N ⇒ CH). The most stable conformers of the bipiperidines are enantiomeric and have the lone pairs of the nitrogen atoms nearly gauche to each other. In these conformations the N−N bond and all alkyl groups are equatorial to each of the piperidine chairs. The barriers (ΔG^*) to the interconversion between these enantiomeric gauche conformations in **199–201** are 52.3 (248 K), 74.1 (345 K), and 79.5 (370 K) kJ/mol, respectively. The barrier for **199** was assigned to the passing inversion of the N atoms,[273] whereas those for **200** and **201** were assigned to the single-passing rotation about the N−N bond.[273] At lower temperatures the signals of **199** split into two ($\simeq$ 85:15), and this process ($\Delta G^* = 38.5$ kJ/mol) was attributed to the ring inversion.[273]

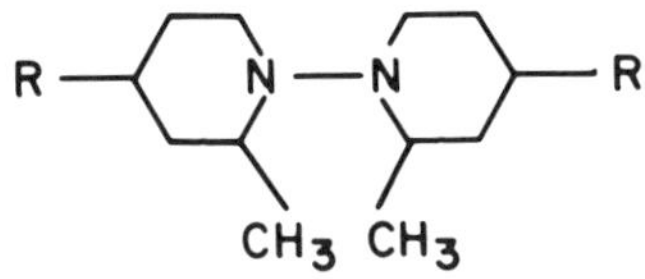

199 R = H
200 R = CH_3
201 R = t − Bu

Riddell and Turner[274] determined the barrier to nitrogen inversion at position 2 in *N,N′*-dimethyltetrahydro-1,2,5-oxadiazine (**202;** $\Delta H^* = 60.2 \pm 0.4$ kJ/mol, $\Delta S^* = -5.0 \pm 1.7$ J/mol K) in toluene-d_5 solvent, and it compares favorably with that in 2-methyltetrahydro-1,2-oxazine (**203;** $\Delta H^* = 63.2 \pm 1.7$ kJ/mol, $\Delta S^* = 9.6 \pm 6.3$ J/mol K) in $CDCl_3$.

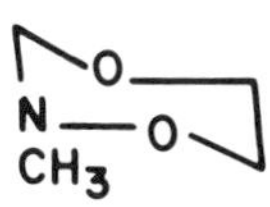

202

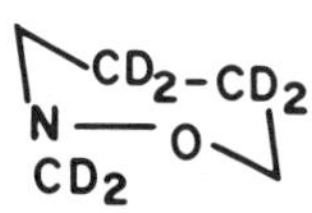

203

c. Conformational Motion in Medium-Sized Rings

Dynamic NMR spectroscopy with both ^{1}H and ^{13}C has been utilized for conformational studies of *trans-syn-trans-* 4,5:9,10-biscyclohexano-1,3,6,8-tetroxecane **(204)** to find a low-energy pathway for boat–chair–boat interconversion ($\Delta G^* = 49.0$ kJ/mol),[276] 5-oxabenzocycloheptene (**205:** $\Delta H^* = 41.4 \pm 0.8$ kJ/mol, $\Delta S^* = 6.3 \pm 4.2$ J/mol K, and $\Delta G^*_{216} = 39.7 \pm 0.4$ kJ/mol for a chair–chair interconversion),[277] *cis-cis*-1,4-cyclooctadiene ($\Delta G^* = 33.5$ kJ/mol for a twist-boat, boat-chair interconversion),[277] *cis-cis*-1,5-cyclononadiene ($\Delta G^* = 43.1 \pm 1.3$ kJ/mol for the conformational enantiomerization of a pair of chiral chairlike conformations),[277] 4,4,7,7-tetramethyl-1,3-dioxepane [**206:** $\Delta G^* = 43$ kJ/mol for the conformational racemization of the symmetrical (C_2) twist–chair forms], and 1,4-bridged cyclooctanes (1:1 mixtures of **207–211** and their mirror images **207′–211′**).[279] Table 5-9 gives the Eyring parameters for the conformational mobility of **207–211.**

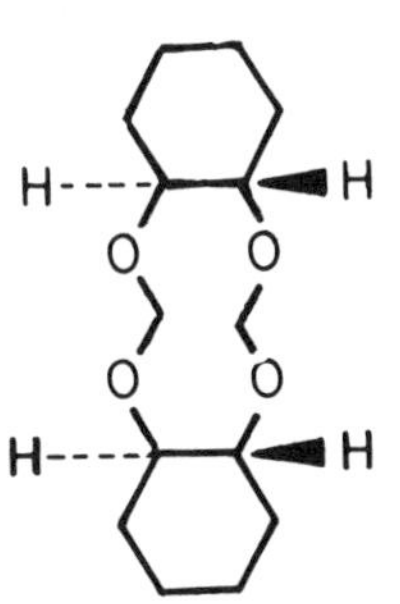

204

205

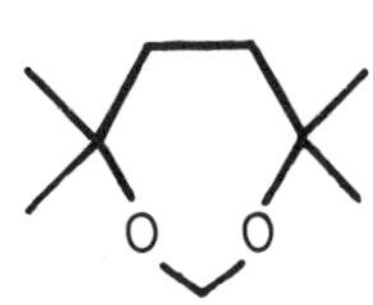

206

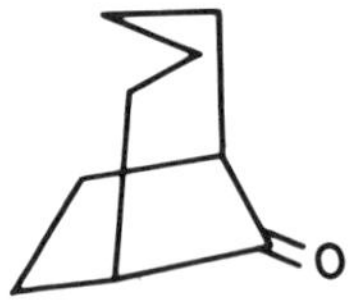

207

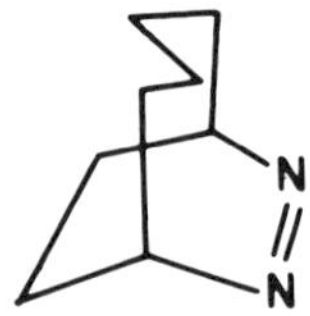

208

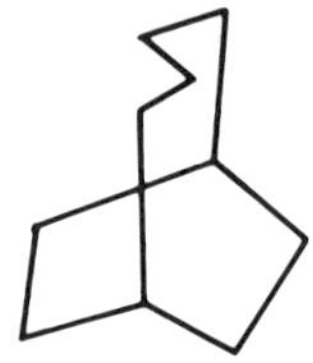

209

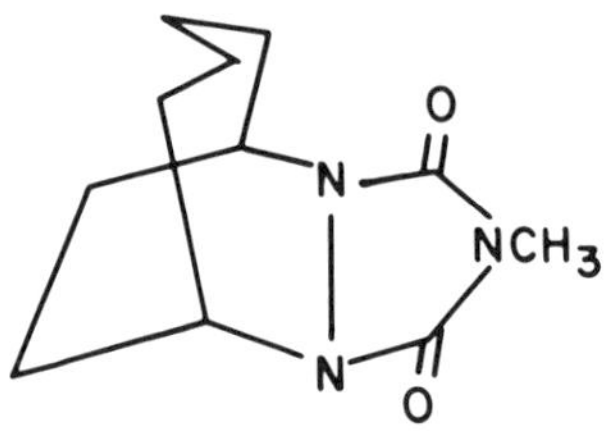

210

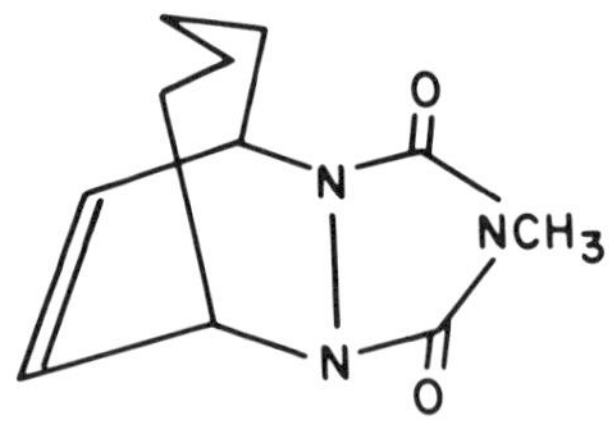

211

TABLE 5-9. Thermodynamic Functions for the Interconversion Processes of **207–211** at 298 K

Rate process	ΔG^* (kJ/mol)	ΔH^* (kJ/mol)	ΔS^* (J/mol K)
207 ⇌ **207′**	25.9 ± 0.4	28.4 ± 0.4	8.4 ± 1.7
208 ⇌ **208′**	36.7 ± 0.8	28.9 ± 1.3	−26.3 ± 6.3
209 ⇌ **209′**	34.8 ± 0.8	32.7 ± 0.8	−7.1 ± 5.4
210 ⇌ **210′**	39.3 ± 0.4	28.9 ± 0.8	−35.1 ± 4.6
211 ⇌ **211′**	46.4 ± 0.4	49.0 ± 0.8	8.4 ± 4.2

9. CONCLUDING REMARKS

In the preceding discussion we have described methods that can yield thermochemically relevant information for structural and stereochemical analysis and related applications. There are, however, many modifications and approaches that were not described or even mentioned in that context. We conclude by briefly listing some of the latter, together with a few results that became available very recently.

Molecular mechanics calculations have been utilized in quite a few of the reports already mentioned.[22–26,28,29,264–267,271,273] The calculations of the late Dr. Burkert on the 1,3-dioxane system[283] deserve, however, a special mention because of their promising agreement with the experiments.[148,236,238]

Most of the methods already illustrated together with infrared techniques and even photoelectron spectra[284] can be employed to clarify the energetics of hydrogen bonding.[285]

A. Photoionization in the Study of Ion Thermochemistry

Although the use of photoionization methods for studying ion thermochemistry could have been dealt with in conventional mass spectrometric counterparts, it has some special merits that speak for treating it separately.

Three reports by Baer and co-workers[286–288] discuss thermochemistry and dissociation dynamics of state-selected $C_4H_8O_2^{+\cdot}$ ions (1,4-dioxane,[286] butanoic acid,[287] and ethyl acetate[288]). The photoionization efficiency curves and appearance energies for the parent ions and several fragment ions are given, and structures and enthalpies of formation for some of the ions are proposed. It can be concluded that the 1,4-dioxane ion does not rearrange near threshold to a more stable structure (ethyl acetate and its enol, and butanoic acid and its enol).[286] Similarly, the results show that butanoic acid does not isomerize before dissociation,[287] although electron impact (EI) ionization produces noninterconverting isomeric structures of $C_4H_8O_2^{+\cdot}$. In contrast to 1,4-dioxane and butanoic acid, ethyl acetate[288] isomerizes possibly to its enol tautomer but not to the 1,4-dioxane or butanoic acid structures ($\Delta_f H°$ at 298 K = 523 ± 4, 569 ± 4, and 506 ± 4 kJ/mol, respectively).

In another series of reports Baer and others[289–291] applied the photoion–photoelectron coincidence (PIPECO) technique to investigate thermochemistry and dissociation dynamics of state-selected C_4H_4X ions where X = S,[289] O,[290] and N.[290] Kutina and co-workers discussed the enthalpies of formation of CH_3S^+, CH_2S^+, CH_2S, and HCS^+ as related to the photoionization mass spectrometry of CH_3SH, CD_3SH, and CH_3SD.[291]

PIPECO mass spectrometry has also proved helpful when studying isomerization and fragmentation of gaseous ions as nitromethane and methylnitrite ions ($\Delta_f H°$ at 298 K = 999 and 938 kJ/mol, respectively),[292] as well as enthalpies and entropies of ionization of alkanes[293] and cycloalkanes.[294]

B. Further Comments on Standard Enthalpies of Formation

Since the first part of this chapter was written, we have become aware of some additional enthalpies of combustion, hence also enthalpies of formation data: for instance, purine,[295] $\Delta_f H_m^\circ$(cr) 169.4 ± 2.7 kJ/mol, and carbon tetrabromide[296] $\Delta_f H_m^\circ$(cr II) 29.4 ± 3.4 and $\Delta_f H°$(g) 83.9 ± 3.4 kJ/mol. In the latter case[296] new standard molar enthalpies of formation were also estimated for gaseous $CHBr_3$ (23.8 ± 4.5 kJ/mol) and CH_2Br_2 (−11.1 ± 5.0 kJ/mol).

The published standard enthalpy of formation of gaseous diphenylcyclopropenone **(212)**[207] has been shown[298] to be wrong by about 210 kJ/mol, and the proper value indicates 88 kJ/mol aromatic stabilization in the cyclopropenone ring.[299] The combustion calorimetric data agree within experimental error with the results of photoacoustic calorimetry[300] when the proper experimental value is employed for diphenylacetylene.

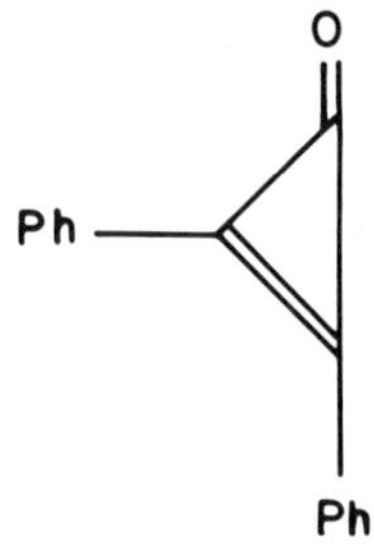

212

Byström and Månsson[301] determined enthalpies of combustion and vaporization of 1,3-dioxane **(213)**, 1,4,7,10-tetraoxacyclododecane **(216)**, and 1,4,7,10,13-pentaoxacyclopentadecane **(217)** and derived their enthalpies of formation in the liquid and gaseous states:

Compound	$-\Delta_f H^\circ_m$(l)(kJ/mol)	$-\Delta_f H^\circ_m$(g)(kJ/mol)
213	377.5 ± 1.1	338.4 ± 1.1
214	355.1 ± 0.9	316.5 ± 0.9
215	516.0 ± 1.2	467.2 ± 1.2
216	696.8 ± 2.1	631.1 ± 2.1
217	879.1 ± 2.0	799.5 ± 2.0

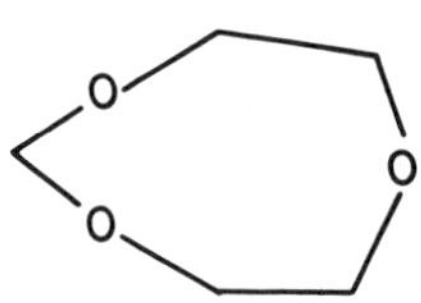

215

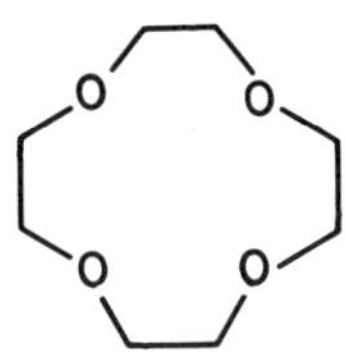

216

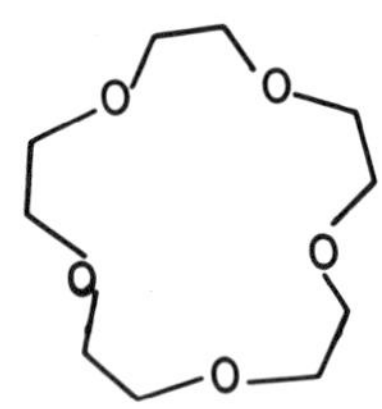

217

These authors have estimated the apparent strain in **213**, 1,3,5-trioxane **(218)**, 1,3,5,7-tetroxane **(219)**, and 1,3,5,7,9-pentoxane **(220)** from

$$[S] = \Delta_f H^\circ_m(\text{g, obs}) - \Delta_f H^\circ_m(\text{g, calc}) - RT \qquad (5\text{-}28)$$

where $\Delta_f H^\circ_m$(g, calc) was calculated using the older Allen parameters[301] and RT is a zero-point correction for a monocyclic ring.[302] We have reestimated the total strain and the strain per 1,3-dioxane unit in **213** and **218–220** using the revised reference values in Tables 5-3 and 5-4 (Table 5-10). From Table 5-10 it is obvious that the best $\Delta_f H^\circ_m$ value for 1,3-dioxane **(213)** is somewhere between the results of Byström and Månsson[301] and this author.[102] In general, all standard enthalpies of formation determined by Månsson have been extremely reliable. In this context it should, however, be emphasized that we put much effort into removing formaldehyde from our 1,3-dioxane sample, a step the other authors[301] do not mention. Whether the 1,3-dioxane

Table 5-10. Apparent Strain Energies (kJ/mol) in Some Cyclic 1,3-Ethers

Compound	$\Delta_f H_m^\circ(g)$	$[S]^a$	$[S]$/1,3-Dioxane unit
213 1,3-Dioxane	338.4,[b] 345.5[c]	10.6,[b] 3.5[d]	7.1[e]
218 1,3,5-Trioxane[b]	465.8	24.2	8.1
219 1,3,5,7-Tetroxane[f]	620.3	33.9	8.5
220 1,3,5,7,9-Pentoxane[f]	779.8	38.5	7.7
			Average 7.8 ± 0.5 kJ/mol

[a] $[S] = \Delta_f H_m^\circ(\text{g, obs}) - \Delta_f H_m^\circ(\text{g, calc}) - RT$.
[b] Reference 301.
[c] From Reference 102, based on ΔH_{vap} given in Reference 301.
[d] Reference 102.
[e] Average value.
[f] Reference 6.

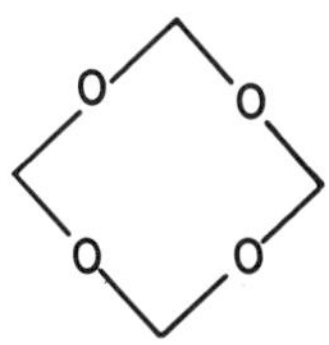

219

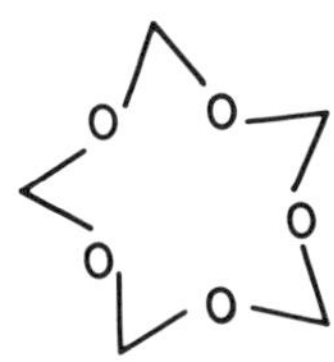

220

ring is significantly strained has also its consequences when discussing its chair–twist enthalpy differences and other nonbonded interactions (cf subsection E of Section 4); hence we will return to this question in a forthcoming paper,[303] since on the basis of previous data[102] the CRSE in **213** has been assumed to be very small.

For the strain increments in **214–217,** we refer to the interesting discussion by Byström and Månsson.[301]

Wiberg and associates have discussed strain energies and enthalpies of formation of propellanes,[304] fused cyclobutanes,[305] and cyclopropane rings[306] in the light of molecular mechanics and ab initio calculations. In their more recent paper[307] they also determined standard enthalpies of formation of *cis*-**(221)** and *trans*-diethylcyclopropanes **(222)** and *cis*- **(223)** and *trans*-bicyclo[6.1.0]nonanes **(224)** by oxygen bomb calorimetry:

Compound	$-\Delta_f H_m^\circ(l)$(kJ/mol)	$-\Delta_f H_m^\circ(g)$(kJ/mol)
221	80.0 ± 1.4	44.6 ± 1.6
222	83.4 ± 1.6	49.1 ± 1.8
223	70.1 ± 1.0	20.8 ± 1.1
224	68.2 ± 1.0	18.8 ± 1.1

223

Finally it is worth mentioning that Wiberg and Squires[307] have continued thermochemical studies of carbonyl reactions and have determined calorimetrically the enthalpies of hydrolysis for a set of aliphatic dimethylketals in addition to those determined earlier to an analogous series of alkyl-substituted dimethylacetals.[97b] The data were utilized to discuss steric effects in acetal and ketal hydrolysis.[307]

Finally, one must emphasize that complete coverage of all relevant literature and even methods (eg, electron diffraction, microwave and infrared spectroscopy) has not been possible in this short review. References that have been omitted are not deemed unimportant. Last but not least, we suggest that the reader become acquainted with the literature cited to learn more about particular methods of investigation. In many cases relevant information also can be found elsewhere in this series.

REFERENCES

1. "Combustion Calorimetry". In "IUPAC: Experimental Chemical Thermodynamics," Vol. 1; Sunner, S.; and Månsson, M. Eds.; Pergamon Press: Oxford, 1979.
2. Cox, J.D.; Pilcher, G. "Thermochemistry of Organic and Organometallic Compounds". Academic Press: London, 1970.
3. "Experimental Thermochemistry", Vol. I, "Measurement of Heats of Reaction", Rossini, F.D., Ed.; Wiley-Interscience: New York, 1956, Chapters 1–10.
4. "Experimental Thermochemistry", Vol. II, Skinner, H.A., Ed.; Wiley-Interscience: New York, 1962, Chapters 1–6, 10, 17, and 19.
5. Pihlaja, K. *Kemia-Kemi* **1980,** *7,* 797.
6. Pedley, J.P.; Rylance, J. "Sussex–N.P.L. Computer Analyzed Thermochemical Data: Organic and Organometallic Compounds". University of Sussex, 1977.
7. Pihlaja, K.; Taskinen, E. In "Physical Methods in Heterocyclic Chemistry", Vol. 6., Katritzky, A.R., Ed.; Academic Press: New York, 1974, Chapter 5, pp. 199–246.
8. McEachern, D.M.; Iniguez, J.C.; Ornelas, H.C. *J. Chem. Eng. Data* **1975,** *20,* 226.
9. Engel, P.S.; Montgomery, R-L.; Månsson, M.; Leckonby, R.A.; Foyt, H.L.; Rossini, F.D. *J. Chem. Thermodyn.* **1978,** *10,* 205.
10. Engel, P.S.; Melaugh, R.A.; Månsson, M.; Timberlake, J.W.; Carner, A.W.; Rossini, F.D. *J. Chem. Thermodyn.* **1976,** *8,* 607.
11. Montgomery, R.L.; Engel, P.S., Leckonby, R.A.; Rossini, F.D. *J. Chem. Eng. Data* **1978,** *23,* 129.
12. Rossini, F.D.; Montgomery, R.L. *J. Chem. Thermodyn.* **1978,** *10,* 465.
13. Sellers, P.; Stridh, G.; Sunner, S. *J. Chem. Eng. Data* **1978,** *23,* 250.
14. Sellers, P. *J. Chem. Thermodyn.* **1977,** *9,* 139.
15. Steele, W.V. *J. Chem. Thermodyn.* **1978,** *10,* 919.
16. Walsh, R.; Wells, J.M. *J. Chem. Thermodyn.* **1976,** *8,* 56.
17. Walsh, R.; Wells, J.M. *J. Chem. Thermodyn.* **1975,** *7,* 149.
18. Kabakoff, D.S.; Bünzli J.-C.; Oth, J.F.M.; Hammond, W.B.; Benson, J.A. *J. Am. Chem. Soc.* **1975,** *97,* 1510.
19. Traeger, J.C.; McLoughlin, R.G. *J. Chem. Thermodyn.* **1978,** *10,* 505.
20. Jochems, R.; Dekker, H.; Mosselman, C.; Somsen, G. *J. Chem. Thermodyn.* **1982,** *14,* 395.
21. Hall, H.K., Jr.; Smith, C.D.; Baldt, J.H. *J. Am. Chem. Soc.* **1973,** *95,* 3197.
22. Parker, W.; Steele, W.V.; Watt, I. *J. Chem. Thermodyn.* **1977,** *9,* 307.
23. Engler, E.M.; Andose, J.D.; Schleyer, P.v.R. *J. Am. Chem. Soc.* **1973,** *95,* 8005.
24. Steels, W.V. *J. Chem. Thermodyn.* **1977,** *9,* 311.
25. Steele, W.V. *J. Chem. Thermodyn.* **1978,** *10,* 585.
26. Allinger, N.L.; Tribble, N.T.; Millar, M.A. *Tetrahedron* **1972,** *28,* 1173.

27. Pittam, D.A.; Pilcher, G. *J. Chem. Soc. Faraday Trans. 1* **1972,** *68,* 2224.
28. Steele, W.V.; Watt, I. *J. Chem. Thermodyn.* **1977,** *9,* 843.
29. Allinger, N.L.; Tribble, M.T.; Millar, M.A.; Wertz, D.W. *J. Am. Chem. Soc.* **1971,** *93,* 1637.
30. Clark, T.; Knox, T. McO.; Mackle, H.; McKervey, M.A.; Rooney, J.J. *J. Am. Chem. Soc.* **1975,** *97,* 3825.
31. Arora, M.; Steele, W.V. *J. Chem. Thermodyn.* **1978,** *10,* 403.
32. Heath, G.A.; Hefter, G.T.; Steele, W.V. *J. Chem. Thermodyn.* **1978,** *10,* 395.
33. Sunner, S.; Wulff, C.A. *J. Chem. Thermodyn.* **1980,** *12,* 505.
34. Månsson, M.; Sellers, P.; Stridh, G.; Sunner, S. *J. Chem. Thermodyn.* **1977,** *9,* 91.
35. Svensson, C. *J. Chem. Thermodyn.* **1979,** *11,* 593.
36. Stridh, G.; Sunner, S.; Svensson, C. *J. Chem. Thermodyn.* **1977,** *9,* 1005.
37. Xu-Wu, A.; Månsson, M. *J. Chem. Thermodyn.* **1983,** *15,* 287.
38. Stridh, G. *J. Chem. Thermodyn.* **1976,** *8,* 895.
39. Stridh, G. *J. Chem. Thermodyn.* **1976,** *8,* 901.
40. Byström, K. *J. Chem. Thermodyn.* **1981,** *13,* 139.
41. Byström, K. *J. Chem. Thermodyn.* **1982,** *14,* 865.
42. McCormick, D.G.; Hamilton, W.S. *J. Chem. Thermodyn.* **1978,** *10,* 275.
43. Hamilton, W.S.; Benton, S.; French, J., McCormick, D.; Pustejovsky, S.; Thompson, P. *J. Chem. Eng. Data* **1978,** *23,* 201.
44. Suradi, S.; El Saiad, N.; Pilcher, G.; Skinner, H.A. *J. Chem. Thermodyn.* **1982,** *14,* 45.
45. Bickerton, J.; Pilcher, G.; Al-Takhim, G. *J. Chem. Thermodyn.* **1984,** *16,* 373.
46. Del Bene, J.E. *J. Am. Chem. Soc.* **1979,** *101,* 6184.
47. Scanlan, M.J.; Hillier, I.H.; McDonald, A.A. *J. Am. Chem. Soc.* **1983,** *105,* 3568.
48. Subach, D.J.; Zwolinski, B.J. *J. Chem. Eng. Data* **1975,** *20,* 232.
49. Subach, D.J.; Zwolinski, B.J. *J. Chem. Thermodyn.* **1975,** *7,* 493.
50. Melaugh, R.A.; Månsson, M.; Rossini, F.D. *J. Chem. Thermodyn.* **1976,** *8,* 623.
51. (a) Good, W.D.; Smith, N.K. *J. Chem. Thermodyn.* **1979,** *11,* 111. (b) Wiberg, K.B.; Wasserman, D.J. *J. Am. Chem. Soc.* **1981,** *103,* 6563. (c) Wiberg, K.B.; Squires, R.R. *J. Chem. Thermodyn.* **1979,** *11,* 773.
52. Månsson, M.; Sunner, S. *J. Chem. Thermodyn.* **1981,** *13,* 671.
53. Smith, N.K.; Stewart, R.C., Jr.; Osborn, A.G.; Scott, D.W. *J. Chem. Thermodyn.* **1980,** *12,* 919.
54. Good, W.D. *J. Chem. Thermodyn.* **1978,** *10,* 553.
55. Good, W.D.; Lee, S.H. *J. Chem. Thermodyn.* **1976,** *8,* 643.
56. Montgomery, R.L.; Rossini, F.D. *J. Chem. Eng. Data* **1978,** *23,* 125.
57. Havel, J.J.; Montgomery, R.L.; Ching-Ching L.; Grissom, M. *J. Chem. Eng. Data* **1978,** *23,* 132.
58. Dekker, H.; Mosselman, C. *J. Chem. Eng. Data* **1978,** *23,* 111.
59. Colomina, M.; Roux, M.V.; Turrión, C. *J. Chem. Thermodyn.* **1976,** *8,* 869.
60. Colomina, M.; Jiménez, P.; Pérez-Ossorio, R.; Turrión, C. *J. Chem. Thermodyn.* **1976,** *8,* 439.
61. Colomina, M.; Jiménez, P.; Roux, M.V.; Turrión, C. *J. Chem. Thermodyn.* **1978,** *10,* 661.
62. Colomina, M.; Jiménez, P.; Pérez-Ossorio, R.; Turrión, C. *J. Chem. Thermodyn.* **1979,** *11,* 1179.
63. Colomina, M.; Jiménez, P.; Roux, M.V.; Turrión, C. *J. Chem. Thermodyn.* **1984,** *16,* 379.
64. Arshadi, M.R. *J. Chem. Thermodyn.* **1980,** *12,* 903.
65. Steele, W.V. *J. Chem. Thermodyn.* **1979,** *11,* 1185.
66. Suradi, S.; Hacking, J.M.; Pilcher, G.; Gümrükcü, I.; Lappert, M.F. *J. Chem. Thermodyn.* **1981,** *13,* 857.
67. (a) Hacking, J.M.; Pilcher, G. *J. Chem. Thermodyn.* **1979,** *11,* 1015. (b) Irving, R.J.; Wadsö, I. *Acta Chem. Scand.* **1970,** *24,* 589. (c) Irving, R.J.; Ribeiro da Silva, M.A.V. *J. Chem. Soc. Dalton Trans.* **1975,** 798.
68. Ferrão, M.L.C.C.H.; Ribeiro da Silva, M.A.V.; Suradi, S.; Pilcher, G.; Skinner, H.A. *J. Chem. Thermodyn.* **1981,** *13,* 567.
69. Stridh, G. *J. Chem. Thermodyn.* **1976,** *8,* 193.
70. Anthoney, M.E., Carson, A.S.; Laye, P.G.; Yürekli, M. *J. Chem. Thermodyn.* **1976,** *8,* 1009.
71. Sabbah, R.; Laffitte, M. *J. Chem. Thermodyn.* **1978,** *10,* 101.
72. Wilson, S.R.; Watson, I.D.; Malcolm, G.N. *J. Chem. Thermodyn.* **1979,** *11,* 911.

73. Gundry, H.A.; Harrop, D.; Head, A.J. *J. Chem. Thermodyn.* **1978,** *10,* 203.
74. Kolesov, V.P.; Slavutskaya, G.M.; Dityat'eva, L.N. *J. Chem. Thermodyn.* **1976,** *8,* 907.
75. Erastov, P.A.; Kolesov, V.P.; Dityat'eva, L.N.; Golovanova, Yu. G. *J. Chem. Thermodyn.* **1981,** *13,* 663.
76. Papina, T.S.; Erastov, P.A.; Kolesov, V.P. *J. Chem. Thermodyn.* **1981,** *13,* 683.
77. Erastov, P.A.; Kolesov, V.P. *J. Chem. Thermodyn.* **1982,** *14,* 103.
78. Pihlaja, K.; Steele, W.V.; Vainiotalo, P. Unpublished results.
79. Pihlaja, K.; Vainiotalo, P. Unpublished results.
80. Hamilton, W.S.; Thompson, P.; Pustejovsky, S. *J. Chem. Eng. Data* **1976,** *21,* 428.
81. Baroody, E.E.; Carpenter, G.A. *J. Chem. Eng. Data* **1976,** *21,* 253.
82. Baroody, E.E.; Carpenter, G.A. *J. Chem. Eng. Data* **1979,** *24,* 3.
83. Rouse, P.E., Jr. *J. Chem. Eng. Data* **1976,** *21,* 16.
84. Prosen, E.J.; Johnson, W.H.; Rossini, F.D. *J. Res. Natl. Bur. Stand.* **1946,** *37,* 51.
85. Pittam, D.A.; Pilcher, G. *Trans. Faraday Soc.* **1967,** *63,* 2357.
86. Pihlaja, K.; Vainiotalo, P. *J. Chem. Eng. Data.* **1985,** *30,* 387.
87. Fenwick, J.O.; Harrop, D.; Head, A.J. *J. Chem. Thermodyn.* **1975,** *7,* 943.
88. Skinner, H.A. *J. Chem. Soc.* **1962,** 4396.
89. (a) Somayajulu, G.R.; Zwolinski, B.J. *J. Chem. Soc. Faraday Trans. 2* **1976,** *72,* 2213. (b) Somayajulu, G.R.; Chao, J.; Reed, C.O.; Kennedy, J.R.; Zwolinski, B.J. *J. Chem. Eng. Data* **1977,** *22,* 299. (c) Kao, J.W.H.; Chung-Phillips, A. *J. Chem. Phys.* **1975,** *63,* 4143. (d) Kao, J.W.H.; Chung-Phillips, A. *J. Chem. Phys.* **1975,** *63,* 4152.
90. Pihlaja, K.; Kankare, J. *Acta Chem. Scand.* **1969,** *23,* 1745.
91. Pihlaja, K. *Acta Chem. Scand.* **1971,** *25,* 451.
92. Kalb, A.J.; Chung, A.L.H.; Allen, T.L. *J. Am. Chem. Soc.* **1966,** *88,* 2938.
93. Månsson, M. *J. Chem. Thermodyn.* **1974,** *6,* 1153.
94. Månsson, M. *J. Chem. Thermodyn.* **1969,** *1,* 141.
95. Smutny, E.J.; Bondi, A. *J. Phys. Chem.* **1961,** *65,* 546.
96. Pihlaja, K.; Tuomi, M.-L. *Acta Chem. Scand.* **1970,** *24,* 366.
97. (a) Hine, J.; Klueppel, A.W. *J. Am. Chem. Soc.* **1974,** *96,* 2924. (b) Wiberg, K.B.; Squires, R.R. *J. Chem. Thermodyn.* **1979,** *11,* 773.
98. Engel, P.S.; Melaugh, P.A.; Page, M.A.; Szilagyi, S.; Timberlake, J.W. *J. Am. Chem. Soc.* **1976,** *98,* 1971.
99. Engel, P.S. *J. Am. Chem. Soc.* **1976,** *98,* 1972.
100. Eliel, E.L.; Allinger, N.L.; Angyal, S.J.; Morrison, G.A. "Conformational Analysis". Wiley-Interscience: New York, 1965.
101. Pihlaja, K. *Suom. Kemistil.* **1966,** *A99,* 223.
102. Pihlaja, K.; Luoma, S. *Acta Chem. Scand.* **1968,** *22,* 2401.
103. Pihlaja, K.; Heikkilä, J. *Acta Chem. Scand.* **1967,** *21,* 2390, 2430.
104. Pihlaja, K.; Kivimäki, M.; Myllyniemi, A.-M.; Nurmi, T. *J. Org. Chem.* **1982,** *47,* 4688.
105. Pihlaja, K. *Kemia-Kemi* **1974,** *1,* 492.
106. Burkert, U. *Tetrahedron* **1979,** *35,* 691.
107. Eliel, E.L. *Acc. Chem. Res.* **1970,** *3,* 1.
108. Pihlaja, K.; Vainiotalo, P.; Anderson, J.E.; Mattinen, J. To be published.
109. Eliel, E.L.; Kandasamy, D.; Yen, C.-Y.; Hargrave, K.D. *J. Am. Chem. Soc.* **1980,** *102,* 3698.
110. Kistiakowsky, G.B.; Ruhoff, J.R.; Smith, H.A.; Vaughan, W.E. *J. Am. Chem. Soc.* **1936,** *58,* 137.
111. Kistiakowsky, G.B.; Ruhoff, J.R.; Smith, H.A.; Vaughan, W.E. *J. Am. Chem. Soc.* **1935,** *57,* 65, 876.
112. (a) Rogers, D.W.; McLafferty, F.J. *Tetrahedron* **1971,** *27,* 3765, and references therein. (b) Roth, W.R.; Lennartz, H.-W. *Chem. Ber.* **1980,** *113,* 1806. (c) Roth W.R.; Klärner, F.-G.; Lennartz, H.-W. ibid, **1980,** *113,* 1818. (d) Roth, W.R.; Kirmse, W.; Hoffmann, W.; Lennartz, H.-W. ibid, **1982,** *115,* 2508. (e) Roth, W.R.; Böhm, M.; Lennartz, H.-W.; Vogel, E. *Angew. Chem. Int. Ed. Engl.* **1983,** *22,* 1007.
113. Turner, R.B.; Jarrett, A.D.; Goebel, P.; Mallon, B.J. *J. Am. Chem. Soc.* **1973,** *95,* 790.
114. Turner, R.B.; Mallon, B.J.; Tichy, M.; Doering, W. von E.; Roth, W.R.; Schröder, G. *J. Am. Chem. Soc.* **1973,** *95,* 8605.
115. Mortimer, C.T. "Reaction Heats and Bond Strengths". Pergamon Press: Oxford, **1962.**
116. Kapteijn, F.; van der Steen, A.J.; Mol, C.J. *J. Chem. Thermodyn.* **1983,** *15,* 146.
117. Good, W.D. *J. Chem. Thermodyn.* **1976,** *8,* 67.

118. Turner, R.B.; Meador, W.B.; Winkler, R.E. *J. Am. Chem. Soc.* **1957,** *79,* 4116.
119. Eliel, E.L. *Pure Appl. Chem.* **1971,** *25,* 509.
120. (a) Taskinen, E. *J. Chem. Thermodyn.* **1973,** *5,* 783; (b) ibid, **1974,** *6,* 271; (c) ibid, **1974,** *6,* 345; (d) ibid, **1974,** *5,* 1021.
121. Taskinen, E.; Liukas, P. *Acta Chem. Scand.* **1974,** *B28,* 114.
122. Taskinen, E.; Mäkinen, A. *Acta Chem. Scand.* **1974,** *B28,* 121.
123. (a) Taskinen, E. *Acta Chem. Scand.* **1974,** *B28* 201; (b) ibid, **1974,** *B28,* 357; (c) **1974,** *B28,* 1234.
124. Taskinen, E.; Ylivainio, P. *Acta Chem. Scand.* **1975,** *B29,* 1.
125. (a) Taskinen, E. *Tetrahedron* **1975,** *31,* 957; (b) ibid, **1976,** *32,* 2327.
126. (a) Taskinen, E. *Acta Chem. Scand.* **1975,** *B29,* 245; (b) ibid, **1980,** *B34,* 203; (c) ibid, **1980,** *B34,* 643.
127. Taskinen, E.; Jokila, K. *Acta Chem. Scand.* **1975,** *B29,* 249.
128. Taskinen, E.; Mustonen, E. *Acta Chem. Scand.* **1976,** *B30,* 1.
129. Taskinen, E.; Sainio, E. *Tetrahedron* **1976,** *32,* 593.
130. Taskinen, E.; Kalva, H. *Finn. Chem. Lett.* **1975,** 208.
131. (a) Taskinen, E. Lähteenmäki, H. *Tetrahedron* **1976,** *32,* 2331; (b) *Finn. Chem. Lett.* **1978,** *47.*
132. Taskinen, E.; Virtanen, R. *J. Org. Chem.* **1977,** *42,* 1443.
133. Taskinen, E.; Anttila, M. *Tetrahedron* **1977,** *33,* 2423.
134. (a) Taskinen, E. *Finn. Chem. Lett.* **1978,** 276; (b) ibid, **1979,** 154.
135. Taskinen, E.; Kuusisto, M. *Finn. Chem. Lett.* **1980,** *B34,* 571.
136. Taskinen, E.; Mukkala, V.-M. *Tetrahedron* **1982,** *38,* 613.
137. Taskinen, E. *Tetrahedron* **1978,** *34,* 425.
138. Taskinen, E. (Thesis) *Ann. Acad. Sci. Fenn. Ser. A2* **1972,** No. 163.
139. Koole. N.J.; de Bie, M.J.A.; Hansen, P.E. *Org. Magn. Reson.* **1984,** *22,* 146, and references therein.
140. Virtanen, R. "Structure–Stability Relationships in Vinyl Sulphides. A Thermodynamic and ^{13}C NMR Spectroscopic Study". Ph.D. thesis, University of Turku, Turku, (Findland), 1984.
141. Gardner, D.V.; McGreer, D.E. *Can. J. Chem.* **1970,** *48,* 2104 and 2110.
142. McGreer, D.E.; Page, B.D.; Kaushal, D.P. *Can J. Chem.* **1973,** *51,* 1239.
143. Virtanen, T.; Nikander, H.; Pihlaja, K.; Rahkamaa, E. *Tetrahedron* **1982,** *38,* 2821.
144. Pihlaja, K.; Nikkilä, A.; Neuvonen, K.; Keskinen, R. *Acta Chem. Scand.* **1976,** *A30,* 457.
145. Keskinen, R., Nikkilä, A.; Pihlaja, K. *J. Chem. Soc. Perkin Trans. 2* **1977,** 343.
146. Pihlaja, K.; Keskinen, R.; Nikkilä, A. *Bull. Soc. Chim. Belg.* **1976,** *83,* 435.
147. Pihlaja, K.; Härkönen, T. *Acta Chem. Scand.* **1978,** *B32,* 769.
148. Pihlaja, K.; Mattinen, J. unpublished results
149. Clay, R.M.; Kellie, G.M.; Riddell, F.G. *J. Am. Chem. Soc.* **1973,** *95,* 4632.
150. Bailey, W.F.; Connon, H.; Eliel, E.L.; Wiberg, K.B. *J. Am. Chem. Soc.* **1978,** *100,* 2202.
151. Lajunen, M.; Hintsanen, A. *Acta Chem. Scand.* **1983,** *A37,* 545.
152. Lajunen, M.; Lyytikäinen, H. *Acta Chem. Scand.* **1981,** *A35,* 131.
153. Pihlaja, K.; Nikkilä, A.; Nikander, H. Unpublished results.
154. Pihlaja, K. *J. Chem. Soc. Perkin Trans. 2* **1974,** 890.
155. Pihlaja, K.; Nikander, H. *Acta Chem. Scand.* **1977,** *B31,* 265.
156. Allinger, N.L.; Pamphilis, N.A. *J. Org. Chem.* **1973,** *38,* 316.
157. Pasto, D.J.; Rao, R.D. *J. Am. Chem. Soc.* **1969,** *91,* 2790.
158. Beak, P.; Woods, T.S.; Müller, D.S. *Tetrahedron* **1972,** *28,* 5507, and references therein.
159. Blackburne, I.D.; Katritzky, A.R.; Read, D.M.; Bodalski, R.; Pietrusiewicz, K. *J. Chem. Soc. Perkin Trans. 2* **1974,** 1155.
160. Burdan, I.J.; Stoddart, J.F. *J. Chem. Soc. Perkin Trans. 1* **1975,** 666, 675.
161. (a) Hussain, S.A.M.T.; Ollis, W.D.; Smith, C.; Stoddart, J.F. *J. Chem. Soc. Chem. Commun.* **1974,** 873. (b) Hussain, S.A.M.T.; Ollis, W.D.; Smith, C.; Stoddart, F. *J. Chem. Soc. Perkin Trans. 1* **1975,** 1480.
162. Jalonen, J.; Pihlaja, K. *Org. Mass Spectrom.* **1973,** *7,* 1203.
163. Pihlaja, K.; Jalonen, J.; Franck, R.W.; Meyerson, S. *Mass Spectrom. Rev.* **1982,** *1,* 107.
164. Mandelbaum, A. *Mass Spectrom. Rev.* **1983,** *2,* 223, and references 5–7 therein.
165. Nikander, H. "2-Oxo-1,3,2-Dioxathianes. Preparation and Structural Analysis with the Aid of Chemical Equilibration, IR and NMR Spectroscopy and Mass Spectra". Ph.D. Thesis, University of Turku, Turku, Finland, 1980.

166. Pihlaja, K.; Nikander, H.; Jordan, D.M. *Adv. Mass Spectrom.* **1980,** *8A,* 821.
167. Morrison, J.D. "MTP Review of Chemistry", Ser. I, Vol. 5., Maccoll, A. Ed.; Butterworths: London, 1972, p. 25.
168. (a) Levsen, K. "Fundamental Aspects of Organic Mass Spectrometry". Verlag Chemie: Weinheim, 1978. (b) Maccoll, A. *Org. Mass Spectrom.* **1982,** *17,* 1.
169. Franklin, J.L.; Sen Sharma, D.K. *Adv. Mass Spectrom.* **1974,** *6,* 947, and references therein.
170. Warren, J.W. *Nature* **1950,** *165,* 810.
171. (a) Haney, M.A.; Franklin, J.L. *J. Chem. Phys.* **1968,** *48,* 4093; *Trans. Faraday Soc.* **1969,** *65,* 1794. (b) Jalonen, J.; Pihlaja, K. *Suom. Kemistil.* **1972,** *A45,* 116. (c) Lossing, F.P.; Tickner, A.W.; Bryce, W.A. *J. Chem. Phys.* **1951,** *19,* 1254. (d) McMillen, D.F.; Golden, D.M. *Annu. Rev. Phys. Chem.* **1982,** *33,* 493.
172. Conrath, K.; Van de Sande, C.; Vandewalle, M. *Org. Mass Spectrom.* **1974,** *9,* 585.
173. Yeo, A.N.H.; Williams, D.H. *Org. Mass Spectrom.* **1971,** *5,* 135, and references therein.
174. (a) Köppel, C.; Schwarz, A.; Bohlmann, F. *Org. Mass Spectrom,* **1974,** *9,* 324; (b) Holmes, J.L.; Fingas, M.; Lossing, F.P. *Can. J. Chem.* **1981,** *59,* 80.
175. (a) Maeda, K.; Semeluk, G.P.; Lossing, F.P. *Int. J. Mass Spectrom. Ion Phys.* **1968,** *1,* 395. (b) Lossing, F.P.; Semeluk, G.P. *Can. J. Chem.* **1979,** *48,* 955.
176. Lossing, F.P.; Maccoll, A. *Can. J. Chem.* **1976,** *54,* 990.
177. Holmes, J.L.; Lossing, F.P. *J. Am. Chem. Soc.* **1980,** *102,* 1591.
178. Holmes, J.L.; Lossing, F.P. *Org. Mass Spectrom.* **1979,** *14,* 512.
179. Holmes, J.L.; Lossing, F.P. *J. Am. Chem. Soc.* **1980,** *102,* 3732.
180. Holmes, J.L. Terlouw, J.K.; Lossing, F.P. *J. Phys. Chem.* **1976,** *80,* 2860.
181. (a) Bouma, W.J.; Poppinger, D.; Radom, L. *J. Am. Chem. Soc.* **1977,** *99,* 6443. (b) Bouma, W.J.; Radom, L. *Aust. J. Chem.* **1978,** *31,* 1167.
182. Lossing, F.P. *J. Am. Chem. Soc.* **1977,** *99,* 7526.
183. Hoppilliard, Y.; Bouchoux, G.; Jaudon, P. *Nouv. J. Chim.* **1982,** *6,* 43.
184. Holmes, J.L.; Burgers, P.C.; Terlouw, J.K. *Can. J. Chem.* **1981,** *59,* 1805.
185. Vajda, J.H.; Harrison, A.G.; Hirota, A.; McLafferty, F.W. *J. Am. Chem. Soc.* **1981,** *103,* 36.
186. Terlouw, J.W.; Heerma, W.; Holmes, J.L.; Burgers, P.C. *Org. Mass Spectrom.* **1980,** *11,* 582.
187. Holmes, J.L.; Lossing, F.P. *J. Am. Chem. Soc.* **1982,** *104,* 2648.
188. Tureĉek, F.; McLafferty, F.W. *Org. Mass Spectrom.* **1983,** *18,* 608.
189. Burgers, P.C.; Holmes, J.L.; Szulejko, J.E.; Mommers, A.A.; Terlouw, J.K. *Org. Mass Spectrom.* **1983,** *18,* 254.
190. (a) Terlouw, J.K.; Wezenberg, J., Burgers, P.C.; Holmes, J.L. *J. Chem. Soc. Chem. Commun.* **1983,** 1121. (b) Apeloig, Y.; Karni, M.; Ciommer, B.; Depke, G.; Frenking, G.; Meyn, S.; Schmidt, J.; Schwarz, H. *J. Chem. Soc. Chem. Commun.* **1983,** 1497. (c) Bouma, W.J.; Nobes, R.H.; Radom, L. *J. Am. Chem. Soc.* **1982,** *104,* 2929. (d) Bouma, W.J.; MacLeod, J.K.; Radom, L. *J. Am. Chem. Soc.* **1982,** *104,* 2930. (e) Holmes, J.L.; Lossing, F.P.; Terlouw, J.K.; Burgers, P.C. *J. Am. Chem. Soc.* **1982,** *104,* 2931.
191. Bouma, W.J.; Nobes, R.H.; Radom, L. *J. Am. Chem. Soc.* **1983,** *105,* 1743.
192. (a) Burgers, P.C.; Holmes, J.L.; Lossing, F.P.; Mommers, A.A.; Povel, F.R.; Terlouw, J.K. *Can. J. Chem.* **1982,** *60,* 2246. (b) Depke, G.; Lifshitz, C.; Schwarz, H.; Tsidony, E. *Angew. Chem. Int. Ed. Engl.* **1981,** *20,* 792.
193. Cassady, C.J.; Freiser, B.S.; Russell, D.H. *Org. Mass Spectrom.* **1983,** *18,* 378.
194. Holmes, J.L.; Burgers, P.C.; Mollah, Y.A. *Org. Mass Spectrom.* **1982,** *17,* 127.
195. Harrison, A.G.; Lin, M.S. *Org. Mass Spectrom.* **1984,** *19,* 67.
196. Herman, J.A.; Harrison, A.G. *Can. J. Chem.* **1981,** *59,* 2133.
197. McLafferty, F.W.; Barbalas, M.P.; Tureĉek, F. *J. Am. Chem. Soc.* **1983,** *105,* 1.
198. Wolkoff, P.; Holmes, J.L.; Lossing, F.P. *Can. J. Chem.* **1980,** *58,* 251.
199. Holmes, J.L.; Burgers, P.C.; Mollah, M.Y.A.; Wolkoff, P. *J. Am. Chem. Soc.* **1982,** *104,* 2879.
200. Wagner-Redeker, W.; Levsen, K.; Schwarz, H.; Zummack, W. *Org. Mass Spectrom.* **1981,** *16,* 361.
201. Herman, J.A.; Li, Y.-H.; Harrison, A.G. *Org. Mass Spectrom.* **1982,** *17,* 143.
202. Selim, E.T.M. *Proc. Indian Acad. Sci. (Chem. Sci.)* **1982,** *91,* 359.
203. Traeger, J.C. *Org. Mass Spectrom.* **1981,** *16,* 193.

204. Burkey, T.J.; Castelhano, A.L.; Griller, D.; Lossing, F.P. *J. Am. Chem. Soc.* **1983,** *105,* 4701.
205. Bouma, W.J.; Dawes, J.M.; Radom, L. *Org. Mass Spectrom.* **1983,** *18,* 12.
206. Holmes, J.L.; Burgers, P.C.; Terlouw, J.K.; Schwarz, H.; Ciommer, B.; Halim, H. *Org. Mass Spectrom.* **1983,** *18,* 208.
207. Apeloig, Y.; Franke, W.; Rappoport, Z.; Schwarz, H.; Stahl, D. *J. Am. Chem. Soc.* **1981,** *103,* 2770.
208. Spalding, T.R. *J. Organometal. Chem.* **1973,** *55,* C 55.
209. Büchler, U.; Vogt, *J. Org. Mass Spectrom.* **1979,** *14,* 503.
210. Günther, H. "NMR Spectroscopy". Wiley: New York, 1980; Chapter VIII, and references therein.
211. Sóhár, P. "NMR Spectroscopy". CRC Press: Boca Raton, Fla., 1983; Vol. I, Section 2.1.5; Vol. II, Sections 3.7.1.1, 4.1.7.
212. Anteunis, M.J.O.; Tavernier, D.; Borremans, F. *Heterocycles* **1976,** *4,* 293.
213. (a) Lambert, J.B.; Shurvell, L.V.; Cooks, R.G.; Stout, G.H. "Organic Structural Analysis". Macmillan; New York, 1976, Chapter 6, and references therein. (b) Nogradi, M. "Stereochemistry. Basic Concepts and Applications". Akadémiai Kiadó: Budapest, 1981. (c) Jackman, L.M.; Cotton, F.A. "Dynamic NMR Spectroscopy". Academic Press: New York, 1975.
214. Pihlaja, K.; Pasanen, P.; Wähäsilta, *J. Org. Magn. Reson.* **1979,** *12,* 331.
215. Karplus, M. *J. Am. Chem. Soc.* **1963,** *85,* 2870.
216. Pihlaja, K.; Nurmi, T.; Pasanen, P. *Acta Chem. Scand.* **1977,** *B31,* 895.
217. Jaime, C.; Osawa, E.; Takeuchi, Y.; Camps, P. *J. Org. Chem.* **1983,** *48,* 4514.
218. Pihlaja, K.; Rossi, K. *Acta Chem. Scand.* **1983,** *B37,* 289.
219. Pihlaja, K.; Jordan, D.M.; Nikkilä, A.; Nikander, H. *Adv. Mol. Rel. processes* **1973,** *5,* 227.
220. Bentrude, W.G.; Day, R.O.; Holmes, J.M.; Quin, G.S.; Setzer, W.N.; Sopchik, A.E.; Holmes, R.R. *J. Am. Chem. Soc.* **1984,** *106,* 106.
221. Giralt, E.; Pericas, M.A.; Riera, A. *Tetrahedron* **1983,** *23,* 3959.
222. Nikander, H.; Mukkala, V.-M.; Nurmi, T.; Pihlaja, K. *Org. Magn. Reson.* **1976,** *8,* 375.
223. Pihlaja, K. *Suom. Kemistil.* **1968,** *B41,* 229.
224. Tavernier, D.; Hosten, N.; Anteunis, M.J.O.; Spiessens, L.I. *Bull. Soc. Chim. Belg.* **1982,** *91,* 839.
225. (a) Spiessens, L.I.; Becu, C.; Hosten, N.; Anteunis-De Ketelaere, F.; Anteunis, M.J.O.; Tavernier, D. *Bull. Soc. Chim. Belg.* **1982,** *91,* 845. (b) Abraham, R.J.; Richards, B.A. *Tetrahedron* **1983,** *39,* 4201.
226. (a) de Leeuw, F.A.A.M.; Altona, C. *Int. J. Peptide Protein Res.* **1982,** *20,* 120. (b) de Leeuw, F.A.A.M.,; Altona, C. *J. Chem. Soc. Perkin Trans. 2* **1982,** 375. (c) de Leeuw, F.A.A.M.; Altona, C.; Kessler, H.; Bermel, W.; Friedrich, A.; Krack, G.; Hull, W.E. *J. Am. Chem. Soc.* **1983,** *105,* 2237.
227. Booth, H.; Lemieux, R.U. *Can. J. Chem.* **1971,** *49,* 777.
228. Schneider, H.-J.; Becker, G.; Freitag, W.; Hoppen, V. *J. Chem. Res.* (S) **1979,** 14–15; (M) **1979,** 0421–0443.
229. Marshall, J.L. "Carbon–Carbon and Carbon–Proton NMR Couplings: Application to Organic Stereochemistry and Conformational Analysis". Methods in Stereochemical Analysis, Vol. 2, Marchand, A.P. Ed.; Verlag Chemie: Deerfield Beach, Fla., 1983.
230. Mursakulov, I.G.; Guseinov, M.M.; Kasumov, N.K.; Zefirov, N.S.; Samoshin, V.V.; Chalenko, E.G. *Tetrahedron* **1982,** *38,* 2213, and references therein.
231. (a) Dalling, D.K.; Grant, D.M. *J. Am. Chem. Soc.* **1967,** *89,* 6612; (b) ibid, **1972,** *94,* 5318.
232. Pihlaja, K.; Björkqvist, B. *Org. Magn. Reson.* **1977,** *9,* 533.
233. Pihlaja, K.; Rossi, K. *Acta Chem. Scand.* **1977,** *B31,* 899.
234. Pihlaja, K.; Nurmi, T. *Isr. J. Chem.* **1980,** *20,* 160.
235. Pihlaja, K.; Eskonmaa, M.; Keskinen, R.; Nikkilä, A.; Nurmi, T. *Org. Magn. Reson.* **1982,** *17,* 245.
236. Pihlaja, K.; Kivimäki, M.; Myllyniemi, A.-M.; Nurmi, T. *J. Org. Chem.* **1982,** *47,* 4688.
237. Pihlaja, K.; Nikander, H.; Rossi, K. *J. Org. Chem.* **1985,** *50,* 644.
238. Nurmi, T. "^{13}C NMR Chemical Shifts of Aliphatic Five- and Six-membered Ring Homologues. Their Parametrization and Dependence on the Ring Geometry and Conformation". Ph.D. thesis, University of Turku, Turku, Finland, 1983.
239. Vierhapper, F.W.; Willer, R.L. *Org. Magn. Reson.* **1977,** *9,* 13.

240. Eliel, E.L.; Pietrusiewicz, K.M. *Top. C-13 NMR Spectrosc.* **1979,** *3,* 171.
241. Eliel, E.L.; Kandasamy, D.; Yen C.-Y.; Hargrave, K.D. *J. Am. Chem. Soc.* **1980,** *102,* 3698, and references therein.
242. Willer, R.L.; Eliel, E.L. *J. Am. Chem. Soc.* **1977,** *99,* 1925.
243. Eliel, E.L.; Manoharan, M.; Pietrusiewicz, K.M.; Hargrave, K.D. *Org. Magn. Reson.* **1983,** *21,* 94.
244. Rossi, K. "Conformational Analysis. ^{13}C NMR Studies of Saturated Heterocycles. Substituent Effects on the ^{13}C Chemical Shifts of Methyl-substituted 2-Oxo- and 4-Oxo-1,3-dioxolanes, 2-Hydroxy-1,3,2-dioxaborolanes, 2-Oxo-1,3-dioxanes, 2-Hydroxy-1,3,2-dioxaborinanes, and 2-Oxo-1,3,2-dioxathianes". Ph.D. thesis, University of Turku, Turku, Finland, 1983.
245. Pihlaja, K. University of Oulu, Report Series in Chemistry, Special Report **1983** (12), 26.
246. Pihlaja, K.; Kettunen, M.-L. To be published.
247. Pihlaja, K.; Aaljoki, A. *Finn. Chem. Lett.* **1982,** 1.
248. Clark, R.A.; Parker, D.C. *J. Am. Chem. Soc.* **1971,** *93,* 7257.
249. Kitching, W.; Olszowy, H.; Adeock, W. *Org. Magn. Reson.* **1981,** *15,* 230.
250. Booth, H.; Everett, J.R. *J. Chem. Soc. Perkin Trans. 2* **1980,** 255, and references therein.
251. Pihlaja, K.; Scheinin, H.; Lindgren, T.; Kivimäki, M. To be published.
252. Stewart, M.A.H.; Schaefer, T.; Hutton, H.M.; Wong, C.M. *Can. J. Chem.* **1971,** *49,* 1085.
253. Peeling, J.; Goodwin, B.W.; Schaefer, T.; Rowbotham, J.B. *Can. J. Chem.* **1973,** *51,* 2110.
254. Peeling, J.; Ernst, L.; Schaefer, T. *Can. J. Chem.* **1974,** *52,* 849.
255. Peeling, J.; Rowbotham, J.B.; Ernst, L.; Schaefer, T. *Can. J. Chem.* **1974,** *52,* 2414.
256. Colebrook, L.D.; Icli, S.; Hund, L.H. *Can. J. Chem.* **1975,** *53,* 1556.
257. Lunazzi, L.; Placueci, G.; Cerioni, G. *J. Chem. Soc. Perkin Trans. 2* **1977,** 1666.
258. Olah, G.A.; Field, L.D.; Watkins, M.I.; Malhotra, R. *J. Org. Chem.* **1981,** *46,* 1761.
259. Müllen, K.; Kotzamani, E.; Schmickler, H.; Frei, B. *Tetrahedron* **1983,** *39,* 3821.
260. Evans, F.E.; Miller, D.W. *J. Am. Chem. Soc.* **1983,** *105,* 4863.
261. Lunazzi, L.; Macciantelli, D.; Grossi, L. *Tetrahedron* **1983,** *39,* 305.
262. Dürst, T.; Gryff-Keller, A.; Terpiński, *J. Org. Magn. Reson.* **1983,** *21,* 657.
263. Boyd, D.R.; McCombe, K.M.; Hamor, T.A.; Jennings B.A.; Wilson, V.E. *J. Chem. Soc. Perkin Trans. 2* **1984,** 95.
264. Beckhaus, H.-D.; Rückhardt, C.; Anderson, J.E. *Tetrahedron* **1982,** *38,* 2299.
265. Anderson, S.; Drakenberg, T. *Org. Magn. Reson.* **1983,** *21,* 602, 730.
266. Tiffon, B.; Lomas, J.S. *Org. Magn. Reson.* **1984,** *22,* 29.
267. (a) Haky, J.E.; Saavedra, J.E.; Hilton, B.D. *Org. Magn. Reson.* **1983,** *21,* 79. (b) Shanin-Atidi, H.; Bar-Eli, E.H. *J. Phys. Chem.* **1970,** *74,* 961.
268. Yamamoto, G.; Ōki, M. *Chem. Lett.* **1984,** 97.
269. Doddrell, D.; Clarrier, C.; Hawkins, B.L.; Crain, W.O., Jr.; Harris, L.; Roberts, J.D. *Proc. Natl. Acad. Sci. USA* **1970,** *67,* 1588.
270. (a) Poupko, R.; Luz, Z. *J. Chem. Phys.* **1981,** *75,* 1675. (b) See References 5 and 6 in 270a.
271. Anderson, J.E.; Jørgensen, F.S.; Thomsen, T. *J. Chem. Soc. Chem. Commun.* **1982,** 333.
272. Sugiura, M.; Takao, N.; Sasaki, Y. *Chem. Pharm. Bull.* **1977,** *25,* 960.
273. Ogawa, K.; Takeuchi, Y.; Suzuki, H.; Nomura, Y. *J. Am. Chem. Soc.* **1984,** *106,* 831, and references therein.
274. Riddell, F.G.; Turner, E.S. *J. Chem. Res.* (S) **1978,** 476.
275. Grindley, T.B.; Szarek, W.A. *Can. J. Chem.* **1974,** *52,* 2566.
276. Canuel, L.; St.-Jacques, M. *Can. J. Chem.* **1974,** *52,* 3581.
277. Anet, F.A.L.; Yavari, I. *J. Am. Chem. Soc.* **1977,** *99,* 6496, 6986.
278. Yavari, I. *Org. Magn. Reson.* **1980,** *14,* 511.
279. Dyllick-Brenzinger, R.; Olsen, H. *J. Am. Chem. Soc.* **1981,** *103,* 704.
280. St.-Amour, R.; St.-Jacques, M. *Can. J. Chem.* **1983,** *61,* 109.
281. Yannoni, C.S.; Clarke, T.C.; Kendrick, R.D.; Macho, V.; Miller, R.D. *Mol. Cryst. Liq. Cryst.* **1983,** *96,* 305.
282. Able, E.W.; King, G.D.; Orrell, K.G.; Šik, V. *Polyhedron* **1983,** *2,* 1363.
283. Burkert, U. *Tetrahedron* **1979,** *35,* 691, 1945.
284. (a) Brown, R.S. *Can. J. Chem.* **1976,** *54,* 1929. (b) Biermann, H.W.; Morton, T.H. *J. Am. Chem. Soc.* **1983,** *105,* 5025.
285. Sherry, A.D. In "The Hydrogen Bond—Recent Developments in Theory and Experiments", Schuster, P., et al., Eds.; North-Holland: Amsterdam, 1976, Chapter 25.

286. Fraser-Monteiro, M.-L.; Fraser-Monteiro, L.; Butler, J.L.; Baer, T.; Hass, J.R. *J. Phys. Chem.* **1982,** *86,* 739.
287 Butler, J.L.; Fraser-Monteiro, M.-L.; Fraser-Monteiro, L.; Baer, T.; Hass, R.J. *J. Phys. Chem.* **1982,** *86,* 747.
288. Fraser-Monteiro, L.; Fraser-Monteiro, M.-L.; Butler, J.L.; Baer, T. *J. Phys. Chem.* **1982,** *86,* 752.
289. Butler, J.L.; Baer, T. *J. Am. Chem. Soc.* **1980,** *102,* 6764.
290. Willet, G.D.; Baer, T. *J. Am. Chem. Soc.* **1980,** *102,* 6769, 6774.
291. Kutina, R.E.; Edwards, A.K.; Goodman, G.L.; Berkowitz, J. *J. Chem. Phys.* **1982,** *77,* 5508.
292. Gilman, J.; Hsieh, T.; Meisels, G.G. *J. Chem. Phys.* **1983,** *73,* 1174.
293. Mautner (Meot-Ner), M.; Sieck, L.W.; Ausloos, P. *J. Am. Chem. Soc.* **1981,** *103,* 5342.
294. Sieck, L.W.; Mautner (Meot-Ner), M. *J. Phys. Chem.* **1982,** *86,* 3646.
295. Kirklin, D.L.; Domalski, E.S. *J. Chem. Thermodyn.* **1984,** *16,* 633.
296. Bickerton, J.; Minas de Piepade, M.E.; Pilcher, G. *J. Chem. Thermodyn.* **1984,** *16,* 661.
297. Hopkins, H.P., Jr.; Bostwick, D.; Alexander, C. *J. Am. Chem. Soc.* **1976,** *98,* 1355.
298. Greenberg, A.; Tomkins, R.P.T.; Dobrovolny, M.; Liebman, J.F. *J. Am. Chem. Soc.* **1983,** *105,* 6855.
299. Steele, W.V.; Gammon, B.E.; Smith, N.K.; Chickos, J.S.; Greenberg, A.; Liebman, J.F. *J. Chem. Thermodyn.* **1985,** *17,* 505.
300. Grabowski, J.J.; Simon, J.D.; Peters, K.S. *J. Am. Chem. Soc.* **1984,** *106,* 4615.
301. Byström, K.; Månsson, M. *J. Chem. Soc. Perkin Trans. 2* **1982,** 565.
302. Nelander, B.; Sunner, S. *J. Chem. Phys.* **1966,** *39,* 35.
303. Pihlaja, K.; Anderson, J.E.; Vainiotalo, P.; Mattinen, J. To be published.
304. Wiberg, K.B. *J. Am. Chem. Soc.* **1983,** *105,* 1227.
305. Wiberg, K.B.; Wendolowski, J.J. *J. Am. Chem. Soc.* **1982,** *104,* 5679.
306. Wiberg, K.B.; Lupton, E.C., Jr.; Wasserman, D.J.; de Meijere, A.; Kass, S.R. *J. Am. Chem. Soc.* **1984,** *106,* 1740.
307. Wiberg, K.B.; Squires, R.R. *J. Am. Chem. Soc.* **1981,** *103,* 4473.

CHAPTER 6

Energetics of Organometallic Species

Henry A. Skinner

University of Manchester, Manchester, England

Joseph A. Connor

University of Kent, Canterbury, Kent, England

CONTENTS

© 1987 VCH Publishers, Inc.
MOLECULAR STRUCTURE AND ENERGETICS, Vol. 2

1. INTRODUCTION

A review of the thermochemistry of organometallic compounds[1] published in 1964 made reference to the known measurements of enthalpies of formation ΔH_f°, of organometallic compounds and listed values for about 100 compounds, almost entirely of main group metals. A more recent review[2] gave available ΔH_f° values for more than 370 organometallic compounds, of which over 150 relate to compounds formed by the transition metals. The thermochemical interest in organic compounds of transition metals began in the early 1970s and has continued with increased effort since. There are now data for a variety of organocompounds of chromium, molybdenum, tungsten, manganese, rhenium, iron, titanium, platinum, and cobalt; but for other transition metals (and for the lanthanides), few reliable enthalpy of formation data have been reported and the thermochemistry remains to be investigated. During the past two decades, spectroscopic and other studies have provided a substantial data bank on the dissociation energies of diatomic molecules containing one or two metal atoms (eg, dimetals, M_2; intermetals, M_1M_2; metal carbides, MC; metal hydrides, MH; and metal oxides, MO), and more recently, of metal molecule-ions (eg, MCH_3^+, $M{=}CH_2^+$, MH^+, MO^+). This chapter examines the bond energies now available for various metal–ligand bonds and seeks trends in bond energy as the metal changes in moving across or down the periodic table.

2. CALORIMETRIC STUDIES

The thermochemical studies have mainly used one or more of three different calorimetric techniques: bomb–combustion calorimetry, solution–reaction calorimetry, and high-temperature "drop" microcalorimetry.

A. Bomb–Combustion Calorimetry

The conventional "static" bomb–combustion calorimeter has been used to measure the energies of combustion in oxygen of more than 100 organometallic compounds, but the method presents difficulties in that the oxidation of the metal on combustion can be incomplete, yielding an ill-defined mixture of oxides and suboxides of the metal. The use of "combustion aids" (eg, benzoic acid, paraffin oil, polyethylene), admixed with the organometallic compound in the combustion crucible, can considerably improve the completeness of combustion, although not for all metals. Rabinovitch, Tel'noi, and co-workers[3] at the Gorky Institute used combustion aids with various transition metal compounds, often mixing the compound with a large excess of auxiliary to achieve "nonexplosive" combustion in the bomb. The static bomb procedure can be satisfactory[4] for combustion of organocompounds of zinc, mercury, boron, tin, phosphorus, arsenic, antimony, and bismuth—and, with combustion aids, for organo compounds of chromium, manganese, iron, cobalt, nickel, titanium, molybdenum, tungsten, yttrium, and the lanthanide metals.

It has generally given unsatisfactory results for combustion of compounds of cadmium, aluminum, silicon, germanium, and lead, but in some of these cases a suitable rotating-bomb combustion procedure has been developed, capable of quality results of high precision. Examples include the combustion of organosilicon compounds by Good,[5] by Iseard,[6] and by Steele,[7] of organogermanium compounds by Bills and Cotton[8] and by Carson,[9] and of alkyl- and aryllead compounds by Good[10] and by Carson.[11]

B. Solution–Reaction Calorimetry

Many organometallic compounds are highly reactive at room temperature and rapidly undergo hydrolysis, oxidative hydrolysis, halogenation, or ligand displacement reactions in solution. Precision solution calorimeters are readily available and are easily adapted to the measurement of the enthalpies of these reactions, enabling heats of formation to be determined in favorable cases. Some examples are summarized below.

1. *Hydrolysis* was used in studies on zinc and cadmium alkyls,[12] butyllithium,[13] metal β-diketonates,[14] and pentamethyltantalum and hexamethyltungsten.[15] The permethyl compounds of tantalum and tungsten are unstable at room temperature in the solid state, and hydrolysis measurements were made on isopentane solutions of the permethyls at 0°C, using aqueous ether for hydrolysis of $TaMe_5$ and aqueous ammonia for hydrolysis of WMe_6. For example:

$$\begin{aligned} &TaMe_5\ (\text{isopentane, soln}) + 2.5H_2O\ (\text{liq}) + \text{excess (aq ether)} \\ &\qquad \rightarrow 0.5Ta_2O_5\ (\text{ppt}) + 5CH_4\ (\text{isopentane/ether}) \end{aligned} \tag{6-1}$$

Adedeji and co-workers[16] measured the enthalpies of acid hydrolysis of the dimethylamido complexes $Ta(NMe_2)_5$ and $W(NMe_2)_6$ to determine their heats of formation; in related studies, Lappert and colleagues[17] obtained ΔH_f° values for the dimethylamido complexes $M(NMe_2)_4$ and diethylamido complexes $M(NEt_2)_4$ of the group IV elements (M = Ti, Zr, Hf), from measurements of the enthalpies of *alcoholysis:*

$$M(NMe_2)_4(l) + 4\ i\text{-PrOH}(l) \rightarrow M(Oi\text{-Pr})_4(l) + 4\ NMe_2H(l) \tag{6-2}$$

Alcoholysis was also used[17] in studies on the group 4 tetraalkyls, $M(CH_2CMe_3)_4$, $M(CH_2SiMe_3)_4$, and $M(CH_2Ph)_4$ (M = Ti, Zr, Hf).

2. *Oxidative hydrolysis* is useful for thermochemical studies on organometallic compounds containing metal–metal bonds in which the metal is in a low oxidation state. The example of hexakis (dimethylamido) ditungsten, $[W_2(NMe_2)_6]$, is typical; this hydrolyzes rapidly to give an ill-defined black

powder precipitate, but in the presence of oxidizing agent the precipitate changes to yellow tungstic acid. Adedeji and co-workers[16] used potassium dichromate in sulfuric acid for oxidative hydrolysis, the measured enthalpy of reaction relating to the overall calorimetric reaction:

$$[W_2(NMe_2)_6, c] + [Cr_2O_7^{2-} + 14H^+ + H_2O](soln) \rightarrow 2H_2WO_4(ppt) + [2Cr^{3+} + 6NMe_2H_2^+](soln) \qquad (6\text{-}3)$$

Ferric chloride in acid solution was used as oxidative hydrolysis agent by Cavell and associates[18] in studies on the tetraacetates of dichromium and dimolybdenum, on the isopropoxy complexes[19] of dimolybdenum, $Mo_2(Oi\text{-}Pr)_6$ and $[Mo_2(Oi\text{-}Pr)_8]$, and on complexes of dimolybdenum and dichromium with 2-hydroxy-6-methylpyridine and *m*-dimethoxybenzene.[20]

3. *Halogenation* by direct reaction with bromine was used by Pedley and co-workers[21] to determine the enthalpy of formation of Me_3SnBr and of $Me_3SnSnMe_3$, from the reactions:

$$SnMe_4(l) + Br_2(g) \rightarrow Me_3SnBr(l) + MeBr(g) \qquad (6\text{-}4)$$

and

$$Me_3SnSnMe_3(l) + Br_2(l) \rightarrow 2Me_3SnBr(l) \qquad (6\text{-}5)$$

Similar studies[22] were made on $SnBu_4$, Me_3PhSn, $Me_3(C_2H_3)Sn$, $Me_3(CH_2Ph)Sn$, and[23] Me_3In, and both bromination and iodination measurements were made on several dialkyls of mercury.[24] Halogen abstraction from CCl_4 or CBr_4 by metal hydrides was used in reactions of the type:

$$[Mo(C_5H_5)_2H_2](c) + 2CCl_4(l) \rightarrow [Mo(C_5H_5)_2Cl_2](c) + 2CHCl_3(l) \qquad (6\text{-}6)$$

by Dias and co-workers[25] in Lisbon.

4. *Redistribution* reactions of the type:

$$Me_2Hg(l) + HgCl_2(c) \rightarrow 2HgMeCl(c) \qquad (6\text{-}7)$$

have been calorimetrically studied to obtain enthalpies of formation of alkyl mercury halides[24] and of alkyl tin halides.[26]

C. Microcalorimetry at Elevated Temperatures

The microcalorimeter has played an increasingly important role in the past decade for thermochemical studies on biochemicals and with organometallic compounds; this is mainly because a microcalorimetric measurement can be made using only a few milligrams of starting material, and the expense and difficulty of preparation of many biochemical and organometallic compounds of current interest is too high to meet the consumption (often as much as 1 g per measurement) of starting material by conventional combustion or reaction calorimeters. In the case of organometallic compounds, the conduction twin-cell microcalorimeter (designed to operate at elevated temperatures) enables study of reactions that occur only on heating.

The Calvet high-temperature microcalorimeter (Setaram, Lyon), used as a "drop-calorimeter,"[27] has been extensively applied in Manchester for studies in the range 100–500°C, mainly on transition metal organic compounds. The sample, usually weighing from 3 to 5 mg and contained in a thin glass capillary tube, is dropped from outside the calorimeter at room temperature down a narrow inlet tube to arrive in the "hot-zone" reaction vessel (maintained at a fixed elevated temperature) about 3 ft below in the center of the microcalorimeter block. An identical (empty) capillary tube is simultaneously allowed to fall into the twin "reference cell" to compensate the thermal disturbance due to the capillary tube container on arrival in the "live cell" of the twin system. The heat flow from the massive microcalorimeter block into the reaction vessel is by conduction through the thermopile surrounding the "live cell" and is recorded by the continuous thermogram trace of heat flow throughout the experiment. The area below the thermogram trace is proportional to the heat flow required to complete the reaction within the "live cell" of the microcalorimeter.

The "drop" microcalorimeter has been used for measurements of the heat of *thermal decomposition* of metal carbonyls (mono- and polynuclear) of the transition metals chromium, molybdenum, tungsten, manganese, rhenium, iron, ruthenium, osmium, cobalt, rhodium, and iridium.[27–31] The carbonyl samples, dropped into the hot zone (argon-filled) at a sufficiently high temperature (usually > 500 K), decompose to form the metal, either as a powder residue in the capillary container and/or as a bright metallic mirror deposit on the wall of the reaction vessel. The measured enthalpy, ΔH_{obs}, refers to the cell reaction, for example,

$$Mo(CO)_6,[c, 298] \rightarrow Mo(metal, T) + 6CO(g, T) \qquad (6\text{-}8)$$

measured at $T = 534$ K,

$$Fe_2Ru(CO)_{12},[c, 298] \rightarrow 2Fe(c, T) + Ru(c, T) + 12CO(g, T) \qquad (6\text{-}9)$$

measured at $T = 545$ K; ΔH_{obs} must be corrected to apply to isothermal disruption at 298 K. Thermal decomposition measurements have also been made on several bisarenes of chromium, molybdenum, and tungsten[32–34] and on carbonyl–arene complexes of the same elements, and cobalt.[32,35–37] Other thermal decomposition studies include those on iron–olefine–carbonyls,[38] carbonyl complexes of chromium, molybdenum, and tungsten with pyridine and acetonitrile,[39] amine–halide complexes of platinum $[Pt(NH_2R)_2X_2]$,[40] and platinum and palladium complexes with phosphine or diene ligands.[41]

In many of these studies, measurements were also made of the heats of thermal decomposition in the presence of *iodine* vapor, the samples being dropped into the hot zone previously charged with excess $I_2(g)$, as, for example, with:

$$[Mn_2(CO)_{10}](c, 298) + 2I_2(g, T) = 2MnI_2(c, T) + 10CO(g, T) \qquad (6\text{-}10)$$

measured[27] at $T = 514$ K. Iodination has an advantage thermochemically compared with thermal decomposition of carbonyl complexes, in that adsorption of CO occurs on freshly deposited metal surfaces but not on the metal iodides. Heats of thermal decomposition may be reduced as a result of adsorption, and it is difficult to correct for this. Accordingly, enthalpies of formation derived from iodination are usually subject to less error than those based solely on thermal decomposition results.

Bromination, as an alternative to iodination, has the further advantage that reaction takes place at lower temperatures, often below the temperature of onset of thermal decomposition of the organometallic compound under investigation. The method was examined in detail by Connor and co-workers[30] in studies on complexes of pentacarbonyl manganese, $[Mn(CO)_5R]$, for R = CH_3, CF_3, Ph, $PhCH_2$, $COCH_3$, COPh, $COCF_3$, Br and $Mn(CO)_5$, for example,

$$[Mn(CO)_5CF_3](c, 298) + 1.5Br_2(g, T) \rightarrow MnBr_2(c, T) + 5CO(g, T) + CF_3Br(g, T) \qquad (6\text{-}11)$$

at $T = 421$ K.

One of the most useful applications of the drop microcalorimeter has been for measurements of enthalpies of *sublimation.* The sample is dropped into the hot zone (argon-filled) held at a temperature *below* that required to induce thermal decomposition; then both live and reference cells are evacuated. The sample sublimes from the hot zone to condense in the cool exit tube, outside the calorimeter. Figure 6-1 shows the thermogram for vacuum sublimation[38] of a sample of $[Fe(C_4H_6)_2CO]$ recorded at 400 K.

$$[Fe(C_4H_6)_2CO](c, 298) \rightarrow [Fe(C_4H_6)_2CO](c, 400) \rightarrow [Fe(C_4H_6)_2CO](g, 400) \qquad (6\text{-}12)$$

Thermal decomposition studies at elevated temperatures on small samples can also be made using *differential scanning calorimetry,* as described

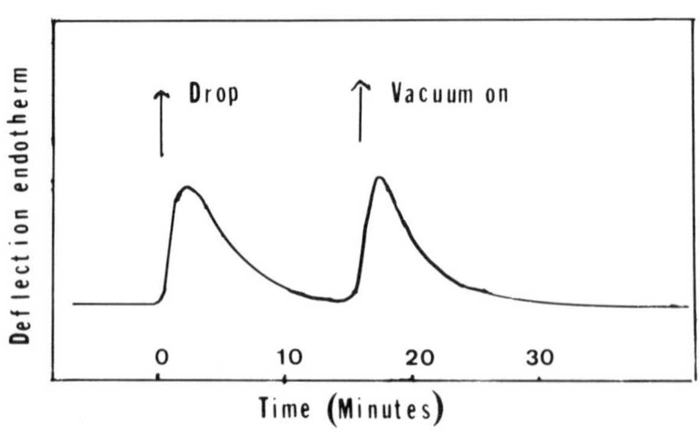

Figure 6-1. Thermogram trace for vacuum sublimation of $[Fe(C_4H_6)_2CO]$ at 400 K.

by Ashcroft and Mortimer.[42] The method has been extensively used on complexes $[MX_2L_2]$, where M is a first-row transition metal, X is a halogen, and the ligand L is pyridine, methylpyridine, quinoline, pyrazine, pyrimidine, or aniline.

3. KINETIC AND SPECTROSCOPIC STUDIES

The high lattice energies of most metals, and of their oxides, halides, carbides, and nitrides, demand high-temperature sources for the observation of the spectra of the diatomic species M_2, MF, MO, MC, and MN. Various arc sources, flames, discharge tubes, microwave discharges, and high-temperature furnaces, and different kinds of flash photolysis and flash heating have been used to generate the high-temperature transient species for spectral studies.[43] More recently, matrix isolation methods have been widely applied to the study of dimetals and small clusters,[44] but despite the technical procedures now available, relatively few optical spectroscopic studies have provided definitive values for the dissociation energies of M_2 and MX molecules. For diatoms of transition metals, one difficulty is the large number of low-lying atomic states, making for ambiguity in the correlation between atomic and molecular states.

Dissociation energies for M_2 and MX have mainly been obtained from high-temperature studies of equilibria involving the gaseous M_2 or MX species using the Knudsen cell–mass spectroscope combination.[45,46] The mass spectroscopic observation and determination of the dissociation energy of gaseous MoNb serves as an example.[47] A graphite-lined tantalum Knudsen cell, precharged with niobium powder and heated to about 3000 K, was then charged with molybdenum cuttings. The effusate from the cell (temperature range 2772–2963 K) was ionized by a 0.3-mA electron beam at 21-eV energy, and the currents due to the ions Nb^+, Mo^+, $MoNb^+$, and Mo_2^+ were measured at several temperatures in this range. (The ions were identified by their m/e ratio and isotopic distribution.) The measured ion intensities, proportional to the partial pressures of the ion precursors (Nb, Mo, MoNb, Mo_2), enabled evaluation of the equilibrium constants K_p for the cell reactions

$$\mathrm{Mo(g) + Nb(g) \rightarrow MoNb(g)} \qquad (6\text{-}13)$$

and

$$\mathrm{Nb(g) + Mo_2(g) \rightarrow MoNb(g) + Mo(g)} \qquad (6\text{-}14)$$

at temperatures T within the range 2772–2963 K. Enthalpy changes were determined from the third-law method,

$$\Delta H^\circ_\theta = -RT \ln K_p - T\Delta \frac{G^\circ_T - H^\circ_\theta}{T} \qquad \theta = 0 \text{ or } 298 \text{ K}$$

for each temperature T.

The third-law treatment of high-temperature equilibria requires the values of the free energy functions $(G^\circ_T - H^\circ_\theta)$ for the atoms and molecules involved, and those for the molecules often need to be estimated. The electronic partition function is difficult to decide if the ground state is unknown and low-lying excited states exist. In the case of MoNb, a doublet ground state was assumed; the average value obtained for ΔH°_{298} of reaction 6-13 from the third-law treatment of the measured K_p values was -452 ± 6 kJ/mol, and for reaction 6-14, $\Delta H^\circ_{298} = -48 \pm 5$ kJ/mol. Combination of the latter with the dissociation energy of Mo_2 gives D_{298} (MoNb) $= 454 \pm 25$ kJ/mol, in very good agreement with the direct measurement from reaction 6-13. Among the advantages of the Knudsen cell–mass spectroscopic method are the abilities to study several equilibria at the same time and to cross-check results.

Stepwise dissociation energies in polyatomic organometallic compounds (usually D_1 values) have been obtained from kinetic studies of pyrolysis, or of iodination, of alkyls of zinc,[48] cadmium,[49] mercury,[50] indium,[51] silicon,[52] and germanium.[53] Several kinetic studies, in solution, of reactions of metal–metal bonded carbonyls have identified the enthalpy of activation with that for homolytic fission of the metal–metal bond,[54] as in $Mn_2(CO)_{10}$, $Tc_2(CO)_{10}$, $Re_2(CO)_{10}$, and $MnRe(CO)_{10}$, in reactions with iodine or oxygen.[55] The values obtained for $D[(CO)_5M-M(CO)_5]$ (M = Tc and Re) of 160 and 166 kJ/mol are in fair agreement with other measurements or estimates, but the value $D = 154$ kJ/mol in $Mn_2(CO)_{10}$ obtained from kinetic studies is much higher than found from electron impact–mass spectroscopic studies,[56] and this result remains controversial.

Recent studies on the thermal decomposition of metal carbonyls [$Fe(CO)_5$, $Cr(CO)_6$, $Mo(CO)_6$, $W(CO)_6$] have used a pulsed laser pyrolysis technique.[57] A slow flow of SF_6/N_2, with small amounts of carbonyl, is passed through a thin cylindrical cell, a portion of which is irradiated by a uniform pulsed CO_2 laser beam. The SF_6 absorbs a fraction of the energy, and this vibrational energy is collision transferred to raise the temperature of the bath gas within milliseconds. Measurement of reaction yields is accomplished by mass spectrometric monitoring of reactant concentrations in the flowing gas mixture from the reaction cell. Arrhenius parameters were determined by comparative rate measurements, relative to the decomposition of dicyclopentadiene. These studies have provided D_1 values in $Fe(CO)_5$, $Mo(CO)_6$, and $W(CO)_6$ and have shown that in $Cr(CO)_6$ the rate-determining step is not the initial disruption, $D_1 \sim 152$ kJ/mol but a subsequent one, [probably D_2 for $Cr(CO)_5 \rightarrow Cr(CO)_4 + CO$], with activation energy ~ 167 kJ/mol.

The bonding of metal cations to various organic ligands has been examined by Staley and co-workers[58] using ion cyclotron resonance techniques for the measurement of equilibrium constants of ligand exchange in the gas phase. An example, $Al(EtOH)^+ + L \rightarrow AlL^+ + EtOH$, was studied for some 30 organic ligands (L = an alcohol, ether, aldehyde, ketone, ester, or alkyl cyanide), providing values for the *relative* bonding energies $D(\mathrm{Al-L})^+ - D(\mathrm{Al-EtOH})^+$, of the different ligands, L. A pulsed laser volatilization–ionization source was used to generate Al^+ ions.

Ion cyclotron resonance experiments have been adapted to the direct measurement of bond dissociation energies $D_0(\mathrm{M^+-X})$, for the transition metals chromium, manganese, iron, cobalt, and nickel, with X = H, CH_3, CH_2, and O, by Beauchamp and co-workers.[59,60] Ion beam reactions:

$$\begin{aligned}
&\mathrm{M^+ + R_2 \rightarrow \overset{+}{M}R + R} \qquad (\mathrm{R_2 = H_2, C_2H_6, C_2H_4, O_2})\\
&\mathrm{M^+ + cyclo\text{-}C_3H_6 \rightarrow \overset{+}{M}CH_2 + C_2H_4}\\
&\mathrm{M^+ + cyclo\text{-}C_2H_4O \rightarrow M^+O + C_2H_4}\\
&\mathrm{M^+ + N_2O \rightarrow M^+O + N_2} \qquad (6\text{-}15)
\end{aligned}$$

were examined, illustrating the versatility and potential of this novel method.

4. DEFINITIONS AND SYMBOLS

For a gaseous diatomic molecule MX, the dissociation process MX(g) → M(g) + X(g) involves an energy change ΔU_D° and an enthalpy change ΔH_D°, both of which are temperature dependent. At the absolute zero, $\Delta U_D^\circ = \Delta H_D^\circ = D_0$. At temperature T, $\Delta U_D^\circ(T) = \Delta H_D^\circ(T) - RT$, and at 298 K, $\Delta U_D^\circ(298) = D = \Delta H_D^\circ - 2.48$ kJ/mol. For a polyatomic molecule MX_n in the gaseous state (X = an atom), we have the enthalpy of *atomization* at 298 K:

$$\Delta H_a^\circ = \Delta H_f^\circ(\mathrm{M, g}) + n\,\Delta H_f^\circ(\mathrm{X, g}) - \Delta H_f^\circ(\mathrm{MX}_n, \mathrm{g})$$

and the energy of atomization at 298 K:

$$\Delta U_a^\circ = \Delta H_a^\circ - nRT = \Delta H_a^\circ - 2.48n$$

The *mean* bond dissociation energy $\overline{D}(\mathrm{M-X}) = \Delta U_a^\circ/n$.

For a molecule MR_n (R = free radical), the enthalpy of the disruption process at 298 K:

$$\mathrm{MR}_n(\mathrm{g}) \rightarrow \mathrm{M(g)} + n\mathrm{R(g)}$$

is given by $\Delta H_{\mathrm{dis}}^\circ = \Delta H_f^\circ(\mathrm{M}) + n\,\Delta H_f^\circ(\mathrm{R}) - \Delta H_f^\circ(\mathrm{MR}_n)$, and the energy of disruption by $\Delta U_{\mathrm{dis}}^\circ = \Delta H_{\mathrm{dis}}^\circ - 2.48\,n$

$$\overline{D}(\mathrm{M-R}) = \textit{mean}\text{ bond disruption energy} = \Delta U_{\mathrm{dis}}^\circ/n$$

For a molecule in which the central atom M is bonded to *different* ligands (eg, MR_aL_b), the energy of disruption at 298 K may be equated to the sum of contributions from the bonds in the molecule:

$$\Delta U_{\mathrm{dis}}^\circ = a\overline{D}(\mathrm{M-R}) + b\overline{D}(\mathrm{M-L})$$

and the evaluation of the $\overline{D}$ values requires an assumed value for one or other of these—for example, that $\overline{D}$(M−R) derived from MR_{a+b} can be transferred unchanged to the "mixed" molecule, MR_aL_b. In some mixed molecules, for example, $[(Et_3P)_2PtCl_2]$, the energies of *partial* disruption $[(Et_3P)_2PtCl_2, g] \rightarrow 2Et_3P(g) + PtCl_2(g)$ are available, to provide mean bond dissociation energies, $\overline{D}(Et_3P - PtCl_2)$.

Thermochemical data are used generally to obtain *mean* bond dissociation energies, but in conjunction with other data from spectroscopic and kinetic studies, the *stepwise* dissociation energies $D_1, D_2, \ldots, D_n$ can sometimes be evaluated, as in CH_4, where D_1 refers to the step $CH_4 \rightarrow CH_3 + H$, D_2 to the step $CH_3 \rightarrow CH_2 + H$, and so on. The mean value $\overline{D}$(C−H) is simply $(D_1 + D_2 + D_3 + D_4)/4$, but the individual stepwise dissociation energies are not equal to one another, nor to $\overline{D}$. Thermochemical data are also commonly used to obtain "bond energy" or "bond enthalpy" *contributions,* which should not be confused with bond dissociation energies. The values are obtained from the energies (or enthalpies) of *atomization* of the molecule, thence to be distributed among the bonds within the molecule. The bond contributions so obtained depend on the distribution scheme adopted and are arbitrary for this reason. As an example, consider tetramethyl silicon, for which $\Delta H_f^\circ(g) = -(233.2 \pm 3.5)$ kJ/mol, $\Delta H_a^\circ = 6166$ kJ/mol, and $\Delta U_a^\circ = 6126$ kJ/mol. The Laidler additivity scheme[61] applied to $SiMe_4$ equates ΔU_a° to the sum of the bond energy contributions in the molecule, $\Delta U_a^\circ = 4E(Si-C) + 12E(C-H, p)$, where C−H is a primary bond [as in C_2H_6 or $C(CH_3)_4$] with the value $E(C-H, p) = 408.6$ kJ/mol, leading to $E(Si-C) = 305.8$ kJ/mol for the Si−C bond energy contribution. This compares closely with the mean bond dissociation energy, $\overline{D}(Si-CH_3) = 312.3 \pm 3$ kJ/mol in $SiMe_4$, but larger differences between E and $\overline{D}$ values occur when the radical in M−R, or the ligand in M−L, undergoes substantial reorganization once liberated from its bonding with M (as, eg, for R = benzyl, or L = CO). $\overline{D}$ values, and stepwise dissociation energies, $D_1, D_2, \ldots, D_n$, are particularly relevant with respect to the thermal stability and chemical reactivity of organometallic compounds. The E values may relate more closely to "in-molecule" properties, such as equilibrium bond lengths.

5. AUXILIARY DATA

The evaluation of E values from thermochemical $\Delta H_f^\circ(g)$ values requires the enthalpies of formation of the gaseous atoms as auxiliary data. The values used here are collected in Table 6-1, chosen mainly from critically evaluated reference tables.[62–66] The evaluation of $\overline{D}$(M−R) requires $\Delta H_f^\circ(g)$ values for the free radicals R, also included in Table 6-1. Values $\overline{D}$(M−L) require $\Delta H_f^\circ(g)$ values of the free ligands L (eg, CO, pyridine, arenes), which were taken mainly from the critical tables by Pedley and Rylance.[67] Error limits

TABLE 6.1 Auxiliary Data: ΔH_f°(g) Values (kJ/mol)[a]

Element	ΔH_f°(g)	Element	ΔH_f°(g)	Element	ΔH_f°(g)	Element	ΔH_f°(g)
H	218.0	V	517	Rh	557	Ho	301
Li	159.3	Cr	397	Pd	375	Er	317
Be	324	Mn	284	Ag	285	Tm	232*
B	565	Fe	417	Cd	112	Yb	152
C	716.7	Co	426.5	In	240.4	Lu	428
N	472.7	Ni	429.5	Sn	301.2	Hf	622
O	249.2	Cu	337.6	Sb	265	Ta	783
F	79.4	Zn	130.5	I	106.8	W	855
Na	107.5	Ga	272	Cs	76.5	Re	775*
Mg	147	Ge	372	Ba	179	Os	790*
Al	330	As	302	La	430	Ir	665*
Si	450*	Br	111.9	Ce	423*	Pt	566
P	316.4	Rb	80.9	Pr	356	Au	369
S	277.0	Sr	160.5	Nd	328	Hg	61.4
Cl	121.3	Y	424	Sm	207	Tl	181
K	89.0	Zr	599	Eu	175	Pb	195
Ca	177.8	Nb	723*	Gd	397.5	Bi	208
Sc	378	Mo	657.5	Tb	389	Th	602
Ti	474	Ru	645*	Dy	290	U	535*

Compound	ΔH_f°(g)	Compound	ΔH_f°(g)	Compound	ΔH_f°(g)
CH_3	144 ± 3	C_5H_5	264 ± 9	OAc	-227 ± 10
C_2H_5	117 ± 5	DMP[b]	20 ± 12	Acac	-202 ± 20
Me_3CCH_2	30 ± 6	OMe	9 ± 8	NMe_2	123.5 ± 8
Ph	332 ± 12	OEt	-26 ± 8	NEt_2	69.4 ± 8
$PhCH_2$	204 ± 4	O*i*-Pr	-59.5 ± 8	mhp[c]	25 ± 10

[a]Asterisks indicate error limits that may reach ±10 kJ/mol (vs < ±4 kJ/mol im most cases).
[b]DMP = *m*-dimethoxyphenyl.
[c]mhp = 2-Oxy-6-methylpyridine.

are not given. They are mostly < ±4 kJ/mol, but where an asterisk is shown could reach ±10 kJ/mol.

6. METAL–METAL BONDS

Bond dissociation energies in homonuclear diatomic metals have been experimentally determined for about 50 metals, mainly by high-temperature studies using the Knudsen cell–mass spectroscopic combination. The $D_0(M_2)$ values in Table 6-2 are as recommended by Gingerich,[68] Huber and Herzberg,[69] and Glushko and Gurvich;[62] some values in parentheses have relatively high uncertainties. For the first- and second-row elements of the periodic table, the D_0 values, moving across the table, rise and fall as the valences of the ground state atoms involved similarly change—but there is

TABLE 6-2. $D_0(M-M)$ Values (kJ/mol) in Homonuclear Dimetals

Dimetal	$D_0(M-M)$	Dimetal	$D_0(M-M)$	Dimetal	$D_0(M-M)$
Li_2	102.0	Sc_2	(159 ± 21)	Nd_2	(80 ± 29)
Na_2	72.1	Ti_2	(126 ± 17)	Sm_2	(50 ± 21)
K_2	(50 ± 2)	V_2	<179	Eu_2	(29 ± 17)
Rb_2	(47 ± 1)	Cr_2	(151 ± 21)	Gd_2	(172 ± 33)
Cs_2	(42 ± 2)	Mn_2	(32 ± 29)	Tb_2	(128 ± 25)
Mg_2	(5 ± 0.5)	Fe_2	(100 ± 21)	Dy_2	(67 ± 29)
Ca_2	(12.5 ± 1)	Co_2	(167 ± 25)	Ho_2	(80 ± 17)
Sr_2	(9.6 ± 2.5)	Ni_2	(230 ± 21)	Er_2	(71 ± 29)
Ba_2	(9.0 ± 2.5)	Cu_2	(192 ± 5)	Tm_2	(50 ± 17)
Al_2	(165 ± 20)	Zn_2	(18.4 ± 6)	Yb_2	(17 ± 17)
Ga_2	(135 ± 8)	Y_2	(156 ± 21)	Lu_2	(138 ± 33)
In_2	(100 ± 8)	Zr_2	(~320)	Hf_2	(~335)
Tl_2	(59 ± 25)	Nb_2	(503 ± 10)	Ta_2	(~440)
Ge_2	(266 ± 17)	Mo_2	(404 ± 20)	W_2	(~550)
Sn_2	(202 ± 0)	Tc_2	(~300)	Re_2	(~400)
Pb_2	(82.5 ± 1)	Ru_2	(~330)	Os_2	(~440)
As_2	(382 ± 1)	Rh_2	(282 ± 21)	Ir_2	(~400)
Sb_2	(299 ± 4)	Pd_2	(105 ± 21)	Pt_2	(358 ± 15)
Bi_2	(195 ± 5)	Ag_2	(160 ± 6)	Au_2	(222 ± 2)
		Cd_2	(8.4 ± 2)	Hg_2	(7.5 ± 2)
		La_2	(243 ± 21)		
Se_2	(329 ± 0.5)	Ce_2	(238 ± 21)	Th_2	(285 ± 21)
Te_2	(258 ± 1)	Pr_2	(126 ± 29)	U_2	(218 ± 21)

no such regularity in passing along the transition metals ($Sc_2 \rightarrow Zn_2$; $Y_2 \rightarrow Cd_2$) nor along the row of lanthanides ($Ce_2 \rightarrow Yb_2$), as shown in Figures 6-2, 6-3, 6-4, and 6-5.

The D_0 values in transition metals and lanthanides have been discussed by Brewer and Wynn[70] in terms of the valence bond model, previously applied to bulk metals.[71] This attempts to identify the "valence state" configuration adopted by the metal and to evaluate the promotion energy needed to achieve it. The group II metals with zero-valent (^1S) ground states, need promotion to a divalent configuration (sp or sd) as a prerequisite to chemical bonding. The promotion energy is seemingly too high to allow other than very weak bonds in ground state dimetals of group II. The Pd atom (d^{10}, ^1S) is also zero valent, but in this case there is a low-lying divalent state (d^9s, $^3D_3$) only 78.5 kJ/mol above ground level. The dissociation energy to yield ground state Pd atoms is $D_0 = 105$ kJ/mol, but with respect to $^3D_3$ Pd atoms, the absolute binding energy D_0^* amounts to about 262 kJ/mol, directly comparable with the dissociation energy ($D_0 \sim 358$ kJ/mol) in Pt_2, where the ground state has the divalent $^3D_3$, d^9s configuration.

The lanthanide metals have ground state configurations $4f^n5d6s^2$ or $4f^n6s^2$

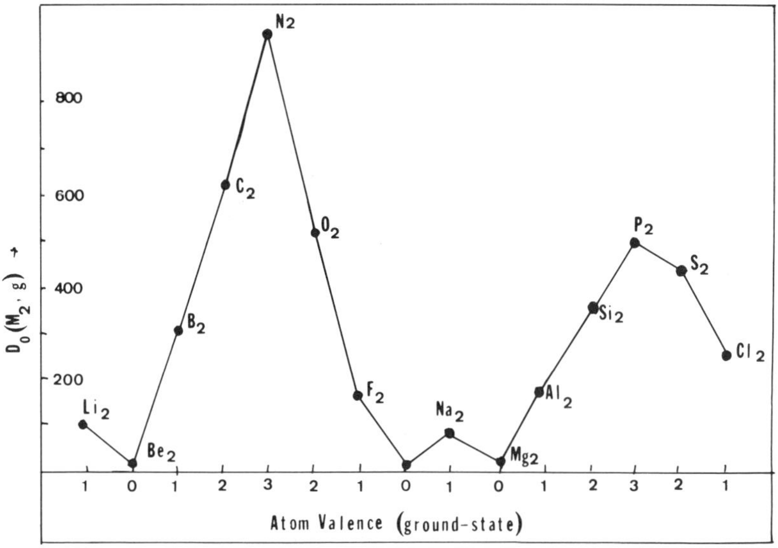

Figure 6-2. Plot of $D_0(M_2)$ values from Li_2 to Cl_2 against valence of the metal atom M in its ground state.

(n = 1–14), but the dimetals are considered to derive from valence states $f^{n-1}d^2s$, f^nsp or f^nds, in which the outermost $6s^2$ pair is promoted to d s or s p. The promotion energies to reach the lowest states of configuration f^nd^2s in the metal atoms from cerium to europium rise steadily from 28 to 155

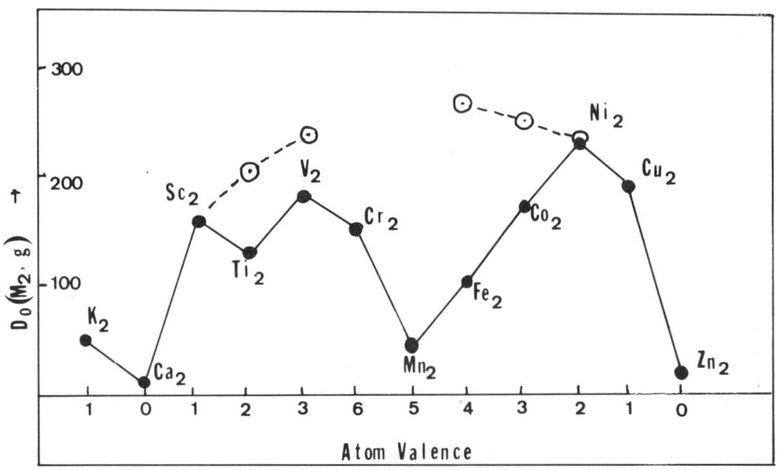

Figure 6-3. Plot of $D_0(M_2)$ values from K_2 to Zn_2 against valence of the metal atom M in its ground state. The dashed line plots D_0^*, measured with respect to the $d^{n+1}s$ valence state of M.

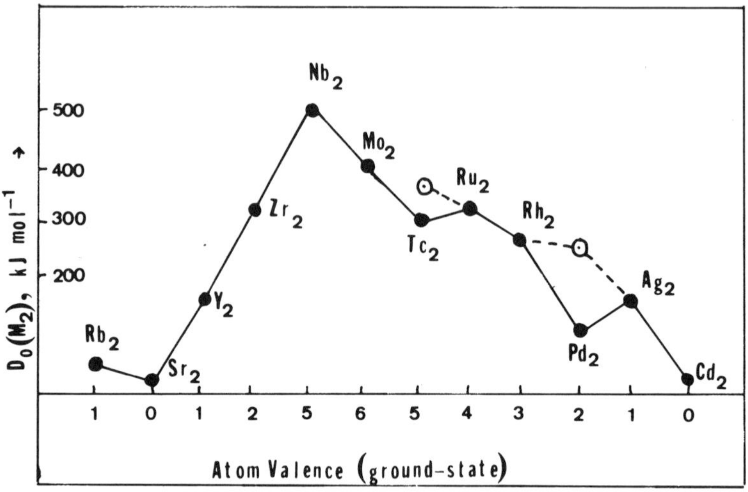

Figure 6-4. Plot of $D_0(M_2)$ values from Rb_2 to Cd_2 against valence of the metal atom M in its ground state. The dashed line plots D_0^*, measured with respect to the $d^{n+1}s$ valence state of M.

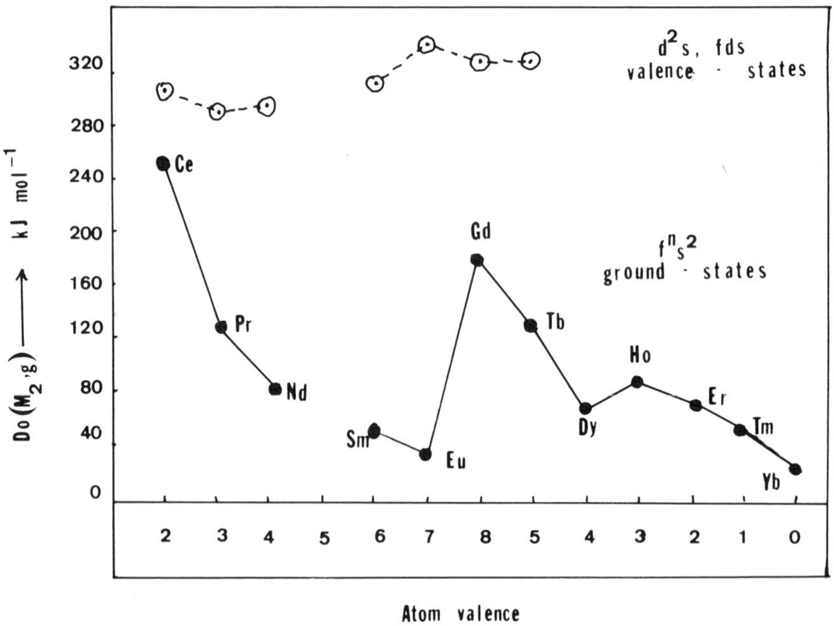

Figure 6-5. Plot of $D_0(M_2)$ values from Ce_2 to Yb_2 against valence of the metal atom M in its ground state.

kJ/mol, while the dissociation energies D_0 simultaneously fall sharply from about 238 to 29 kJ/mol; the dissociation energies $D_0^* = D_0 + 2x$, where x = promotion energy remain virtually constant, increasing gradually from $(294 \pm 20)^*$ in Ce_2 to $(338 \pm 17)^*$ kJ/mol in Eu_2. The sharp rise in D_0 at Gd_2 to about 172 kJ/mol coincides with a sharp fall at Gd for promotion to the f^7d^2s configuration (76 kJ/mol), corresponding to a valence state dissociation energy $D_0^* = 324 \pm 33$ kJ/mol.

A similar correlation may be noted with the first-row transition dimetals from Mn_2 to Ni_2, for which the ground states of the metal atoms are all d^ns^2. The promotion energies $d^ns^2 \rightarrow d^{n+1}s$ fall dramatically from Mn(204 kJ/mol) to Ni(2.5 kJ/mol), whereas the D_0 values rise from about 32 in Mn_2 to about 230 in Ni_2. The second-row transition metals from Nb to Pd, have $d^{n+1}s$ ground states, except in Tc(d^5s^2) and Pd(d^{10}).

Detailed discussion is hampered by lack of knowledge of the multiplicity of the ground states of most M_2 dimetals. Some transition dimetals are reported to have ground states of high multiplicity—notably Fe_2 (Mössbauer studies by Montano[72] indicate the septuplet, $^7\Sigma$), Sc_2 (the quintet $^5\Sigma$ is identified by Knight and co-workers[73]) and V_2, for which spectral studies have shown the triplet $^3\Sigma_g^-$ as ground state. The spectral studies on Cr_2 by Gurvich,[75] confirmed by Michalopoulos[76] indicate the $^1\Sigma_g^+$ configuration as ground state, consistent with a sextuple bond, and with the very short bond length (1.685 Å) in this molecule. The bond energy ($D_0 \sim 151$ kJ/mol) in Cr_2 is surprisingly low for sextuple bonding, and it has been suggested that d orbitals have only weak bonding power.[70,77] This is hardly borne out by other transition dimetals (eg, Nb_2, Mo_2), but it may be that the very shortness of the Cr—Cr bond in Cr_2 (and in V_2, where $r_e = 1.77$ Å) forces the σ component of the multiple bond into strain. (The σ bond is shortened from its normal length by at least 0.7 Å in Cr_2.)

The dashed lines plotted in Figures 6-3 and 6-4 are drawn through D_0^* values, measured with respect to $d^{n+1}s$ valence states as opposed to the actual d^ns^2 ground states (d^{10} in Pd) of the atoms concerned. These lines follow atom valence changes across the periodic table in a more regular fashion, as in Figure 6-2. But the point for Cr_2 (unchanged, since the ground state of Cr is already d^5s) is less than one-half expectation, emphasizing that there is a weakening factor, not yet properly identified.

The power of a metal atom to bind itself to a neighbor is employed to the maximum in the condensed state of the metal, wherein it surrounds itself (in most cases) with 12 near neighbors. The energy of *atomization* ΔU_a° is a measure of this *global* bonding power and may be compared with the *linear* bonding power, as measured by the energy of atomization of the dimer [ie, $\frac{1}{2}D_0^\circ(M_2)$]. This straight comparison is made in Table 6-3, in terms of a ratio, $G/L = 2\Delta U_a^\circ/D_0$. The high values of G/L for the group II metals emphasize the weakness of linear relative to global bonding in these metals—consistent with the view that promotion to the divalent state *is* achieved in the metallic state, but *not* in the dimetals. Most metals show G/L values in the range 3–

TABLE 6-3. G/L Values for Metals[a]

Metal	G/L	Metal	G/L	Metal	G/L	Metal	G/L
Li	3.1	Ga	4.0	Nb	2.9	Pt	3.2
Na	3.1	In	4.0	Ta	(3.5)	Ce	3.6
K	3.6	Tl	6.1	Cr	5.2	Pr	5.7
Rb	3.5	Ge	2.8	Mo	3.2	Nd	8.2
Cs	3.8	Sn	3.2	W	(3.1)	Sm	8.2
Cu	3.5	Pb	4.8	Mn	18	Eu	12
Ag	3.6	As	1.6	Tc	(4.2)	Gd	4.6
Au	3.3	Sb	1.8	Re	(3.9)	Tb	6.2
Mg	60	Bi	2.1	Fe	8.3	Dy	8.8
Ca	28	Sc	4.7	Ru	(3.9)	Ho	7.6
Sr	34	Y	5.4	Os	(3.6)	Er	9.0
Ba	40	La	3.5	Co	5.1	Tm	9.4
Zn	14	Ti	7.5	Rh	4.0	Yb	18
Cd	27	Zr	(3.7)	Ir	(3.3)	Lu	6.2
Hg	16	Hf	(3.7)	Ni	3.7	Th	4.2
Al	4.0	V	5.7	Pd	7.1	U	4.9

[a]Values in parentheses are for numbers having higher uncertainties.

5, but ytterbium, manganese, europium, iron, palladium, and titanium and several rare earths stand out as unusually weak in linear bonding power.

A. The *G*/*L* Ratio

The ratio G/L relates to *ground* state metal atoms. A more realistic measure would relate to the *valence* states of the metal when condensed, and when involved in dimetal bonding. Thus, in the case of Mg, the metallic state presumably originates from divalent Mg, sp, V_2, and ΔU_a° is to be replaced by $(\Delta U_a^\circ + x)$, where x is the required promotion energy. The D_0 value remains unchanged for van der Waals-type bonding, so that here $(G/L)^* > G/L$. For alkali metals, promotion is not an obvious necessity either in the metal or in the dimetal.

In diamond (and in the diamond structures of Si, Ge, and Sn), promotion from the divalent ^3P ground state to a tetravalent sp^3 state is a minimal requirement and ΔU_a^* exceeds ΔU_a, whereas the dimetals derive from ^3P ground state atoms, so that again $(G/L)^*$ is greater than G/L. For metals in general, promotion is necessary in the condensed state ($\Delta U_a^* > \Delta U_a$), and may be so, although to a lesser degree, in the gaseous dimetals ($D_0^* > D_0$). It is arguable to what extent promotion occurs, but in general G/L^* values are expected to be larger than the corresponding G/L values.

B. Bond Energy and Coordination Number

In most metal structures, the coordination number is 12 (face-centered cubic, fcc; hexagonal close-packed, hcp), or 8 (body-centered cubic, bcc). The average bond energy in fcc metals is directly obtained from $\overline{E}(M-M) = 2\Delta U_a^\circ/12$; in bcc structures, each metal is coordinated to eight near neigh-

TABLE 6-4. Bond Energy Contributions in Metallic Bonds

Metal	Structure	r (Å)	Coordination number	$\overline{E}$ (kJ/mol)
Li	bcc	3.04	8/6	33.5
Na	bcc	3.72	8/6	23
K	bcc	4.54	8/6	19
Rb	bcc	4.95	8/6	17.5
Cs	bcc	5.31	8/6	16.5
Cu	fcc	2.56	12	56
Ag	fcc	2.89	12	47.5
Au	fcc	2.88	12	61.5
Be	fcc	2.25	12	53
Mg	fcc	3.20	12	24
Ca	fcc	3.95	12	29.5
Sr	fcc	4.30	12	27
Ba	bcc	4.35	8/6	38
Hg	Rhombohedral	3.00	6	20
Al	fcc	2.86	12	54.5
Sc	hcp	3.28	6/6	63
Y	hcp	3.60	6/6	70.5
La	hcp	3.75	6/6	72
Lu	hcp	3.46	6/6	71
Si	Diamond	2.35	4	223
Ge	Diamond	2.45	4	184.5
Sn	Diamond	2.81	4	150.5
Pb	fcc	3.50	12	33
Ti	hcp	2.92	6/6	78.5
Zr	hcp	3.20	6/6	100
Hf	hcp	3.16	6/6	104
V	bcc	2.62	8/6	109
Nb	bcc	2.86	8/6	153
Ta	bcc	2.86	8/6	166
Cr	bcc	2.50	8/6	84
Mo	bcc	2.72	8/6	139
W	bcc	2.74	8/6	182
Mn(γ)	fcc	2.73	12	(47)
Tc	hcp	2.72	6/6	116
Re	hcp	2.75	6/6	129
Fe	bcc	2.48	8/6	88
Ru	hcp	2.67	6/6	107.5
Os	hcp	2.70	6/6	131.5
Co	fcc	2.51	12	71
Rh	fcc	2.69	12	92.5
Ir	fcc	2.71	12	111.5
Ni	fcc	2.49	12	71
Pd	fcc	2.75	12	62
Pt	fcc	2.77	12	94

bors at distance r_8, but also to six others at $r_6 \sim 1.15\ r_8$, and it is necessary to distribute the total bonding energy between the immediate and next-nearest coordination spheres. The values for $\overline{E}$ given in Table 6-4, which are based on a distribution that allocates 85% of the total energy to the eight nearest-neighbor bonds, refer to the bond lengths r_8 of these bonds. In hcp

TABLE 6-5. Intermetal D_0 Values (kJ/mol)[a]

M_AM_B	D_0	ΔD	M_AM_B	D_0	ΔD
LiNa	(86 ± 2)	—	AuSc	(277 ± 17)	89
LiK	(76 ± 2)	—	AuY	(304 ± 8)	118
LiRb	(74 ± 2)	—	AuLa	(335 ± 21)	103
LiCs	(72 ± 2)	—	AuLu	(328 ± 17)	153
NaK	(62 ± 1)	—	CuGe	(197 ± 17)	—
NaRb	(55 ± 4)	—	AgGe	(171 ± 21)	—
NaCs	(56 ± 2)	—	AuGe	(270 ± 5)	62
KCs	(46 ± 2)	—	CuSn	(168 ± 10)	—
CuLi	(189 ± 9)	50	AgSn	(134 ± 21)	−21
AgLi	(174 ± 6)	47	AuSn	(251 ± 8)	68
AuLi	(281 ± 7)	131	AuPb	(126 ± 42)	—
CuNa	(172 ± 17)	54	CuSi	(218 ± 6)	—
AgNa	(136 ± 11)	29	AgSi	(174 ± 10)	—
AuNa	(212 ± 13)	86	AgBi	(192 ± 42)	—
CuAg	(170 ± 11)	—	AuBi	(293 ± 8)	85
AuCu	(224 ± 5)	—	AuV	(238 ± 12)	38
AuAg	(201 ± 11)	—	AuNb	(297 ± 21)	−37
AlLi	(172 ± 15)	43	AuCr	(209 ± 17)	26
GaLi	(129 ± 15)	—	AgMn	(96 ± 21)	24
InLi	(87 ± 14)	—	AuMn	(188 ± 13)	104
CuAl	(170 ± 11)	—	AuFe	(188 ± 21)	39
AgAl	(172 ± 17)	—	AuCo	(218 ± 17)	25
AuAl	(322 ± 6)	131	AuRh	(229 ± 29)	−21
AgGa	(178 ± 6)	31	AuNi	(251 ± 21)	25
AuGa	(230 ± 38)	57	AuPd	(151 ± 21)	—
AgIn	(163 ± 6)	37	RhSc	(440 ± 11)	228
AuBa	(251 ± 10)				
AuMg	(243 ± 42)		RhY	(442 ± 11)	232
AuBe	280		RhLa	(525 ± 17)	263
AuCa	238		RhTi	(387 ± 15)	199
HgLi	10		RhV	(360 ± 29)	135
HgNa	5		RhEu	(232 ± 34)	142
HgK	6		RhTh	(547 ± 42)	264
HgRb	5		RhU	(516 ± 17)	268
HgCs	5				
			RuCe	(527 ± 25)	247
AuCe	(335 ± 21)	105	OsCe	(503 ± 33)	179
AuPr	(305 ± 21)	138	RhCe	(546 ± 25)	287
AuNd	(297 ± 21)	164	IrCe	(581 ± 25)	272
AuEu	(239 ± 11)	159	PdCe	(318 ± 17)	160
AuTb	(289 ± 33)	120	PtCe	(551 ± 25)	259
AuDy	(254 ± 20)	132	PtY	(470 ± 8)	234
AuHo	(264 ± 33)	131	PtLa	(496 ± 15)	201
AgEu	(123 ± 13)	55	PtLu	(398 ± 34)	176
AgDy	(124 ± 19)	20	PtTi	(394 ± 11)	182
AgHo	(120 ± 17)	—	PtTh	(547 ± 42)	227
GeCr	(166 ± 29)	—	IrY	(453 ± 16)	203
GeCo	(230 ± 21)	54	IrLa	(573 ± 12)	261
GeNi	(276 ± 13)	70	IrTh	(571 ± 42)	233
GePd	(259 ± 17)	120	RuV	(410 ± 29)	167

TABLE 6-5. Intermetal D_0 Values (kJ/mol)[a], contd.

M_AM_B	D_0	ΔD	M_AM_B	D_0	ΔD
BiGa	(155 ± 17)	—	RuTh	(588 ± 42)	281
BiTl	(117 ± 13)	—	LaY	(197 ± 17)	—
BiSn	(206 ± 8)	—	MoNb	(448 ± 25)	—
BiPb	(134 ± 8)	—	PdAl	(251 ± 12)	119

[a]Data mainly from Reference 68.

structures, there are usually small differences between the bond lengths to the six nearest and six next-nearest neighbors, and $\overline{E}$ values calculated from $\Delta U_a^\circ/6$ refer to a bond length intermediate between the extremes. The metallic bond lengths are usually decidedly longer than expected for a covalent M–M bond, and the $\overline{E}$ values are less than would apply to a bond of unit order. But in a number of organometallic compounds (the polynuclear carbonyls of the transition metals provide a good example), metal–metal separations as large as those in the metallic state do occur, and the $\overline{E}$ values in metals become relevant in considerations of the various bond energy contributions in such molecules.

C. Intermetal Bonds

Experimental values are available for the dissociation energies of more than 100 intermetal diatomic molecules, of which more than 30 contain gold as one of the metals. The values in Table 6-5 are mainly from Gingerich.[68] The values ΔD are differences $(D_0 - D_{calc})$, where $D_{calc} = [D_0(AA) \cdot D_0(BB)]^{1/2}$, using the $D_0^\circ(AA)$, $D_0^\circ(BB)$ values listed in Table 6-2 except for intermetals of tin and germanium, where single-bond $\overline{E}$ values (Table 6-4) are used. Small ΔD values (<20 kJ/mol) are not recorded.

The alkali metals have ΔD positive when combined with copper, silver, or gold. The group 1B metals usually show positive ΔD, particularly for the electronegative Au atom when combined with the more electropositive rare earth metals. The *negative* ΔD in AuNb and AuRh probably arise from using D_0 values from multibonded Nb_2 and Rh_2 when single-bonded M–M values are actually required. The large positive ΔD for intermetals of rhodium, cerium, platinum, and iridium point to multiple bonding in the intermetals of a higher order than in the parent dimetals. The high values of D_0 in the dimetals of gold with the alkali earths suggest ion pair bonding M^+Au^-, and the low D_0 values of the dimetals of mercury with the alkali metals indicate "van der Waals" bonding.

7. METAL–HYDROGEN AND METAL–ALKYL BONDS

Thermochemical, kinetic, and spectroscopic studies on the hydrides MH_n and alkyls MR_n(R = Me, Et, Pr, Bu) of the main group and subgroup B

TABLE 6-6. Mean Bond Energies in MH_n and MR_n (kJ/mol)

Compound	$\Delta H_f^\circ(g)^a$	ΔU_a°	$\bar{D}(M-H)$	$\bar{D}(M-R)$
LiH	139	235	235	
LiEt	58 ± 6	216		216 ± 6
LiBu	−(25 ± 5)	255		255 ± 5
MgH	230 ± 6	135	135 ± 6	
BH	446 ± 7	334	334 ± 7	
BH_3	92 ± 10	1120	373 ± 3	
BMe_3	−(123 ± 11)	1113		371 ± 4
BEt_3	−(149 ± 6)	1058		353 ± 2
BPr_3	−(237 ± 13)	1073		360 ± 4
BBu_3	−(291 ± 5)	1067		356 ± 2
AlH	249 ± 4	296	296 ± 4	
$AlMe_3$	−(81 ± 11)	835		279 ± 4
$AlEt_3$	−(164 ± 4)	837		279 ± 2
$AlPr_3$	−(239 ± 15)	846		282 ± 5
GaH	214 ± 5	273	273 ± 5	
$GaMe_3$	−(42 ± 6)	738		246 ± 2
$GeEt_3$	−(61 ± 8)	677		226 ± 3
InH	215 ± 2	241	241 ± 2	
InMe				~170
$InMe_3$	173 ± 7	492		164 ± 3
TlH			100 ± 5	
CH	597.4 ± 1	335	335 ± 1	
CH_4	−74.5	1653	413	
CMe_4	−167.4	1450		362.5
SiH	369 ± 8	297	297 ± 8	
SiH_4	35 ± 2	1277	319 ± 1	
$SiMe_4$	−(233 ± 4)	1249		312 ± 1
$SiEt_4$	−(265 ± 15)	1173		293 ± 4
GeH			309 ± 10	
GeH_4	91 ± 2	1143	286	
$GeMe_4$	−(71 ± 10)	1009		252 ± 3
$GeEt_4$	−(162 ± 8)	991		248 ± 2
$GePr_4$	−(227 ± 7)	969		242 ± 2
SnH			251 ± 12	
SnH_4	163 ± 2	1001	250	
$SnMe_4$	−(19 ± 2)	886		222
$SnEt_4$	−(45 ± 3)	804		201 ± 1
PbH			<154	
PbH_4	~268	789	197	
$PbMe_4$	136 ± 4	625		156 ± 1
$PbEt_4$	110 ± 5	544		136 ± 2
PH	232 ± 13	300	300 ± 13	
PH_3	5 ± 1	965	322	
PMe_3	−(101 ± 5)	842		281 ± 3
AsH			347	
AsH_3	66 ± 1	883	294	
$AsMe_3$	13 ± 10	714		238 ± 4
SbH_3	145 ± 1	764	255	
$SbMe_3$	32 ± 25	655		218 ± 8
BiH			280	

TABLE 6-6. Mean Bond Energies in MH_n and MR_n (kJ/mol) contd.

Compound	$\Delta H_f^\circ(g)^a$	ΔU_a°	$\bar{D}(M-H)$	$\bar{D}(M-R)$
$BiMe_3$	194 ± 14	437		146 ± 5
ZnH			83	
ZnMe				≃142
$ZnMe_2$	50 ± 2	363		182 ± 1
$ZnEt_2$	57 ± 7	303		151 ± 4
$ZnPr_2$	−(14 ± 23)	329		165 ± 12
$ZnBu_2$	−(52 ± 24)	323		162 ± 12
CdH			66	
CdMe				≃97
$CdMe_2$	106 ± 2	287		144 ± 1
$CdEt_2$	105 ± 3	234		117 ± 2
HgH			37	
HgMe				≃36
$HgMe_2$	94 ± 1	250		125 ± 1
$HgEt_2$	72 ± 2	218		109 ± 1
$HgPr_2$	36 ± 6	211		106 ± 3

[a]The $\Delta H_f^\circ(g)$ values refer in all cases to monomeric gaseous species. Values in parentheses are for numbers having higher uncertainties.

metals of groups II, III, IV, and V have provided values for M−H and M−R bond energies of these particular metals, but little is known for bonds of this type formed by transition metals or lanthanides. The available data (298 K) are listed in Table 6-6, taken mainly from Pilcher and Skinner,[2] Glushko and Gurvich,[62] Kondratiev,[64] and Huber and Herzberg.[69]

Comparison of the *mean* bond dissociation energies $\bar{D}(M-H)$ in MH_n and $\bar{D}(M-R)$ in MR_n shows that the M−H values are usually slightly larger than the corresponding M−R values. The *stepwise* bond dissociation energies $D_1, D_2, \ldots, D_n$ fluctuate above and below the mean $\bar{D}$ values, as shown in Table 6-7.

TABLE 6-7. Fluctuation of Stepwise Bond Dissociation Energies

	Dissociation energies (kJ/mol)				
Compound	D_1	D_2	D_3	D_4	$\bar{D}$
CH_4	434	462	422	335	413
SiH_4	396	245	339	297	319
GeH_4	346	—	—	309	286
BH_3	419	367	334	—	373
$SiMe_4$	355	—	—	—	312
$GeMe_4$	322	—	—	—	252
$SnMe_4$	290	—	—	—	222
$InMe_3$	198	124	170	—	164
$ZnMe_2$	197	166	—	—	182
$CdMe_2$	190	97	—	—	144
$HgMe_2$	214	36	—	—	125

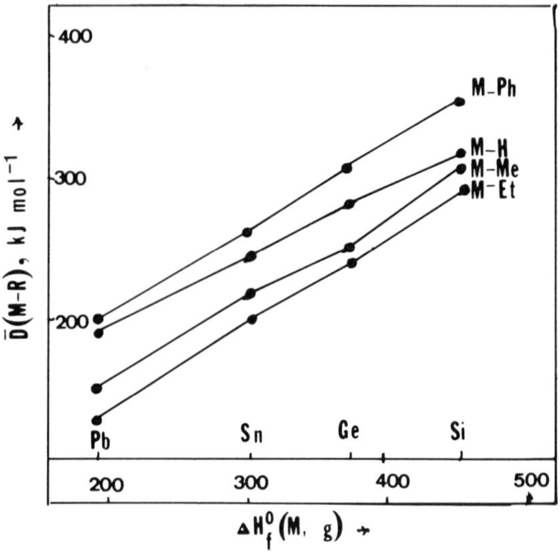

Figure 6-6. Plot of $\overline{D}$(M−R) values against ΔH_f°(M, g) for group IV metals, (R = H, Me, Et, Ph).

It is noteworthy that the D_1 values are in all cases larger than the mean, $\overline{D}$. The low values of D_2 for the group II metals reflect the need to promote from ^{1}S ground states to a divalent configuration (the minimal ^{3}P–^{1}S separations are 242, 386, 360, and 450 kJ/mol in Mg, Zn, Cd, and Hg). In like manner, promotion to a trivalent configuration from the monovalent ground states of group III metals is likely to lead to a low value for D_2—as appears to be the case in $InMe_3$ (the minimal ^{4}P–^{2}P separation in In is 418 kJ/mol, and 345 kJ/mol in B). Promotion from the divalent ^{3}P ground state to the tetravalent ^{5}S state in group IV elements should result in a relatively low value for D_2—as seen in SiH_4 (the ^{5}S–^{3}P separation in Si is 399 kJ/mol). The CH_4 molecule is anomolous in that the ground state of the methylene radical has the triplet $^3B_1$ configuration.

For the alkyls of the main group metals, there is a correlation between the mean bond dissociation energies $\overline{D}$(M−R) and the enthalpies of atomization ΔH_f°(M, g), of the metals.[2] Figure 6-6 shows this for $\overline{D}$(M−H), $\overline{D}$(M−Me), $\overline{D}$(M−Et), and $\overline{D}$(M−Ph), plotted against ΔH_f° (M, g) for the group IV elements (M = Si, Ge, Sn, and Pb). In the case of silicon, germanium, and tin (all of which have the "diamond" lattice in the solid state), ΔH_f°(M, g) is twice the M−M single-bond energy, so that the plot in Figure 6-6 simply states that $\overline{D}$(M−H) and $\overline{D}$(M−R) values increase as E(M−M) increases. It is interesting to note that the $\overline{D}$ values for lead alkyls fit the same plot, although the metal has coordination number 12 and does not adopt the diamond lattice structure. Figure 6-7 shows a similar plot for the group V elements (M = P, As, Sb, Bi); in the solid state, arsenic, antimony, and bismuth have a rhombohedral structure (coordination number 3,3), whereas P

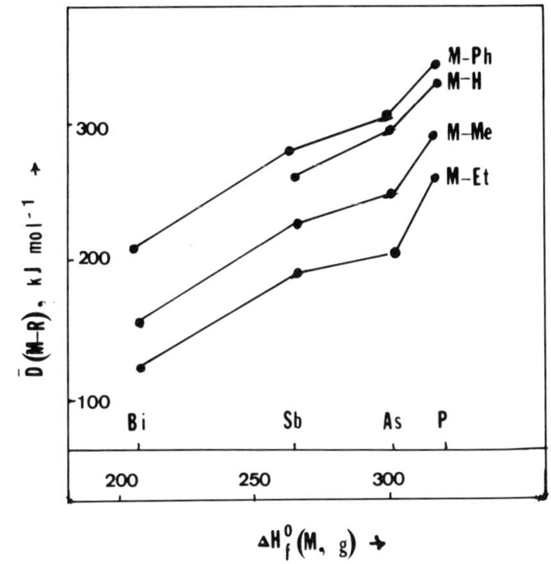

Figure 6-7. Plot of $\overline{D}(M-R)$ values against $\Delta H_f^\circ(M, g)$ for group V metals (R = H, Me, Et, Ph).

adopts the orthorhombic form (coordination number 3), and $\Delta H_f^\circ(g)$ of the metal is approximately 1.5 $E(M-M)$.

Thermochemical studies on transition metal alkyls are limited to the measurements by Adedeji and co-workers[16] on the heats of hydrolysis of $TaMe_5$ and WMe_6, and of Lappert and co-workers[17] on the heats of reaction with isopropanol of $M(CH_2CMe_3)_4$, $M(CH_2SiMe_3)_4$, and $M(CH_2Ph)_4$ (M = Ti, Zr, and Hf). Results are summarized in Table 6-8.

The $\overline{D}(W-Me)$ value is influenced by steric strain in this highly unstable molecule, and the relatively low values in the neopentyls of titanium and zirconium also suggest steric strain. The $\overline{D}$ values in the benzyls of these metals are reduced by virtue of the resonance stabilization of the benzyl

TABLE 6-8. Transition Metal–Alkyl Bonds (kJ/mol)

Compound	$\Delta H_f^\circ(g)$	ΔU_a°	$\overline{D}(M-R)$
$TaMe_5$	212 ± 30	1278	256 ± 6
WMe_6	772 ± 36	928	155 ± 6
$Ti(CH_2SiMe_3)_4$	−745 ± 34	1081	270 ± 9
$Ti(CH_2CMe_3)_4$	−157 ± 34	751	185 ± 10
$Ti(CH_2Ph)_4$	413 ± 34	867	217 ± 9
$Zr(CH_2SiMe_3)_4$	−786 ± 34	1288	322 ± 9
$Zr(CH_2CMe_3)_4$	−174 ± 34	883	221 ± 10
$Zr(CH_2Ph)_4$	353 ± 34	1052	263 ± 9
$Hf(CH_2CMe_3)_4$	−228 ± 34	960	240 ± 10

radical and extrapolate to approximate $\overline{D}$ values of 280 kJ/mol in $TiMe_4$ and 325 kJ/mol in $ZrMe_4$.

Metal–methyl bond energy contributions for titanium, zirconium, hafnium, molybdenum, tungsten, and platinum can be evaluated from thermochemical studies on biscyclopentadienyl complexes of these metals (Cp_2MX_2; Cp = C_5H_5), by Tel'noi and Rabinovitch[3] (heats of combustion) and by Dias, Calado, and co-workers[25] (reactions in solution). The evaluation requires a division of the *total* energy of the disruption reaction:

$$MCp_2X_2(g) \rightarrow M(g) + 2Cp(g) + 2X(g) \tag{6-16}$$

into contributions from the M—X and M—Cp bonds. The assumption made is that the M—Cl contributions in the dichloro complexes (M = Ti, Zr, Hf) are the same as in MCl_4. (Structural analysis[78] shows that the Ti—Cl bond length in Cp_2TiCl_2 is closely similar to that in $TiCl_4$.) This allows evaluation of the contribution from M—Cp which, on transfer to other complexes, enables the M—X contributions to be obtained. For the biscyclopentadienyl complexes of molybdenum and tungsten, the assumption made is that $\overline{D}$(Mo—Cl) and $\overline{D}$(W—Cl) in the hexachlorides $MoCl_6$ and WCl_6 apply in the compounds [Cp_2MoCl_2] and [Cp_2WCl_2]; similar values for M—Cp result by transfer of $\overline{D}$(Mo—Br) and $\overline{D}$(W—Br) from the hexabromides to the [Cp_2MBr_2] complexes. Comparison of $\overline{D}$(M—R) values in Tables 6-8 and 6-9 (M = Ti, Zr, W) shows only "moderate" agreement, which is not

TABLE 6-9. Biscyclopentadienyl Derivatives of Titanium, Zirconium, Hafnium, Molybdenum, Tungsten, and Platinum

Compound	$\Delta H_f^\circ(g)$	ΔU_a°	Basis	$\overline{D}$(M—X)
$(Cp)_2TiCl_2$	−(266 ± 9)	1501	$\overline{D}$(TiCl) = 428	Ti—Cp = 322
$(Cp)TiCl_3$	−(506 ± 12)	1598	$\overline{D}$(TiCl) = 428	Ti—Cp = 314
$(Cp)_2TiMe_2$	134 ± 12	1147	$\overline{D}$(TiCp) = 318	Ti—Me = 255
$(Cp)_2TiPh_2$	382 ± 12	1274	$\overline{D}$(TiCp) = 318	Ti—Ph = 319
$(Cp)_2Ti(CH_2Ph)_2$	279.5 ± 10	1121	$\overline{D}$(TiCp) = 318	Ti—C_7H_7 = 242
$(Cp)_2ZrCl_2$	−(433 ± 4)	1794	$\overline{D}$(ZrCl) = 486	Zr—Cp = 411
$(Cp)_2ZrMe_2$	37 ± 3	1369	$\overline{D}$(ZrCp) = 411	Zr—Me = 274
$(Cp)_2ZrPh_2$	368 ± 12	1414	$\overline{D}$(ZrCp) = 411	Zr—Ph = 296
$(Cp)_2HfCl_2$	−(429 ± 4)	1813	$\overline{D}$(HfCl) = 495	Hf—Cp = 412
$(Cp)_2MoCl_2$	4.6 ± 5	1414	$\overline{D}$(MoCl) = 304	Mo—Cp = 403
$(Cp)_2MoBr_2$	109 ± 20	1291	$\overline{D}$(MoBr) = 244	Mo—Cp = 402
$(Cp)_2MoI_2$	170 ± 9	1220	$\overline{D}$(MoCp) = 403	Mo—I = 207
$(Cp)_2MoH_2$	303 ± 6	1310	$\overline{D}$(MoCp) = 403	Mo—H = 252
$(Cp)_2MoMe_2$	354 ± 6	1110	$\overline{D}$(MoCp) = 403	Mo—Me = 152
$(Cp)_2WCl_2$	34 ± 5	1579	$\overline{D}$(W—Cl) = 343	W—Cp = 446
$(Cp)_2WBr_2$	112 ± 18	1482	$\overline{D}$(W—Br) = 292	W—Cp = 449
$(Cp)_2WI_2$	162 ± 9	1422	$\overline{D}$(W—Cp) = 447	W—I = 264
$(Cp)_2WMe_2$	359 ± 6	1299	$\overline{D}$(W—Cp) = 447	W—Me = 203
$(Cp)_2WH_2$	311 ± 6	1495	$\overline{D}$(W—Cp) = 447	W—H = 300
$(Cp)PtMe_3$	246 ± 3	1006		Pt—Cp = 366
$(Cp)_2PtMe_2$	214 ± 8	1158		Pt—Me = 213

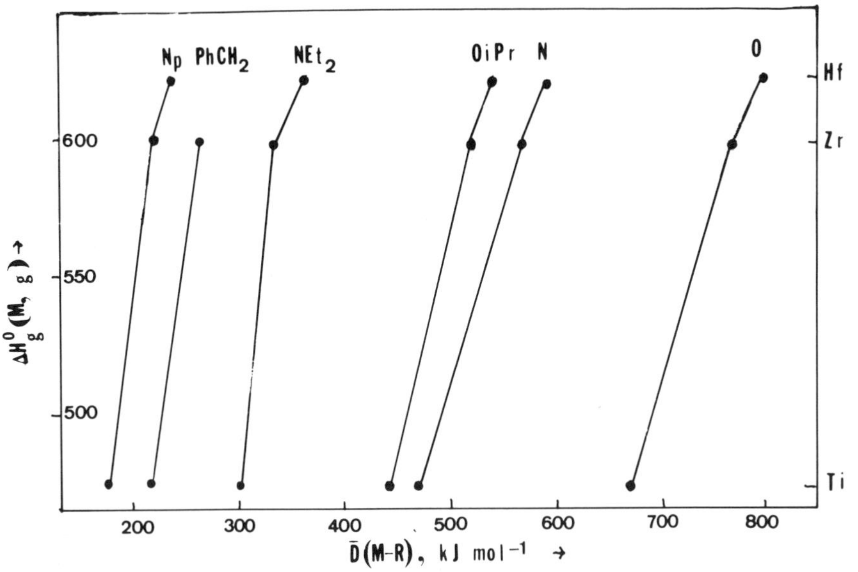

Figure 6-8. Plot of $\overline{D}(M-R)$ values against $\Delta H_f^\circ(M, g)$ (M = Ti, Zr, Hf; R = Me_3CCH_2, $PhCH_2$, NEt_2, O*i*-Pr, N, O).

surprising in view of assumptions made and combined experimental uncertainties.

The mean bond dissociation energies $\overline{D}(M-R)$ in MR_4 for the group IVA transition metals (M = Ti, Zr, Hf), with R = neopentyl (Np), $PhCH_2$ (Bz), NEt_2, and O*i*-Pr, are plotted against $\Delta H_f^\circ(g)$ of the metals in Figure 6-8. These energies (kJ/mol) are as follows:

Ti—Np	185	Ti—Bz	217	Ti—NEt_2	307	Ti—O*i*-Pr	447
Zr—Np	221	Zr—Bz	263	Zr—NEt_2	337	Zr—O*i*-Pr	517
Hf—Np	240			Hf—NEt_2	364	Hf—O*i*-Pr	535

Figure 6-8 also shows dissociation energies in the diatomic MO and MN molecules, and these follow the same general pattern. The group IVA metals have the same crystal form (hcp) at room temperature.

8. METAL CARBONYLS

The enthalpies of formation of the mononuclear carbonyls $M(CO)_n$, of chromium, molybdenum, tungsten, iron, and nickel lead directly to $\overline{D}(M-CO)$ values for these metals. The pulsed laser pyrolysis technique, recently applied to study the gas phase thermal decomposition kinetics of mononuclear carbonyls,[57] has provided *first* bond dissociation energies D_1 that differ significantly from the mean $\overline{D}$ values, namely:

	$Cr(CO)_6$	$Mo(CO)_6$	$W(CO)_6$	$Fe(CO)_5$	$Ni(CO)_4$
$\overline{D}$	105	149	176	116	144
D_1	152	167	190	171	102

in an irregular fashion. The D_1 and $\overline{D}$ values often differ by such magnitudes, or even larger. In the dialkyls of zinc, cadmium, and mercury, $D_1 > \overline{D} > D_2$, and in CH_4, D_1 exceeds D_4 by nearly 100 kJ/mol. Such differences are to be expected in MR_n, when the valence state of M differs from that of the ground state of the atom, since the valence state excitation energy is recovered during the stepwise disruption, but not uniformly at each step. The valence states of chromium, iron, and nickel in $Cr(CO)_6$, $Fe(CO)_5$, and $Ni(CO)_4$ are considered to derive from zero-valent d^6, d^8, and d^{10} configurations of these atoms. The excitation energies to reach these states in chromium and iron are quite high,[79] consistent with $D_1 > \overline{D}$ in $Cr(CO)_6$ and $Fe(CO)_5$. The excitation energy required to reach zero valent d^6 in molybdenum is almost certainly less than in chromium (^1I, d^6 in molybdenum lies 473 kJ/mol above ground level, but the same state in chromium has not been observed and probably lies above the ionization limit). The finding that $(D_1 - \overline{D})$ is smaller in $Mo(CO)_6$ than in $Cr(CO)_6$ may reflect this. In $Ni(CO)_4$, the reverse finding (ie, $\overline{D} > D_1$) suggests that the valence state change from zero-valent d^{10} to divalent d^8s^2 occurs at the first disruption step, $Ni(CO)_4 \rightarrow Ni(CO)_3 + CO$, coincident with a drop in excitation energy[79] from 176 to 88 kJ/mol.

The thermochemical data on polynuclear carbonyls [eg, $Ru_3(CO)_{12}$, $Ir_4(CO)_{12}$] are insufficient to lead directly to $\overline{D}(M-CO)$ values; Connor and co-workers,[28] on the basis of a subdivision of metal–carbonyl bonds into "terminal" (M—CO) and "bridging" (M—CO—M) types, and with estimated metal–metal contributions, have evaluated the "terminal" M—CO bond contributions in these molecules. A plot of such bond energies against ΔH_f° (M, g) in Figure 6-9 follows the pattern of similar plots in Figures 6-6, 6-7, and 6-8.

9. METAL ARENES AND CYCLOPENTADIENYLS

Enthalpies of formation have been determined[33,34] for a number of bisarenes of chromium and for bisbenzene molydenum and bistoluene tungsten. The derived mean bond dissociation energies, $\overline{D}$(Cr—arene) depend on the arene, and range from about 164 kJ/mol in $[Cr(C_6H_6)_2]$ to 144 kJ/mol in bisnaphthalene Cr. The $\overline{D}$ values along the series $[Cr(C_6H_6)_2] \rightarrow [Mo(C_6H_6)_2] \rightarrow [W(C_6H_5CH_3)_2]$ increase with increasing ΔH_f° of the metal (Figure 6-9). The enthalpies of formation of triscyclopentadienyls of scandium, yttrium, lanthanum, praseodymium, thulium, and ytterbium and of biscyclopentadienyls of transition metals from the first row, have been obtained from

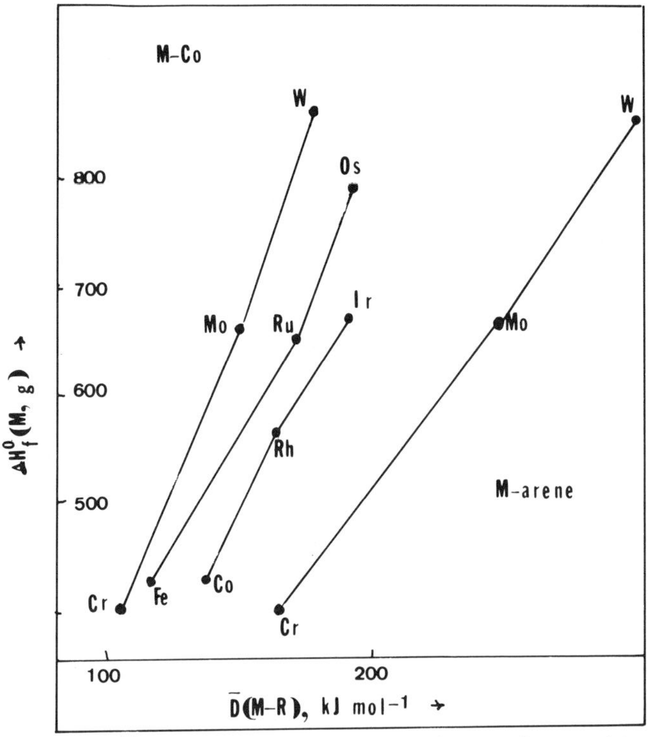

Figure 6-9. Plot of $\overline{D}$(M–CO) values against ΔH°_f(M, g) for transition metals.

measurements of their energies of combustion.[80,81] The derived $\overline{D}$(M–Cp) values are included in Table 6-10, where available bond dissociation energy values for metal–carbon bonds of different types are collected for comparison. The gaps in this table outnumber the data items, some of which are subject to considerable uncertainty (values in parentheses). Nevertheless, a clearly discernible pattern indicates that the strongest metal–carbon bonds occur in the metal carbide diatomic molecules, followed by M–Cp > M–arene > M–CH_3 > M–CO. This pattern matches expectation in that the bond orders of the metal–carbon bonds diminish in passing from MC to MCH_3; the bond order rises again at M–CO, and it is the large reorganization energy in the CO ligand on dissociation that is responsible for the lowering of $\overline{D}$(M–CO).

10. LIGAND REPLACEMENT ENTHALPIES

The enthalpies of formation of a variety of ligand-substituted metal carbonyls [$LM(CO)_{n-1}$, $L_2M(CO)_{n-2}$, $L_3M(CO)_{n-3}$, from $M(CO)_n$] have been determined, from which the heats of ligand replacement are readily derived. Examples include the replacement of benzene in [$(C_6H_6)Cr(CO)_3$] by a dif-

TABLE 6-10. Metal–Carbon Bond Energies (kJ/mol)[a]

Metal	D_0(MC)	$\bar{D}$(M−Cp)	$\bar{D}$(M−C_6H_6)	$\bar{D}$(M−CH_3)	$\bar{D}$(M−CO)
Sc	<440	360	—	—	—
Y	~417	381	—	—	—
La	453	355	—	—	—
Ti	<432	(318)	—	(255–280)	—
Zr	<556	(411)	—	(274–325)	—
Hf	<536	(412)	—	(340)	—
V	419	419	286	—	(156)
Nb	564	—	—	—	—
Ta	—	—	—	256	—
Cr	—	340	164	—	105
Mo	478	(403)	245	(152)	149
W	—	(447)	298	155 (203)	176
Mn	—	263	—	(153)	(93)
Tc	561	—	—	—	—
Re	—	—	—	(218)	(179)
Fe	—	350	—	—	116
Ru	645	—	—	—	(170)
Os	645	—	—	—	(189)
Co	—	323	—	—	(133)
Rh	580	—	—	—	(160)
Ir	622	—	—	—	(189)
Ni	—	298	—	—	144
Pd	430	—	—	—	—
Pt	606	(366)	—	260 (213)	—

[a]Values in parentheses are for numbers having higher uncertainties.

ferent arene[37] and the replacement of CO in a mononuclear carbonyl by a different ligand, such as pyridine or acetonitrile.[82,57] The displacement enthalpies ΔH_1, ΔH_2, ΔH_3 for the following reactions:

$$[(C_6H_6)Cr(CO)_3,g] + \text{arene}(g) \rightarrow [(\text{arene})Cr(CO)_3,g] + C_6H_6(g) \qquad \Delta H_1$$
$$[M(CO)_n, g] + L(g) \rightarrow [LM(CO)_{n-1}, g] + CO(g) \qquad \Delta H_2$$
$$[M(CO)_6, g] + 3L(g) \rightarrow [L_3M(CO)_3, g] + 3CO(g) \qquad \Delta H_3 \qquad (6\text{-}17)$$

are negative if the replacement ligand is more strongly bonded than the displaced ligand and are positive when the reverse is the case. Values of ΔH_1, ΔH_2, and ΔH_3 are listed in Table 6-11.

The weak bonding of methylbenzoate and acetophenone (relative to benzene) is consistent with electron withdrawal from the ring by the $COOCH_3$ and $COCH_3$ substituent groups; conversely, the stronger binding by the methyl-substituted benzenes and by dimethylaniline may be attributed to electron donation to the ring by alkyl and $-NMe_2$ substituents. The negative ΔH_2 for Ph_3P shows that phosphine donors are stronger than nitrogen donors and the ΔH_2, ΔH_3 values indicate that CO is at least as strong a donor as pyridine, and stronger than methyl cyanide and ethylene. Recent studies

TABLE 6-11. Ligand Replacement Enthalpies (kJ/mol)

Arene	ΔH_1	Ligand, L	ΔH_2	ΔH_3
C_6Me_6	−26	Ph_3P [into $W(CO)_6$]	−123	
Mesitylene	−12	$C_5H_{11}N$ [into $W(CO)_6$)]	+5	
$PhNMe_2$	−3	C_5H_5N [into $W(CO)_6$]	+27	+15
Toluene	+4	CH_3CN [into $W(CO)_6$]		+26
$PhOCH_3$	+14	C_5H_5N [into $Mo(CO)_6$]		+18
PhCl	+19	CH_3CN [into $Mo(CO)_6$]		+39
Naphthalene	+26	$C_5H_{11}N$ [into $Cr(CO)_6$]	+8	
$PhCOCH_3$	+44	C_5H_5N [into $Cr(CO)_6$]	+28	
$PhCOOCH_3$	+64	C_2H_4 [into $Fe(CO)_5$]	+22	

on complexes $[L_2PtCl_2]$ and $[L_2PdCl_2]$ have also indicated that phosphine donors are appreciably stronger than nitrogen donors,[40,41] the values $\bar{D}(L-PtCl_2)$ diminishing along the series L = i-Pr_3P > $(C_6H_{11})_3P$ > Et_3P > Ph_3P > olefin (in cycloocta-1,5-diene) > pyridine > methylamine > NH_3. The $\bar{D}(L-PdCl_2)$ values for L = Ph_3P, olefin, and MeCN are of similar magnitude to one another.

11. TRANSITION METAL MOLECULE-IONS

The ion beam studies by Beauchamp and co-workers[59,60] of reactions of first-row transition metal cations with H_2, C_2H_6, C_2H_4 and O_2 (Equations 6-15) have provided values for the dissociation energies $D(M^+-H)$, $D(M^+-CH_3)$, $D(M^+-CH_2)$, and $D(M^+-O)$. The D values relate to ground state ions, Cr^+, d^5; Mn^+, d^5s; Fe^+, d^6s; Co^+, d^8; and Ni^+, d^9.

Beauchamp has argued that the configurations d^n are less able to form strong σ bonds than are $d^{n-1}s$, so that the values obtained for binding to Cr^+, Co^+, and Ni^+ are relatively low. Accordingly, Table 6-12 gives both the experimental D_0 *and* adjusted values D_0^* (measured with respect to the *lowest* M^+, $d^{n-1}s$, configuration). Comparison of available $D_0(MX)$ of isoelectronic neutral molecules with $D_0^*(M^+X)$ shows these to be of similar magnitude.

12. METAL–OXYGEN AND METAL–NITROGEN BONDS

The dissociation energies $D_0(MO)$ of the monoxides of most metals are reported,[83] but relatively little is known of the energies of metal–oxygen bonds in the alkoxides $M(OR)_n$. The available data are listed in Table 6-13, where comparison is made in terms of the ratio $D_0(MO)/\bar{D}$ (M−OR). The average ratio of about 1.8 is consistent with a higher bond order in monoxide and a shorter bond length than in the corresponding alkoxide. Similar comparison is made, through the ratio D_0 (MN)/$\bar{D}(M-NR_2)$ of M−N bond

Table 6-12. $D_0(\overset{+}{M}-X)$ and $D_0^*(\overset{+}{M}X)$ Values (kJ/mol) for Transition Metals

Compound	$D_0(\overset{+}{M}-X)$	$D_0^*(\overset{+}{M}-X)$	Comparison
Cr^+H	146 ± 17	289*	
Cr^+CH_3	155 ± 29	298*	
$Cr^+{=}CH_2$	272 ± 29	415*	VC 419
Mn^+H	222 ± 12	222	CrH (275 ± 50)
Mn^+CH_3	297 ± 29		
$Mn^+{=}CH_2$	393 ± 25	393	
Fe^+H	243 ± 21	243	MnH (241)
Fe^+CH_3	285 ± 17	285	
$Fe^+{=}CH_2$	402 ± 21	402	
Co^+H	218 ± 17	256*	
Co^+CH_3	255 ± 12	293*	
$Co^+{=}CH_2$	356 ± 29	393*	
Ni^+H	180 ± 8	280*	CoH (309)
Ni^+CH_3	201 ± 21	301*	
$Ni^+{=}CH_2$	360 ± 25	460*	
Cu^+H	<60	<320*	NiH 297
Zn^+H	241	241	CuH 264
Zn^+CH_3	280	280	
Cd^+H	203	203	AgH 220
Cd^+CH_3	226	226	
Hg^+H	289	289	AuH 311
Hg^+CH_3	293		

[a]Asterisks indicate D_0 values measured with respect to the lowest M^+, $d^{n-1}s$, configuration.

Table 6-13. $\overline{D}(M-O)$ Values (kJ/mol) in Metal Alkoxides

Compound	$\Delta H_f^\circ(g)$	ΔH_{dis}°	$\overline{D}(M-R)$	$D_0(MO)$	Ratio, $D_0/\overline{D}$
$B(OMe)_3$	−900 ± 2	1492	495	805	1.63
$Si(OEt)_4$	−1338 ± 10	1684	419	794	1.89
$Ge(OEt)_4$	−964 ± 5	1232	306	654	2.14
$Ti(O\textit{i}\text{-}Pr)_4$	−1561 ± 10	1797	447	668	1.49
$Zr(O\textit{i}\text{-}Pr)_4$	−1715 ± 10	2076	517	772	1.49
$Hf(O\textit{i}\text{-}Pr)_4$	−1768 ± 10	2152	535	797	1.49
$P(OMe)_3$	−706 ± 6	1049	347	596	1.72
$As(OMe)_3$	−550 ± 2	879	291	480	1.65
$V(O\textit{t}\text{-}Bu)_4$	−1367 ± 8	1500	373	621	1.67
$Nb(OMe)_5$	−1333 ± 12	2101	418	765	1.83
$Ta(OMe)_5$	−1316 ± 12	2144	426	795	1.87
$Cr(O\textit{t}\text{-}Bu)_4$	−1276 ± 8	1289	320	436	1.36
$Mo(O\textit{i}\text{-}Pr)_4$	−986 ± 26	1406	349	556	1.59
$Ti(NEt_2)_4$	−485 ± 9	1237	307	465[a]	1.52
$Zr(NEt_2)_4$	−482 ± 8	1359	337	561[a]	1.66
$Hf(NEt_2)_4$	−566 ± 25	1466	364	590[a]	1.62
$Ta(NMe_2)_5$	−237 ± 15	1638	325	~607[a]	1.87
$W(NMe_2)_6$	268 ± 14	1328	219	—	—
$Mo(NMe_2)_4$	131 ± 8	1021	253	—	—

[a]$D_0(MN)$.

TABLE 6-14. $\overline{D}$(M—O) Values (kJ/mol) in Metal Acetylacetonates

Compound	ΔH_f°(g)	ΔH_{dis}°	$\overline{D}$(M—O)	D_0(MO)	Ratio, $D_0/\overline{D}$
$Al(acac)_3$	-1672 ± 5	1396	231	507	2.19
$Ga(acac)_3$	-1354 ± 6	1020	169	349	2.07
$In(acac)_3$	-1284 ± 6	918	152	<316	<2.08
$Cr(acac)_3$	-1431 ± 7	1222	202	436	2.15
$Mo(acac)_3$	-1202 ± 8	1254	208	556	2.68
$Mn(acac)_3$	$(-1253)^a$	931	156	399	2.55
$Fe(acac)_3$	$(-1189)^a$	1000	164	386	2.35
$Co(acac)_3$	$(-1240)^a$	1065	176	380	2.16
$Cu(acac)_2$	-700 ± 4	634	157	266	1.69
$Be(acac)_2$	-1156 ± 3	1076	268	431	1.61
$Ca(acac)_2$	$(-1210)^b$	984	245	436	1.78
$Zn(acac)_2$	$(-833)^b$	560	139	<267	<1.92
$Mn(acac)_2$	$(-937)^b$	817	203	399	1.97
$Pt(acac)_2$	-573 ± 6	735	183	387	2.11

$^a\Delta H_{sub}$ = 120 kJ/mol (assumed).
$^b\Delta H_{sub}$ = 100 kJ/mol (assumed).

energies in metal nitrides and in dialkylamino-metals, $M(NR_2)_n$, for the few examples for which data are known.

The enthalpies of formation have been determined for several metal acetylacetonates,[14] from which the enthalpies of disruption, $[M(acac)_n]$ (g) $\rightarrow$ M(g) + n(acac)(g), and the mean $\overline{D}$(M—O) values in Table 6-14 are derived. The values of the ratio $D_0(MO)/\overline{D}$(M—O) in the trisacetylacetonate complexes are generally greater than 2 and are larger than in the alkoxides. The ratio values are smaller (<2) in the bisacetylacetonates of copper, beryllium, calcium, and zinc—possibly because the zero-valent ground state of atoms of these elements are not parents of the biscomplexes—but the ratio goes back above >2 in $Pt(acac)_2$, which derives directly from the divalent ground state of Pt.[84]

Metal–nitrogen and metal–oxygen $\overline{D}$ values can also be evaluated in complexes $[M_2L_6]$ and $[M_2L_4]$ formed by dichromium, dimolybdenum, and ditungsten (Table 6-15). As an example, in hexakis(dimethylamino) dimolybdenum, the *partial* disruption to form the dimetal:

$$[Mo_2(NMe_2)_6, g] \rightarrow Mo_2(g) + 6\cdot NMe_2(g) \tag{6-18}$$

has $\Delta H_{dis}^\circ = (1524 \pm 25)$ kJ/mol, corresponding to $\overline{D}(Me_2N-Mo_2) = (252 \pm 4)$ kJ/mol. The similar disruption reaction in $[Mo(NMe_2)_4]$:

$$[Mo(NMe_2)_4, g] \rightarrow Mo(g) + 4\cdot NMe_2(g) \tag{6-19}$$

has $\Delta H_{dis}^\circ = (1020 \pm 18)$ kJ/mol, and $\overline{D}(Me_2N-Mo) = (253 \pm 4)$ kJ/mol. The virtual equivalence of the $\overline{D}$(Mo—N) values in the two compounds is matched by that of the $\overline{D}$(Mo—O) values in hexakis(isopropoxy) dimolybdenum and $[Mo(Oi\text{-}Pr)_4]$, where $\overline{D}(i\text{-}PrO-Mo_2) = 355 \pm 4$ kJ/mol and $\overline{D}(i\text{-}PrO-Mo) = 349 \pm 8$ kJ/mol. The Mo—N bond length in

TABLE 6-15. $\overline{D}(L-M_2)$ Values (kJ/mol) in M_2L_6 and M_2L_4

Complex[a]	$\Delta H_f^\circ(g)$	ΔH_{dis}°	$\overline{D}(L-M_2)$
$Cr_2(OAc)_4$	−1984 ± 28	1719	427
$Cr_2(mhp)_4$	−798 ± 10	1541	383
$Cr_2(DMP)_4$	−769 ± 22	1492	371
$Mo_2(NMe_2)_6$	128 ± 13	1524	252
$Mo_2(O\textit{i}\text{-}Pr)_6$	−1549 ± 14	2145	355
$Mo_2(OAc)_4$	−1806 ± 12	1809	450
$Mo_2(mhp)_4$	−597 ± 10	1608	400
$W_2(NMe_2)_6$	133 ± 11	(1768)[b]	(292)[b]

[a]mhp = 2-Oxy-6-methylpyridine; DMP = *m*-dimethoxyphenyl.
[b]Values having higher uncertainties.

$[Mo_2(NMe_2)_6]$ is only slightly longer (1.97 Ă) than in $[Mo(NMe_2)_4]$ (1.93 Å), and similar values for the bond energy contributions, $E(Me_2N-Mo)$, in these compounds are to be anticipated. But the extension of this equivalence to the mean bond dissociation energies, $\overline{D}(Me_2N-Mo_2)$ and $\overline{D}(Me_2N-Mo)$, requires no significant change in metal–metal bonding energy as a result of the disruption reaction (6-18)—despite the resulting sharp contraction in Mo−Mo bond length from the triple bond (2.21 Å) in $[Mo_2(NMe_2)_6]$ to the sextuple bond (1.93 Å) in the dimetal Mo_2. Cavell and co-workers[20] have constructed smooth energy–length plots for Cr−Cr and Mo−Mo bonds, using the metals themselves and complexes $[M_2L_6]$ and $[M_2L_4]$ as sources of data; the points for the dimetals Cr_2 and Mo_2 fall short of these, illustrating that the sextuple bonds are not noticeably stronger than the longer triple and quadruple bonds formed by chromium and molybdenum in dimetal complexes.

The value $\overline{D}(Me_2N-W_2)$ in $[W_2(NMe_2)]$ exceeds $\overline{D}(Me_2N-W)$ in $[W(NMe_2)_6]$ by about 74 kJ/mol. The W−N bond length is shorter by 0.05 Å in $[W_2(NMe_2)_6]$, and this, combined with the lower coordination number, is consistent with the higher $\overline{D}$ value. (Comparison with $\overline{D}$ in $[W(NMe_2)_4]$ cannot be made, since thermochemical data are not available.) The value $\overline{D}(Me_2N-W_2)$ is linked to the estimate $D_0(W_2)$ = 550 kJ/mol (Table 6-2); a lower estimate would increase the $\overline{D}(Me_2N-W_2)$ value, but only by one-sixth of the change made to $D_0(W_2)$.

REFERENCES

1. Skinner, H.A. *Adv. Organomet. Chem.* **1964,** *2,* 49.
2. Pilcher, G.; Skinner, H.A. In "The Chemistry of the Metal–Carbon Bond", Hartley, F.R.; and Patai, S., Eds.; Wiley: New York, 1982, Chapter 2.
3. Tel'noi, V.I.; Rabinovitch, I.B. *Usp. Khim.* **1977,** *46,* 1337.
4. Pilcher, G. In "Thermochemistry and Its Applications to Chemical and Biochemical Systems", NATO AS1 Series. Reidel: Dordrecht, Holland, 1984, pp. 353–382.
5. Good, W.D.; Lacina, J.L.; De Prater, B.L.; McCullough, J.P. *J. Phys. Chem.* **1964,** *68,* 579.

6. Iseard, B.S.; Pedley, J.B.; Treverton, J.A. *J. Chem. Soc. A* **1971,** 3095.
7. Steele, W.V. *J. Chem. Thermodyn.* **1978,** *10,* 445; **1983,** *15,* 595.
8. Bills, J.L.; Cotton, F.A. *J. Phys. Chem.* **1964,** *68,* 806.
9. Adams, G.P.; Carson, A.S.; Laye, P.G. *Trans. Faraday Soc.* **1969,** *65,* 113. Carson, A.S.; Carson, E.M.; Laye, P.G.; Spencer, J.A.; Steele, W.V., ibid, **1970,** *66,* 2459.
10. Good, W.D.; Scott, D.W.; Waddington, G. *J. Phys. Chem.* **1956,** *60,* 1090. Good, W.D.; Scott, D.W.; Lacina, J.L.; McCullough, J.P. ibid **1959,** *63,* 1139.
11. Carson, A.S.; Laye, P.G.; Spencer, J.A.; Steele, W.V. *J. Chem. Thermodyn.* **1972,** *4,* 783.
12. Carson, A.S.; Hartley, K.; Skinner, H.A. *Trans. Faraday Soc.* **1949,** *45,* 1159. *Proc. R. Soc. London Ser. A* **1949,** *195,* 500.
13. Fowell, P.A.; Mortimer, C.T. *J. Chem. Soc.* **1961,** 3793.
14. Ribeiro da Silva, M.A.V. In "Thermochemistry and Its Applications to Chemical and Biochemical Systems," NATO ASI Series. Reidel: Dordrecht, Holland, 1984.
15. Adedeji, F.A.; Connor, J.A.; Skinner, H.A.; Galyer, L.; Wilkinson, G. *J. Chem. Soc. Chem. Commun.* **1976,** 159.
16. Adedeji, F.A.; Cavell, K.J.; Cavell, S.; Connor, J.A.; Pilcher, G.; Skinner, H.A.; Zafarani-Moattar, M.T. *J. Chem. Soc. Faraday Trans. 1* **1979,** *75,* 603.
17. Lappert, M.F.; Patil, D.S.; Pedley, J.B. *J. Chem. Soc. Chem. Commun.* 1975, 830.
18. Cavell, K.J.; Garner, C.D.; Pilcher, G.; Parkes, S. *J. Chem. Soc. Dalton Trans.* **1979,** 1914.
19. Cavell, K.J.; Connor, J.A.; Pilcher, G., Ribeiro da Silva, M.A.V.; Ribeiro da Silva, M.D.M.C.; Skinner, H.A.; Virmani, Y.; Zafarani-Moattar, M.T. *J. Chem. Soc. Faraday Trans. 1* **1981**, *77,* 1585.
20. Cavell, K.J.; Garner, C.D.; Martinho-Simoes, J.A.; Pilcher, G.; Al-Samman, H.; Skinner, H.A.; Al-Takhin, G.; Walton, I.B.; Zafarani-Moattar, M.T. *J. Chem. Soc. Faraday Trans. 1* **1981,** *77,* 2927.
21. Pedley, J.B.; Skinner, H.A.; Chernick, C.L. *Trans. Faraday Soc.* **1957,** *53,* 1612.
22. Pedley, J.B.; Skinner, H.A. *Trans. Faraday Soc.* **1959,** *55,* 544.
23. Clark, W.D.; Price, S.J.W. *Can. J. Chem.* **1968,** *46,* 1633.
24. Hartley, K.; Pritchard, H.O.; Skinner, H.A. *Trans. Faraday Soc.* **1950,** *46,* 1019; **1951,** *47,* 254. Mortimer C.T.; Pritchard, H.O.; Skinner, H.A. ibid, **1952,** *48,* 220.
25. Dias, A.R.; Salema, M.S.; Martinho-Simoes, J.A. *J. Organomet. Chem.* **1981,** *222,* 69. Calado, J.C.G.; Dias, A.R.; Martinho-Simoes, J.A. ibid, **1980,** *195,* 203.
26. Nash, G.A.; Skinner, H.A.; Stack, W.F. *Trans. Faraday Soc.* **1965,** *61,* 640.
27. Connor, J.A.; Skinner, H.A.; Virmani, Y. *J. Chem. Soc. Faraday Trans. 1* **1972,** *68,* 1754.
28. Connor, J.A.; Skinner, H.A.; Virmani, Y. *Faraday Symp. Chem. Soc.* **1973,** *8,* 18.
29. Baev, A.K.; Connor, J.A.; El-Saied, N.I.; Skinner, H.A. *J. Organomet. Chem.* **1981,** *213,* 151.
30. Connor, J.A.; Zafarani-Moattar, M.T.; Bickerton, J.; El-Saied, N.I.; Suradi, S.; Carson, R.; Al-Takhin, G.; Skinner, H.A. *Organometallics* **1982,** *1,* 1166.
31. Al-Takhin, G.; Connor, J.A.; Skinner, H.A. *J. Organomet. Chem.* **1983,** *259,* 313.
32. Connor, J.A.; Skinner, H.A.; Virmani, Y. *J. Chem. Soc. Faraday Trans. 1* **1973,** *69,* 1218.
33. Connor, J.A.; Martinho-Simoes, J.A.; Skinner, H.A.; Zafarani-Moattar, M.T. *J. Organomet. Chem.* **1979,** *179,* 331.
34. Connor, J.A.; El-Saied, N.I.; Martinho-Simoes, J.A.; Skinner, H.A. *J. Organomet. Chem.* **1981,** *212,* 405.
35. Adedeji, F.A.; Brown, D.L.S.; Connor, J.A.; Leung, M.L.; Paz-Andrade, M.I.; Skinner, H.A. *J. Organomet. Chem.* **1975,** *97,* 221.
36. Brown, D.L.S.; Connor, J.A.; Demain, C.P.; Leung, M.L.; Martinho-Simoes, J.A.; Skinner, H.A.; Zafarani-Moattar, M.T. *J. Organomet. Chem.* **1977,** *142,* 321.
37. Al-Takhin, G.; Connor, J.A.; Skinner, H.A.; Zafarani-Moattar, M.T. *J. Organomet. Chem.* **1984,** *260,* 189.
38. Brown, D.L.S.; Connor, J.A.; Leung, M.L.; Paz-Andrade, M.I.; Skinner, H.A. *J. Organomet. Chem.* **1976,** *110,* 79. Connor, J.A.; Demain, C.P.; Skinner, H.A.; Zafarani-Moattar, M.T. ibid, **1979,** *170,* 117.
39. Adedeji, F.A.; Connor, J.A.; Demain, C.P.; Martinho-Simoes, J.A.; Skinner, H.A.; Zafarani-Moattar, M.T. *J. Organomet. Chem.* **1978,** *149,* 333.
40. Al-Takhin, G.; Skinner, H.A.; Zaki, A.A. *J. Chem. Soc. Dalton Trans.* **1983,** 2323.
41. Al-Takhin, G.; Skinner, H.A.; Zaki, A.A. *J. Chem. Soc. Dalton Trans.* **1984,** 371.
42. Ashcroft, S.J.; Mortimer, C.T. In "Thermochemistry of Transition Metal Complexes". Academic Press: London, 1970. Mortimer, C.T. In "Thermochemistry and Its Applications

to Chemical and Biochemical Systems", NATO ASI Series. Reidel: Dordrecht, Holland, 1984, pp. 289–315.
43. Cheetham, C.J.; Barrow, R.F. *Adv. High Temp. Chem.* **1967,** *1,* 7 (Academic Press: New York).
44. Kundig, E.P.; Moskovits, M.; Ozin, G.A. *Angew. Chem. Int. Ed. Engl.* **1975,** *14,* 292. Klotzbucher, W.; Ozin, G.A.; Norman, J.G.; Kolari, H.J. *Inorg. Chem.* **1977,** *16,* 2871.
45. Drowart, J.; Goldfinger, P. *Angew. Chem. Int. Ed. Engl.* **1967,** *6,* 581.
46. Cocke, D.L.; Gingerich, K.A. *J. Phys. Chem.* **1971,** *75,* 3264.
47. Gupta, S.K.; Gingerich, K.A. *J. Chem. Phys.* **1978,** *69,* 4318.
48. Price, S.L.W.; Trotman-Dickenson, A.F. *Trans. Faraday Soc.* **1957,** *53,* 1208.
49. Price, S.L.W.; Trotman-Dickenson, A.F. *Trans. Faraday Soc.* **1957,** *53,* 939.
50. Gowenlock, B.G.; Polanyi, J.C.; Warhurst, E. *Proc. R. Soc. London Ser. A* **1969,** *218,* 269. Kominar, R.J.; Price, S.J.W. *Can. J. Chem.* **1969,** *47,* 999. Pritchard, H.O. *J. Chem. Phys.* **1956,** *25,* 267.
51. Jacko, M.G.; Price, S.J.W. *Can. J. Chem.* **1964,** *42,* 1198.
52. Walsh, R. *Acc. Chem. Res.* **1981,** *14,* 246.
53. Dzarnoski, J.; Ring, M.A.; O'Neal, H.E., *Int. J. Chem. Kinet.* **1982,** *14,* 279.
54. Poe, A.J. In "Reactivity of Metal–Metal Bonds", Chisholm, M.H., Ed.; American Chemical Society Symposium Series No. 155. ACS: Washington, D.C., 1981, Chapter 7.
55. Fawcett, J.P.; Poe, A.J.; Twigg, M.V. *J. Organometer. Chem.* **1973,** *51,* C17; *J. Am. Chem. Soc.* **1976,** *98,* 1401.
56. Bidinosti, D.R.; McIntyre, N.S. *J. Chem. Soc. Chem. Commun.* **1966,** 555; *Can. J. Chem.* **1970,** *48,* 593.
57. Lewis, K.E.; Golden, D.M.; Smith, G.P. *J. Am. Chem. Soc.* **1984,** *106,* 3905.
58. Uppal, J.S.; Staley, R.H. *J. Am. Chem. Soc.* **1982,** *104,* 1229, 1235, 1238. Kappes, M.M.; Staley, R.H. ibid, **1982,** 1813, 1819. Jones, R.W.; Staley, R.H. ibid, **1982,** *104,* 2296; *J. Phys. Chem.* **1982,** *86,* 1387, 1669.
59. Armentrout, P.B.; Beauchamp, J.L. *Chem. Phys.* **1980,** *50,* 37; *J. Chem. Phys.* **1981,** *74,* 2819; *J. Am. Chem. Soc.* **1981,** *103,* 784, 6628.
60. Armentrout, P.B.; Halle, L.F.; Beauchamp, J.L. *J. Am. Chem. Soc.* **1981,** *103,* 6501, 6624; *J. Chem. Phys.* **1982,** *76,* 2449.
61. Laidler, K.J. *Can. J. Chem.* **1956,** *34,* 626.
62. "Thermodynamic Properties of Individual Substances", Vols. 1–4, Glushko, V.P.; and Gurvich, L.V., Eds.; Akad. Nauk: Moscow, 1978–1982.
63. "Thermochemical Constants of Substances", Vols. 1–10; Glushko, V.P.; Medvedev, V.A.; Bergman, G.A.; and Gurvich, L.V., Eds.; Akad. Nauk: Moscow, VINITI, 1965–1982.
64. Gurvich, L.V.; Karachevtsev, G.V.; Kondratiev, V.N.; Lebedev, Yu.A.; Medvedev, V.A.; Potapov, V.K.; Hodiev, Yu. S. "Energies of Dissociation of Chemical Bonds". Akad. Nauk: Moscow, 1974.
65. NBS Tables of Chemical Thermodynamic Properties, *J. Phys. Chem. Ref. Data,* **1982,** *11,* Suppl. 2.
66. Jones, M.N.; Skinner, H.A. *Annu. Rep. Chem. Soc. C,* **1982,** Chapter 2.
67. Pedley, J.B.; Rylance, J., "Sussex–NPL Computer-Analysed Thermochemical Data, Organic and Organometallic Compounds", University of Sussex, Brighton, UK, 1977.
68. Gingerich, K.A. *Faraday Symp. Chem. Soc.* **1980,** *14,* 109.
69. Huber, K.P.; Herzberg, G. Molecular Spectra and Molecular Structure, Vol. 4, "Constants of Diatomic Molecules". Van Nostrand, Reinhold: New York, 1979.
70. Brewer, L.; Winn, J.S. *Faraday Symp. Chem. Soc.* **1980,** *14,* 126.
71. Brewer, L. In "Phase Stability in Metals and Alloys", Rudman, P.; Stringer, J.; and Jaffee, R.L., Eds.: McGraw-Hill: New York, 1967, pp. 39–61, 241–249, 344–346, 560–568.
72. Montano, P.A. *Faraday Symp. Chem. Soc.* **1980,** *14,* 79.
73. Knight, L.B.; Van Zee, R.J.; Weltner, W. *Chem. Phys. Lett.* **1983,** *94,* 296.
74. Langridge-Smith, P.R.R.; Morse, M.D.; Hansen, G.P.; Smalley, R.E.; Merer, A.J. *J. Chem. Phys.* **1984,** *80,* 593.
75. Efremov, Yu.M.; Samoilova, A.N.; Kozhukhovsky, V.B.; Gurvich, L.V. *J. Mol. Spectrosc.* **1978,** *73,* 430.
76. Michalopoulus, D.L.; Geusic, E.; Hansen, S.G.; Powers, D.E.; Smalley, R.E. *J. Phys. Chem.* **1982,** *86,* 3914.
77. Wood, C.; Doran, M.; Hillier, I.H.; Guest, M.F. *Faraday Symp. Chem. Soc.* **1980,** *14,* 159.
78. Ponova, I.A.; Alekseev, N.B. *Dokl. Akad. Nauk. SSSR* **1967,** *174,* 614.

79. Skinner, H.A.; Sumner, F.H. *J. Inorg. Nucl. Chem.* **1957,** *4,* 245.
80. Devyatykh, C.G.; Rabinovitch, I.B.; Tel'noi, V.I.; Boisov, G.K.; Zyazina, L.F., *Dokl. Akad. Nauk. SSSR* **1974,** *217,* 673.
81. Tel'noi, V.E.; Kirynov, K.V.; Ermolaev, V.I.; Rabinovitch, I.B. *Dokl. Akad. Nauk. SSSR* **1974,** *209,* 197.
82. Al-Takhin, G.; Connor, J.A.; Pilcher, G.; Skinner, H.A.; *J. Organomet. Chem.* **1984,** *265,* 263.
83. Pedley, J.B.; Marshall, E.M. Thermochemical Data for Gaseous Monoxides. In "Industrial Uses of Thermochemical Data", Barry, T.I., Ed.; Royal Society of Chemistry: London, 1980, p. 83. *J. Phys. Chem. Ref. Data,* **1983,** *12,* 967–1032.
84. Al-Takhin, G.; Pilcher, G.; Bickerton, J.; Zaki, A.A. *J. Chem. Soc. Dalton Trans.* **1983,** 2657.

CHAPTER 7

Absolute Values of Gas Phase Proton Affinities and Basicities of Molecules: A Comparison Between Theory and Experiment

David A. Dixon

E.I. du Pont de Nemours & Company, Wilmington, Delaware

Sharon G. Lias

National Bureau of Standards, Gaithersburg, Maryland

CONTENTS

1. INTRODUCTION

Because proton transfer reactions are ubiquitous in chemistry, the study of acid–base reactions in solution has an extensive history. The interpretation

© 1987 VCH Publishers, Inc.
MOLECULAR STRUCTURE AND ENERGETICS, Vol. 2

of results from these studies, however, has not always been straightforward, since the acidity or basicity of a compound may be modified by the solvent so that experimentally observed thermochemical quantities may differ from those that can be attributed strictly to intrinsic molecular properties. In the past 20 years experimental techniques have become available that permit the quantitative study of proton transfer reactions in the gas phase, thereby allowing elegantly detailed interpretations of the effects of molecular structure on acid–base properties. Particularly in the years since 1971 when the first gas phase ion–molecule equilibrium studies on proton transfer appeared,[1] there has been a burgeoning of data on the thermochemistry of proton transfer reactions, and consequently, a great interest in using information from such gas phase studies to distinguish intrinsic molecular acidities and basicities from the chemical effects of solvation, and to interpret the thermochemistry of proton transfer in terms of molecular structure and bonding.[2]

Data on the thermochemistry of proton transfer processes in the gas phase have usually been presented in terms of two related quantities, the proton affinity (PA) and the gas basicity. Both the gas phase basicity and the proton affinity of a molecule M are defined in terms of the hypothetical reaction:

$$M + H^+ \rightarrow MH^+ \tag{7-1}$$

The gas phase basicity is the negative of the free energy change associated with this reaction; the proton affinity is the negative of the corresponding enthalpy change, or the $M-H^+$ bond strength. Thus, since:

$$-RT \ln K_{eq} = \Delta G_{Rn} = \Delta H_{Rn} - T\Delta S_{Rn} \tag{7-2}$$

the determination of an equilibrium constant for a proton transfer reaction:

$$MH^+ + N \rightleftarrows NH^+ + M \tag{7-3}$$

$$K_{eq} = \frac{[NH^+]}{[MH^+]} \cdot \frac{[M]}{[N]} \tag{7-4}$$

directly leads to the difference in the gas basicities of molecules M and N. If the entropy change of the reaction is determined or can be reliably estimated, a value for the enthalpy change of reaction, or relative proton affinity, is obtained. Most of the quantitative experimental data on PA values available at this writing have been generated through the determination of equilibrium constants of proton transfer reactions in mass spectrometers of various types in which the ratio $[NH^+]/[MH^+]$ can be directly observed in a mixture of known composition of M and N. When an interrelated set of such data exists, it is known as a thermochemical "ladder."

This chapter provides two related reviews on the current state of knowledge of the absolute values of gas phase proton affinities of neutral molecules derived from experiment and from theory. The extensive thermochemical ladders that have been derived from proton transfer equilibrium constant measurements provide relative gas phase basicities and relative proton affinities, but do not directly give any information about absolute values of proton affinities. Therefore, it is important to give attention to details of the assignment of absolute values to the proton affinity scale. Subsection B of Section 2 presents a brief review of a recent comprehensive evaluation[3] of the data on proton affinities, with particular attention to the rationale for the assignment of absolute values to the scale, based on heats of formation of relevant MH^+ ions and the corresponding neutral M molecules or molecular fragments. Since highly accurate gas phase values for heats of formation of cations are difficult to obtain, it is true in some cases (particularly for small M) that quantum chemical calculations can provide more accurate values for heats of formation of MH^+ ions than can experiment.[4] In Section 3, the available high-level theoretical determinations of proton affinities are compiled, reviewed, and evaluated.

2. THE EXPERIMENTAL PROTON AFFINITY SCALE

A. History

In the mid-1950s mass spectrometric instrumentation was developed sufficiently to permit the investigation of the elementary bimolecular reactions of positive ions with neutral molecules in ion sources of the conventional sector-field type of mass spectrometer by using the simple expedient of increasing the pressure in the instrument so that the ions generated in the gas could collide with neutral molecules before being collected. The earliest reactions studied[5] were, in fact, proton transfer reactions, such as:

$$CH_4^+ + CH_4 \rightarrow CH_5^+ + CH_3 \qquad (7\text{-}5)$$

In the early studies of proton transfer processes, the thermochemical information derived was necessarily crude, consisting mainly of the assignment of upper and/or lower limits to the relative values that could be assigned to proton affinities based on the assumption that only exothermic reactions would be fast enough to be observed in the ion source of a mass spectrometer. That is, the preferred direction of reaction 7-3 was assumed to be the exothermic direction. Furthermore, if:

$$MH^+ + N \rightarrow NH^+ + M \qquad (7\text{-}6)$$

was observed, it was assumed to be exothermic, and if it was not observed, one concluded that the reaction was probably endothermic. For example, in 1956 Tal'roze and Frankevich[5c] observed the occurrence of the reaction:

$$H_2S^+ + H_2O \rightarrow H_3O^+ + HS \qquad (7\text{-}7)$$

and the nonoccurrence of the reaction:

$$C_2H_2^+ + H_2O \rightarrow H_3O^+ + C_2H \qquad (7\text{-}8)$$

From these observations and known heats of formation of various relevant species, they concluded that the proton affinity of water lay between 163 and 172 kcal/mol. In 1965 Munson[6] used the approach (called bracketing) of observing the direction of proton transfer reactions in a series of molecules to establish the relative ordering of the gas basicities of the methylamines.

By the late 1960s mass spectrometric instrumentation had become sufficiently varied and sophisticated that several different experimental approaches were routinely being used for the investigation of ion–molecule reactions. These included high-pressure mass spectrometers equipped with photon sources for better energy resolution or with pulsed electron sources for time-resolved experiments, flowing afterglow and drift tube techniques in which the reactant ions are generated in a bath gas that ensures thermalization before reaction, and various ion cyclotron resonance (ICR) instruments in which the ions are trapped at low pressures for long times. Kebarle and Godbok[7] had been studying ion–molecule clustering equilibria of the type:

$$MH^+ + M \rightleftarrows M_2H^+ \tag{7-9}$$

since 1963 in a high-pressure mass spectrometer, but it was not until 1971 that simple bimolecular proton transfer equilibria (reaction 7-3) were studied in an ICR spectrometer.[1] Within months, numerous papers reported the observation of proton transfer equilibria in the gas phase in ICR spectrometers,[8] high-presssure mass spectrometers,[9] and flowing afterglow instruments.[10] Although initially there were in some quarters reservations about whether a thermodynamic equilibrium could actually be established at the low pressures of an ICR spectrometer, the results obtained under the different conditions of these various instruments were found to be in excellent agreement. Furthermore, studies of charge transfer equilibria carried out in an ICR generated scales of relative ionization potentials that reproduced energy intervals known independently from well-established ionization potentials of the relevant molecules.[11]

By the mid-1970s extensive interlocking ladders of free energy changes for proton transfer reactions had been published; in about one-fourth of these studies, equilibrium constants had been determined as a function of temperature, so that the entropy changes, and therefore the enthalpy changes, for the proton transfer reactions were known. In most cases, however, the experimentally determined gas basicity scales were converted to scales of proton affinities by making the simplifying assumption that the only significant contribution to the entropy change would arise from the ratios of rotational partition functions:

$$\begin{aligned}\Delta S_{\text{rot}} &= R\ln\frac{(q_{\text{NH}^+})(q_{\text{M}})}{(q_{\text{MH}^+})(q_{\text{N}})} \\ &= R\ln\left[\frac{(I_{\text{NH}^+})(I_{\text{M}})}{(I_{\text{MH}^+})(I_{\text{N}})}\right]^{1/2}[(\sigma_{\text{MH}^+})(\sigma_{\text{N}})/(\sigma_{\text{NH}^+})(\sigma_{\text{M}})]\end{aligned} \tag{7-10}$$

(where the q's are the rotational partition functions, the I's are the moments of inertia, and the σ's are the symmetry numbers of the designated species from reaction 7-3). It is usually further assumed that the ratio of the moments of inertia is close to unity, so that the entropy change can be estimated from an examination of the symmetry numbers alone.

The composite proton affinity scale resulting from work in several laboratories covered compounds of low proton affinity (ie, the scale from O_2 to N_2O for compounds having proton affinities between 100 and 137 kcal/mol), and hundreds of compounds whose proton affinities lay in the range 160–250 kcal/mol. Absolute values must be assigned to this scale on some comparison standard incorporated in the thermochemical ladder for which heats of formation of both M and MH^+ are independently available. These species are necessarily limited to those for which an MH^+ ion of known structure can be generated in a mass spectrometer so that a reliable independent gas phase heat of formation is known. During most of its existence, the upper portion of the proton affinity scale has traditionally been assigned absolute values relative to the proton affinity of ammonia. In some studies[9] the PA of ammonia was taken to be 207 kcal/mol, based on a determination of the heat of formation of NH_4^+ from the threshold energy for the fragmentation:[12]

$$C_2H_7N \rightarrow NH_4^+ + C_2H_2 + H + e \qquad (7\text{-}11)$$

[in which C_2H_7N is $C_2H_5NH_2$ and $(CH_3)_2NH$]. After the proton affinity difference between ammonia and isobutene had been determined through proton transfer equilibrium measurements, the value for the proton affinity of ammonia was effectively based on the proton affinity of isobutene (ie, based on the heats of formation of isobutene and the tertiary butyl ion). That is, although absolute values assigned to experimentally determined relative scales of proton affinities were usually cited relative to the proton affinity of ammonia, that value became in fact a secondary standard, and the value for isobutene, defined by:

$$iso\text{-}C_4H_8 + H^+ \rightarrow tert\text{-}C_4H_9^+ \qquad (7\text{-}12)$$

actually served as the primary standard for the scale. Unfortunately, in the years between 1975 and 1981, the value assigned to the heat of formation of the tertiary butyl ion, *tert*-$C_4H_9^+$, varied between 163 and 169 kcal/mol, and absolute values assigned to proton affinity determinations varied considerably from year to year and from paper to paper. (The value assigned to ammonia varied between 202.3 and 208.0 kcal/mol.) Furthermore, the proton affinity difference between isobutene and ammonia is not well established, since results from various laboratories show considerable variation in proton transfer equilibrium constant measurements involving ammonia.

Some of the early published studies on proton affinities derived from equilibrium constant measurements also suffered from uncertainties in the temperature at which the experiments were carried out. Since several of these data sets were very extensive and covered energy ranges of up to 72 kcal/mol, uncertainties of only a few percent in the operating temperature used in the original experiments translated into differences of several kilocalories per mole in the absolute values assigned to gas basicities or proton affinities when applied across such an energy range relative to a single absolute standard. For instance, early determinations of the thermochemical ladder connecting the proton affinities of H_2O and NH_3 reported that the PA difference of these species was 32 kcal/mol; corrections to the originally published values of ΔG, brought about by the realization that the operating temperatures of the instruments were slightly higher than originally reported, have led to a revision of this difference to 37 kcal/mol.

As a result of these problems, anyone consulting the original literature on determinations of proton affinities from measurements of gas phase proton transfer equilibrium constants will find a number of massive inconsistencies regarding the absolute values assigned to particular proton affinities (even in cases of actual equilibrium constant data that are in excellent agreement). Those working in the field are usually acutely aware of the nature of the arbitrary fluctuations in absolute values assigned to gas phase proton affinities, but workers in related fields who have sought to use the data have often fallen unwarily into the trap of using absolute proton affinity values from the literature without realizing that the cited values were based on relative thermochemical relationships and that the assigned absolute values were subject to change because of subsequent reevaluations of the thermochemistry of a primary standard.

Recently, the entire body of published data on proton affinities has been compiled and evaluated to provide a complete and internally consistent set of gas phase proton affinity values.[3] Absolute values were assigned to the resulting proton affinity scale using the best current information about the thermochemistry of positive ions, imposing the requirement of internal consistency. Recent studies have firmly established the positions of propylene and ethylene in the scale, thereby providing additional primary standards since the 298 K heats of formation of the ethyl and *sec*-propyl ions are well known. The discussion that follows briefly reviews the procedures used in that evaluation and explains how absolute values were assigned to the overall scale and how internal consistency in the data was achieved.

B. Evaluation of Experimental Proton Affinity Data

In the recent comprehensive evaluation of the proton affinity scale,[3] about 86% of the 780 compounds included had been studied through determinations of equilibirum constants for reaction 7-3. Thus, there were available several extensive overlapping thermodynamic ladders from various laboratories. These notably included the work of Bohme and co-workers[10,13] (flowing afterglow), Kebarle and coworkers[9e–f,14] (high-pressure mass spectrometry), the research group at the University of California at Irvine notably including Taft, McIver, and Hehre, and co-workers[8a,c–d,15] (ICR), Aue and Bowers and coworkers[1,8b,e,16] (ICR), and starting in the late 1970s, Meot-Ner and collaborators[17] (high-pressure mass spectrometry) and McMahon and co-workers[18] (ICR). The relative gas phase basicities of several series of compounds from these thermodynamic ladders were made the initial basis for the evaluation, using consistency between sets of data from different laboratories and matching of the intervals of the scale with those predicted from widely separated primary standards as the criteria for judgment. To minimize problems due to uncertainties in the temperature at which measurements were made, the thermodynamic ladders were related to as many local standards over the course of the scale as possible. If the original authors had reported that the initially reported temperatures needed to be revised, the revisions were carried out, multiplying the reported values of ΔG by T_{rev}/T_{orig}.

There are certain species for which measurements of equilibrium constants for proton transfer reactions are difficult or impossible. These include free radicals and species for which the MH^+ ion undergoes fast reactions with M or alternate reaction channels with N. In these cases, which constitute about 12% of the data in the evaluated compilation, the basicity is usually estimated using the technique known as "bracketing." With the availability of the extensive thermodynamic ladder that furnishes accurate gas basicity values for reference bases, results obtained from this approach have become somewhat more quantitative than those from the early "bracketing" studies described above. Essentially, one reacts MH^+ with a series of bases, observing whether proton transfer occurs:

$$MH^+ + B_1 \rightarrow B_1H^+ + M \tag{7-13}$$

$$MH^+ + B_2 \rightarrow \text{no proton transfer} \tag{7-14}$$

The basicity of M is assumed to lie between those of B_1 and B_2, and where these species have known gas basicity values, a value can be assigned to M. In the evaluation, after the standardization of the backbone thermodynamic ladder, results obtained from "bracketing" experiments were related to the evaluated scale, assuming that the gas basicity of the bracketed species lay between those of the two bracketing compounds. Results obtained through bracketing experiments are generally less reliable than those obtained through other methods because of numerous possible complications. For instance, exothermic proton transfer reactions sometimes do not occur when there is an energetically favorable alternate channel open to the reactants. Furthermore, endothermic proton transfer reactions can be observed if the reaction has a negative Gibbs free energy change, or if the reaction is not endothermic by more than a few kilocalories per mole. [Note, however, that the limiting rate constant for an observable reaction depends on the number of collisions the reactant ion undergoes during the experimental determination. In an ICR at temperatures in the 300–400K range, the effective lower limit for an observable reaction would be one with a rate constant of 10^{-11} cm^2/molecule-s, which corresponds approximately to a lowering of the probability of reaction from the collision rate by an energy barrier of 2.5–3 kcal/mol. In a flowing afterglow instrument or high-pressure mass spectrometer, where the reactant ion may undergo many more collisions during the determination, the effective limit on observable reactions would correspond to an energy barrier of 8–10 kcal/mol in most instruments.]

Finally, if there are several isomeric structures of B_1H^+, the observed proton transfer reaction may involve a rearrangement of the B_1 or B_1H^+ species in the reaction complex to a more stable structure, so that the observed "bracketing" does not necessarily reflect the thermochemistry of a straightforward or even unique proton transfer reaction. For example, it has been shown[19] that the straightforward proton transfer reaction:

$$MH^+ + trans\text{-}2\text{-}C_4H_8 \rightarrow sec\text{-}C_4H_9^+ + M \qquad (7\text{-}15)$$

occurs in competition with the proton transfer/rearrangement process:

$$MH^+ + trans\text{-}2\text{-}C_4H_8 \rightarrow tert\text{-}C_4H_9^+ + M \qquad (7\text{-}16)$$

for all MH^+ for which proton transfer is less exothermic than about 10 kcal/mol. When reaction 7-15 becomes endothermic, reaction 7-16 continues to be observed, with a probability that is greater, the greater the dipole moment of M. The interpretation of this effect in terms of the relevant potential surfaces for several such rearrangement processes occurring in ion–molecule collision complexes has recently been discussed.[19b]

Among the 780 compounds included in the compilation[3] are a few for which proton affinities were derived from other types of experiment. For example, heats of formation of a number of alkyl carbonium ions relative to the heat of formation of the tertiary butyl ion were available from studies in which hydride transfer equilibirum constants were determined in alkane mixtures:[20]

$$C_nH_{2n+1}^+ + C_mH_{2m+2} \rightleftarrows C_nH_{2n+2} + C_mH_{2m+1}^+ \qquad (7\text{-}17)$$

The heat of formation of the $C_mH_{2m+1}^+$ carbonium ion can be used to calculate an absolute proton affinity value for the corresponding olefin, C_mH_{2m}. Quantitative information about relative proton affinities has also been obtained through the determination of the energy barrier associated with endothermic proton transfer reactions.[21] Also, determinations of the equilibrium constants of processes such as:

$$B + AH^+ \rightleftarrows ABH^+ \qquad (7\text{-}18)$$

lead to values of heats of formation of the condensation or association ions, ABH^+. If the heats of formation of AH^+ and B are known, this information can then be used to derive the proton affinity of AB.[22]

In the final step of the evaluation, the experimentally determined scale of relative gas basicities was translated into a scale of proton affinities by evaluating the entropy change associated with reaction 7-3. For most polyatomic species, the entropy change of the half-reaction ($M \rightarrow MH^+$) was estimated by making the usual simplifying assumption, discussed above, that this quantity can be adequately approximated by the expression:

$$\Delta S = R \ln \left[\frac{\sigma(M)}{\sigma(MH^+)} \right] \tag{7-19}$$

The use of this expression is obviously inexact, in some cases enough so that a more exact calculation of the entropy change from the complete partition function is warranted. This is especially true for species in which internal rotations are lost or gained upon protonation [eg, $(CH_3)_2C{=}CH_2 + H^+ \rightarrow (CH_3)_3C^+$,] or in which there is a significant difference in the moments of inertia of the protonated and unprotonated species (eg, $NH_3 + H^+ \rightarrow NH_4^+$). In such cases, the entropy change was calculated from the complete partition function. In all cases for which experimental entropy change determinations of reaction 7-3 had been made, the experimental results became the basis of the evaluation of the relevant entropy changes.

C. Assignment of Absolute Values to the Scale: Reference Standards

Absolute values were assigned to the proton affinity scale based on the proton affinities that can be assigned to several primary standard species whose positions in the PA scale are known and for which well-established values of the heats of formation of both M and MH^+ are available. Experimentally, most independently available values for heats of formation of MH^+ species have been obtained from determinations of the ionization potential of MH:

$$MH \rightarrow MH^+ + e \tag{7-20}$$

or the appearance potential of MH^+ from a larger molecule:

$$N \rightarrow MH^+ + A + e \tag{7-21}$$

Of course, such values are known only in cases for which MH is a species stable enough to be introduced into a mass spectrometer or generated in situ, or for which an MH^+ ion of known structure exists as a fragmentation product from some larger molecular species. Such absolute values for proton affinities are available for only a very few species of the 780 compounds on which data have been reported, largely because in most cases the neutral MH molecule does not exist, and the energetics of formation in ionic fragmentation processes of an MH^+ ion of the required structure is not reliably established.

In the discussion that follows, we present details of the evaluations of the proton affinities of the compounds that served as the primary standards for assigning absolute values to the gas basicity and proton affinity scales. For the lowest part of the scale (H_2 to C_2H_6), several primary standards—namely, carbon monoxide, carbon dioxide, the oxygen atom, and the oxygen molecule—are available. For the part of the scale above ethylene (PA = 162.6 kcal/mol), the absolute values assigned to the scale were standardized relative to the proton affinities of ethylene, propylene, and isobutene. The heats of formation of $C_2H_5^+$ and $CH_3C^+HCH_3$ have been carefully studied in numerous laboratories and are now rather well established. The heat of formation of the *tert*-butyl ion is somewhat less well established, but recent evaluations of this heat of formation give a proton affinity value for isobutene that is in fairly good agreement with information from other comparison standards in the immediate vicinity in the scale (ketene and ammonia). Although the proton affinity of ammonia was not utilized as a standard in the evaluation,[3] we begin with an analysis of that quantity, since many relative PA data quoted in the literature are given relative to it. The discussion also presents details of the evaluation of the proton affinities of some other compounds (ketene, formaldehyde, water) for which independently determined heats of formation of MH^+ are available. These data serve to confirm the assignment of absolute values to the proton affinity scale; the compounds also served as secondary standards in the evaluation. The relevant heats of formation and evaluated 300 K proton affinity values of the standard compounds are summarized in Table 7-1.

It should be noted that the heats of formation of positive ions cited here are given using the so-called stationary electron convention commonly used by mass spectrometrists.[23] According to this convention, the integrated heat capacity of the electron in Equations 7-20 and 7-21 is taken as zero at all temperatures, and the thermal energy of the electron is not taken into

TABLE 7-1. Summary of Thermochemical Data (kcal/mol) on Comparison Standards for Proton Affinity Scale

Standard (M)	ΔH_f(M) at 298 K	$\Delta H_f(MH^+)$[a]	Proton affinity[b]	Gas basicity
NH_3	−11.02	150.7	204.0 ± 3	195.6
Primary standards				
$(CH_3)_2C{=}CH_2$	−4.04	165.8	195.9 ± 1.5	187.3
$CH_3CH{=}CH_2$	4.83	191.0	179.5 ± 0.8	171.7
From thermochemical ladder, relative to ethylene:			179.7	
$CH_2{=}CH_2$	12.48	215.6	162.6 ± 1	155.6
CO	−26.4	197.4	141.9 ± 2	134.4
CO_2	−94.05	140.8	130.9	124.4
O	59.56	308.96	116.3 ± 1	110.7
O_2	0.00	262.4	100.9 ± 0.5	95.0
Secondary and confirming standards				
$CH_2{=}C{=}O$	−11.4	156.0	198.3	
From thermochemical ladder, relative to isobutene			199.0	189.5
H_2O	−57.8	141.4	166.5 ± 2	
From thermochemical ladder, relative to ethylene			166.3	159.0

[a]Heats of formation using "stationary electron convention."
[b]The heat of formation (298 K) of H^+ in "stationary electron convention" = 365.7 kcal/mol.

account in calculating the heat of formation of the ion at temperatures other than 0 K. Obviously, since ionic species occur on both sides of Equation 7-1, the absolute value for the proton affinity does not change with the convention, but one must take care to use consistent data in calculating a proton affinity from ionic heats of formation. To translate "stationary electron convention" heats of formation of positive ions to the convention used in commonly consulted compilations of thermodynamic data (eg, the NBS tables of chemical thermodynamic properties[24] or the JANAF thermochemical tables[25]) in which the electron is treated as a conventional chemical element, the 289 K "stationary electron" values must be increased by 1.48 kcal/mol. For the defining Equation 7-1, the 298 K heat of formation of the proton according to the stationary electron convention is 365.7 kcal/mol. Relevant heats of formation of reference ions used to standardize the scale were taken at 298 K.

D. Ammonia

For reasons that appear to have more to do with tradition than with science, published scales of gas phase proton affinities have often been related to the proton affinity of ammonia as a standard. This proton affinity is defined by the reaction:

$$NH_3 + H^+ \rightarrow NH_4^+ \tag{7-22}$$

An experimental determination of the heat of formation of NH_4^+ based on its appearance energy from the $(NH_3)_2$ dimer[26]:

$$(NH_3)_2 \rightarrow NH_4^+ + NH_2 + e \tag{7-23}$$

led to a reported value of the proton affinity of ammonia of 202.3 kcal/mol at 0 K, which was estimated to correspond to a value of 203.6 kcal/mol for the PA of ammonia at 298 K. Although error limits of 1.3 kcal/mol were cited for this value of the proton affinity, an examination of the error limits associated with the appearance energy measurement ($\pm$0.5 kcal/mol), the heat of formation of NH_2 ($\pm$3 kcal/mol), and the assumed well depth for formation of the $(NH_3)_2$ dimer ($\pm$1 kcal/mol) leads one to assign error limits of at least $\pm$3.2 kcal/mol to this value. Substitution of the dimer well depth used in that determination[26] (3.5 kcal/mol) with a value from a more recent ab initio calculation[27] (1.9 kcal/mol) leads to a modification of the proton affinity derived from this experimental result to a value of 202.0 kcal/mol. A more exact value for this proton affinitiy is actually obtained from ab initio calculations, as discussed in Subsection A of Section 3 ("Theory"). An arbitrary value of 204.0 kcal/mol was selected for the proton affinity of ammonia in the recently published evaluation.[3] This is an intermediate value taken from the several thermochemical ladders that have been reported relating the PA of ammonia to that of the primary standard isobutene.[14n,15j,28]

The exact value for the proton affinity of ammonia is of less importance than it is usually accorded, since the position of ammonia in the scale of relative gas phase basicities is not well established. In fact, results from various laboratories show considerable variation in measurements involving this molecule. For instance, relative to the evaluated proton affinity of isobutene (see below), one would assign values for the PA of ammonia ranging from 203.5 to 205.0 kcal/mol, depending on the set of measurements selected. (Although this range is not really large in terms of absolute quantities, the reproducibility of measurements of proton transfer equilibrium constants often allows one to assign error limits of ± 0.1 kcal/mol to any particular directly measured interval in the free energy scale. The apparently poor quality of the measurements involving this compound is easy to rationalize, since ammonia tends to adsorb on surfaces, making accurate pressure measurements difficult.) However, because the agreement between the different thermochemical scales is generally good except for the position of ammonia, the exact value of the proton affinity of ammonia was given little weight in carrying out the evaluation. This proton affinity was treated as a secondary standard, and, in fact, was allowed to "float" relative to the rest of the scale in evaluating different data sets to maintain the overall agreement between the different experimentally determined thermodynamic ladders.

The entropy change associated with the half-reaction ($NH_3 \rightarrow NH_4^+$) is -1.8 cal/mol K, calculated from the complete partition function. This leads to a value of 195.6 kcal/mol for the 298 K gas phase basicity.

E. Primary Standards for the Proton Affinity Scale

a. Isobutene

The proton affinity of isobutene is defined by reaction 7-12. The heat of formation of the *tert*-$C_4H_9^+$ ion is less well established than those of the ethyl and *sec*-propyl ions discussed below. A recent analysis[29] of the 298 K heat of formation of this ion as derived from appearance energy measurements led to a value of 166.2 ± 0.8 kcal/mol. A value of 166.5 kcal/mol is obtained for this heat of formation taking a value of 12.0 ± 1.0 kcal/mol for the heat of formation of the *tert*-butyl radical[30] and a value of 6.7 eV for the ionization potential.[31] However, the latter value cannot be considered to be well established without additional corroborating evidence, since other experimentally determined values for the heat of formation of the radical ranging from 10.5 to 8.4 kcal/mol have also been reported,[32] and a value of 6.58 eV has been reported for the ionization potential.[33] The corroborating evidence is derived from measurements[34] of the equilibrium constant of the reaction:

$$C_6H_5CH_2^+ + (CH_3)_3CCl \rightleftarrows C_6H_5CH_2Cl + tert\text{-}C_4H_9^+ \qquad (7\text{-}24)$$

which leads to a value of -1.3 kcal/mol for the enthalpy change at 298 K; from these data the heat of formation of the *tert*-butyl ion would be estimated to be 165.8 kcal/mol. Based on this result, the proton affinity of isobutene is 195.9 kcal/mol.

Taking into account changes in external rotational symmetry numbers, the gain of an internal rotation upon protonation, and the change in the moments of inertia, the entropy change for the half-reaction (i-$C_4H_8 \rightarrow$ *tert*-$C_4H_9^+$) is about +0.7 cal/mol K. This leads to a value of 187.3 kcal/mol for the 298 K gas phase basicity.[28]

b. Propylene

The proton affinity of propylene is defined by the reaction:

$$CH_3CH{=}CH_2 + H^+ \rightarrow sec\text{-}C_3H_7^+ \qquad (7\text{-}25)$$

The heat of formation of the *sec*-$C_3H_7^+$ ion at 298 K has been determined to be 191.0 kcal/mol,[29,35] leading to a value for the proton affinity of propylene of 179.5 kcal/mol.

A calculation of the rotational entropy change associated with changes in the moments of inertia and rotational symmetry numbers for the half-reaction ($CH_3CH{=}CH_2 \rightarrow$ *sec*-$C_3H_7^+$) leads to a value of −0.9 cal/mol K; this estimate fails to take into account the predicted positive entropy change brought about by the loss of the double bond. (The methyl group in the ion can be taken to be essentially a free rotor.[36]) In the evaluation of the gas basicity of propylene, it is assumed that the entropy change of this half-reaction is zero. On this basis, the gas basicity of propylene is taken to be 171.7 kcal/mol.

The gas basicity of propylene has been related through thermochemical ladders to several of the other comparison standards listed here.[28] If one estimates the proton affinity through such data using ethylene as the primary standard for the proton affinity scale,[3,37] one obtains a value of 179.7 kcal/mol for the PA of propylene, in excellent agreement with the conclusion presented above. On the other hand, the experimentally determined interval in the thermochemical ladder between isobutene and propylene appears to be slightly constricted, an effect that may result from small inaccuracies in temperature measurement as discussed above or may possibly reflect uncertainties in the heat of formation of the *tert*-butyl ion, also discussed above.

c. Ethylene

The proton affinity of ethylene is defined by the reaction:

$$C_2H_4 + H^+ \rightarrow C_2H_5^+ \qquad (7\text{-}26)$$

The heat of formation of $C_2H_5^+$ at 298 K has been determined to be 215.6 kcal/mol,[29,35] leading to a value of 162.6 kcal/mol for the proton affinity of ethylene. Assuming the $C_2H_5^+$ ion to have a nonclassical bridged structure of C_{2v} symmetry (which has been calculated to be 5.2 kcal/mol lower in energy than the classical structure[14b,38]), the entropy change calculated from the complete partition function for the half-reaction ($C_2H_4 \rightarrow C_2H_5^+$) is −2.4 cal/mol K. The gas basicity of this compound is 155.6 kcal/mol.

d. Carbon Monoxide

The heat of formation of HCO^+ derived from appearance potential measurements in formaldehyde is 196.9 kcal/mol, corresponding to a proton affinity of 142.4 kcal/mol,[39] or 197.6 kcal/mol, corresponding to a PA of 141.7 kcal/mol.[40] The selected value, 141.9 kcal/mol, is the average of these two and some additional values derived from appearance potential measurements. Accepting a recent recommended value for the heat of formation of HCO[32d] of 8.9 $\pm$1.2 kcal/mol, an ionization potential determination for this radical of 8.27 $\pm$ 0.01 eV[41] leads to a heat of formation of the HCO^+ ion of 199.6 kcal/mol and a corresponding proton affinity of 139.7 $\pm$ 1.4 kcal/mol, in reasonably good agreement with the values derived above.

Following the recommendation of Bohme and co-workers,[13a] the entropy change associated with the half-reaction ($CO \rightarrow HCO^+$) is taken as 0.8 cal/mol K.

e. Carbon Dioxide

From the appearance potential of HCO_2^+ in formic acid (12.29 eV),[42] one can derive a heat of formation of the ion of 140.8 kcal/mol, which corresponds to a proton affinity of 130.9 kcal/mol.

Following the previously cited recommendation of Bohme,[13a] the entropy change associated with the half-reaction ($CO_2 \rightarrow HCO_2^+$) is taken as 4.5 cal/mol K.

f. Oxygen Atom

The heat of formation of the OH^+ ion has been determined from its appearance potential in H_2O (18.115 $\pm$ 0.008 eV) to be 308.96 kcal/mol,[43] in reasonably good agreement with values derived from other experimental results.[44] A value for the proton affinity of the O atom of 116.3 kcal/mol results.

Bohme and co-workers[13a] recommend a value of 6.5 cal/mol K for the entropy change associated with the half-reaction ($O \rightarrow OH^+$). On this basis, the gas basicity of the O atom at 298 K is taken as 110.7 kcal/mol.

g. Molecular Oxygen

The ionization potential of O_2H has been determined to be 11.35 $\pm$ 0.01 eV.[45] Accepting values for the heat of formation of O_2H of 3.2 kcal/mol at 0 K, or 2.5 kcal/mol at 298 K,[22] one obtains a value for the heat of formation of O_2H^+ ion of 264.9 or 262.4 kcal/mol at 0 or 298 K, respectively, leading to a value for the proton affinity of oxygen of 100.5 kcal/mol. This result is in good agreement with a value of 100.9 kcal/mol, which has been cited for this proton affinity, based on the appearance potential of O_2H^+ in H_2O_2.[46]

The entropy change associated with the half-reaction ($O_2 \rightarrow O_2H^+$) is taken to be 6.5 cal/mol K.[13a] The gas basicity of O_2 is taken as 95.0 kcal/mol.

F. Major Secondary Standards for the Proton Affinity Scale

a. Ketene

The proton affinity of ketene is defined by the reaction:

$$CH_2{=}C{=}O + H^+ \rightarrow CH_3CO^+ \qquad (7\text{-}27)$$

A recent determination of the heat of formation of CH_3CO^+ at 298 K led to a value* of 156 kcal/mol.[47] Accepting a value of -11.4 kcal/mol for the heat of formation of ketene, this would correspond to a proton affinity of 198.3 kcal/mol. The position of ketene in the thermochemical ladder of proton affinities has been established[23a,14a] and, relative to the value chosen for isobutene, leads to a PA for ketene of 199.0 kcal/mol, in reasonably good agreement with the value deduced from the threshold determination of the heat of formation of CH_3CO^+.

From experimentally determined entropy changes,[23a,14a] it is known that the entropy change associated with the half-reaction ($CH_2C{=}O \rightarrow CH_3CO^+$) is -2 cal/K-mol, which leads to a value for the gas basicity of ketene of 189.5 kcal/mol.

b. Formaldehyde

Approximately following the recent discussion of Collyer and McMahon,[37] an analysis of the proton affinity of formaldehyde is given. The proton affinity is defined by:

$$H_2CO + H^+ \rightarrow CH_2OH^+ \qquad (7\text{-}28)$$

Values for the heat of formation of CH_2OH^+ derived from the appearance energy of this ion from CH_3OH range from 170.0 to 170.7 kcal/mol,[48] corresponding to proton affinity values of 170–171 kcal/mol. Since the PA of formaldehyde is 5.2 kcal/mol above that of water,[13i] the value predicted from the evaluated proton affinity scale is 171.7 kcal/mol, in reasonably good agreement with that obtained from threshold measurements.

c. Water

A value for the heat of formation of H_3O^+ has been obtained from a measurement of the appearance energy of this ion from a hydrogen-bonded dimer[49]:

$$(H_2O)_2 \rightarrow H_3O^+ + OH + e \qquad (7\text{-}29)$$

From this experiment, one obtains a value for the proton affinity of H_2O of 165.8 kcal/mol at 0 K, which corresponds to a value of 166.5 kcal/mol for the proton affinity of water at 298 K. The error limits cited by the authors for this PA are ± 1.8 kcal/mol, with the major uncertainty being the well

*Value reevaluated in Reference 3.

depth for formation of the water dimer. A recent state-of-the-art ab initio calculation[50] verifies the well depth of −4 kcal/mol assumed in the interpretation of those results.

Recent equilibrium studies have tied the proton affinity of H_2O to that of ethylene.[37] The experimental results from those studies indicate that the Gibbs free energy change of the reaction:

$$C_2H_5^+ + H_2O \rightarrow H_3O^+ + C_2H_4 \tag{7-30}$$

is −3.4 kcal/mol. Accepting a value for the proton affinity of C_2H_4 of 162.6 kcal/mol, this result leads to a PA of water of 166.7 kcal/mol if one calculates the entropy change for reaction 7-27 taking into account only the changes in the rotational symmetry numbers of the reactants and products; a more complete calculation of the entropy change for this reaction leads to a value of 166.3 kcal/mol. Both estimates of the entropy change give proton affinity values that are in agreement with the value derived from the threshold determination of the heat of formation of H_3O^+.

The entropy change associated with the half-reaction ($H_2O \rightarrow H_3O^+$) is −1.03 cal/mol K.[51] The evaluated proton affinity of water, 166.5 kcal/mol, thus corresponds to a gas phase basicity of 159.0 kcal/mol.

3. CALCULATIONS OF PROTON AFFINITIES

Molecular orbital calculations offer certain advantages for determining proton affinities. As pointed out in Section 1, the accurate determination of the heat of formation of the MH^+ cations of interest through ionization threshold measurements is not always possible. Furthermore, although the measurement of equilibrium constants for proton transfer reactions permits the accurate determination of relative values of proton affinities in ranges of the scale for which a sufficient number of reference compounds with established proton affinities are available, such reference standards may not be available for comparison, particularly in portions of the high and low ends of the scale. Although PA calculations, at present, have been carried out only for atoms[52,53] and for some simple hydrocarbon radicals,[54] it should be possible to calculate the proton affinity of radicals, a quantity that is quite difficult to obtain experimentally. Experimental measurements of proton affinities do not, in general, provide structural information. However, theoretical determinations do provide structural information on the ion and can be used to quantitatively determine the difference in energy for protonation at two different sites in the same molecule (eg, protonation at C or at O in CO). Theoretical calculations of proton affinities can also be used to establish heats of formation of gas phase ions from a combination of Equation 7-1 and the experimental heats of formation of neutrals. A more detailed understanding of the energetics can be gained by examining various energy components, for example, using the charge decomposition analysis of Morokuma.[55] Finally, more detailed information about the intrinsic nature of the proton affinity can be determined from theory, for example, the separation of chemical from physical effects (eg, zero-point energy differences).

One drawback of the computational approach to proton affinities is that the calculations are so sophisticated that restrictions may be imposed on the size of the molecule that can be studied due to the available computer resources. Furthermore, significant amounts of computer (the theoretician's apparatus) time are required for these studies.

In this section, we focus on how to calculate accurate proton affinities and present a summary of high-quality determinations of PAs from theory.

A. Theory

The definition of the proton affinity is $-\Delta H_{Rn}$ for reaction 7-1. If one were to simply determine the total energy of B and BH^+ (the electronic energy of H^+ is zero), one would calculate a ΔE for reaction 7-1 related to the bottom of the well for each molecule. Let us consider the calculation of $\Delta H_{Rn}(298)$ for reaction 7-1 from first principles. Thermodynamics gives:

$$\Delta H_{Rn}(T) = \Delta E_{Rn}(T) + \Delta(PV) \qquad (7\text{-}31a)$$

$$= \Delta E_{Rn}(T) + \Delta nRT \qquad (7\text{-}31b)$$

where the conversion from one reaction to the next (ie, 7-31a to 7-31b) assumes ideal gas behavior, and T is the temperature. For reaction 7-1, $n = -1$ and:

$$\Delta H_{Rn}(T) = \Delta E_{Rn}(T) - RT \qquad (7\text{-}31c)$$

The quantity ΔE_{Rn} can be written as a sum of ΔE values of the degrees of freedom and is dependent on temperature:

$$\Delta E_{Rn} = \Delta E^{elec}_{Rn}(T) + \Delta E^{vib}_{Rn}(T) + \Delta E^{rot}_{Rn}(T) + \Delta E^{trans}_{Rn}(T) \qquad (7\text{-}32)$$

where ΔE_{Rn} has been broken into sums of electronic, vibrational, rotational, and translational components. The term ΔE^{trans}_{Rn}, assuming classical behavior, is $-3/2RT$. The term ΔE^{rot}_{Rn} can have different values depending on the nature of the geometry of M and MH^+. Assuming classical behavior again, if M and MH^+ have the same type of structure, both linear or both nonlinear, then $\Delta E^{rot}_{Rn}(T) = 0$. If M is linear and MH^+ nonlinear, $\Delta E^{rot}_{Rn}(T) = 0.5$ RT; if the opposite is true (less common), then $\Delta E^{rot}_{Rn}(T) = -0.5\ RT$.

The term $\Delta E^{vib}_{Rn}(T)$ can be considered to be a sum of two terms:

$$\Delta E^{vib}_{Rn}(T) = \Delta E'^{vib}_{Rn}(T) + \Delta ZPE \qquad (7\text{-}33)$$

where the first term contains the temperature dependence of the vibrations and ΔZPE contains the difference in zero-point energies of M and MH^+. The term $\Delta E'^{vib}_{Rn}(T)$ is approximately zero at room temperature if there are no low-lying vibrationally excited states that are significantly populated at this temperature. If such states are populated, as in the case for $MH^+ = H_3O^+$, the appropriate Boltzmann averaging should be done. In general, this term would contribute less than 1 kcal/mol. The term ΔZPE cannot be neglected in determining absolute proton affinities because there are three more degrees of freedom in MH^+ than in M, and thus there is more zero-point energy in MH^+ than in M. Since an isolated MH^+ is always more stable than the asymptotic limit $M + H^+$, this has the consequence of lowering the absolute value of the proton affinity. An estimate of ZPE for the reaction:

$$NH_3 + H^+ \rightarrow NH_4^+ \tag{7-34}$$

based on empirical estimates of the frequencies is 8 kcal/mol.[56,57] This large value for ΔZPE arises because of the strong binding of the proton (large force constants) and the large value of the reduced mass of the proton. For example, in first-row neutral hydrides, the ZPE of an A—H bond is 4–6 kcal/mol. This demonstrates that the force fields for both the neutral and the ion are required to determine accurate absolute proton affinities. Because of the paucity of experimental data on the force fields of isolated gas phase molecular ions,[58] the appropriate force fields must be calculated. Such calculations are discussed below.

The final term $\Delta E_{Rn}^{\text{elec}}$ can also be written as a sum of two terms:

$$\Delta E_{Rn}^{\text{elec}}(T) = \Delta E'^{\text{elec}}_{Rn}(T) + \Delta E_0^{\text{elec}} \tag{7-35}$$

where the first term again carries the temperature dependence. At room temperature, both B and BH^+ are, in general, in the ground electronic state and $\Delta E'^{\text{elec}}_{Rn}(T) = 0$. The ab initio calculations themselves deal with the determination of ΔE_0^{elec}. It is now clear that for even qualitatively correct absolute values, a basis set size of at least double zeta plus polarization functions on all atoms (DZ + P) is required for the reference Hartree–Fock (HF) wave function. For very accurate absolute values, an even larger basis set is required as discussed below. It is then necessary to perform some type of calculation beyond Hartree–Fock to account for corrections due to the correlation energy. As discussed below, such correlation corrections cannot be neglected for absolute values, although for relative values of certain types such a neglect may be reasonable.

The two standard approaches to obtaining the correlation energy correction are configuration interaction (CI) and perturbation theory expansions. The CI calculations should be done at the minimum level of all single and double excitations (CI-SD) from the valence space to the virtual space.[59] The orbitals for such a calculation can be obtained from Hartree–Fock wave functions, provided a single determinantal wave function gives a good description of M and MH^+. We also recommend inclusion of a correction for quadruple excitations (CI-SDQ) as the final step.[60] For the perturbation theory calculations, it is probably appropriate to do the calculations at the level of fourth-order Møller–Plesset (MP4)[61] if such a calculation is possible, although in a number of cases calculations at the MP3 level seem to be adequate. We note that there are a variety of ways of performing these MP calculations and recommend that authors be as specific as possible in their descriptions of the technique. Two other approximate techniques for treating the correlation energy that have been used in PA calculations are the independent electron pair approximation (IEPA) and the coupled electron pair approximation (CEPA).[62,63] The IEPA treatment is based on a sum of pair correlation energies. The CEPA treatment is better than IEPA because it allows for interactions of the pairs. Both techniques usually do provide a reasonable estimate for the correlation energy. Neither technique provides a rigorous bound to the energy. In fact, one problem with the IEPA formalism is that the estimate of the correlation energy at this level may be greater than the actual correlation energy. To provide a balanced calculation, one should use the same basis set for the correlation calculations as used for the self-consistent field (SCF) calculations. An estimate of correlation corrections can be obtained with a smaller basis set, although this is not recommended.

In many cases, correlation energy corrections reduce the absolute value of the proton affinity. A simple physical argument suggests that in such cases, the proton usually binds to a lone pair of M to form MH^+. Thus the electrons in the lone pair will be pulled further away from the nucleus in MH^+ than in M. Because the electrons in the $M-H^+$ bond have a larger spatial extent, the intrapair correlation energy can decrease, and consequently the correlation energy of MH^+ is less than that of M. This leads to decrease in the proton affinity.

B. Computational Techniques

There are a wide variety of molecular orbital programs available for use. However, two program systems are more generally available and these are HONDO(GAMESS)[64] and GAUSSIAN 82.[65] The major differences are (a) the typical standard basis set and (b) the typical standard correlation energy correction. GAUSSIAN 82 uses primarily Gaussian basis sets of the type developed by Pople and co-workers[66]: STO-3G, 4–31G, 3–21G, 6–31G and 6–31G**. The correlation energy calculations are primarily done using the Møller-Plesset perturbation techniques up through MP4. Calculations with the program HONDO(GAMESS) generally involve more general basis sets,[67] and correlation corrections are handled by CI techniques which are based on the GUGA-CI formalism.[68] The GUGA (graphical unitary group approach)-CI formalism is a very computationally efficient technique for performing standard CI calculations, including all single and double excitations from the valence space to the virtual space (see also Del Bene, Vol.1, Chapter 9).

There are a number of possible basis sets to employ in such calculations, beginning with a minimum basis set such as STO-3G. As discussed below, this basis set is totally inadequate for absolute values and often will not even give good relative values. A variety of DZ basis sets are also available. These

DZ basis sets come in two types: (a) full double zeta, which has double-zeta inner shells and a double-zeta valence space, and (b) single-zeta inner shells with a double-zeta valence space. A variety of DZ basis sets of the latter type have been widely implemented: 4–31G, 6–31G, 3–21G, and Dunning and Hay's[67] contraction (9,5)/[3,2]. The convention that we are using for defining the basis set is (ns,mp,od) primitive functions contracted to [n′s,m′p,o′d] functions for the SCF and CI calculations. This second type of basis set is very useful for CI calculations, since the virtual space is somewhat smaller and there are fewer virtual orbitals with large eigenvalues. We recommend the use of at least Dunning and Hay's basis sets, although the 6–31G basis set is also often acceptable for qualitative values. For these energy calculations the DZ basis set should be augmented by a set of polarization functions on each atom.

A variety of larger basis sets have also been used in proton affinity calculations. One such basis set is Dunning's[69] [5,3] contraction of Huzinaga's[70] (10,6) primitive set for first-row atoms. For second-row atoms the [6,4] or [6,5] contracted basis set from a (12,9) primitive set of McLean and Chandler[71] is recommended. Dunning's[69] (5)/[3] basis set for hydrogen is recommended; this will alleviate the need for employing different scale factors. It is recommended that the extended s,p basis set be augmented by two sets of d polarization functions contracted from a primitive set of four sets of d functions on heavy atoms and, if possible, two sets of p functions on hydrogen.[72] One set of p functions on hydrogen also works quite well. For accurate absolute values, we recommend the use of the foregoing basis set, although basis sets with fewer polarization functions may also be adequate. Even the DZ + P basis set often provides useful values, and a DZ basis set can sometimes provide useful qualitative information about relative proton affinities.

In all such calculations the geometries employed are an important feature. With the advent of gradient techniques,[73] it is now possible to obtain good geometries at the DZ + P level for molecules of average size. Since there is very little available information on gas phase ion structures, it is important to determine the molecular geometry of the ion to obtain accurate results. It is also recommended that the structure for the neutral be determined with the same level of calculation as employed for determining the structure of the ion. In this way systematic errors will cancel each other. We recommend, if possible, determination of the structure at least at the DZ + P level (DZ + D if necessary). For some systems accurate geometries at the CI level are available, usually from grid searches, although derivative techniques are now available.[74] At the Hartree–Fock level with a DZ + P basis set, systematic errors in geometry are well understood and good structures are usually obtained. The recipe above, of course, relies on the fact that a single determinental wave function provides an adequate description of the basic electronic structure. Although the DZ basis set usually provides adequate geometries, it does tend to make sp^3 N centers approximately planar[75] and it leads to angles at O that are too large.[76] This can lead to errors, especially if the DZ geometry is used with a larger basis set.

The availability of gradient techniques[74c,77] and faster computers also allows one to determine molecular frequencies for molecules and ions. New techniques[78] promise to make this approach even faster, and it is possible that the force field will come at approximately the same computational price as the determination of the structure. As discussed above, the force field is required for the zero-point correction. Again, it is preferable to calculate these frequencies at a better level of basis set. However, it is usually adequate to calculate the force field with a double-zeta basis set such as 3–21G or 4–31G. There are usually errors of 10–15% in the calculated frequencies as compared to the experimentally observed values. These differences are due to the neglect of correlation corrections and to the fact that harmonic frequencies are calculated as compared to the anharmonic frequencies measured experimentally. However, at present there is no method other than calculation to obtain the required force fields for ions, since there are only a few experimentally measured gas phase ion frequencies. It is thus often useful to scale the frequencies either by a factor (usually 0.9) times the diagonal force constants or by a factor determined by the ratio of the experimental zero-point energy of B to the calculated zero-point energy of B. If necessary, for larger systems, an STO-3G calculation of the frequencies is adequate, especially since differences between two sums of frequencies are being used, which leads to cancellation of errors. For these STO-3G calculations, some form of scaling is required, usually the ratio of the experimental zero-point energy of the base to the calculated value. An added bonus from the calculations is that the calculated frequencies can be used to guide the experimental search for such quantities.

C. Basis Set Dependence Studies

Before beginning our discussion of the accurate determinations, we discuss some of the work at the Hartree–Fock level on basis set and geometry effects on the determination of relative proton affinities. The most extensive studies of the basis set dependence of proton affinities calculated at the Hartree–Fock level have been made by Del Bene[79–81] and Dixon and co-workers.[82,83] We also discuss for the carbonyl bases the effect of the MP2 and MP3 levels of correlation with different basis sets on the relative PAs.[81] The experimental PA values are all taken from the recent compilation of Lias, Liebman, and Levin.[3]

Del Bene has focused primarily on carbonyl bases. In her initial work on RCHO for R = H, CH_3 and OH, she performed four sets of calculations: STO-3G basis/rigid monomer geometry for the ion, STO-3G/optimized ion geometries, 4–31G/rigid monomer geometry for the ion, and 4–31G/optimized ion geometries. These results are summarized in Table 7-2 for the ΔE_0^{elec} values relative to that for H_2CO. For conciseness, we present only absolute values of ΔE_0^{elec} for H_2CO in Table 7-3. It is clear that the absolute ΔE_0^{elec} for H_2CO at the STO-3G level is far too large even if an estimate of ΔZPE ($\backsim$10 kcal/mol) were made, and by analogy the values of ΔE_0^{elec} for

TABLE 7-2. Values (kcal/mol)[a] for $\delta\Delta E_0^{elec}$ for RHCO Relative to ΔE_0^{elec} (CH_2O)

Method[b]	R=CH_3	NH_2	OH	F
STO-3G/rigid	−14.0	−31.2	−9.5	4.8
STO-3G/STO-3G	−15.3	−41.9	−17.7	0.8
4-31G/rigid	−11.8	—	+2.3	23.9
4-31G/4-31G	−13.1	−30.3	−4.0	20.1
6-31G*/4-31G	−12.6	−28.9	−5.7	14.1
6-31G**/4-31G	−12.8	−29.0	−6.0	14.0
6-31G*/6-31G*	−12.6	−28.7	−5.7	13.1
6-31G**/6-31G*	−12.6	−28.6	−5.8	13.3
6-311G**/6-31G*	−12.7	−28.3	−5.8	14.1
6-31G*-MP3/4-31G	−13.6	−32.3	−7.3	12.9
6-31G**-MP3/4-31G	−13.4	−32.3	−7.9	12.4
6-31G*-MP2/6-31G*	−13.0	−30.9	−6.7	12.5
6-31G*-MP3/6-31G*	−13.5	−31.9	−7.2	12.5
6-31G**-MP2/6-31G*	−13.0	−31.0	−7.3	12.0
6-31G**-MP3/6-31G*	−13.3	−31.9	−7.7	12.0
Experimental (ΔPA)	−14.9	−26.7	−7.1	—

[a]Negative values imply an exoergic reaction, $\Delta E_0^{elec}(RHCO) < \Delta E_0^{elec}(CH_2O)$. Positive values imply the opposite.
[b]Computational method for determination of energy/geometry determined at this level of calculation.

the larger molecules are also too large. Furthermore, the relative proton affinity differences are too large with either the rigid or relaxed ion geometries. In fact, worse agreement is found with the relaxed ion structures at the STO-3G level. Qualitatively correct results were found for the relative PAs if a relaxed geometry was used at the 4–31G level. In more recent work[81]

TABLE 7-3. Dependence of ΔE_0^{elec} (CH_2O) on Basis Set

Method	$-\Delta E_0^{elec}$ (kcal/mol)
STO-3G/rigid	217.2
STO-3G/STO-3G	221.3
4–31G/rigid	178.9
4–31G/4–31G	181.1
6–31G*/4–31G	181.4
6–31G**/4–31G	186.0
6–31G*/6–31G*	182.0
6–31G**/6–31G*	186.8
6–31G**/6–31G*	185.7
6–31G**–MP3/4–31G	175.6
6–31G**–MP3/4–31G	181.7
6–31G*–MP2/6–31G*	174.2
6–31G*–MP3/6–31G*	177.0
6–31G**–MP2/6–31G*	180.3
6–31G**–MP3/6–31G*	183.2
Experimental (ΔPA)	171.7

Del Bene has extended the values of R to NH_2 and F and the type of basis set to 6–31G*, 6–31G**, and 6–311G**. Very similar results are found for relative PAs at the 4–31G level if either the STO-3G or 4–31G geometries are used. Comparisons with the larger basis sets, 6–31G* and 6–31G**, were made only with the 4–31G and 6–31G* geometries. Remarkably good agreement is found for all basis sets independent of the geometry and independent of the basis set except for R = F, where the 4–31G basis set leads to an error in PA.

In the latter study,[81] MP3 energies were compared at both geometries with the 6–31G* and 6–31G** basis sets, and again little change was observed. Furthermore, comparison of HF, MP2, and MP3 relative energies at the 6–31G* basis set geometry with either 6–31G* or 6–31G** showed little variation.

Although the relative values of ΔE_0^{elec} show little variation, there are much larger variations in the absolute ΔE's, since the absolute value for ΔE_0^{elec} varies for H_2CO. At the 4–31G geometry, there is a significant variation in the absolute values of ΔE_0^{elec} that is dependent on the basis set; the 4–31G energies are the lowest and the largest basis set 6–31G** gives the largest values of ΔE_0^{elec}. A similar trend in going from the 6–31G* to the 6–311G** basis set is found at the 6–31G* geometry. Little difference in the absolute values of ΔE_0^{elec} at the level of 6–31G* or 6–31G** using either the 4–31G or 6–31G* geometry was found. We also note that a comparison of ΔE_0^{elec} at the 6–31G* geometry using MP2 and MP3 shows some interesting effects on the absolute values. The MP2 corrections always show a much larger correlation effect than do the MP3 corrections and the effect is always negative, leading to a decrease in the value of ΔE_0^{elec}. The trends in the MP2 and MP3 ΔE_0^{elec}'s are the same as at the HF level when comparing between the 6–31G* and 6–31G** basis sets. For more recent work by Del Bene in this area, see Vol. 1, Chapter 9.

At the Hartree–Fock level Dixon and co-workers[82] have carried out two studies on proton affinities at N. In the first study, the basis set dependence of the relative proton affinity at N was studied for the methyl-substituted amines. Values for $\delta\Delta E_0^{elec}$ relative to ΔE_0^{elec} (NH_3) are given in Table 7-4. The values for ΔE_0^{elec} for NH_3 are given in Table 7-5. In this study experimental structures were employed for the neutral bases. Geometries optimized at the level of partial retention of diatomic differential overlap (PRDDO) were employed for the ions with scaled N^+H distances taken from calculations on NH_4^+ at the appropriate basis set level. This is because the N^+H distance is significantly longer with either the PRDDO or STO-3G minimum basis set calculations than it is with larger basis sets. The PRDDO method[84] is an approximate molecular orbital technique employing a minimum basis set of Slater-type orbitals that is computationally fast and gives results comparable to ab initio results using a minimum basis set. The absolute values of ΔE_0^{elec} for NH_3 from the PRDDO or STO-3G calculations are again far too large (Table 7-5). PRDDO does not even yield correct relative proton affin-

Table 7-4. Values for $\delta\Delta E_0^{elec}$ for Simple Substituted Amines and Imines Relative to $\Delta E_0^{elec}(NH_3)$ at the SCF Level[a]

Method	NH_2CH_3	$NH(CH_3)_2$	$N(CH_3)_3$	CH_2NH	CH_3CHNH	CH_2CHNH_2(N)[b]	CH_2CHNH_2(C)[b]
PRDDO	−3.1	−2.7	−0.6	6.8	−4.4	1.8	−17.7
STO-3G	−8.0	−13.4	−14.5	1.9	−9.8	−3.8	−20.6
STO-3G/rigid	−7.7	−12.9	−16.4	—	—	—	—
DZ	−9.4	−16.2	−18.3	−7.0	−16.6	1.7	−15.7
DZ + D	−11.1	−17.8	−20.5	−4.2	−13.9	−2.1	−20.8
DZ + P	−11.3	−18.5	−21.8	−4.6	−14.9	−1.8	−20.1
DZ + "ghost"	−9.9	−17.0	−19.2	—	—	—	—
DZ + D + "ghost"	−11.4	−18.3	−21.0	—	—	—	—
DZ + P + "ghost"	−11.5	−18.8	−22.1	—	—	—	—
Experimental (ΔPA)	−10.1	−16.6	−21.1	—	−10.2	—	—

[a]Negative values imply an exoergic reaction $\Delta E_0^{elec}(\text{base}) < \Delta E_0^{elec}(NH_3)$. Positive values imply the opposite.
[b] (N) implies protonation at N. (C) implies protonation at C.

TABLE 7-5. Dependence of $\Delta E_0^{\text{elec}}(NH_3)$ (kcal/mol) on Basis Set

Method	$-\Delta E_0^{\text{elec}}(1)^a$	$-\Delta E_0^{\text{elec}}(2)^b$
PRDDO	259.1	258.4
STO-3G	260.3	259.4
STO-3G/rigid	258.3	—
DZ	221.1	218.2
DZ + D	215.8	215.6
DZ + P	217.7	217.5
DZ + "ghost"	219.0	—
DZ + D + "ghost"	214.9	—
DZ + P + "ghost"	217.1	—
Extended GTO	—	216.9
Extended GTO + "ghost"	—	216.7
Extended STO	—	216.2
Extended STO/CI-SD	—	213.9
Extended STO/CI-SDQ	—	213.2
Experimental (PA)	—	204.0

[a]Evaluated using optimized geometry for NH_4^+ and experimental geometry for NH_3.
[b]Evaluated using optimized NH_4^+ and NH_3 geometries.

ities for these systems. Although STO-3G is adequate for providing relative PAs for CH_3NH_2 and $(CH_3)_2NH_2$, it fails for $(CH_3)_3N$. A comparison with the older work of Hehre and Pople[85] shows the effect of geometry relaxation on the relative PAs at the STO-3G level. Again the rigid monomer restriction gives better agreement with experiment at the STO-3G level. The results with the larger basis sets DZ, DZ + D, and DZ + P are quite similar. (The DZ + D basis set includes only d functions on heavy atoms.) The basis sets in this case are from Dunning and Hay.[67] The latter two basis sets show very similar results and differ somewhat from the DZ results. All the larger basis sets show reasonable agreement with experiment.

The inclusion of "ghost orbitals" was also evaluated, and only small changes are observed in the relative PAs. "Ghost orbitals" are orbitals added to the neutral at approximately the same geometrical position as in the ion except that no nucleus is placed at this position. This helps to account for the fact that a larger basis set is used in the ion as compared to the neutral, even though the number of electrons is the same. The largest changes are observed at the DZ level, where the basis set deficiency is being accounted for by the ghost orbitals. At the level of the DZ + P basis, only a small effect on the Hartree–Fock absolute energy is found with ghost orbitals. This demonstrates the small effect such orbitals have at the HF level if the basis set is large enough.

In other terminology, the inclusion of ghost orbitals is equivalent to inclusion of a correction for the "counterpoise" or superposition error. This is most pronounced for small basis sets. For a system A—B, the interaction energy between A and B will be overestimated because the fragment A(B) improves the quality of its basis set by using basis functions on fragment B(A). This leads to an energy for A−B that is lower than it should be when compared to the sum of the energies for isolated A and B.

The values for absolute ΔE_0^{elec} show some interesting trends for NH_3. The DZ and DZ + P values are quite similar, while the DZ + D value is somewhat smaller. This is different from the results observed by Del Bene for H_2CO.

A similar comparison of the dependence on basis set for the relative proton affinities of CH_2NH, CH_3CHNH, and CH_2CHNH_2 (C or N) protonation has been made.[83] All geometries were determined at the PRDDO level, and the N–H distances were scaled to those in NH_3 or NH_4^+ at the same basis set level. For these systems, it is only at the DZ + D level and DZ + P level that consistent results are found. The absolute ΔE_0^{elec} values in this case show a decrease in ΔE_0^{elec} between the DZ and DZ + D level and a slight increase in going from the DZ + D to the DZ + P level. The difference in the DZ and DZ + P values is significant.

The results above demonstrate that at the level of DZ + D or 6–31G*, qualitatively accurate relative PAs can be obtained at the Hartree–Fock level by approximating ΔPA as $\delta\Delta E_0^{elec}$ for systems of specific atom types. We note that these results are based on the fact that protonation was occurring at the same type of atomic center. Thus, differences in zero-point corrections and, apparently, correlation corrections approximately cancel, leading to good values for relative PAs. We also note that it is important to calculate relative PAs with the same basis set due to the variation in absolute ΔE_0^{elec}.

D. Accurate Calculated Absolute Proton Affinities

For our discussion of accurately calculated PAs, we have chosen values that were determined with at least extended basis sets with polarization functions and some form of correlation correction. We have also included zero-point corrections and the appropriate thermodynamic corrections required to calculate ΔH. The values are then compared to the best experimental values. Where appropriate, we also compare to lower level calculations as long as some form of correlation correction is included. The discussion is arranged as follows: atoms, diatomics, and polyatomics.

a. Atoms

The proton affinities of all atoms from H to Ar and that of Kr have been calculated very accurately and are given in Table 7-6. The earliest work in electronic structure theory dealt with the simplest diatomic, H_2^+.[52] This problem of one electron on two nuclei has an analytical solution, since the wave function is separable in terms of two confocal elliptic coordinates and an azimuthal angle. Since the exact energy of H is known, this is the most precisely calculated proton affinity. The most recent theoretical value is a

TABLE 7-6. Proton Affinities of Atoms

Atom	Frequency, $\nu(A-H^+)(cm^{-1})$[a]	Proton affinities (kcal/mol) SCF	CI	Experimental
H	2321[b]	62.0	—	62.0[b]
He	2911	40.6	43.4	43.4
Li	443	193.7	193.0	192.2
Be	2227	174.1	169.5	171.0[c]
B	2530	180.5	169.7	168.1[c]
C	2851	148.5	147.6	148.0[c]
N	3059	81.6	81.1	( )[d]
O	3124	115.7	116.9	116.3
F	3116	78.4	80.5	81.0
Ne	2670	47.0	49.4	48.1
Na	302	201.5	198.8	( )[d]
Mg	1703	183.1	184.2	186.6
Al	1695	198.1	191.8	( )[d]
Si	2154	202.4	201.3	200.3
P	2376	150.6	148.9	143.3
S	2566	158.6	161.6	158.3
Cl	2688	122.0	124.5	123.6
Ar	2611	87.1	90.6	>61.0
Kr	2463	99.4	108.1	101.6

[a]Determined at the CEPA level.
[b]Reference 52e
[c]The ions MH^+ dissociate to $M^+ + H$. These values are calculated from $D_0^0(M-H^+) = D_0^0(M^+-H) + IP(H) - IP(M)$, where D_0 is the appropriate dissociation energy and IP is the ionization potential. The D_0^0 values are then converted to ΔH's. The values for $D_0^0(M^+-H)$ are from Reference 53. The IPs are from Moore, C.E. NSRDS-NBS 34, National Bureau of Standards: Washington, D.C., 1970.
[d]The values for $D_0^0(M^+-H)$ are not known.

nonadiabatic calculation and includes relativistic and Lamb shift corrections.[52e] The proton affinity can be calculated by applying the proper thermodynamic considerations to convert $D_0\ (H_2^+)$ to a PA.

The proton affinity of He can also be calculated essentially exactly using a many-term correlated wave function.[86,87] Again this should correspond to a very precise theoretical determination of the proton affinity, which can be determined from the value for $D_0\ (HeH^+)$.

The proton affinities of the first- and second-row atoms excluding the rare gases have been calculated by Rosmus and Meyer.[53] These authors calculated both $D_0\ (AH^+)$ and the dissociation channel for $AH^+ \rightarrow A + H^+$. Only for A = O, F are these two the same, and indeed most of the AH^+ molecules will dissociate to $A^+ + H$, since the ionization potential of A is lower than that of H. These calculations were done with very large basis sets and correlation was accounted for by the CEPA technique. Potential curves were determined both at the SCF and correlated levels. For the first-row atoms an (11,6,2,1)/[9,5,2,1] basis set was employed. For the second-row atoms a [13,10,2,1] basis set was used. For Na, a set of p functions was added, while for Mg two sets of p functions and a set of d functions were added. The hydrogen basis set for LiH^+ and NaH^+ was of the form (7,4,2)/[5,4,2]. For the next two atoms in each row, the basis on H was (7,2,2)/[5,2,2], while for the remaining molecules the H basis set was (6,2,1)/[4,2,1]. For LiH^+ and BeH^+, all electrons were correlated, while for NaH^+, MgH^+, and AlH^+ the 2s and 2p orbitals were also correlated to account for core correlation effects. This is required for good bond distances for NaH^+ and MgH^+. For the remaining hydrides, only the valence electrons were correlated. For most of these PAs, only small correlation corrections ($\sim$ 0.1 eV) are found. This correction can be either positive or negative. For Be, B, and Al, larger negative correlation corrections are found. These authors quote an accuracy of ± 0.05 eV (± 1 kcal/mol) on the dissociation energies; thus, these values should be quite accurate. The conversion from ΔE to ΔH for these systems requires a correction of only 1.5 RT, since the ΔE_{Rn}^{Rot} (T) term of $+RT$ cancels the nRT term. Since these are hydrides, the frequencies are quite high, and only for LiH^+ or NaH^+ would vibrational corrections possibly be required. Where comparison with experiment is possible, the calculated values of D_0 show excellent agreement and are always within the experimental error limits.

Rosmus and co-workers have also calculated the proton affinities of the rare gases neon, argon, and krypton, again with the CEPA method and large basis sets.[88–90] For NeH^+ and KrH^+ all valence electrons are correlated and all single and double excitations are included using the self consistent electron pair (SCEP) formalism. The effect of higher order substitutions are approximated by CEPA. For ArH^+, the CEPA formalism was used. The basis set for Ne in NeH^+ is [12,8,3,1] and for Kr in KrH^+ is [16,12,5,1]; the hydrogen basis set has the form [7,2,1]. For ArH^+, a [14,10,3,1] basis set was used with a [6,2,1] basis set on H. Again the errors in D_0 are quoted as ± 0.05 eV. For NeH^+ and ArH^+, very good agreement with the experimental (spectroscopic) measurements is found. For KrH^+, the calculated value is preferred over the value derived from scattering experiments. (See Reference 90 for a discussion of this point.)

b. Diatomics

Hydrogen

The proton affinities for the diatomics are given in Table 7-7. The simplest diatomic is H_2 and the proton affinity of this molecule can be determined from the literature. The best calculations on H_3^+ that also reported a value for $E(H_2)$ are those of Dykstra and co-workers,[91] who used an (8,3,1)/[6,3,1] basis set and full CI (H_2 and H_3^+ are two-electron systems). Their value for H_2 is within 0.0011 au of the exact result of Kolos and Wolniewicz[92] for $E(H_2)$. The surface for H_3^+ was optimized at the CI level. The frequencies for H_3^+ can be taken from an extensive fit of an SCF-CI surface by Carney

TABLE 7-7. Accurate Calculated Proton Affinities for Diatomics

			$-\Delta E_0^{elec}$(kcal/mol)		Proton affinities (kcal/mol)	
Molecule	Ref	ΔZPE[a]	SCF	CI	Calculated[b]	Experimental
H_2	91	6.1	—	106.3	101.4*	101.3
H_2	95	6.8	104.1	106.6	101.0	—
CO(C)	95	7.2	143.7	147.1	141.4*	141.9
CO(C)	96	7.9	144.5	150.7	144.3	—
CO(C)	97	(7.9)	144.6	149.7	143.3	—
CO(O)	95	5.7	111.9	106.1	101.9*	—
CO(O)	96	5.3	115.2	112.5	108.7	—
CO(O)	97	(5.3)	114.9	113.4	109.6	—
N_2	99	6.6	—	123.3	118.2*	118.2
N_2	100	7.0	120.9	124.9	119.4*	—
N_2	96	7.4	120.6	125.2	119.3	—
N_2	97	(7.4)	120.8	125.1	119.2	—
HF	101	6.0	122.8	123.9	119.1*	117.0
HF	102	(6.0)	122.5	120.8	116.0	—
HF	96	6.3	127.4	131.0	125.9	—

[a]Values in parentheses not calculated in the literature; each one taken from the ΔZPE above it in the list.
[b]Recommended values are indicated by asterisks.

and Porter,[93] which includes anharmonic effects. They report a zero-point energy of 4345 cm^{-1}, and we estimate the error in this value to be ± 50 cm^{-1} (0.14 kcal/mol). Since the ZPE for H_3^+ is quite accurate, we have used the experimental ZPE for H_2 for the calculation of PA(H_2).[94] Correcting ΔE to ΔH [assuming $\Delta E_{Rn}^{rot}(T) = \Delta E_{Rn}^{vib}(T) = 0$], we obtain a value within 0.1 kcal/mol of the experimental value. Because of the precision in the energy of H_2, we feel that the calculated value is very accurate.

With a smaller basis set (5,1)/[3,1] Dixon and associates[95] found a ΔE_0^{elec} value within 0.3 kcal of that of Dykstra at the full CI level. The geometries for this study were taken from optimized SCF structures and the frequencies were evaluated at the SCF level. The effect of correlation energy is to raise ΔE_0^{elec}. These authors find PA(H_2) = 101.0 kcal/mol, in excellent agreement with the experimental value and the other theoretical value.

Carbon Monoxide

The proton affinity of CO has attracted a significant amount of interest because it is a small molecule with two binding sites for the proton and because of its importance in interstellar clouds. The most extensive calculations on this system employed a basis set of the form (11,7,1/6,1)/[5,4,1/3,1], and correlation corrections were made at the CI-SDQ level.[95] The geometries and vibrational frequencies were taken from CI-SDQ calculations. Anharmonic corrections to ΔZPE would be on the order of 0.2 kcal/mol. The proton affinity of CO at C is significantly larger than PA(CO) at

O. Correlation at the CI-SD level raises the proton affinity for reaction at C; at the CI-SDQ level this increase is somewhat smaller. The best value for PA(CO) at C is 141.4 kcal/mol, in excellent agreement with the experimental value of 141.9 kcal/mol. At the level of 6–31G**/MP-SDQ the PA(CO) at C is 144.3 kcal/mol,[96] and this value is clearly too high. For this second study the geometry was taken from 6–31G* SCF calculations, and frequencies were evaluated with the 3–21G basis set. Since the 3–21G frequencies will clearly overestimate ΔZPE, it is clear that ΔE_0^{elec} is too high. A 6–31G**/MP-3 calculation[97] gives PA(CO) = 143.3 kcal/mol, which is again too high. The 6–31G** geometry was used in this latter calculation and the set of 3–21G frequencies from Del Bene and co-workers[96] was employed. These calculations do show the same trend of an increase in PA when correlation is included, as do the large basis set CI-SDQ calculations.

For protonation of CO at O, the best results give a value 101.9 kcal/mol. For this site a large decrease in PA is found when correlation corrections are included. The vibrational correction is smaller here than for HCO^+, since the bending mode is very low. This demonstrates that errors in relative PAs evaluated at the ΔE_0^{elec} level can have errors due to ΔZPE effects if the sites are not the same. This low bending frequency shows that a small temperature correction of 0.1–0.2 kcal/mol for $\Delta E_{Rn}^{\text{vib}}(T)$ may be required. The value for PA(CO) at O is, at present, unknown from experiment. The only experimental measurement related to PA(CO) at O is that of Bowers and associates,[98] who found that HOC^+ was formed 6 ± 5% of the time in reactions of $H_3^+ + CO$. The dominant channel was formation of HCO^+, as predicted by the calculations. The low value for PA(CO) at O is important in determining the thermodynamics of H^+ transfer in dense interstellar clouds. The value for PA(CO) at O is essentially the same as that for PA(H_2) evaluated with this type of calculation. In fact, comparison of the ΔE_0^{elec} values for PA(CO) at O and PA(H_2) shows that PA(H_2) is actually higher at this level and is only lower because of zero-point effects. Since the proton transfer reagent in these clouds is H_3^+, then HCO^+ will be predominantly formed. The value for PA(CO) at O has also been calculated at the level 6–31G**/MP4-SDQ using the same techniques as described above and is again found to be significantly too high.[96] In fact comparison with H_2 at this level would lead to an incorrect conclusion.

Nitrogen

The proton affinity of N_2 has been investigated by a number of groups. There have been two high-quality calculations of PA(N_2). Using a very large basis set (11,7,2/6,2)/[8,5,2/4,2], Botschwina[99] has calculated PA(N_2) using CEPA for determining the correlation correction. No SCF value was given in this work. A very accurate determination of the stretching frequencies was made in this work, and the energy of the bending vibrations was taken from a separate set of CI-SDQ calculations.[100b] The value for PA(N_2) from this calculation is 118.2 kcal/mol (although Botschwina reports a value of 117.4 kcal/mol).

With a somewhat smaller basis set of the same form as used in the accurate study on CO (11,7,1/6,1)/[5,4,1/3,1], Kraemer and colleagues[100] found that the correlation correction raised PA(N_2) by 4.0 kcal/mol over the SCF value. For this study the correlation correction was determined at the CI-SDQ level, and the geometries and harmonic frequencies were also determined at this level. This gives a value for PA(N_2) that is 1.6 kcal/mol higher than Botschwinna's value for ΔE_0^{elec}. Kraemer predicts a slightly larger zero-point effect of 7 kcal/mol. Converting to T = 298 K gives 119.4 kcal/mol. The best experimental value is 118.2 kcal/mol, which is in very good agreement with Botschwina's value. Based on these results, we would recommend a value of 118.5 $\pm$ 1 kcal/mol for PA(N_2); however, see the discussion of hydrogen fluoride, below. Two other calculations of N_2 using 6–31G** basis sets at the MP3[97] or MP4[96] level give values within 0.3 kcal/mol of that found by Kraemer. These studies employed SCF geometries at the 6–31G* or 6–31G** level, and frequencies were evaluated with the 3–21G basis set.

Hydrogen Fluoride

There are one high-quality[101] and two reasonable theoretical determinations[96,102] of the proton affinity of HF. The best calculations have been done by Dixon and Komornicki[101] with a basis set of the form (10,6,4/5,2)/[5,3,2/3,2]; final energies were determined at the CI-SDQ level. Geometries were optimized at the Hartree–Fock level and zero-point energies were also determined at this level. The calculated value for ZPE (HF) was scaled to the experimental value, and this scale factor was used to scale ΔZPE. The calculated value is somewhat above the reported experimental value. However, this experimental value is based on PA(N_2). Foster and Beauchamp[103] showed that PA(HF) is slightly less than PA(N_2). Comparison of the value for PA(HF) calculated by Dixon and Komornicki[101] with the higher calculated value for PA(N_2) given above shows this result. This suggests that the higher of the two good theoretical values for PA(N_2) may be the likely value. This value for PA(HF) is clearly much larger than the value of 95 kcal/mol obtained by Lee and co-workers[104] from the photoionization of an HF molecular beam that contains $(HF)_2$.

The work by Lischka[102] was done with a large basis set, (11,7,2/6,2)/[7,4,2/4,2], and the IEPA technique was used for treating electron correlation. Although the IEPA technique is not what would be recommended now, this was an impressive early calculation. A slight correlation correction, decreasing PA(HF), was found. The geometry in this work was determined at the correlated level. Although zero-point corrections were not made (only the symmetric frequencies were determined), an estimate can be made from that of Del Bene and colleagues, who give ΔZPE = 6.3 kcal/mol.[96] This gives PA(HF) = 116 kcal/mol, which is in reasonable agreement with the results of Foster and Beauchamp,[103] who determined PA(HF) relative to PA(N_2). The other theoretical value employed a much smaller basis set, 6–31G**, and MP4 for the correlation correction.[96] As often found with the smaller basis sets, ΔE_0^{elec} increased with the correlation correction, and the calculated value of 126.2 kcal/mol for PA(HF) is clearly too high.

C. Polyatomics

Ammonia

The accurate proton affinities for the polyatomics are given in Table 7-8. For historical purposes, we discuss PA(NH_3) as our first example of a polyatomic calculation of the PA. Much of the relative PA data from ICR measurements employed NH_3 as the reference base and PAs were given relative to PA(NH_3). The best calculation for PA(NH_3)[105] employed a basis set of Slater-type orbitals of the form [5,4,2/2,1]. Correlation was accounted for at the CI-SDQ level in the valence space. The geometries for B and BH^+ were determined from very large Gaussian-type orbital (GTO) calculations and the ZPEs were calculated with a large GTO basis: (11,7,2/5,1)/[7,6,2/3,1].

TABLE 7-8. Accurate Calculated Polyatomic Proton Affinities

Molecule	Ref.	ΔZPE (kcal/mol)	$-\Delta E_0^{elec}$ (kcal/mol)		Proton affinities (kcal/mol)	
			SCF	CI	Calculated[a]	Experimental
NH_3	105	9.1	216.2	213.2	205.6*	204.0
NH_3	96	10.2	219.6	220.8	212.1	—
H_2O	101	7.2	176.0	174.3	168.6*	166.5
H_2O	109	6.3	174.3	172.8	168.0*	—
H_2O	102	(7.2)[b]	176.6	174.0	168.3	—
H_2O	108	8.3	174.6	171.8	164.8	—
H_2O	110	9.2	177.4	177.8	170.1	—
H_2O	96	8.6	179.8	180.8	173.7	—
PH_3	111	7.1	200.4	193.7	188.1*	188.6
H_2S	111	6.9	177.3	177.1	171.7*	170.2
CH_4	112	4.4	130.7	131.7	128.8*	132.0
H_2CO	113	8.0	185.0	179.3	172.8*	171.7
H_2CO	95	9.0	185.4	182.0	174.5	—
H_2CO	81[c]	8.5	185.7	182.2	175.2	—
H_2CO	81[d]	8.5	186.8	183.2	176.2	—
H_2CO	96	9.2	183.6	181.8	174.1	—
CH_2NH	113	8.6	221.8	216.9	209.8*	—
CH_2NH	115[c]	9.6	223.7	219.6	211.5	—
CH_2NH	115[d]	9.6	225.3	222.4	214.3	—
CH_2NH	96	9.5	225.3	221.4	213.4	—

[a]Recommended values are indicated by asterisks.
[b]From Reference 101.
[c]6–311G**/MP3.
[d]6–31G**/MP3.

The ΔZPE term was scaled by a factor of 0.89 as determined by a comparison of the calculated and experimental values for NH_3. The STO and GTO values for ΔE_0^{elec} at the SCF level differed by 0.7 kcal/mol, with the GTO being higher. "Ghost" orbitals were shown to have a very small effect on the GTO energies at the SCF level (0.2 kcal/mol). Correlation was shown to decrease ΔE_0^{elec} and the effect of quadruples could not be ignored. These STO CI-SDQ calculations yield a value of 205.6 kcal/mol and an estimated error of ± 1 kcal/mol. The recommended experimental value is slightly lower, 204 kcal/mol. Again the photoionization molecular beam experiments yield too low a value for $PA(NH_3)$.[106]

A study[107] of the basis set dependence of the correlation corrections using perturbation theory suggested that quite large basis sets were required before the correlation corrections for $PA(NH_3)$ showed consistent behavior.[4] These results are summarized in Table 7-9. This point requires further study. We note that the smaller GTO basis sets used in this study do not give quite as low a value for ΔE_0^{elec} at the SCF level as the large GTO or STO basis sets. The MP4 method for determining the correlation corrections does appear to provide consistent results with the large CI-SDQ calculations.

Water

The proton affinity of H_2O has been studied by a number of groups.[101,102,107–109] The ion H_3O^+ has a very low inversion barrier, and thus a small vibrational correction must be included to account for this. There are three high-quality calculations for $PA(H_2O)$. The first one employed a large GTO basis (11,7,1/6,1)/[5,4,1/3,1] of the same type as used for CO and N_2 and a full CI-SD calculation. The geometries and frequencies were determined with the correlated wave functions. Diercksen and co-workers[109] also treated the vibrational motions carefully and included coupling between the inversion and stretching motions. The value for $PA(H_2O)$ including a small correction for $\Delta E'^{vib}_{Rn}(T)$ is 168.0 kcal/mol. In the second study of reasonably high quality,[108] an extended 6–31G basis set, 6–311+G**(2d), was employed with single, double, triple, and quadruple substitutions at the MP4 level for correlation effects. This study gave a large correlation correction and a value for ΔE_0^{elec} that is 1 kcal/mol smaller than that found by Diercksen.[109] For this study the geometry and frequencies were taken from 6–31G* calculations. A simpler treatment of the vibrations by Del Bene and colleagues[108] at the 3–21G level gave a larger zero-point energy of 8.3 kcal/mol. Because of the somewhat better treatment of the vibrations and the overall quality of the calculations, we prefer the higher value. Following the same prescription as in the most accurate calculations on PA(HF), Dixon and Komornicki[101] have determined $PA(H_2O)$ with the large [5,3,2/3,2] basis set at the CI-SDQ level. Vibrational frequencies were determined at the SCF level in the harmonic approximation and the ratio $ZPE_{exp}(H_2O)$/$ZPE_{calc}(H_2O)$ was used to scale ΔZPE. Because of the difficulty in determining the value of ZPE for ν_2 (the mode describing molecular inversion) an

TABLE 7-9. Correlation and Basis Set Effects on ΔE_0^{elec} for NH_3 and H_2O

Method	$-\Delta E_0^{elec}$ (kcal/mol)	
	NH_3	H_2O
6–31G**–HF	219.0	179.8
6–31G**–MP4	220.9	180.9
6–31G**–CI-SD	221.1	180.9
6–311G**–HF	218.6	178.6
6–311G**–MP4	218.2	179.6
6–311G**–CI-SD	218.9	179.9
6–311G**(2d)–HF	218.0	178.1
6–311G**(2d)–MP4	215.9	177.3
6–311G**(2d,2p)–HF	218.5	177.9
6–311G**(2d,2p)–MP2	215.4	175.9
6–311G**(2d,2p)–MP4	—	176.6

estimate of this term as 0.1 kcal/mol was made based on other literature values.[109] This gives a value for PA(H_2O) in good agreement with the value of Diercksen[109] value and slightly higher. The value of ΔE_0^{elec} determined in this later calculation is somewhat better than the earlier result of Diercksen, and we recommend the higher value for PA(H_2O). Our preferred value for H_2O is higher than the evaluated experimental value by 2.1 kcal/mol. For H_2O, the molecular beam–photoionization technique[49] gives somewhat better agreement for PA(H_2O) = 166.5 kcal/mol, but this value is still slightly low as compared to the best calculated values. We note that the calculated value for PA(NH_3) is higher than the experimental value by 2.4 kcal/mol. Thus the difference between PA(NH_3) and PA(H_2O) is essentially the same using theoretical or experimental PA's. Again we note that the error limits on the theoretical value for PA(H_2O) are probably 1.0 kcal/mol. As in the case of NH_3, Frisch and co-workers[107] have carried out an extensive set of calculations of the dependence of PA(H_2O) on the basis set and correlation correction (see Table 7-9). Again only for large basis sets are consistent correlation corrections found.

There are two other calculations of PA(H_2O). The earlier calculation of Lischka[102] used the IEPA method for treating the electron correlation and a large GTO basis: (11,7,1/6,1)/[7,4,1/4,1]. These calculations gave values for ΔE_0^{elec} that are similar to those of Dixon and Komornicki.[101] Lischka's corrected PA is in good agreement with that given by the largest calculations. The more recent calculations of Kello and associates[110] employed a decent-sized GTO basis (9,5,1/4,1)/[4,3,1/2,1] and many-body Rayleigh-Schrödinger perturbation theory (MBRSPT) to third order for treating the correlation correction. The value for ΔE_0^{elec} (SCF) is somewhat too high and the correlation correction is in the wrong direction. These authors also employed "ghost" orbitals (a counterpoise correction) to decrease the correlation energy correction effect. Because of other deficiencies in their calculation, however, this effect is probably exaggerated here.

Phosphine and Hydrogen Sulfide

The values for PA(PH_3) and PA(H_2S) have been determined by a technique analogous to that used for determining the best value of PA(NH_3).[111] The geometries were taken from CI-SD calculations at the DZ + P level (STO basis) for PH_3, PH_4^+, and H_3S^+ and from experiment for H_2S.* The final calculations of the energy were done at the CI-SDQ level using a DZ-core and TZ-valence space of STOs augmented by two sets of d functions for P and S; for H, a [3,1] STO basis was used. The ZPE calculations were done with a DZ + P basis of GTOs and the ZPE term was scaled by 0.91 for both. For PA(PH_3) a large correlation effect was observed, leading to a decrease in PA(PH_3). There was even a significant effect of 2 kcal/mol due to the estimate of the effect of quadruple excitations. For H_2S no CI correction to the Hartree–Fock value was found. The value for PA(PH_3) is in excellent agreement with the experimental value of 188.6 kcal/mol. The calculated value for PA(H_2S) is 1.5 kcal/mol higher than the experimental value. The error limits on the calculation are probably ± 1 to ± 1.5 kcal/mol, and we would thus expect a slightly higher experimental value for PA(H_2S). We note that PA(H_2S) − PA(H_2O) = 2.9 and 3.7 kcal/mol from theory and from experiment, respectively.

Methane

Komornicki and Dixon[112] have calculated the proton affinity with a large basis set and the correlation correction was accounted for at the CI-SDQ level. The basis set is of the form described previously: (10,6,4/5,2)/[5,3,2/3,2]. The geometries and vibrational frequencies were determined analytically at the SCF level with the large basis set. The value for ZPE was scaled to the experimental value for ZPE(CH_4). The structure of CH_5^+ is shown for the most stable form (**1**) and for the rotamer (**2**), which is less than 0.1 kcal/mol higher in energy.

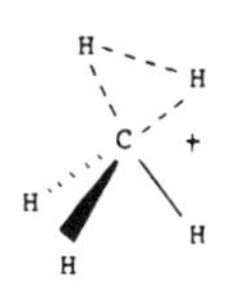

1

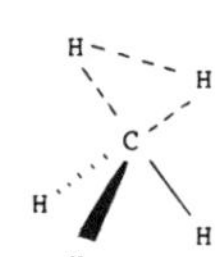

2

The fact that **2** is a transition state on the rotation surface was verified by a determination of the force field, which showed one imaginary frequency.

*The experimental parameters for H_2S were used because large calculations were already available at this geometry. This is reasonable, since the optimized structure for PH_3 is in excellent agreement with the experimental structure.

There is actually some uncertainty in calculating the ΔZPE term for this system due to the fluxional nature of the CH_5^+. In the harmonic approximation, analytical gradient techniques yield a value of 128 cm^{-1} for the lowest frequency in this system. Since **2** is missing this frequency (it is a transition state), **2** actually has a lower ZPE content, making it more stable. What this demonstrates is the inadequacy of the harmonic approximation for describing the very low energy torsional mode. Thus, the molecule CH_5^+ at room temperature has an almost free internal rotation and we have an uncertainty of ±0.25 kcal/mol in our ΔZPE term. The calculated value for $PA(CH_4)$ is lower than experiment by 3.2 kcal/mol. This is a surprising result in light of most of the other molecular PA calculations, where the calculated PA is slightly above the experimental value. This is even more striking for CH_5^+, since a consideration of the internal vibrational energy contribution for CH_5^+ at T = 298 K is certainly not zero. The $\Delta E'^{\mathrm{vib}}_{Rn}(T)$ term raises the value of the energy for CH_5^+, thereby decreasing $PA(CH_4)$. We thus suggest that the experimental value for $PA(CH_4)$ is too high [see below for a discussion of how this affects $PA(CO_2)$].

Formaldehyde

The proton affinity of H_2CO has been determined at a number of levels. Dixon and co-workers[95] calculated $PA(H_2CO)$ using a (9,5,1/5,1)/[4,2, 1/3,1] basis set and evaluated the correlation corrections at the CI-SDQ level. Geometries and zero-point energies were determined at the Hartree–Fock level with this basis set. These investigators found that correlation energy decreases $\Delta E_0^{\mathrm{elec}}$ and that H^+ bonds to oxygen. Their value of 174.5 kcal/mol is somewhat higher than the experimental value of 171.7 kcal/mol. This discrepancy probably exists because the basis set is still somewhat too small.

The best calculation for $PA(H_2CO)$ follows the technique employed in determining the PAs of the first-row hydrides.[113] In this case a basis set of the form (10,6,4/5,1)/[5,3,2/3,1] was employed. Geometries and frequencies were taken from the foregoing work. The value for ΔZPE was scaled as described earlier by the ratio $ZPE_{\mathrm{exp}}(H_2CO)/ZPE_{\mathrm{theor}}(H_2CO)$. The SCF value for $PA(H_2CO)$ is similar to that determined with the smaller basis set. However, the correlation effect is much larger, and the best calculated value for $PA(H_2CO)$ is 1.1 kcal/mol above the experimental value. Since the experimental value is probably low, we prefer the theoretical value of 172.8 kcal/mol.

Del Bene has calculated the PA for H_2CO at the level of 6–31G**/MP3. The geometry was determined at the 6–31G* level and zero-point corrections were made at the 3–21G level. For H_2CO, this gives a value of 176.2 kcal/mol for $PA(H_2CO)$, which is too high as compared to the calculation with the larger basis set or with experiment.

Methyleneimine

Although the proton affinity of the compound CH_2NH is not known accurately from experiment because of the difficulty in generating CH_2NH and

$CH_2NH_2^+$, it has been treated by reasonably sophisticated theoretical calculations. Dixon[113] has determined PA(CH_2NH) using a (9, 5, 1/4, 1)/[3, 2, 1/2, 1] basis set at the CI-SDQ level. Geometries were evaluated with this basis set and zero-point energies were determined with the 3–21G basis set. At this level PA(CH_2NH) = 212.4 kcal/mol. Calculations with the same basis set as used for the best H_2CO calculations were used to determine PA(CH_2NH), again at the CI-SDQ level. The geometries were taken from the Hartree–Fock calculations above and the ΔZPE term from above was scaled by 0.9. (The value of 0.9 seems to be a reasonable scale factor based on a variety of calculations as described elsewhere in this chapter.) The final value for PA(CH_2NH) is 209.8 kcal/mol, and we would suggest a value for PA(CH_2NH) between 209 and 210 kcal/mol. Although no good experimental data are available for PA(CH_2NH), an estimate near PA(NH_3) has been made. We prefer our somewhat larger calculated value. This would give a smaller CH_3 effect on the PA based on the experimental measurements for PA(CH_2NH) − PA(CH_3CHNH).[114]

Del Bene[115] has also calculated PA(H_2CNH) with the prescription given above for H_2CO. For H_2CNH the basis set was also extended to 6–311G** and 6–31+G** at the MP3 level. The values are 214.3, 211.5, and 211.0 kcal/mol with the 6–31G**, 6–311G**, and 6–31+G** basis sets, respectively. The latter two values are in reasonable agreement with the more accurate calculations above.

E. Less Accurate Proton Affinity Calculations

In this section we describe less accurate calculations on molecular proton affinities (Table 7-10). Although the results may not be quantitative within ±1 kcal/mol, the calculations reported are still quite sophisticated. Correlation corrections were included and the basis sets are mostly of DZ + P quality.

a. H_mABH_n: A,B = First-Row Atoms

The most extensive study of molecular proton affinities with correlated wave functions has been carried out by Del Bene and co-workers[96] for H_mABH_n, where A or B may be carbon, nitrogen, oxygen, or fluorine. Some of these calculations have been discussed previously in comparison with larger calculations. All calculations were done at the 6–31G**/MP4(SDQ) level with the effect of triple excitations included from 6–31G*/MP4(SDTQ) calculations. All calculations were done with geometries determined at the 6–31G*/HF level. Force fields were determined at the 6–31G* level. The results are given in Table 7-10, and comparison with experiment and other calculated values shows that this level of calculation gives proton affinities that are too large by about 5 kcal/mol. Exceptions are H_2O_2, H_2CO, CO, N_2, and HCN, where better agreement with experiment or other theoretical values is found. The PAs calculated by Del Bene[96] provide qualitatively correct PAs, but they are certainly not quantitative. For CH_3OH,

TABLE 7-10. Less Accurate Calculated Polyatomic Proton Affinities

Molecule	Ref.	ΔZPE (kcal/mol)	$-\Delta E_0^{elec}$ (kcal/mol)		Proton affinities (kcal/mol)	
			SCF	CI	Calculated	Experimental
C_2H_2	96	4.5	164.8	162.8	159.8	155.2
C_2H_4	96	6.4	177.0	172.8	167.9	162.6
HCN	96	7.9	180.8	179.2	172.8	171.4
CH_3NH_2	96	10.2	230.3	230.8	222.1	214.1
CH_3OH	96	8.3	194.9	194.1	187.3	181.9
CH_3OH	116	7.9	194.9	194.4	188.0	—
CH_3F	96	5.6	157.0	154.6	150.5	150.0
N_2H_2	96	9.2	198.4	197.5	189.8	—
N_2H_4	96	9.8	221.4	222.3	214.0	204.7
HNO(N)	96	9.7	177.5	178.6	170.5	—
H_2NOH(N)	96	9.6	208.6	209.1	201.0	—
NH_2F(N)	96	10.0	195.7	196.7	188.2	—
H_2O_2	96	8.5	172.7	173.4	166.4	162.0
HOF(O)	96	8.6	145.6	147.9	140.8	—
CH_3CHO	81	8.0	199.4	196.5	190.0	186.6
NH_2CHO	81	8.8	215.4	215.1	207.8	198.4
HOCHO	81	8.0	192.6	190.9	184.4	178.8
FCHO	81	7.9	173.5	171.2	164.8	—
CH_3CHNH	115	9.3	234.4	231.0	223.0	214.2
NH_2CHNH	115	10.2	246.7	244.7	236.0	—
HOCHNH	115	9.1	228.1	225.7	218.1	—
FCHNH	115	9.3	212.3	209.7	201.9	—
H_2SiO	113	7.2	219.4	211.7	206.0	—
H_2SiNH	113	9.4	244.8	238.7	230.8	—
CO_2	118	6.2	133.8	135.6	130.6	130.9
HCP	120	5.9	179.7	174.5	170.1	—
H_2CPH(P)	120	5.9	190.8	189.9	185.5	—
Li_2O	121	4.9	304.8	298.9	295.5	—
C_2H_4O	122	7.4	195.8	192.5	186.5	187.9
C_2H_4S	122	5.3	200.6	204.6	200.6	194.6
C_2H_6	123	5.4	146.5	147.7	143.4	143.6
PH_2CH_3	125	6.0	272.3	267.9	263.4	—

Nobes and Radom[116] have performed similar calculations at the 6–31G**/MP3 level and have evaluated the ΔZPE term with a 4–31G basis set. Their result of 188.0 kcal/mol is in excellent agreement with the value of 187.3 kcal/mol found by Del Bene. Both are too high as compared to the experimental value of 181.9 kcal/mol.

b. RCHO and RCHNH

Del Bene has calculated the PAs for R = H, CH_3, NH_2, OH, and F for both the carbonyl[81] and imine[115] compounds at the 6–31G**/MP3 level. The geometries were determined at the 6–31G* level and zero-point corrections were made at the 3–21G level. The PAs for R = H have been discussed above. Based on the foregoing comparisons, we would expect that the PAs calculated at this level would be too large by about 5 kcal/mol. As expected, the value for PA(HCOOH) is too high as compared to the experimental value, as is the calculated value for PA(NH_2CHO). The new value recommended for PA(CH_2NH) given above suggested that the CH_3/H effect in the imine is somewhat smaller than the effect calculated at this level, which is somewhat surprising.

c. H_2SiO and H_2SiNH

Compounds containing multiple bonds between silicon and other atoms are under study by a number of experimental and theoretical groups. Dixon[113] has calculated PA(H_2SiO) and PA(H_2SiNH) with a (11,7,1/9,5,1)/[6,4,1/3,2,1/2,1] basis set at the CI-SDQ level. In part, this work was done to examine the change in proton affinity arising from substitution of silicon for carbon in compounds with double bonds to heteroatoms. Geometries were determined with this basis set at the SCF level, and the term ΔZPE was evaluated at the 3–21G level. Based on previous comparisons, the calculated PAs are probably too high by a 2–3 kcal/mol on an absolute scale. The PAs for these two compounds are unknown, but PA[$(CH_3)_2SiO$] has been measured. Comparison of PA[$(CH_3)_2SiO$)] − PA[$(CH_3)_2CO$)] (measured) and PA(H_2SiO) − PA(H_2CO) (calculated) shows excellent agreement.[117] For these cases, substitution of silicon for carbon significantly increases the PA.

d. Carbon Dioxide

The proton affinity of CO_2 has also been calculated using a DZ + P basis set of the form (9,5,1/4,1)/[4,2,1/2,1] and with correlation corrections evaluated at the CI-SD level.[118] The geometry was determined at the CI-SD level. Since no frequencies were evaluated at this level, we have calculated them at the 3–21G level. No scale factor was employed because the ratio $ZPE_{exp}(CO_2)/ZPE_{calc}(CO_2)$ was roughly 1. The ΔE_0^{elec} term is 135.8 kcal/mol, which gives a value of 130.6 for PA(CO_2). Protonation is predicted to occur at oxygen. This is in good agreement with the experimental value of 130.9 kcal/mol. We do note one problem with this agreement. A number of experiments[119] demonstrate that PA(CH_4) is about 1.5 kcal/mol greater than PA(CO_2) and, indeed, the experimental value for PA(CH_4) is based on these measurements. We have recommended that PA(CH_4) be lowered by 3 kcal/mol, to about 129 kcal/mol. This would necessitate lowering PA(CO_2) by a comparable amount. As seen above, the DZ + P basis set at the CI-SD level tends to give PAs that are 2–3 kcal/mol too high. A decrease of 3 kcal/mol would give good agreement for ΔPA between CH_4 and CO_2.

e. HCP and H_2CPH

The proton affinities of HCP and H_2CPH have been determined by Lohr and co-workers[120] at the 6–31G**/MP4 level. The geometries were determined at the 6–31G* level and the ZPE correction was evaluated at the 3–

21G level. For HCP, the proton is predicted to bind only at carbon and PA(HCP) = 170.0 kcal/mol. For H_2CPH, protonation is predicted to occur at P, and this process is favored by 11 kcal/mol over protonation at C. The proton affinity of H_2CPH is calculated to be 185.5 kcal/mol. There are no known values for these compounds, and the calculated values are probably high by about 5 kcal/mol.

f. OLi_2

The proton affinity of OLi_2 has been calculated with a large STO basis set of the form [4,2,1] on Li, [7,5,2] on O, and [3,1] on H.[121] The correlation correction was estimated at the CI-SDQ level with a smaller basis set of the form [3,1] on Li, [4,2,1] on O, and [2,1] on H. The ZPE corrections were determined at the STO-3G level and were scaled to the experimental quantity for the neutral. The value for $PA(Li_2O)$ is 302.3 kcal/mol, and we expect an error of up to ± 3 kcal/mol because of the way in which the correlation correction was determined. However, these results do demonstrate that OLi_2 has the highest PA known for a neutral molecule.

g. C_2H_4O, C_2H_4S, and C_2H_6

Lischka and Kohler[122,123] have calculated the PAs for C_2H_4O, C_2H_4S and C_2H_6 using GTO basis sets and CEPA for the correlation correction. The ΔE_0^{elec} values at the SCF level for C_2H_4O and C_2H_4S were determined with a basis set of the form (7,3,1/3,1)/[4,2,1/2,1] for carbon, oxygen, and hydrogen, while the correlation energy correction was determined with the same basis set without polarization functions on hydrogen. For sulfur, the basis set was of the form (10,6,1)/[6,4,1]. Reasonable agreement with experiment is found for oxirane, with the calculated value being somewhat low. For thiirane, the agreement is much worse; the experimental value is much lower. At this point, we do not know which value to recommend.

For C_2H_6, Kohler and Lischka[123] employed the same techniques as described above. Geometries were either taken from MINDO/3 or STO-3G calculations. The ΔZPE term was calculated at the MINDO/3 level. The $PA(C_2H_6)$ is calculated to be 143.4 kcal/mol for a product ion of the form shown in **3**.

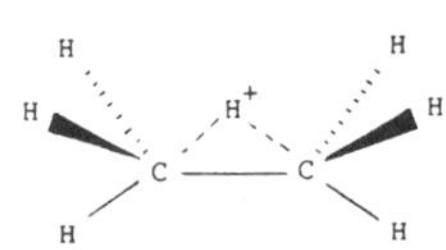

3

Considering the method employed for calculating the PA, this number is in remarkably, albeit fortuitous, good agreement with the experimented value of 143.6 kcal/mol. The C-protonated structure $CH_3CH_4^+$, analogous to CH_5^+, was found to be 5.1 kcal/mol higher in energy than the C—C bond-protonated structure; this corresponds to a lower proton affinity. This difference in energy is in reasonable agreement with the experimental energy difference of 7.8 kcal/mol.[124]

h. PH_3CH_2

The energy of protonation of the ylid PH_3CH_2 has been calculated by Lischka[125] in a manner similar to that employed above. A basis set of the form (10,6,1/8,4,1/3,1)/[6,4,1/5,3,1/2,1] was employed, and correlation was accounted for in the CEPA framework. The geometries were determined at the CEPA level. No force constants were determined, so we estimated ΔZPE as 6 kcal/mol based on the results on similar phosphorus systems of Lohr and co-workers.[120] The ΔE_0^{elec} terms are 272.3 kcal/mol at the SCF level and 267.9 kcal/mol at the CEPA level. For information, we also note that the IEPA method gives 263.4 while Pseudonatural Orbital Configuration Interaction (PNO-CI) gives 269.7 kcal/mol for ΔE_0^{elec}. The proton affinity is calculated to be 263.4 kcal/mol and is not known from experiment.

REFERENCES

1. Bowers, M.T.; Aue, D.H.; Webb, H.M.; McIver, R.T. *J. Am. Chem. Soc.* **1971,** *93,* 4314.
2. For reviews see: (a) Bohme, D.K., In "Interactions Between Ions and Molecules," Ausloos, P., Ed.; Plenum Press: New York, 1975, pp. 489–504. (b) Taft, R.W. In "Proton Transfer Reactions," Caldin, E.F., Gold, V., Eds.; Wiley: New York, 1975, pp. 31–78. (c) Arnett, E.M. In "Proton Transfer Reactions," Caldin, E.F., Gold, V., Eds.; Wiley: New York, 1975, pp. 79–101. (d) Kebarle, P., *Ann. Rev. Phys. Chem.* **1977,** *28,* 445. (e) Aue, D.H.; Bowers, M.T. In "Gas Phase Ion Chemistry," Bowers, M.T., Ed.; Academic Press: New York, 1979, pp. 1–51. (f) Taft, R.W., *Prog. Phys. Org. Chem.* **1983,** *14,* 248. (g) Moylan, C.R.; Brauman, J.I. *Ann. Rev. Phys. Chem.* **1983,** *34,* 187.
3. Lias, S.G.; Liebman, J.F.; Levin, R.D. *J. Phys. Chem. Ref. Data* **1984,** *13,* 695.
4. (a) "Methods of Electronic Structure Theory", Schaefer, H.F., III, Ed.; Plenum Press: New York, 1977. (b) "Applications of Electronic Structure Theory", Schaefer, H.F., III, Ed.; Plenum Press: New York, 1977. (c) "Computational Methods in Chemistry", Bargon, J., Ed.; Plenum Press: New York, 1980.
5. (a) Tal'roze, V.L.; Lyubimova, A.K. *Dokl. Akad. Nauk SSSR,* **1952,** *86,* 909. (b) Stevenson, D.P.; Schissler, D.O. *J. Chem. Phys.* **1955,** *23,* 1353. (c) Tal'roze, V.L.; Frankevich, E.L., *Izv. Akad. Nauk SSSR, Otd. Khim. Nauk* **1959,** No. 7, 1351.
6. Munson, M.S.B. *J. Am. Chem. Soc.* **1959,** *87,* 2332.
7. Kebarle, P.; Godbole, E.W. *J. Chem. Phys.* **1965,** *424,* 798.
8. (a) McIver, R.T.; Eyler, J.R. *J. Am. Chem. Soc.* **1971,** *93,* 4314. (b) Aue, D.H.; Webb, H.M.; Bowers, M.T. *J. Am. Chem. Soc.* **1972,** *94,* 4726. (c) Henderson, W.G.; Taagepera, M.; Holtz, D.; McIver, R.T., Jr.; Beauchamp, J.L.; Taft, R.W., *J. Am. Chem. Soc.* **1972,** *94,* 4728. (d) Arnett, E.M.; Jones, F.M., III; Taagepera, M.; Henderson, W.G.; Beauchamp, J.L.; Holtz, D.; Taft, R.W. *J. Am. Chem. Soc.* **1972,** *94,* 4724. (e) Aue, E.M.; Webb, H.M.; Bowers, M.T., *J. Am. Chem. Soc.* **1973,** *95,* 2699. (f) Taft, R.W.; Taagerpera, M.; Summerhays, K.D.; Mitsky, J. *J. Am. Chem. Soc.* **1973,** *95,* 3811.
9. (a) Chong, S.-L.; Myers, R.A., Jr.; Franklin, J.L. *J. Chem. Phys.* **1971,** *56,* 2427. (b) Chong, S.-L.; Franklin, J.L. *J. Am. Chem. Soc.* **1972,** *94,* 6347. (c) Chong, S.-L.; Franklin, J.L. *J. Am. Chem. Soc.* **1972,** *94,* 6630. (d) Kaspar, S.F.; Franklin, J.L. *J. Chem. Phys.* **1972,** *56,* 1156. (e) Briggs, J.P.; Yamdagni, R.; Kebarle, P. *J. Am. Chem. Soc.* **1972,** *94,* 5128. (f) Yamdagni, R.; Kebarle, P. *J. Am. Chem. Soc.* **1973,** *95,* 3504. (g) Hopkins, J.M.; Bone, L.I. *J. Chem. Phys.* **1973,** *58,* 1473.

10. (a) Fennelly, P.F.; Hemsworth, R.S.; Schiff, H.I.; Bohme, D.K. *J. Chem. Phys.* **1973,** *59,* 6405. (b) Hemsworth, R.S.; Rundle, H.W.; Bohme, D.K.; Schiff, H.I.; Dunkin, D.B.; Fehsenfeld, F.C. *J. Chem. Phys.* **1973,** *59,* 61. (c) Bohme, D.K.; Hemsworth, R.S.; Rundle, H.W.; Schiff, H.I. *J. Chem. Phys.* **1973,** *58,* 3504.
11. Lias, S.G.; Ausloos, P. *J. Am. Chem. Soc.* **1978,** *100,* 6027.
12. Haney, M.A.; Franklin, J.L. *J. Chem. Phys.* **1969,** *50,* 2028.
13. (a) Bohme, D.K.; Mackay, G.I.; Schiff, H.I. *J. Chem. Phys.* **1980,** *73,* 4976. (b) Bohme, D.K.; Mackay, G.I. *J. Am. Chem. Soc.* **1981,** *103,* 2173. (c) Bohme, D.K.; Stone, J.A.; Mason, R.S.; Stradling, R.S.; Jennings, K.R. *Int. J. Mass Spectrom. Ion Phys.* **1981,** *37,* 283. (d) Mackay, G.I.; Bohme, D.K. *Int. J. Mass Spectrom. Ion Phys.* **1978,** *26,* 327. (e) Mackay, G.I.; Schiff, H.I.; Bohme, D.K. *Can. J. Chem.* **1981,** *59,* 1771. (f) Payzant, J.D.; Schiff, H.I.; Bohme, D.K. *J. Chem. Phys.* **1975,** *63,* 149. (g) Rakshit, A.B.; Bohme, D.K. *Int. J. Mass Spectrom. Ion Phys.* **1983,** *49,* 275. (h) Schiff, H.I.; Bohme, D.K. *Int. J. Mass Spectrom. Ion Phys.* **1975,** *16,* 167. (i) Tanaka, K.; Mackay, G.I.; Bohme, D. *Can. J. Chem.* **1978,** *56,* 193. (j) Tanner, S.D.; Mackay, G.I.; Hopkinson, A.C.; Bohme, D.K. *Int. J. Mass Spectrom. Ion Phys.* **1979,** *29,* 153.
14. (a) Davidson, W.R.; Lau, Y.K.; Kebarle, P. *Can. J. Chem.* **1978,** *56,* 1016. (b) French, M.; Kebarle, P. *Can. J. Chem.* **1975,** *53,* 2668. (c) Hiraoka, K.; Kebarle, P. *Can. J. Chem.* **1975,** *53,* 970. (d) Hiraoka, K.; Kebarle, P. *Can. J. Chem.* **1976,** *54,* 6119. (e) Hiraoka, K.; Kebarle, P. *J. Am. Chem. Soc.* **1977,** *99,* 360. (f) Ikuta, S.; Kebarle, P.; Bancroft, G.M.; Chan, T.; Puddephatt, R.J. *J. Am. Chem. Soc.* **1982,** *104,* 5899. (g) Kebarle, P.; Yamdagni, R.; Hiraoka, K.; McMahon, T.B. *Int. J. Mass Spectrom. Ion. Phys.* **1976,** *19,* 71. (h) Lau, Y.K.; Ph.D. thesis, University of Alberta, **1979.** (i) Lau, Y.K.; Kebarle, P. *J. Am. Chem. Soc.* **1976,** *98,* 7452. (j) Lau, Y.K.; Nishizawa, K.; Tse, A.; Brown, R.S.; Kebarle, P. *J. Am. Chem. Soc.* **1981,** *103,* 6291. (k) Lau, Y.K.; Saluja, P.P.S.; Kebarle, P.; Alder, R.W. *J. Am. Chem. Soc.* **1978,** *100,* 7328. (l) Saluja, P.P.S.; Kebarle, P. *J. Am. Chem. Soc.* **1979,** *101,* 1084. (m) Sharma, R.B.; Blades, A.T.; Kebarle, P. *J. Am. Chem. Soc.* **1984,** *106,* 510. (n) Yamdagni, R.; Kebarle, P. *J. Am. Chem. Soc.* **1976,** *98,* 1320.
15. (a) Hehre, W.J.; McIver, R.T., Jr.; Pople, J.A.; Schleyer, P.v.R. *J. Am. Chem. Soc.* **1974,** *96,* 7162. (b) Wolf, J.F.; Harch, P.G.; Taft, R.W. *J. Am. Chem. Soc.* **1975,** *97,* 2904. (c) Cook, M.J.; Katrisky, A.R.; Taagepera, M.; Singh, T.D.; Taft, R.W. *J. Am. Chem. Soc.* **1976,** *98,* 6048. (d) Devlin, J.L., III; Wolf, J.F.; Taft, R.W.; Hehre, W.J. *J. Am. Chem. Soc.* **1976,** *98,* 1990. (e) Wolf, J.F.; Abboud, J.L.M.; Taft, R.W. *J. Org. Chem.*, **1977,** *42,* 3316. (f) DeFrees, D.J.; McIver, R.T., Jr.; Hehre, W.J. *J. Am. Chem. Soc.*, **1977,** *99,* 3853. (g) Summerhays, K.D.; Pollack, S.K.; Taft, R.W.; Hehre, W.J. *J. Am. Chem. Soc.*, **1977,** *99,* 4585. (h) Staley, R.H.; Taagepera, M.; Henderson, W.G.; Koppel, I.; Beauchamp, J.L.; Taft, R.W. *J. Am. Chem. Soc.*, **1977,** *99,* 326. (i) Pollack, S.K.; Devlin, J.L., III; Summerhays, K.D.; Taft, R.W.; Hehre, W.J. *J. Am. Chem. Soc.*, **1977,** *99,* 4583. (j) Wolf, J.F.; Staley, R.H.; Koppel, I.; Taagepera, M.; McIver, R.T.; Jr.; Beauchamp, J.L.; Taft, R.W. *J. Am. Chem. Soc.* **1977,** *99,* 5417. (k) Taft, R.W.; Wolf, J.F.; Beauchamp, J.L.; Scorrano, G.; Arnett, E.M. *J. Am. Chem. Soc.* **1978,** *100,* 1240. (l) Locke, M.J.; Hunter, R.L.; McIver, R.T., Jr. *J. Am. Chem. Soc.* **1979,** *101,* 272. (m) Alder, R.W.; Arrowsmith, R.J.; Casson, A.; Sessions, R.B.; Heilbronner, E.; Kovac, B.; Huber, H.; Taagepera, M. *J. Am. Chem. Soc.* **1981,** *103,* 6137. (n) Bromilow, J.; Abboud, J.L.M.; Lebrilla, C.B.; Taft, R.W.; Scorrano, G.; Lucchini, V. *J. Am. Chem. Soc.* **1981,** *103,* 5448. (o) Taagepera, M.; Summerhays, K.D.; Hehre, W.J.; Topsom, R.D.; Pross, A.; Radom, L.; Taft, R.W. *J. Org. Chem.* **1981,** *46,* 891.
16. (a) Aue, D.H.; Betowski, L.D.; Davidson, W.R.; Bowers, M.T.; Beak, P.; Lee, J. *J. Am. Chem. Soc.* **1979,** *101,* 1361. (b) Aue, D.H.; Davidson, W.R.; Bowers, M.T. *J. Am. Chem. Soc.* **1976,** *98,* 6700. (c) Aue, D.H.; Webb, H.M.; Bowers, M.T. *J. Am. Chem. Soc.* **1975,** *97,* 4136. (d) Aue, D.H.; Webb, H.M.; Bowers, M.T. *J. Am. Chem. Soc.* **1975,** *97,* 4137. (e) Aue, D.H.; Webb, H.M.; Bowers, M.T. *J. Am. Chem. Soc.*, **1976,** *98,* 311. (f) Aue, D.H.; Webb, H.M.; Bowers, M.T.; Liotta, C.L.; Alexander, C.J.; Hopkins, H.P., Jr. *J. Am. Chem. Soc.*, **1976,** *98,* 854. (g) Aue, D.H.; Webb, H.M.; Davidson, W.R.; Vidal, M.; Bowers, M.T.; Goldwhite, H.; Vertal, L.E.; Douglas, J.E.; Kollman, P.A.; Kenyon, G.L. *J. Am. Chem. Soc.*, **1980,** *102,* 5151.
17. (a) Meot-Ner (Mautner), M. *J. Am. Chem. Soc.* **1979,** *101,* 2396. (b) Meot-Ner (Mautner), M. *J. Phys. Chem.*, **1980,** *84,* 2716. (c) Meot-Ner (Mautner), M. *J. Am. Chem. Soc.*, **1983,** *105,* 4906. (d) Meot-Ner (Mautner), M.; Nelsen, S.F.; Willi, M.F.; Frigo, T.B., ibid, **1984,** *106,* 7384. (e) Meot-Ner (Mautner), M.; Field, F.H. *J. Chem. Phys.*, **1977,** *66,* 4527. (f)

Meot-Ner (Mautner), M.; Hamlet, P.; Hunter, E.P.; Field, F.H. *J. Am. Chem. Soc.,* **1980,** *102,* 6393. (g) Meot-Ner (Mautner), M.; Hunter, E.P.; Field, F.H. *J. Am. Chem. Soc.,* **1980,** *102,* 686. (h) Meot-Ner (Mautner), M.; Hunter, E.P.; Hamlet, P.; Field, F.H., "Proceedings of the 28th Annual Conference on Mass Spectrometry and Allied Topics", May 25–30, 1980, p. 233. (i) Meot-Ner (Mautner), M.; Sieck, L.W. *J. Am. Chem. Soc.,* **1983,** *105,* 2956.

18. (a) Doiron, E.; McMahon, T.B. *Can. J. Chem.,* **1981,** *59,* 2689. (b) Drummond, D.F.; McMahon, T.B. *Int. J. Mass Spectrom. Ion Phys.* **1982,** *42,* 265. (c) Doiron, E.; Grein, F.; McMahon, T.B.; Vasudevan, K. *Can. J. Chem.* **1979,** *57,* 1751.
19. (a) Lias, S.G.; Shold, D.M.; Ausloos, P. *J. Am. Chem. Soc.,* **1980,** *102,* 2540. (b) Ausloos, P.; Lias, S.G. *Int. J. Mass Spectrom. Ion Proc.* **1984,** *58,* 165.
20. (a) Meot-Ner (Mautner), M.; Solomon, J.J.; Field, F.H. *J. Am. Chem. Soc.,* **1976,** *98,* 1025. (b) Solomon, J.J.; Field, F.H., ibid, **1975,** *97,* 2625. (c) Solomon, J.J.; Field, F.H., ibid, **1976,** *98,* 1567. (d) Goren, A.; Munson, B. *J. Phys. Chem.,* **1976,** *80,* 2848.
21. Hiraoka, K. *Int. J. Mass Spectrom. Ion Phys.,* **1978,** *27,* 139.
22. (a) Hiraoka, K.; Kebarle, P. *Can. J. Chem.,* **1975,** *53,* 970. (b) Hiraoka, K.; Kebarle, P. *J. Am. Chem. Soc.,* **1976,** *98,* 6119. (c) Hiraoka, K.; Kebarle, P., ibid, **1977,** *99,* 360.
23. (a) Lias, S.G., In "Kinetics of Ion-Molecule Reactions", Ausloos, P., Ed.; Plenum Press: New York, 1979, p. 223. (b) Rosenstock, H.M., In "Kinetics of Ion-Molecule Reactions", Ausloos, P., Ed.; Plenum Press: New York, 1979, p. 246.
24. Wagman, D.; Evans, W.H.; Parker, V.B.; Schumm, R.H.; Halow, I.; Bailey, S.M.; Churney, K.L.; Nuttall, R.L. *J. Phys. Chem. Ref. Data,* **1982,** *11,* Suppl. 2.
25. (a) Stull, D.R.; Prophet, H., NSRDS-NBS 37 National Bureau of Standards, Washington, D.C. (1971). (b) Chase, M.W.; Curnutt, J.L.; Prophet, H.; McDonald, R.A.; Syverud, A.N. *J. Phys. Chem. Ref. Data,* **1975,** *4,* 1. (c) Chase, M.W., Jr.; Curnutt, J.L.; Downey, J.R., Jr; Mcdonald, R.A.; Syverud, A.N.; Valenzuela, E.A. *J. Chem. Phys. Ref. Data,* **1982,** *11,* 695.
26. Ceyer, T.; Tiedemann, P.W.; Mahan, B.H.; Lee, Y.T. *J. Chem. Phys.* **1979,** *70,* 14.
27. Frisch, M.J.; Pople, J.A.; Del Bene, J.E. Unpublished results, personal communication.
28. Lias, S.G.; Shold, D.M.; Ausloos, P. *J. Am. Chem. Soc.,* **1980,** *102,* 2540.
29. Traeger, J.C.; McLoughlin, R.G., *J. Am. Chem. Soc.,* **1981,** *103,* 3647.
30. Tsang, W., *Int. J. Chem. Kinet.,* **1978,** *10,* 41.
31. Houle, A.; Beauchamp, J.L. *J. Am. Chem. Soc.,* **1979,** *101,* 4067.
32. (a) Canosa, E.; Marshall, R.M. *Int. J. Chem. Kinet.,* **1981,** *13,* 303. (b) Castelhano, L.; Marriott, P.R.; Griller, D. *J. Am. Chem. Soc.,* **1981,** *103,* 4262. (c) Holmes, J.L.; Lossing, F.P. *Int. J. Mass Spectrom. Ion Phys.* In press. (d) McMillen, F.; Golden, D.M. *Annu. Rev. Phys. Chem.,* **1982,** *33,* 493.
33. Dyke, J.M.; Jonathan, N.; Lee, E.; Morris, A.; Winter, M. *Phys. Sci.,* **1977,** *16,* 197.
34. (a) Abboud, J.-L.M.; Hehre, W.J.; Taft, R.W. *J. Am. Chem. Soc.,* **1976,** *98,* 6072. (b) Jackson, J.-A.A.; Lias, S.G.; Ausloos, P. *J. Am. Chem. Soc.,* **1977,** *99,* 7515. (c) Sen Sharma, K.; Kebarle, P. *Can. J. Chem.,* **1981,** *59,* 1592.
35. (a) Baer, T. *J. Am. Chem. Soc.,* **1980,** *102,* 2482. (b) Rosenstock, H.M.; Buff, R.; Ferreira, M.A.A.; Lias, S.G.; Parr, A.C.; Stockbauer, R.L.; Holmes, J.L. *J. Am. Chem. Soc.,* **1982,** *104,* 2337.
36. Cremer, D.; Binkley, J.S.; Pople, J.A.; Hehre, W.J. *J. Am. Chem. Soc.,* **1974,** *96,* 6900.
37. Collyer, S.M.; McMahon, T.B. *J. Phys. Chem.,* **1983,** *87,* 909.
38. Del Bene, J.E.; Frisch, M.J.; Raghavachari, K.; Pople, J.A. *J. Phys. Chem.,* **1982,** *86,* 1529.
39. Guyon, M.; Chupka, W.A.; Berkowitz, J. *J. Chem. Phys.,* **1976,** *64,* 1419.
40. Matthews, C.; Warneck, P. *J. Chem. Phys.,* **1969,** *51,* 854.
41. Dyke, J.M.; Jonathan, N.; Morris, A.; Winter, M.J. *Mol. Phys.,* **1980,** *39,* 629.
42. Warneck, P. *Z. Naturforsch. A,* **1974,** *29,* 350.
43. McCulloh, K.E. *Int. J. Mass Spectrom. Ion Phys.,* **1976,** *21,* 333.
44. (a) Berkowitz, J.; Appelman, E.H.; Chupka, W.A. *J. Chem. Phys.,* **1973,** *58,* 1950 (b) Katsumata, S.; Lloyd, D.R. *Chem. Phys. Lett.,* **1977,** *45,* 519.
45. Dyke, J.M.; Jonathan, N.; Morris, A.; Winter, M.J. *Mol. Phys.,* **1981,** *44,* 1059.
46. McCulloh, K.E., Proceedings of the 25th Annual Conference on Mass Spectrometry and Allied Topics, Washington, D.C., 1977. Also cited as data "To be published" in Hartman, K.N.; Lias, S.; Ausloos, P.; Rosenstock, H.M.; Schroyer, S.S.; Schmidt, C.; Martinsen, D.; Milne, G.W.A. "A Compendium of Gas Phase Basicity and Proton Affinity Measurements", NBSIR 79-1777, National Bureau of Standards, Washington, D.C. 1979.
47. Traeger, C.; McLoughlin, R.G.; Nicholson, A.J.C. *J. Am. Chem. Soc.,* **1982,** 104, 5318.

48. (a) Refaey, K.M.A.; Chupka, W.A. *J. Chem. Phys.,* **1968,** *48,* 5205. (b) Omura, I.; Kaneko, T.; Yamada, Y.; Tanaka, K. *J. Phys. Soc. Japan,* **1981,** *27,* 981. (c) Lossing, F.P. *J. Am. Chem. Soc.,* **1977,** *99,* 7526.
49. Ng, C.Y.; Trevor, D.J.; Tiedemann, P.W.; Ceyer, S.T.; Kronebusch, P.L.; Mahan, B.H.; Lee, Y.T. *J. Chem. Phys.,* **1977,** *67,* 4235.
50. Del Bene, J.E.; Mettee, H.D.; Frisch, M.J.; Luke, B.T.; Pople, J.A. *J. Phys. Chem.,* **1983,** *87,* 3279.
51. Clair, R.; McMahon, T.B. *Int. J. Mass Spectrom. Ion Phys.,* **1981,** *39,* 27.
52. (a) Burrau B., *Det. K. Dan. Videnskapselsk.* **1927,** *7,* 1. (b) Hylleraas, E.A. *Z. Phys.* **1931,** *71,* 739. (c) Jaffe, G. *Z. Phys.* **1934,** *87,* 535. (d) Pauling, L.; Wilson, E.B., Jr. "Introduction to Quantum Mechanics". McGraw-Hill: New York, 1935, pp. 327–340. (e) Bishop, D.M. *Mol. Phys.* **1974,** *28,* 1397.
53. Rosmus, P.; Meyer, W. *J. Chem. Phys.* **1977,** *66,* 13.
54. Nicholas, A.M. de P.; Boyd, R.J.; Arnold, D.R. *Can. J. Chem.* **1982,** *60,* 3011.
55. (a) Noell, J.O.; Morokuma, K. *J. Phys. Chem.* **1978,** *80,* 2675. (b) Morokuma, K. *J. Chem. Phys.* **1971,** *56,* 1236. (c) Morokuma, K.; Kitaura, K. *Int. J. Quantum Chem.* **1976,** *10,* 325.
56. (a) Herzberg, G. "Electronic Spectra and Electronic Structure of Polyatomic Molecules". Van Nostrand-Reinhold: New York, 1966. (b) Huber, K.P.; Herzberg, G. "Constants of Diatomic Molecules". Van Nostrand-Reinhold: New York, 1979.
57. Durig, J.R.; Antion, D.J. *J. Chem. Phys.* **1969,** *51,* 3639.
58. For example, see: (a) Crofton, M.W.; Oka, T. *J. Chem. Phys.* **1983,** *79,* 3157. (b) Schafer, E.; Begemann, M.H.; Gudeman, C.S.; Saykally, R.J. ibid, **1983,** *79,* 3159. (c) Gudeman, C.S.; Begemann, M.H.; Pfaff, J.; Saykally, R.J. *Phys. Rev. Lett* **1983,** *50,* 727. (d) Gudeman, C.S.; Begemann, M.H.; Pfaff, J.; Saykally, R.J. *J. Chem. Phys.* **1983,** *78,* 5837. (e) Foster, S.C.; McKellar, A.R.W.; Sears, T.J. ibid, **1984,** *81,* 578. (f) Haese, N.N.; Oka, T. ibid, **1984,** *80,* 572. (g) Begemann, M.H.; Gudeman, C.S.; Pfaff, J.; Saykally, R.J. *Phys. Rev. Lett.* **1983,** *51,* 554; (h) Amano, T.J. *Chem. Phys.* **1983,** *79,* 3595; (i) Gudeman, C.S.; Saykally, R.J. *Annu. Rev. Phys. Chem.* **1984,** *83,* 387.
59. Shavitt, I. In Reference 4a, p. 189.
60. Langhoff, S.R.; Davidson, E.R. *Int. J. Quantum Chem.* **1974,** *8,* 61.
61. (a) Pople, J.A.; Krishnan, R.; Schlegel, H.B.; Binkley, J.S. *Int. J. Quantum Chem. Symp.* **1978,** *14,* 545; (b) Krishnan, R.; Pople, J.A. ibid, **1978,** *14,* 91; (c) Krishnan, R.; Frisch, M.J.; Pople, J.A. *J. Chem. Phys.* **1980,** *72,* 4244. (d) Pople, J.A.; Binkley, J.S.; Seeger, R. *Int. J. Quantum Chem. Symp.* **1976,** *10,* 1.
62. Kutzelnigg, W. In Reference 4a, p. 129.
63. (a) Meyer, W. In Reference 4a, p. 413. (b) Meyer, W. *Int. J. Quantum Chem.* **1971,** *5,* 341. (c) Meyer, W. *Theor. Chim. Acta* **1974,** *35,* 277.
64. (a) Dupuis, M.; Rys, J.; King, H.F. *J. Chem. Phys.* **1976,** *65,* 111. (b) King, H.F.; Dupuis, M.; Rys, J. "National Resource for Computer Chemistry Software Catalog", Vol. 1; Program QH02 (HONDO), 1980. HONDO will handle any but generally contracted GTO basis sets. It will not do MPn calculations. For an MP2 version, see Carksy, P.; Hess, B.A., Jr.; Schaad, L.J.; Urban, M.; Kello, V. Quantum Chemistry Program Exchange (QCPE), Indiana University: Bloomington, 1984, Program No. 482.
65. (a) GAUSSIAN 80: Binkley, J.S.; Whiteside, R.A.; Raghavachari, K.; Seeger, R.; DeFrees, D.J.; Schlegel, H.B.; Topiol, S.; Kahn, L.R.; Pople, J.A. Carnegie–Mellon University, Pittsburgh. (b) GAUSSIAN 82: Binkley, J.S.; Frisch, M.J.; DeFrees, D.J.; Raghavachari, K.; Whiteside, R.A.; Schlegel, H.B.; Fluder, E.M.; Pople, J.A. Carnegie–Mellon University, Pittsburgh. GAUSSIAN 82 will handle more general basis sets and will do CI-SD calculations.
66. (a) STO-3G: Hehre, W.J.; Stewart, R.F.; Pople, J.A. *J. Chem. Phys.* **1969,** *51,* 2657. (b) 4–31G: Ditchfield, R.; Hehre, W.J.; Pople, J.A. *J. Chem. Phys.* **1971,** *54,* 724. (c) 6–31G* and 6–31G**: Hariharan, P.C.; Pople, J.A. *Theor. Chim. Acta* **1973,** *28,* 213. (d) 3–21G: Binkley, J.S.; Pople, J.A.; Hehre, W.J. *J. Am. Chem. Soc.* **1980,** *102,* 939. (e) 6–311G**: Raghavachari, K.; Binkley, J.S.; Seeger, R.; Pople, J.A. *J. Chem. Phys.* **1980,** *72,* 650. (f) 6–31+G**: Clark, T.; Chandrasekhar, J.; Spitznagel, W.G.; Schleyer, P.v.R. *J. Comput. Chem.* **1983,** *4,* 294.
67. Dunning, T.H., Jr.; Hay, P.J. In Reference 6a, p. 1.
68. (a) Shavitt, I. *Int. J. Quantum Chem. Symp.* **1977,** *11,* 131; ibid, **1978,** *12,* 5. (b) Brooks, B.R.; Schaefer, H.F., III *J. Chem. Phys.* **1979,** *70,* 5092; (c) Brooks, B.R.; Laidig, W.D.; Saxe, P.; Handy, N.C.; Schaefer, H.F., III *Phys. Scripta* **1980,** *21,* 312.

69. Dunning, T.H., Jr. *J. Chem. Phys.* **1971,** *55,* 716.
70. Huzinaga, S. *J. Chem. Phys.* **1965,** *42,* 1293.
71. McLean, A.D.; Chandler, G.S. *J. Chem. Phys.* **1980,** *72,* 5639.
72. Komornicki, A.; Dixon, D.A. Unpublished results.
73. (a) Pulay, P. In Reference 4b, p. 153. (b) Komornicki, A.; Ishida, K.; Morokuma, K.; Ditchfield, R.; Conrad, M. *Chem. Phys. Lett.* **1977,** *45,* 595. (c) Komornicki, A.; Pauzat, F.; Ellinger, Y. *J. Phys. Chem.* **1983,** *87,* 3847.
74. (a) Brooks, B.R.; Laidig, W.D.; Saxe, P.; Goddard, J.D.; Yamaguchi, Y.; Schaefer, H.F., III *J. Chem. Phys.* **1980,** *72,* 4652. (b) Krishnan, R.; Schlegel, H.B.; Pople, J.A. ibid, **1980,** *72,* 4654. (c) Pople, J.A.; Krishnan, R.; Schlegel, H.B.; Binkley, J.S. *Int. J. Quantum Chem. Symp.* **1979,** *13,* 225. (d) Brooks, B.R.; Laidig, W.D.; Saxe, P.; Goddard, J.S.; Schaefer, H.F., III In "Lecture Notes in Chemistry", Vol. 22; Hinze, J., Ed., Spring-Verlag: Berlin, 1981, p. 158. (e) Schlegel, H.B. In "Computational Theoretical Organic Chemistry", Csizmadia, I.G.; and Daudel, R., Eds., Reidel: Dordrecht, Holland, 1981, p. 129.
75. Eades, R.A.; Weil, D.A.; Dixon, D.A.; Douglass, C.H., Jr. *J. Phys. Chem.* **1981,** *85,* 976.
76. Dunning, T.H., Jr; Pitzer, R.M.; Aung, S. *J. Chem. Phys.* **1972,** *57,* 5044.
77. (a) Osamura, Y.; Yamaguchi, Y.; Saxe, P.; Vincent, M.A.; Gaw, J.F.; Schaefer, H.F., III *Chem. Phys.* **1981,** *72,* 131. (b) Osamura, Y.; Yamaguchi, Y.; Schaefer, H.F., III *J. Chem. Phys.* **1982,** *77,* 383. (c) Yamaguchi, Y.; Osamura, Y.; Fitzgerald, G.; Schaefer, H.F., III *J. Chem. Phys.* **1983,** *78,* 107.
78. Komornicki, A.; King, H.F. Private communication.
79. Del Bene, J.E. *Chem. Phys. Lett.* **1978,** *55,* 235.
80. Del Bene, J.E. *J. Am. Chem. Soc.* **1978,** *100,* 1673.
81. Del Bene, J.E. *Chem. Phys. Lett.* **1983,** *94,* 213.
82. Eades, R.A.; Weil, D.A.; Dixon, D.A.; Douglass, C.H., Jr. *J. Phys. Chem.* **1981,** *85,* 981.
83. Eades, R.A.; Weil, D.A.; Ellenberger, M.R.; Farneth, W.E.; Dixon, D.A.; Douglass, C.H., Jr. *J. Am. Chem. Soc.* **1981,** *103,* 5372.
84. (a) Halgren, T.A.; Lipscomb, W.N. *J. Chem. Phys.* **1973,** *58,* 1569. (b) Halgren, T.A.; Kleier, D.A.; Hall, J.H., Jr; Brown, L.D.; Lipscomb, W.N. *J. Am. Chem. Soc.* **1978,** *100,* 6595.
85. Hehre, W.J.; Pople, J.A. *Tetrahedron Lett.* **1970,** 2959.
86. (a) Kolos, W.; Peek, J.M. *J. Chem. Phys.* **1976,** *12,* 381. (b) Bishop, D.M.; Cheung, L. *J. Mol. Spectrosc.* **1979,** *75,* 462.
87. Peyerimhoff, S. *J. Chem. Phys.* **1965,** *43,* 998.
88. Rosmus, P.; Reinsch, E.-A. *Z. Naturforsch. A* **1980,** *35,* 1066.
89. Rosmus, P. *Theor. Chim. Acta* **1979,** *51,* 359.
90. Rosmus, P.; Reinsch, E.-A.; Werner, H.S. *NATO Adv. Sci. Inst.* **1983,** *90,* 407.
91. Dykstra, C.E.; Gaylord, A.S.; Gwinn, W.D.; Swope, W.C.; Schaefer, H.F., III. *J. Chem. Phys.* **1978,** *68,* 3951.
92. Kolos, W.; Wolniewicz, L. *J. Chem. Phys.* **1964,** *41,* 3663; *J. Mol. Spectrosc.* **1975,** *54,* 303.
93. Carney, G.D.; Porter, R.N. *J. Chem. Phys.* **1976,** *65,* 3547.
94. Huber, K.P.; Herzberg, G. "Constants of Diatomic Molecules". Van Nostrand-Reinhold: New York, 1979.
95. Dixon, D.A.; Komornicki, A.; Kraemer, W.P. *J. Chem. Phys.* **1984,** *81,* 3603.
96. Del Bene, J.E.; Frisch, M.J.; Raghavachari, K.; Pople, J.A. *J. Phys. Chem.* **1982,** *86,* 1529.
97. Ikuta, S. *Chem. Phys. Lett.* **1984,** *109,* 550.
98. Illies, A.J.; Jarrold, M.F.; Bowers, M.T. *J. Chem. Phys.* **1982,** *77,* 5847.
99. Botschwina, P. *Chem. Phys. Lett.* **1984,** *107,* 535.
100. (a) Kraemer, W.P.; Komornicki, A.; Dixon, D.A. To be published. (b) Hennig, P.; Kraemer, W.P.; Diercksen, G.H.F. MPI/PAE Astro 135, Max Planck Institut für Physik und Astrophysik: Munich, Germany, 1977.
101. Dixon, D.A.; Komornicki, A. To be published.
102. Lischka, H. *Theoret. Chim. Acta* **1973,** *31,* 39.
103. Foster, M.S.; Beauchamp, J.L. *Inorg. Chem.* **1975,** *14,* 1229.
104. Tiedemann, P.W.; Anderson, S.L.; Ceyer, S.T.; Hirooka, T.; Ng, C.Y.; Mahan, B.H.; Lee, Y.T. *J. Chem. Phys.* **1979,** *71,* 605.
105. Eades, R.A.; Scanlon, K.; Ellenberger, M.R.; Dixon, D.A.; Marynick, D.S. *J. Phys. Chem.* **1980,** *82,* 2840.
106. Ceyer, S.T.; Tiedemann, P.W.; Mahan, B.H.; Lee, Y.T. *J. Chem. Phys.* **1970,** *70,* 14.
107. Frisch, J.J.; Del Bene, J.E.; Raghavachari, K.; Pople, J.A. *Chem. Phys. Lett.* **1981,** *83,* 240.

108. Del Bene, J.E.; Mettee, H.D.; Frisch, M.J.; Luke, B.T.; Pople, J.A. *J. Phys. Chem.* **1983,** *87,* 3279.
109. Diercksen, G.H.F.; Kraemer, W.P.; Roos, B.O. *Theor. Chim. Acta* (Berlin) **1975,** *36,* 249.
110. Kello, V.; Urban, M.; Hubac, I.; Carsky, P. *Chem. Phys. Lett.* **1978,** *58,* 83.
111. Marynick, D.S.; Scanlon, K.; Eades, R.A.; Dixon, D.A. *J. Phys. Chem.* **1981,** *85,* 3364.
112. Komornicki, A.; Dixon, D.A. To be published.
113. Dixon, D.A. To be published.
114. Ellenberger, M.R.; Dixon, D.A.; Farneth, W.E. *J. Am. Chem. Soc.* **1981,** *103,* 5377.
115. Del Bene, J.E. *J. Comput. Chem.* **1984,** *5,* 381.
116. Nobes, R.H.; Radom, L. *Org. Mass Spectrom.* **1982,** *17,* 340.
117. Weil, D.A. Ph.D. thesis, University of Minnesota, January 1984.
118. Green, S.; Schor, H.; Sieghahn, P.; Thaddeus, P. *Chem. Phys.* **1976,** *17,* 479.
119. (a) Staley, R.H.; Beauchamp, J.L. *J. Chem. Phys.* **1975,** *62,* 1998. (b) Kaspar, S.F.; Franklin, J.L. ibid, **1972,** *56,* 1156. (c) Meot-Ner (Mautner), M.; Field, F.H. ibid, **1977,** *66,* 4527. (d) Hemsworth, R.S.; Rundle, H.W.; Bohme, D.K.; Schiff, H.I.; Dunkin, D.B.; Fehsenfeld, F.C. ibid, **1973,** *59,* 61; (e) Bohme, D.K.; Hemsworth, R.S.; Rundle, H.W.; Schiff, H.I. ibid, **1973,** *58,* 3504.
120. Lohr, L.L.; Schlegel, H.B.; Morokuma, K. *J. Phys. Chem.* **1984,** *88,* 1981.
121. Marynick, D.S.; Dixon, D.A. *J. Phys. Chem.* **1983,** *87,* 3430.
122. Lischka, H.; Kohler, H.-J. *Chem. Phys. Lett.* **1979,** *63,* 326.
123. Kohler, H.-J.; Lischka, H. *Chem. Phys. Lett.* **1978,** *58,* 175.
124. Hiraoka, H.; Kebarle, P. *J. Am. Chem. Soc.* **1976,** *98,* 6119.
125. Lischka, H. *J. Am. Chem. Soc.* **1977,** *99,* 353.

CHAPTER 8

Universality: The Differences and Equivalences of Heats of Formation, Strain Energy, and Resonance Energy

Joel F. Liebman

University of Maryland Baltimore County, Catonsville, Maryland

Deborah Van Vechten*

Naval Research Laboratory, Washington, D.C.

CONTENTS

*Current Address: Sachs/Freeman Associates, Code 4671, Naval Research Laboratory, Washington, D.C.

© 1987 VCH Publishers, Inc.

MOLECULAR STRUCTURE AND ENERGETICS, Vol. 2

≪Statements in angular brackets, ≪ . . . ≫, are intended to supplement the text by offering details of arguments, additional examples, and extensions of the ideas presented. They may be omitted in an initial reading without loss of continuity.≫

1. INTRODUCTION

In this chapter we introduce the ideas, methodology, and applications of "conceptual vectors" to problems of chemical interest. The material presented provides both a conceptual structure for the general understanding of chemical phenomena and relatively concrete models we have found useful in investigations of rather broad areas of molecular structure and energetics. We discuss applications to the varied topics enumerated in the chapter contents; in addition, three appendices extend the concept of "diagonal" logic employed in the topical sections. The order of the various applications arises from the historical order in which they were studied. In particular, most of the specialized vocabulary associated with "conceptual vectors" was introduced at the same time as the topic of Section 2. Thus we made it the first applications section and wrote it more explicitly and pedagogically than the others. We recommend that readers read it first, although they may choose to read any or all of the sections in any personal order desired. This chapter is explicitly cross-referenced internally, as well as externally to other chapters in this series that discuss a class of compounds, phenomenon, or general principle more extensively than we have.

Before beginning the applications sections, we present the salient mathematical properties of vectors that are employed in this chapter. Readers who feel adequately comfortable with the concept of vectors and their mathematical description are encouraged to bypass this introduction (Section 1). As used here, vectors are essentially mathematical shorthand, and little is to be gained by recalling that vectors have magnitude and direction. Indeed, the sole reason for calling our "objects" vectors is that the mathematical properties are the same, not that they have any physical reality as found for velocity, force, or molecular dipole moment. As such, we often refer to them as conceptual vectors.

The first property for "real" and "conceptual" vectors alike is that an n-dimensional vector **V** can be written as an ordered set of n numbers (V_1, V_2, . . . , V_n). These numbers are referred to as the vector **V**'s components. While the number n is usually 3 in most physical applications because our everyday environment has three spatial dimensions, for our purposes the dimension is arbitrary and is determined solely by the number of components V_i we wish to consider. As such, depending on the application, we may speak of 2-vectors, 3-vectors, and so on. Most applications in this chapter involve 3-vectors, for which readers' prior experience on vectors such as dipole moments and force will be directly transferable.

Although many mathematical operations on vectors have been devised, surprisingly few are of relevance here, and so we list these few and consider them as the defining properties of vectors. We start with addition. Two vectors may be added only when they have the same dimensionality or number of components and $\mathbf{U} + \mathbf{V} = \mathbf{W}$ is taken to mean:

$$(U_1, U_2, \ldots) + (V_1, V_2, \ldots) = (W_1, W_2, \ldots) \tag{8-1}$$

where $W_i = U_i + V_i$. As with ordinary addition on ordinary numbers, $\mathbf{U} + \mathbf{V} = \mathbf{V} + \mathbf{U}$.

Turning to multiplication, for any number (scalar) c and vector **V**, the quantity $c\mathbf{V}$ is defined by:

$$c(V_1, V_2, \ldots) = (cV_1, cV_2, \ldots). \tag{8-2}$$

As such $c\mathbf{V} = \mathbf{W}$ only when $cV_1 = W_1$, $cV_2 = W_2$, . . . One can also multiply two vectors to form a single number s (a scalar) if the two vectors have the same dimensionality:

$$\mathbf{U} \cdot \mathbf{V} = U_1V_1 + U_2V_2 + \cdots = s \tag{8-3}$$

(This is called the inner or dot product of the two vectors, where for mathematical reasons the vector on the left, **U** in this case, is referred to as a "row" vector and the vector on the right, **V**, is referred to as a "column" vector. There are also the cross, wedge, and outer or direct products of vectors, which are seemingly of no consequence to our analysis.) Quite clearly, $\mathbf{U} \cdot \mathbf{V} = \mathbf{V} \cdot \mathbf{U}$. When the product $\mathbf{U} \cdot \mathbf{V} = 0$, and neither **U** nor **V** has all its components equal to zero, the two vectors are said to be orthogonal.

The final property of vectors that we will mention is that of linear independence. A set of vectors {**U**, **V**, **W**, . . . } is said to be linearly independent if there are no numbers a, b, c, . . . such that $a\mathbf{U} + b\mathbf{V} + c\mathbf{W} + \cdots = \mathbf{0}$, except for the trivial and irrelevant case of $a = b = c \cdots = 0$. If the vectors are independent, one cannot derive any one vector in the set from a combination of the others. Likewise, if a set of conceptual vectors are independent, one cannot derive the information content of one from knowledge of the predictions of the other conceptual vectors.

2. THE STRAIN ENERGY OF UNSUBSTITUTED ALICYCLIC HYDROCARBONS

This section presents a review of the authors' initial application[1] of the current conceptual framework to a problem of molecular structure and energetics. Unsubstituted alicyclic hydrocarbons are a large class of organic compounds that may all be described by the general formula $C_\alpha(CH)_\beta(CH_2)_\gamma$ for which C, CH, and CH_2 are quaternary carbons, methine or tertiary carbons, and methylene or secondary carbons, respectively. For scientific, esthetic, and numerous but generally personal reasons, these parent species or unsubstituted compounds, compared to their substituted derivatives,[2] have historically been of much greater interest to both the synthetic and theoretical organic chemistry communities.[3]

≪Following from an intuitive point of view rather than the textbook definition, substituents are taken as the groups or atoms other than hydrogen that are connected to the alicyclic framework by but one bond at one site. That is, breaking this bond divides the molecule into two pieces. The substitutent is thus bound to the rest of the molecule rather peripherally or weakly, at least in a conceptual way. This usage parallels the literature convention of the nomenclature of such species. For example, cyclo-$C_6H_{11}CH_3$ is named methylcyclohexane to indicate that it is a derivative of hexane, while its isomer $(CH_2)_7$ is named cycloheptane to indicate that it is a derivative of heptane. We do not know of the alternative names, 1,6- and 1,7-didehydroheptane respectively, ever being suggested. Numerous syntheses of these species also reflect this substituent–framework dichotomy; for example, methylcyclohexane is logically made from a 6-carbon species, cyclohexanone, whereas cycloheptane syntheses more often start with 7-carbon species such as 1,7-dibromoheptane.≫

Our interest in unsubstituted alicyclic hydrocarbons is predominantly in their strain energy, a concept to be discussed in reasonable length in this chapter. Suffice it to say for now that strain energy is the difference of the "real" heat of formation the species has (see Chapters 4 and 5 of this volume and 5 and 7 of Volume 3 of this series) and that for an often hypothetical model species in which such distorted and destabilizing features as stretched bonds and compressed internuclear angles are absent. As in most literature studies of strained compounds, we will also presume these hydrocarbons to be gases at 25°C, even though most are either liquids or solids under conventional experimental conditions.[4] This nearly universal presumption is no doubt due in large part to an appreciation of the inherent "pristineness" and conceptual simplicity of gas phase species, but it is reinforced by the current inability of theory to adequately describe phase transition energies or condensed phase species outright. (There are, however, numerous methods for measuring—and, with varying degrees of accuracy, generality, and complexity—estimating, heats of vaporization and of sublimation, see Chapter 3, this volume.) For such gaseous, unsubstituted alicyclic hydrocarbons, the strain energy may be naturally defined as follows:

$$SE_{cpd} = \Delta H_f^\circ(\text{g, cpd}) - \alpha\Delta H_f^\circ(\text{g, >C<}) - \beta\Delta H_f^\circ(\text{g, >CH}-) - \gamma\Delta H_f^\circ(\text{g, }-\text{CH}_2-) \tag{8-4}$$

One notational simplification arises from deleting the "g" and the superscript ° due to our presumptions.

Another notational and conceptual simplification arises upon recalling the general equivalence of a number triplet and a three-dimensional vector or 3-vector and recognizing in Equation 8-3 the form of a dot product. To see this, one rewrites the set of numbers (α, β, γ) as $(\nu_1, \nu_2, \nu_3) = \mathbf{v}$ and the set of heats of formation: $[\Delta H_f(>\text{C}<), \Delta H_f(>\text{CH}-), \Delta H_f(-\text{CH}_2-)]$ as $(\eta_1, \eta_2, \eta_3) = \boldsymbol{\eta}$. The strain energy may then simply be recast as:

$$SE_{cpd} = \Delta H_f^\circ(\text{g, cpd}) - (\nu_1\eta_1 + \nu_2\eta_2 + \nu_3\eta_3) \tag{8-5a}$$
$$= \Delta H_{f(\text{cpd})} - (\nu_1\eta_1 + \nu_2\eta_2 + \nu_3\eta_3) \tag{8-5b}$$
$$= \Delta H_{f(\text{cpd})} - \Sigma_k \nu_k \eta_k \tag{8-5c}$$
$$= \Delta H_{f(\text{cpd})} - \mathbf{v} \cdot \boldsymbol{\eta} \tag{8-5d}$$

The ν and η number triplets, now rewritten as $\mathbf{v}$ and $\boldsymbol{\eta}$ vectors, are different in a fundamental way: the components of the $\mathbf{v}$ vector are defined solely and uniquely by the structure of the compound, inasmuch as they are the number of $>$C$<$, $>$CH$-$, and $-CH_2-$ groups from which the molecule is composed. In contrast, $\boldsymbol{\eta}$ is independent of the molecule and depends only on the scheme for determining the heats of formation of these groups. Many schemes exist (see the literature citations in Reference 1).

Thus a major fraction of this section must be devoted to a discussion of how $\boldsymbol{\eta}$ is determined. Because $\boldsymbol{\eta}$ is three-dimensional for unsubstituted hydrocarbons, only three choices for $\boldsymbol{\eta}$ are linearly independent. That is, given the predictions of three schemes for the strain energy of a compound, the predictions of any fourth scheme for the same compound can be determined solely by considering the four schemes' $\boldsymbol{\eta}$ vectors, that is, without any reference to any other detail of the compound other than its (total) heat of formation. Because three groups are being considered, $\mathbf{v}$ must be three-dimensional and thus so must $\boldsymbol{\eta}$ for the dot product $\mathbf{v} \cdot \boldsymbol{\eta}$ (and strain energy) to be well defined. Strain energy schemes differ by at least one of the values they assign to the heats of formation of each of the groups. Thus there is a one-to-one correspondence between the reference states defining a strain energy scheme and the $\boldsymbol{\eta}$ vector associated with it. Moreover, there is no objective (ie, mathematical) way of deciding which of the literature approaches is to be preferred.

Our solution to the question of which **η** to use was to devise our own. Effectively, we opted to define the strain energy as zero for three well-defined species having the property that their **v** vectors ν_{R1}, ν_{R2}, ν_{R3} are multiples of one of the orthogonal, unit vectors (1,0,0), (0,1,0), and (0,0,1), namely, $\nu_{R1} = a(1,0,0)$, $\nu_{R2} = b(0,1,0)$, $\nu_{R3} = c(0,0,1)$. Notice that in a diagonal scheme, any species' **v** vector (α,β,γ) can be written as $(\alpha/a)(\nu_{R1}) + (\beta/b)(\nu_{R2}) + (\gamma/c)(\nu_{R3})$. (It is rarely possible to find reference species so that any, no less all, of a, b, and $c = 1$.) Moreover, for a diagonal scheme, if ΔH_{f1}, ΔH_{f2}, and ΔH_{f3} represent the heats of formation of the reference species, the scheme's **η** vector equals $\text{col}(\Delta H_{f1}/a, \Delta H_{f2}/b, \Delta H_{f3}/c)$. That is, each reference species is composed of only a single type of group. The idea of such idealized "diagonal" reference states need not be limited to unsubstituted alicyclic hydrocarbons. Diagonal reference states definitionally possess the following advantages, some chemical and some more mathematical:

1. The three structural (**v**) vectors corresponding to these three reference species are mathematically the simplest set possible. That is, more components in these **v** vectors are 0 than in any other independent set of vectors. This presumably also means that their chemical realizations are simpler, having more symmetry and less structural ambiguity. Of course, the syntheses of such diagonal species may be more difficult than those of substituted derivatives, since substituents provide chemical handles for such problems as the control of the site of ring closure. But this problem is not unique to diagonal species—the history of strained species is filled with examples of the synthesis of substituted derivatives antedating that of the parent bare hydrocarbons.[2] In compensation, once made, the various forms of spectroscopy are generally far simpler for parent hydrocarbons than for substituted derivatives. For example, the proton and carbon NMR spectra become almost barren—there is but a single peak for the protons and carbons, respectively—for strictly diagonal hydrocarbons.
2. The value of each component of the **η** vector is determined immediately and unambiguously from the heat of formation of a single reference species and its associated **v** vector if the reference species is diagonal. Calculational, conceptual, and even copying or transcriptional errors in the determination of one component do not affect any other. This should also simplify matters when a new value for an old group or component, or extension of the model to a new group or component, is desired.
3. The chemical (but not mathematical) independence of the value for one group to that of another described in item 2 requires the reference states' **v** vectors to be orthogonal, and this feature is unique to diagonal reference states. Mathematically this orthogonality is shown by showing that their dot product is zero.

≪We do not mean to imply that only when two vectors are orthogonal are they mathematically independent. That is unequivocally false. Chemical independence is used to express the lack of effect that changing the value for one group has on the value of any other.≫

A proof of uniqueness begins by noting the chemical constraint that groups may be only either there or absent, so that the components of **v** may take on integer values starting at zero only. Thus all the terms in the dot product must be either zero or positive, and the sum can be zero only if every term is zero. This cannot happen unless each of the three vectors has only one nonzero component; that is, each must be a multiple of a different one of the orthonormal unit vectors (1,0,0), (0,1,0), and (0,0,1). Diagonal reasoning cannot determine the numerical value of these multiplicative factors; the choice must be based on chemical reasoning from among the possible diagonal candidates. That is, if X and Y are both diagonal but Z is not, chemical logic focusing on the plausibility of the compound having zero strain energy tells us whether to use X or Y for the group of interest, but Z is to be eliminated from consideration because it is nondiagonal.

4. Many conceptual models in the literature are most simply applied to diagonal species and indeed, numerous well-defined and already synthesized molecules may be recognized as diagonal. For example, consider the discussion of the bonding of cyclopropane in terms of the Walsh model in Reference 5.

5. The same mathematical framework is applicable for numerous properties. As such, comparison of different properties as well as different molecules is simplified.

6. "Eventually" one will want to study substituted derivatives of general compounds. There are intrinsically fewer substituted derivatives of diagonal species than of isomeric nondiagonal species. Indeed, there is but one monosubstituted derivative of a strictly diagonal species for each possible substituent. This is true because if all the groups of the initial molecule are identical, there is no way to distinguish one substitution site from another.

For the alicyclic hydrocarbons, the choice of cyclohexane for the $-CH_2-$ components was unequivocal, resulting in $\eta_3 = -4.9$ kcal/mol (-20.6 kJ/mol)* and that of diamond for >C< was quite unavoidable, resulting[6] in $\eta_1 = +0.45$ kcal ($+1.9$ kJ).

≪Do not be dismayed by the difference in signs for η_1 and η_3. Remember that these numbers are referenced to graphite and hydrogen, which are set

*Note that we give energies in terms of both the "conventional" kilocalories per mol and the "thermochemical" kilojoules per mol for which there is the exact conversion factor of 4.184. Whereas the latter (kJ/mol) values are strictly speaking more proper, the former allow for easier recognition of the numbers for most readers. To compensate for the "wasted" space arising from using both sets of numbers, the "/mol" is deleted, since we are always giving quantities per mol. Experimental uncertainties and error bars are rarely given. However, when employed, they appear in parentheses after the quantity they refer to; for example, 21.0(0.5) means 21.0 ± 0.5 and 1.6(2.2) means 1.6 ± 2.2.

precisely equal to 0. That diamond is less stable than graphite means that η_1 is positive. That an unstrained $>CH_2$ group is stable against decomposition to the elements means that η_3 is negative. We also admit that the diagonal standard for quaternary carbons is higher in energy than that obtained by subtracting two ethanes $[(CH_3)_2]$ from neopentane, $C(CH_3)_4$. Since the energy difference is small—about 0.2 kcal (1 kJ)—we feel permitted to opt for the diagonal reference species much as we had opted for cyclohexane instead of cycloheptadecane as the reference state for $>CH_2$ groups in Reference 1.≫

Both values are comparable to those suggested in the literature: the average of the seven literature values, and thus definitionally more conventional, approaches were −5.0(0.1) and +0.4(0.5) kcal (21.0(0.5) and +1.6(2.2) kJ.)

However, the choice for the $>CH-$ component, η_2, presented a dilemma, since no intuitively strainless, diagonal species for this component is known. Two "finessed" solutions were given for $>CH-$: the use of cubane (**1**), with bond energy values corrected by quantum chemical calculations, and the use of a "bond-diagonal" species, adamantane (**2**), with all twelve C−C

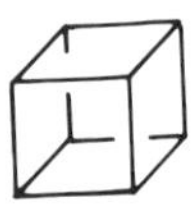

1

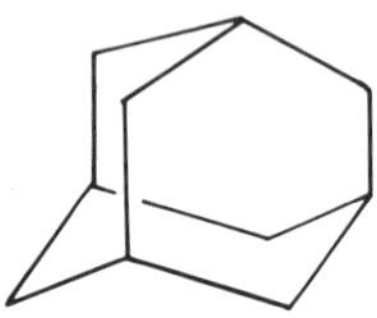

2

bonds equivalent even if there are two different types of carbon. Encouragingly, the two computed values were comparable, −2.3 and −2.0 kcal (−9.7 and −8.3 kJ) and nearly identical to the average of the seven more standard heats of formation of this increment collected from the literature methods, −1.9(0.4) kcal (−7.8(1.6) kJ). Either value of η_2 may be used.

It should be emphasized that the equivalence of models and vectors always implies that the user has the conceptual freedom, if not obligation, to make subjective choices as to which model to employ. Equivalently, there are limitations on the objective usefulness of the model chosen. We recognize that the diagonal approach, and its accompanying virtues, represent but one guiding principle in the study of molecular structure and energetics. However, because of the explicitness of its underlying structure of conceptual vectors, it is a powerful methodology, and its usefulness is heavily documented in the subsequent sections of this chapter.

3. STRAIN ENERGY "PER"—ATOMS, BONDS . . . .

In the preceding section it was shown that the numerical values for the heats of formation of the $>C<$, $>CH-$, and $-CH_2-$ increments were almost independent of the choice of reference states (conceptual model) used to generate them. Is this independence of model a general phenomenon? Our experience has shown that the answer is no. In this section we consider the strain energy of unsubstituted, alicyclic hydrocarbons from a perspective different from that of Section 2. We do this because these species are well understood and so serve to show strong model dependence.

First let us consider how the numerical value of strain energy depends on the size and structure of the species. As a test case, consider the strain energy of the cycloalkanes $(CH_2)_n$. It turns out to vary nonmonotonically with n starting in the mid-twenty kilocalorie ($\simeq$100 kJ) range for $n = 3$ and 4 because of internuclear angle compression, decreasing to a (near-zero) minimum at $n = 6$, and having some additional rather marked maxima and minima as n increases further, due to varying transannular or intraring nonbonded repulsions before, for large n, it decreases asymptotically to zero. (Presumably a large enough cycloalkane is indistinguishable from a large enough alkane.) Numerous authors[7] have commented on the near equality of strain energies of cyclopropane and cyclobutane (despite the considerably greater degree of internuclear angle compression in the former) and the contrasting fact that cyclopropane is much more reactive. We add that the hypothetical (and really not that interesting) 1,6,11,16-tetrakis(cyclopentyl)hexadecane should have about the same strain energy as cyclopropane and cyclobutane. (This conclusion was simply derived from multiplying the strain energy of cyclopentane by 4.) Yet we doubt that most readers would consider this hypothetical substance particularly strained. The differences among these three species of essentially identical strain energy becomes more intelligible when we divide the total destabilization energy (which the strain energy is supposed to represent, after all) by the number of carbon atoms in the strained ring or rings that can "share" this energy. That is, we divide ("normalize") the strain energies of cyclopropane and cylcobutane by 3 and 4, respectively, whereas that for the cyclopentane derivative is divided by 20 (not 36), since we assume that the aliphatic hexadecane chain does not contribute to the strain energy. This new normalized quantity, the strain energy per carbon atom, SE_C, reflects quite correctly our biases and our experience: cyclopropane, cyclobutane, and the hypothetical but predictable tetrakis(cyclopentyl)hexadecane have $SE_C \simeq 9$, 7, and 1 kcal (38, 29, and 4 kJ), respectively.

Some probes of strain energy reflect the atomic environment; others reflect that of the bonds.[8] For the three species discussed above, the strain energy per atom (SE_C) and per bond (SE_B) are identical, and so no new information can be gleaned by considering SE_B. This equality is not, however, a general phenomenon. We recall that α, β, and γ are the numbers of quaternary, tertiary, and secondary carbons and note that to avoid double counting of bonds, these groups contain $\frac{4}{2}$, $\frac{3}{2}$, and $\frac{2}{2}$ C—C bonds, respectively. From these definitions, we derive the quantities:

$$SE_{\mathrm{C}} = \frac{SE}{[\alpha + \beta + \gamma]} \tag{8-6}$$

and

$$SE_{\mathrm{B}} = \frac{SE}{[\frac{1}{2}(4\alpha + 3\beta + 2\gamma)]} = \frac{SE}{[2\alpha + \frac{3}{2}\beta + \gamma]} \tag{8-7}$$

Both may be rewritten as $SE/(a\alpha + b\beta + c\gamma) = SE/(a\nu_1 + b\nu_2 + c\nu_3)$, where $(a, b, c) = (1, 1, 1)$ and $(2, \frac{3}{2}, 1)$ for the per-carbon and per-bond analyses, respectively. That these last two 3-vectors are mathematically independent suggests that conclusions from their use must be assumed to be independent as well, in general. (The cycloalkanes represent the special case where $\alpha = \beta = 0$ and thus $SE_{\mathrm{C}} = SE/\gamma = SE_{\mathrm{B}}$.) Indeed it is only because there are many alicylic hydrocarbon rearrangements (eg, most yielding adamantane, its alkylated derivatives, and its higher homologues[11]) that do not affect the individual values of α, β, and γ that the predictions of SE_{C}, SE_{B} (and the total strain energy SE) are indistinguishable for numerous sets of compounds.

Let us now consider a case where the predictions of SE_{C}, SE_{B}, and SE are distinguishable: contrast bicyclobutane (**3**) and prismane (**4**). The strain

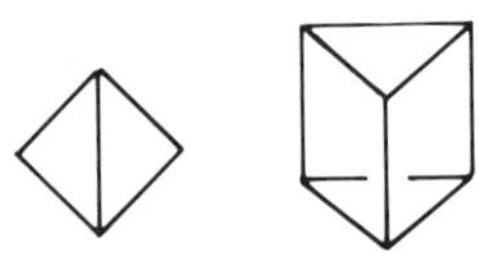

3 **4**

energy of the former is well chronicled as 66 kcal (278 kJ), whereas the latter has been only approximately determined to be about 135 kcal (560 kJ).

≪The former value was taken from Reference 3 (p. 72, and references cited therein). The latter was obtained by taking the difference of 6 times η_2, the heat of formation of the strainless >CH− increment and the heat of formation of prismane from Reference 9. In Reference 9, it is noted that the macroincrementation value for the heat of formation of prismane, 128 kcal (534 kJ), is inexplicably different from a suggested value in the literature, 119 kcal (496 kJ). Part of the discrepancy may arise from our ignoring the conjugative stabilization in the prismane inherent in its bicyclopropyl substructure[10]; however, more work remains to be done on this interesting and important molecule. No qualitative nor quantitative conclusion is affected by this ambiguity.≫

While the strain energies per bond for prismane, 15 kcal (63 kJ), and for bicyclobutane, 13 kcal (56 kJ), are comparable, the strain energy per atom of prismane, 22.5 kcal (94 kJ), far exceeds that of bicyclobutane, 17 kcal (70 kJ). This no doubt contributes to the difference in relative safety (ie, metastability) of the two species. Prismane is explosive when pure[12] (a very rare feature for unsubstituted hydrocarbons) and yet has a half-life of about 11 h in solution at 90°C. The metastability of prismane arises from requiring an archetypal symmetry-forbidden reaction to rearrange, and so prismane has been compared to "an angry tiger unable to break out of a paper cage."[13] Bicyclobutane, while labile owing to its C_1-C_3 bond, is much more benign, even if we "utilize" a "pussycatalyst" to encourage its exothermic rearrangement to butadiene.

≪This has been somewhat simplistic, since the total amount of energy released per unit weight or volume from the isomerization reaction of bicyclobutane to form butadiene is considerably less than that from the reaction of prismane to form benzene. Furthermore, we do not know what triggers the explosive decomposition of prismane and have surmised rather than proved the connection with its high strain energy.≫

It would take us too far astream to compare other pairs of species, their reactivities, and the predictions from the use of SE, SE_C, and SE_B. We may, however, inquire what is the most helpful remaining 3-vector ("strain per") to consider. Reference 5 speaks of strain energy per ring and discounts it on chemical grounds. We note that the number of rings cannot be written as $a\nu_1 + b\nu_2 + c\nu_3$ for any a, b, and c; the number of rings is $\nu_1 + \frac{1}{2}\nu_2 + 1$. The presence of the "1" takes this expression outside the realm of 3-vectors, while in its absence, there would result an inadmissible dividing by zero for cycloalkanes. Nonetheless, we note that polycyclic species usually have greater strain than a corresponding assemblage of monocyclic species that has the identical total number of rings of the same size in their rings as the polycycle. (For example, the strain energy of bicyclobutane is greater than that of two cyclopropanes and that of prismane is greater than that of two cyclopropanes and two cyclobutanes.) This suggests that $>C<$ and $>CH-$ groups contribute more to the strain energy than $-CH_2-$ groups. This in turn suggests that any third, ring-related "strain energy per" scheme trying to consider the rings should have larger a and b than c. Before attempting to derive such a new approach, however, note that there are surprisingly few unsubstituted alicyclic hydrocarbons with a nonzero number of quaternary carbons and fewer yet for which we have data regarding their gas phase energetics. Indeed, spiropentane and its two higher homologues (5,4)spirodecane and (5,5)spiroundecane[14] are seemingly the only "normal" compounds known to the authors for which such data exist! It is quite clear that more experimental, synthetic, and calorimetric, as well as theoretical, both thermochemical and conceptual, chemistry needs to be done. Our efforts await such data.

4. AROMATICITY AND ANTIAROMATICITY: BENZENE VERSUS CYCLOBUTADIENE

In this section we discuss the nature of aromaticity in hydrocarbons, commencing with a comparison of the realtive π energies of cyclobutadiene and benzene derivatives. It is easily established that benzene is considerably more stable than cyclobutadiene: the former may be heated without decomposition far above its boiling point at 80°C under atmospheric pressure, while cryogenic conditions are needed to isolate the latter.[15] Moreover, derivatives generally have analogous relative stabilities, and so we are not surprised that it is only in the past decade or so that highly hindered, substituted cyclobutadienes have been isolated at room temperature.[16] Furthermore, with the obvious exception of biphenylene (**5**) and its derivatives, few annelated cyclobutadienes are known[17] (eg, **6**), in contrast to the extraordinary number of annelated benzene derivatives.[18]

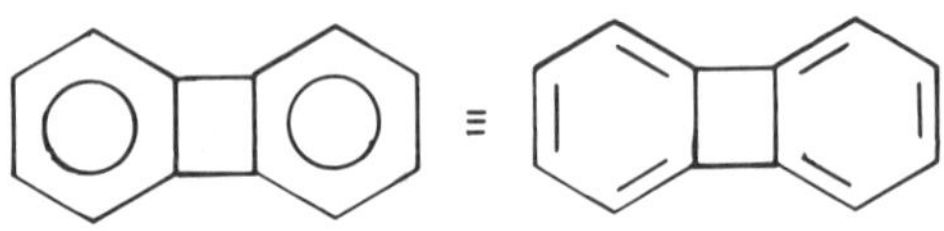

5

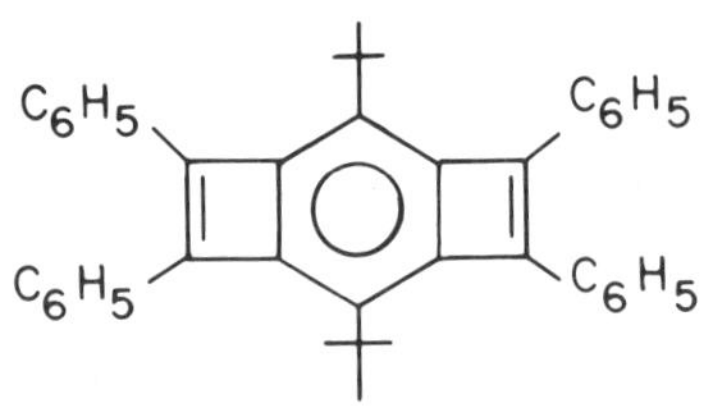

6

How may all these findings be understood? The generally phenomenal disparity in stability of formally analogous cyclobutadiene and benzene derivatives has usually been explained in terms of the relative stability of the parent hydrocarbons. Indeed, cyclobutadiene has often been referred to as antiaromatic[19] to explicitly counterpoint its marked instability with respect to acyclic polyenes and to contrast the extreme stability of the aromatic benzene. The simplest explanation of this difference in stability involves Hückel's $4n + 2$ rule for one-ring cyclic polyenes: counting π electrons, aromaticity exists only when there are $N = 4n + 2$ electrons with n integral. However, this rule cannot be immediately extended to many-ring systems: we suspect few would want to suggest that the parent species of **6**, benzo-1,2:4,5-cyclobutadiene, is aromatic because it has 10 ($n = 2$) π electrons and that biphenylene (**5**) is antiaromatic because it has 12. Numerous extensions of Hückel's rule exist, ranging from the automatic assumption of destabilization due to a cyclobutadiene ring to the total exclusion of such species from discussion.

Despite material in this series relating directly to hydrocarbons and associated aromaticity (Chapters 2 in this volume, 2 in Volume 3, and 1 in Volume 4, respectively), we nonetheless explicitly compare in this section the π-electron energies associated with four- and six-membered rings. For this, we have considered two homologous series of compounds. The first is the "acenes"[20] or "hexagonal ladders": benzene (**7**), naphthalene (**8**), anthracene (**9**), naphthacene or tetracene (**10**), and so on. The first seven members of

7

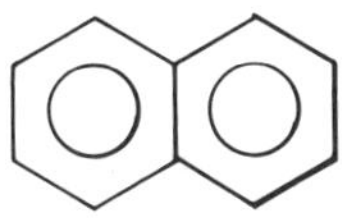

8

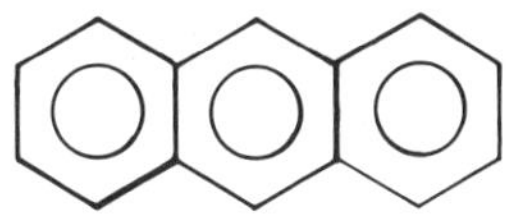

9

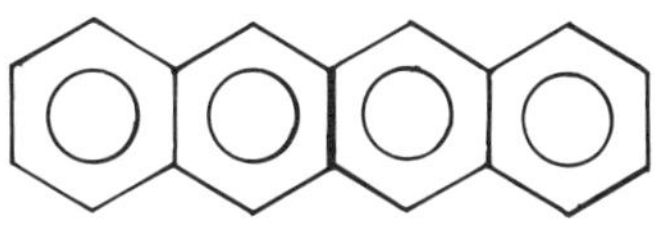

10

this series have been isolated.[20] In addition, the heats of formation in both the condensed and gas phases have been reported in the experimental literature[21] for the first four compounds (**7–10**). The second series is that of corresponding "quadragonal ladders" commencing with cyclobutadiene (**11**), bicyclo(2.2.0)hexatriene (**12**; also called[22] "butalene"), tricyclo(4.2.0.$0^{2,5}$)octatetraene (**13**, also called[23] "bicyclobutadienylene").

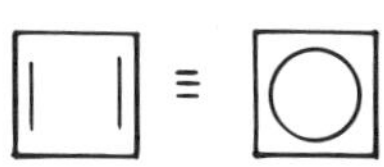

11

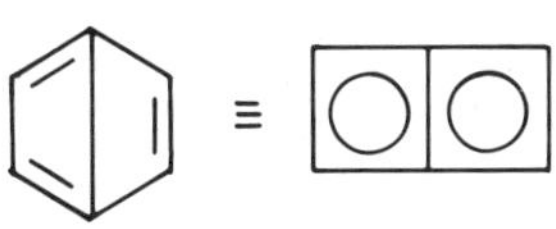

12

13

In this latter series, only the first member has been characterized,[15] although there is good evidence for the intermediacy of the second.[22] While the intermediacy of a substituted three-square compound, **14**, has been suggested[23] in the degenerate rearrangement of a quite complicated organometallic, and a potential precursor, **15**, of the dibenzo analogue has been synthesized,[24] the parent species is still an object of strictly theoretical investigation.[25,26]

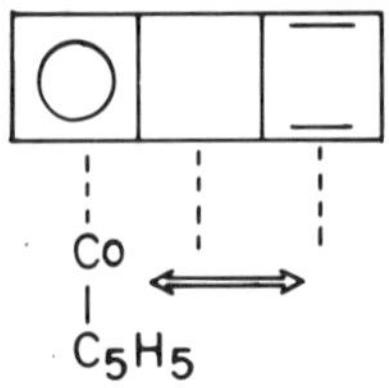

14

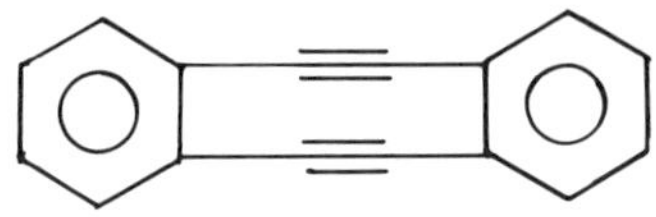

15

Higher analogues have been essentially totally ignored[26]—where theory has been reported, it has almost always been "only" at the Hückel molecular orbital (HMO) level.[27] Thus for our comparison of the energetics of the hexagonal and quadragonal ladders, we continue to make use of the superbly simple HMO approach. We admit that the omission of σ-electron effects such as the ring system or framework strain energy[28] is serious. However, Hückel theory is precisely the level of rigor at which arguments such as aromaticity and antiaromaticity are generally made. Since HMO theory gives far more information than just molecular energetics,[27] it would appear that still further simplifications are possible if all one wants to know is the total π-electron energy of a molecule. Indeed, arduous but well-established procedures allow one to derive directly the individual orbital energies, π orbital by π orbital, of both the hexagonal[29] (Equation 8-8) and quadragonal[26] (Equation 8-9) ladders merely as a function of the number of (carbon) atoms N. For the hexagonal ladder, looking solely at the positive energies and letting r be the number of rings, $N = 4r + 2$ and s an integer such that $1 \leq s \leq r$, the individual orbital energies are:

$$\lambda_{+}^{(s)} = \frac{1}{2}\left\{\left[9 + 8\cos\left(\frac{s\pi}{(r+1)}\right)\right]^{1/2} - 1\right\} \tag{8-8a}$$

$$\lambda_{-}^{(s)} = \frac{1}{2}\left\{\left[9 + 8\cos\left(\frac{s\pi}{(r+1)}\right)\right]^{1/2} + 1\right\} \tag{8-8b}$$

$$\lambda_{-}^{(r+1)} = +1 \tag{8-8c}$$

So summing these, this results on the energy of $4r + 2$ carbon species,

$$E^{(r)} = 2\left\{1 + \sum_{s=1}^{r}\left[9 + 8\cos\left(\frac{s\pi}{r+1}\right)\right]^{1/2}\right\} \tag{8-8d}$$

For the quadragonal ladder, summing just the positive energies is not so direct. Letting r be the number of rings, $N = 2r + 2$, we find the energy levels. Defining $N = 2p$, and s as an integer such that $1 \leq s \leq p$, these energies are:

$$\lambda_+^{(s)} = 1 + 2\cos\left(\frac{s\pi}{(p+1)}\right) \tag{8-9a}$$

$$\lambda_-^{(s)} = -1 + 2\cos\left(\frac{s\pi}{(p+1)}\right) \tag{8-9b}$$

Defining some new terms, $m(x) \equiv$ greatest integer component of x [eg, $m(9.9) = 9$], $\theta_p \equiv \pi/(p+1)$, so results in the energy of a $2p$ carbon species.

$$E^{(p)} = m\left[\frac{2(p+1)}{3}\right] - m\left[\frac{p+1}{3}\right] + 2\left[\sin\left(\frac{\theta_p}{2}\right)\right]^{-1}\sin\left\{\theta_p\frac{m[(p+1)/3]}{2}\right\}$$
$$\cos\left\{\theta_p\frac{[m[(p+1)/3]+1]}{2}\right\} + 2\left[\sin\left(\frac{\theta_p}{2}\right)\right]^{-1}\sin\left\{\theta_p\frac{m[2(p+1)/3]}{2}\right\}$$
$$\cos\left\{\theta_p\frac{[m[2(p+1)/3]+1]}{2}\right\} \tag{8-10}$$

(All HMO energies in this chapter are taken in units of β where 1 β equals about 18 kcal/mol = 75 kJ/mol.)

We turn immediately to the cases of an infinite[30] number of rings, and briefly discuss the finite number case afterward. For the hexagonal and quadragonal ladder cases, species **16** and **17**, the total π energy per electron

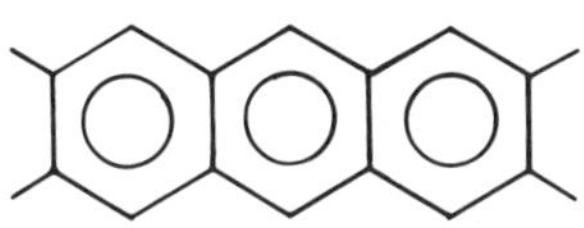

16

17

may be directly found from Equations 8-8d and 8-10. For these species, the delocalization energies per electron are 0.403β (Reference 29) and 0.436β (Reference 26), respectively. The difference is quantitatively small: 0.033β or about 0.6 kcal or 2 kJ. Qualitatively, however, it is significant—at least in the simplest level of understanding of the Hückel molecular orbital picture of conjugated π species and its resulting omission of strain effects, it would seem that cyclobutadienes are *not* always less stable than the formally related benzene derivatives.

At this time it must be admitted that simple consideration of the π energy per electron is often fraught with conceptual difficulties. It is generally conceded to be more reasonable to speak of the resonance energy per electron, a concept that automatically requires a reference state to which to compare the species of interest. For almost all aromatic species with an even number of carbons, one may draw resonance structures with precisely half that number of double bonds. (See Chapter 2, this volume). Although isolated double bonds with their two carbons and two π electrons apiece are understandable in terms of their archetype, ethylene, it is likewise generally acknowledged that simple comparison of a compound with N atoms and N π electrons with one with $N/2$ ethylenes is also too simple to mimic conventional chemical understanding, if not chemical "reality." In the current case, comparison with ethylenes would return us to the same (Hückel-level) qualitative conclusion as before—cyclobutadienes are not always less stable than the formally related benzene derivatives. This is a result we are reluctant to accept.

We nonetheless abstain from dealing with these difficulties and from discussing the numerous literature reference states and accompanying definitions of resonance energy (RE) and resonance energy per electron (REPE)—such a discussion would be too lengthy to include here. Instead we note that a casual review of the experimental and theoretical literature on the various aromatic, antiaromatic, and nonaromatic hydrocarbons shows there are three different types of carbon environment: those with one carbon neighbor or "partner," $=CH_2$ (eg, as found in fulvene, **18**, and 3,4-dimethylenecyclobutene, **19**), those with two carbon neighbors, $=CH-$, and those with three, $=C<$.

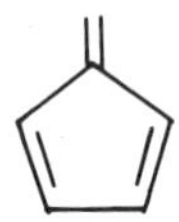

18

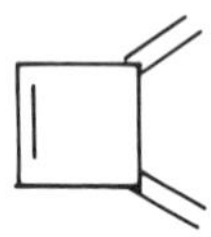

19

≪This is not strictly a true statement in that there is the trivial case of methyl (CH_3) with no carbon neighbors, and to date experimentally unrealized candidates of molecules containing a square planar carbon bonded to four different carbon atoms of an annulene periphery.[31] Both these cases are ignored in this chapter. Having so ignored these species, one may easily prove some of the earlier cited relationships. First of all, for $=CH_2$, $=CH-$, and $=C<$, there is one carbon apiece and but "half" a double bond. Since double bonds have two π electrons, this means that each group contributes $\frac{2}{2}$ electrons and so the number of carbon atoms equals the number of π electrons. Furthermore, if the species is composed only of these groups, then since one cannot have "half" a double bond, the number, N, of carbon atoms and π electrons must be even.≫

After choosing the appropriate reference states and deriving their associated energy contributions for the three types of carbon environment (ie, for $=CH_2$, $=CH-$, and $=C<$), we may define the resonance or stabilization energy, RE or STE, as the difference of the total π energy and the contributions of its components, $E_\pi - \mathbf{v} \cdot \boldsymbol{\varepsilon}$. We admit that as in the earlier case of strain energies of alicyclic hydrocarbons, there is an intrinsic subjectivity to the choice of the prototype carbon environments (the reference states) that define the $\boldsymbol{\varepsilon}$ vector. Following our earlier prejudices, we select diagonal reference states. In particular, the following three theoretically and experimentally well-characterized species serve as reference states: ethylene, benzene, and graphite (species **20**).

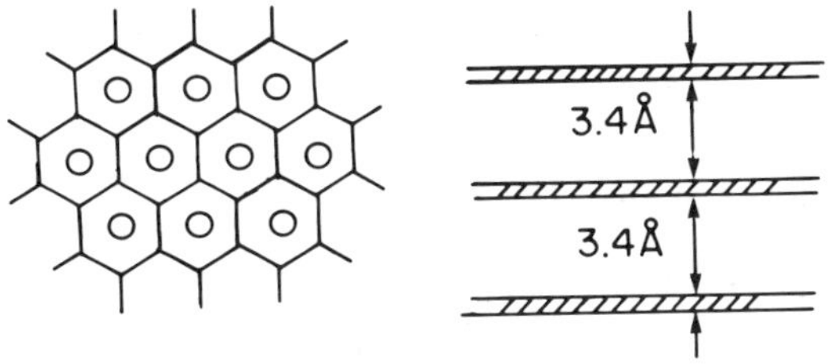

20

In doing this, we arbitrarily assign the value of zero to their resonance or stabilization energies. We wish to emphasize now that this does not mean we are assuming that the π electrons of benzene and graphite are localized; rather, we assert that reference states may refer explicitly to delocalized species.

In order to compare hexagonal and quadragonal ladders, let us talk first in terms of HMO theory. The total π energies per electron for ethylene, benzene, and graphite, equivalently ϵ_1, ϵ_2, and ϵ_3, are 1.0000β, 1.3333β, and 1.5746β.[32] What species do we consider? We could proceed through all the thermochemically characterized polynuclear hydrocarbons and those quadragonal ladders related by the same total number of either atoms or rings (eg, butalene and benzene could be compared because both have six carbons; butalene and naphthalene could be compared because both have two rings). Instead, we limit our attention to three cases: the one-ring benzene and cyclobutadiene, the case calculated for the two series to be where they have comparable stabilities and the case of the infinite limits of both ladders. As a consequence of being taken as a reference state, the total resonance energy of benzene is taken as 0: recall we are using inherently delocalized reference states. Cyclobutadiene has a total π energy, E_π of 4β. Since it is composed of four $=$CH$-$ groups, the π energy "should" have been $4(1.5746) = 5.3333\beta$, and so there is a total resonance energy of -1.3333β. Since destabilization relative to the reference states is the norm, we define an accompanying destabilization energy by reversing the sign of RE. Furthermore, to compensate ("normalize") for the compound-dependent number of π electrons (carbon atoms) N, we divide by N. The resulting number is herein called DSTP. For cyclobutadiene, $N = 4$ so it is $+1.3333\beta/4$ or DSTP $= 0.3333\beta$. Since DSTP is positive and, as will be seen, relatively large, cyclobutadiene is shown to be destabilized in the current model relative to benzene, for which DSTP $= 0/6 = 0\beta$ by definition.

Not at all surprisingly, there is no chemically realizable value of N for which the energies of the two ladders are exactly equal—the functional dependences on N of the total π energies or π energies per electron are much too complicated to have had more than a vague hope for another outcome.

The case of 26 atoms and electrons is, however, a good approximation.[33] For the hexagonal ladder with 6 rings and an associated $\mathbf{v}$ = (0,10,16), the total π energy is 36.156β, while for the quadragonal ladder with 12 rings and an associated $\mathbf{v}$ = (0,4,22) the π energy is 36.250β. Normalizing, these correspond to 1.391β and 1.394β for the π energy per electron. Close enough. The hexagonal ladder has a resonance energy of -0.832β and associated DSTP of 0.032β, while the quadragonal ladder has the corresponding values of -3.819β and 0.147β. It is thus seen that relative to the suggested standard states, the quadragonal ladder is destabilized considerably more than the hexagonal.

Turning to the case of infinite numbers of rings, carbon atoms, and electrons, the total π energy is clearly infinite as well. Immediately shifting attention over to the energies per electron returns us to the earlier mentioned values of 1.403β and 1.436β. In this limiting case, the hexagonal ladder is easily seen to be composed of equal "quantities" of the =CH− and =C< components, giving a reference π energy of ½(1.3333β + 1.5746β) = 1.454β. The net resonance energy (per electron) is -0.051β and the associated DSTP is 0.051β. Likewise, in the limiting case, the quadragonal ladder is composed solely of =C< components; that is, it is as "diagonal" as graphite, which results in a reference π energy, resonance energy (per electron), and DSTP of 1.575β, -0.139β, and 0.139β respectively.

A lot of numbers were just presented. The chemical conclusions, however, are quite clear. First and foremost, for the three cases discussed—and presumably in all cases—the hexagonal ladder is more stable than the corresponding quadragonal ladder case with the same number of π electrons. Thus we recover the intuitively desired conclusion: cyclobutadiene rings are inherently destabilizing. Second, in accord with experimental experience,[20] we find that the hexagonal ladder structure becomes less stable as the number of rings is increased. In contrast, for the quadragonal ladders the stability increases. That is, both aromaticity and antiaromaticity are lessened as the size of the molecule is increased.

5. THE AROMATICITY OF POLYNUCLEAR HYDROCARBONS

As noted in Section 4, an ancillary benefit of the choice of ethylene, benzene, and graphite as reference states for consideration of aromaticity is that these species correspond to theoretically well-defined and experimentally well-characterized substances. The graphite case must, however, be further analyzed before it may be used as a reference state. Unlike diamond, which we argued[1] could be considered to be a gas of quaternary carbon units, graphite must be treated as a solid because its sheetlike structure[34] forces on us the consideration of intermolecular interactions. These were addressed in a recent analysis of the thermochemistry of polynuclear hydrocarbons,[35] which concludes that the effective heat of sublimation of graphite is 1.45 kcal or 6.1 kJ per mol-C. Adding this number to the definitionally zero experimental heat of formation of solid graphite, we find the heat of formation of "gaseous" graphite to be 1.45 kcal (6.1 kJ) per mol-C. Thus ϵ_3 is taken to be 15 kcal or 6.1 kJ. The experimental heats of formation of ethylene [ie, $(CH_2)_2$)] and gaseous benzene [ie, $(CH)_6$)] are 12.5 and 19.8 kcal (52.2 and 82.9 kJ), respectively. We thus derive the following experimental values for the $\boldsymbol{\varepsilon}$ vector: $\epsilon_1 = 12.5/2 = 6.3$ kcal (26.1 kJ), $\epsilon_2 = 19.8/6 = 3.3$ kcal (13.8 kJ), and $\epsilon_3 = 1.5$ kcal (6.1 kJ). (While theoretical values for the same $\boldsymbol{\varepsilon}$ vector could be calculated, nonaromaticity issues such as strengths of C—H σ bonds intrude significantly. We opt to avoid these issues and so do not attempt to compare theoretically and experimentally derived $\boldsymbol{\varepsilon}$ vectors.)

Let us now look at the experimental resonance energies of the more general polynuclear hydrocarbons. Commencing with naphthalene (**8**) with an associated $\mathbf{v}$ vector of (0,8,2), the predicted heat of formation would be 29.4 kcal (122.8 kJ), while the experimental value is 35.9 kcal (150.4 kJ), corresponding to a destabilization (as always, relative to the reference states chosen) of 6.5 kcal (27.6 kJ). Anthracene (**9**) and phenanthrene (**21**), both with

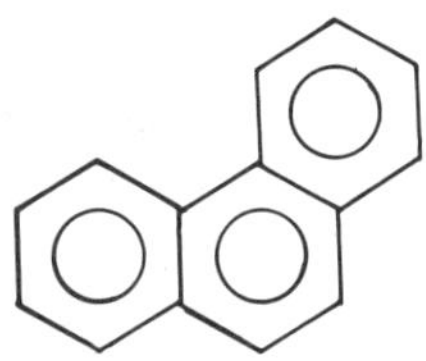

21

the $\mathbf{v}$ vector (0.10,4), have a predicted heat of formation of 39.0 kcal (162.4 kJ) and experimental heats of formation of 55.0 and 49.5 kcal (230.1 and 206.9 kJ). The corresponding destabilization energies are 16.0 and 10.5 kcal (67.9 and 44.5 kJ). As with theoretical numbers, it seems appropriate to divide these destabilization energies by the number of electrons or carbon atoms to derive the value of DSTP, and so we find values of DSTP of 0.7, 1.1, and 0.8 kcal (2.8, 4.9, and 3.2 kJ). These findings support the conventional observation, assertion, and prejudice that aromaticity increases in the order: anthracene, phenanthrene, naphthalene, benzene.

Likewise, consider pyrene (**22**), with its accompanying $\boldsymbol{\eta}$ vector of

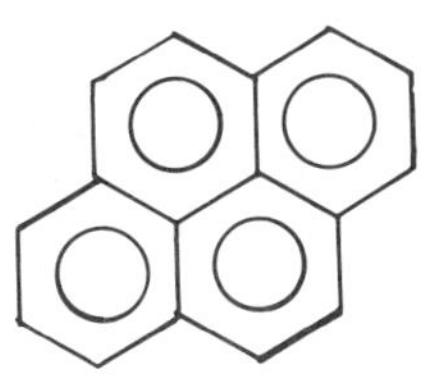

22

(0,10,6). This pericondensed species has a predicted heat of formation of 42.0 kcal (175 kJ).

≪There are several essentially interchangeable definitions of "pericondensed." The first focuses on the =C< and =CH− components. If at least one =C< component is bonded to three other =C< and no =CH− components, the species is pericondensed. This is equivalent to the presence of at least one carbon found in common with three rings. The second definition defines a test procedure: put a dot in the center of each ring and draw a line between the dots of each ring that share an edge. If this new diagram of dots and lines has any rings in it, the species is pericondensed. By inspection, pyrene is pericondensed by the first definition. Using the second procedure, a dot/line structure results that is equivalent to bicyclobutane (**3**), and so pyrene is again found to be pericondensed. Species that are not pericondensed are called "catacondensed," a term that is employed below. Finally, we note an additional definition of pericondensation applicable exclusively to polynuclear hydrocarbons composed solely of six-membered rings: If the empirical formula of the n-ring hydrocarbon is $C_{4n+2}H_{2n+4}$, it is *catacondensed.* If the species has any other stoichiometry, it may be assumed to be *pericondensed.*≫

A value of 51.6 kcal (216 kJ) is reported for **22**'s heat of formation in the archives; more recently, however, a value of 54.2 kcal (227 kJ) was reported[36] with no discussion of the disparity. These two experimental values result in DSTPs of 0.6 and 0.8 kcal (2.6 and 3.2 kJ), respectively. The former value of DSTP unequivocally confirms the assertion that pericondensed hydrocarbons are more stable than catacondensed.[18] The latter value appears to be less definitive but in fact also corroborates the same conclusion. The DSTP numbers presented for the acenes increase with the number of rings and with the total number of π electrons and carbon atoms. Since this finding is also shown by a more thorough analysis of the hexagonal ladder HMO equation,[25c,29] we are encouraged but not surprised that the trend continues for the experimental data on the 18-electron, isomeric four-ring catacondensed hexagonal ladders with their accompanying (0,12,6) vectors. There are five such species: tetracene (**10**), benz[*a*]anthracene (**23**), chrysene (**24**), benz[*c*]phenanthrene (**25**), and triphenylene (**26**).

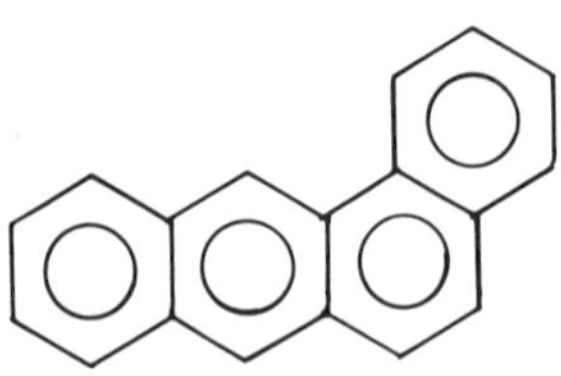

23

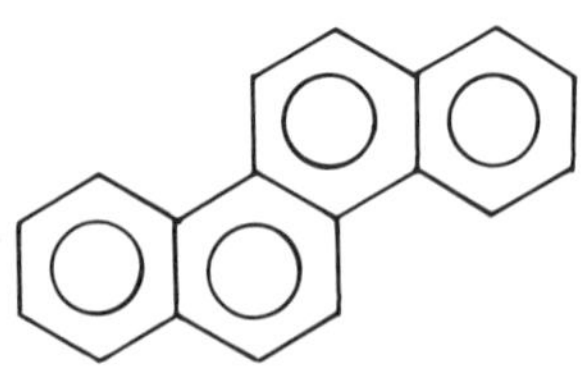

24

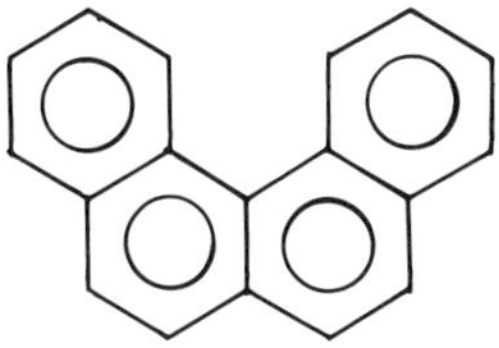

25

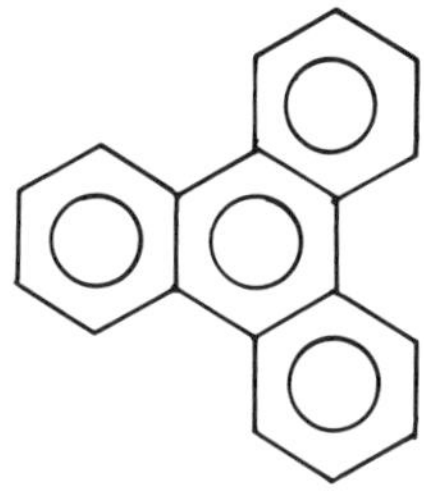

26

The predicted heats of formation are automatically the same, numerically 48.6 kcal (202.4 kJ), since all five have the same **v** vector. The experimental values[37] on these five species show a rather wide spread: tetracene, 67.8(1.2) kcal (284(5) kJ); benz[*a*]anthracene, 70.2(0.9) kcal (294(4) kJ); chrysene, 63.0(1.2) kcal (263(5) kJ); benz[*c*]phenanthrene, 69.9(1.2) kcal (291(5) kJ); and triphenylene, 66.5(1.0) kcal (278(4) kJ). These values result in associated destabilizations of 19.2, 21.6, 14.4, 21.3, 17.9 kcal (82, 90, 61, 89, and 68 kJ) and DSTPs of 1.1, 1.2, 0,8, 1.2, and 1.0 kcal (4.6, 5.0, 3.4, 4.9, and 3.8 kJ).

If cata- and pericondensed species could be treated uniformly, one could interpolate the 16-electron pyrene between the 14-electron isomeric pair of phenanthrene and anthracene and the 18-electron set above. The DSTP of pyrene would be expected to lie between approximately 0.9 kcal (4 kJ), the average of phenanthrene and anthracene, and 1.1 kcal (4.4 kJ), the average of the five four-ring catacondensed hydrocarbons. That is, it "should" be around 1.0 kcal (4.1 kJ). That it is 0.8 kcal (3.2 kJ) strongly suggests that pyrene, and by inference other pericondensed species, remain comparatively stabilized relative to catacondensed hydrocarbons with comparable numbers of carbons. DSTP analysis thus quantifies another element of the "folklore" of aromatic hydrocarbons.[18]

We now briefly turn to nonbenzenoid hydrocarbons. The first is cyclooctatetraene, **27**, diagonal much as benzene is, for which the predicted heat

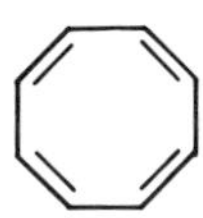

27

of formation is 8(3.3) = 26.4 kcal (110 kJ). This is to be compared with the experimental value of 71.1 kcal (298 kJ), resulting in an associated DSTP of (71.1 − 26.4)/8 = 5.6 kcal (23.4 kJ). This DSTP or discrepancy between predicted and experimental values may appear to be exceptionally large. It reflects the difference between aromatic and nonaromatic species corresponding, in the case of benzene versus (6/8)cyclooctatetraene, to a total net stabilization of benzene of more than 33.6 kcal (140 kJ), a figure comparable to that normally given for the delocalization, aromatic, or resonance energy of benzene. However, to make the desired comparison between aromatic and antiaromatic species, the DSTP value must be augmented by the "normalized" planarization energy of cyclooctatetraene which is about 16/8 kcal (8.3 kJ).

≪See Chapter 2 in Volume 3 of this series for references and a related discussion on the aromaticity (antiaromaticity) and planarization energies of cyclooctatetraenes.≫

The resulting value of DSTP, 7.6 kcal (33 kJ), is so large that perhaps we should expand **v** and **ε** from 3-vectors to 4-vectors to be able to include as the fourth component a =CH− group derived from a nonbenzenoid, nonaromatic source: planar, antiaromatic cyclooctatetraene is not then so destabilized. Furthermore, this would allow us to more naturally study a wider class of compounds such as butadiene (**28**), fulvene (**18**), 3,4-dimethylenecyclobutene (**19**), cyclooctatetraene (**27**), styrene (**29**), and *trans*-stilbene (**30**).

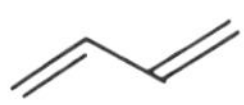

28

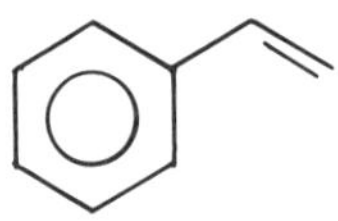

29

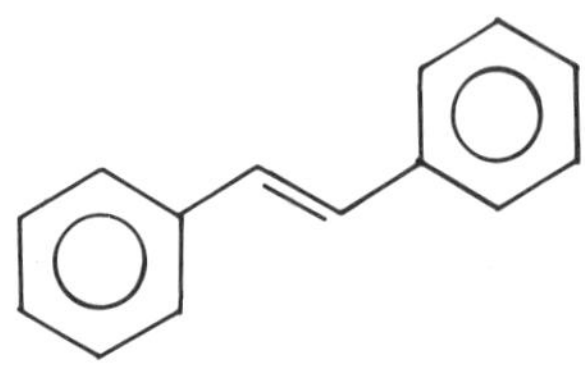

30

Alternatively, one may reduce the number of species that may be studied by reducing the dimensionality of the vectors from 3 to 2 by removing the $=CH_2$ component. This would eliminate all comparisons with nonaromatic acyclic species and, in a formal way, would mimic the earlier treatment of alicyclic species wherein no compound bearing a carbon bonded to but one other carbon was considered. For now we do neither, and so leave this as an avenue for future research.

Our second nonbenzenoid species is the naphthalene isomer, azulene (**31**). Since it has the same **v** vector as naphthalene (**8**), the predicted heat

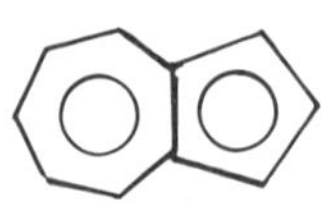

31

of formation is the same—29.4 kcal (123 kJ), to be contrasted with the experimental value of 69.1 kcal (289 kJ). The DSTP is thus (69.1 − 29.4)/10 = 4.0 kcal (16.6 kJ). While hardly as bleak as for cyclooctatetraene, the size of DSTP suggests that azulene is considerably less aromatic than naphthalene. It has been said that azulene is "merely" a mildly perturbed [10]annulene or cyclodecapentaene,[38,39] a description that has also been used for the 1,6-methano derivative of [10]annulene, bicyclo(4.4.1)undeca-1,3,5,7,9-pentaene (**32**).

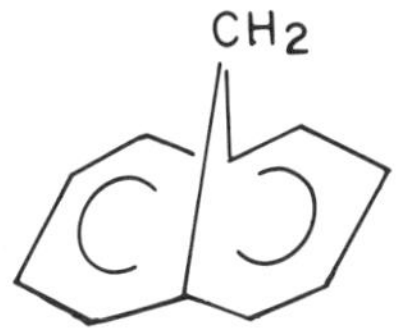

32

An admittedly crude model for **32** describes it with the same components as naphthalene plus one $-CH_2-$ group from our earlier cited alicyclic "collection." So doing results in a predicted heat of formation of 29.3 + (−4.9) = 24.4 kcal (102 kJ), to be compared with the experimental value of 75.3 kcal (315 kJ). The destabilization is thus 50.9 kcal (213 kJ). As far as normalization is concerned, it is preferable to divide this number by 10, the number of π electrons, rather than by 11, the total number of carbon atoms, if we wish to make comparisons with other polynuclear hydrocarbons. The resulting for DSTP, 5.1 kcal (21.3 kJ), is comparable to the value for azulene 4.0 kcal (16.6 kJ). While we are not surprised that the value for methano[10]annulene is larger than that for azulene, because of the effects of twisting the π framework, the results for both species are suggestive of their description as perturbed [10]-annulenes. (These values of about 4.5 kcal (19 kJ) are no doubt upper bounds because we have ignored the strain inherent in the σ framework or skeleton,[39] which destabilizes both species relative to our reference compounds.)

We conclude this section with an application of the current model to cyclobutadiene. There being no thermochemical data on it, we will employ biphenylene (**5**), the sole cyclobutadiene derivative for which experimental thermochemical data exist. Because the central ring in biphenylene has four members, biphenylene may also be identified as a cyclobutane derivative. As such, even in the absence of effects due to double bonds or π electrons, the additional strain due to this four-membered ring alone would result in destabilization of about 27 kcal (113 kJ). Replacing an alicyclic $>CH_2$ carbon in cyclobutane by a tricoordinate carbon increases the strain energy by about 1 kcal (~4 kJ).

≪This number is rather approximate and may be derived by taking the difference of the strain energy of cyclobutane and those of methylenecyclobutane (with its one tricoordinate carbon) or half the difference of cyclobutane and cyclobutene (with its two such carbons).[40]≫

Four such replacements are needed for cyclobutadiene and biphenylene and (assuming additivity), we thus expect them to be destabilized by 4(1) = 4 kcal (4(4) = 16 kJ) relative to cyclobutane. Thus in the absence of any additionally destabilizing antiaromaticity effects, biphenylene is automatically expected to have destabilization energy of 27 + 4(1) = 31 kcal (129 kJ).

The experimental[41] heat of formation of biphenylene is 99.7 kcal (417 kJ), while the unstrained model species, as befits a compound with a (0,8,4) **v** vector, is 32.4 kcal (135 kJ), resulting in an approximate total destabilization of 99.7 − 32.4 = 67 kcal (≃280 kJ). This destabilization is in part due to the already discussed σ effects associated with ring strain of cyclobutane and introduction of the four tricoordinate carbons. Given our just-derived estimate of the strain energy of 31 kcal (129 kJ), the net destabilization attributable to additional π effects, such as whatever the antiaromaticity of cyclobutadiene is, is 67 − 31 = 36 kcal (135 kJ). One contribution relates to the total number of π electrons. Recall that the "normal" naphthalene and anthracene are destabilized by 6.5 and 16.0 kcal (28 and 68 kJ), respectively, relative to the model species. Naphthalene has 10 π electrons, anthracene 14, and biphenylene 12. If we pretend that biphenylene were likewise "normal" (ie, that one could linearly interpolate between naphthalene and anthracene, in analogy to what we earlier did regarding pyrene), we would expect a destabilization about 12 kcal (~50 kJ) due solely to the number of π electrons. We thus conclude that the presence of the cyclobutadiene ring causes an additional destabilization of about 36 − 12 = 24 kcal (~100 kJ). We ascribe this energy to the antiaromaticity.

Is the last number, 24 kcal (100 kJ), an index of the antiaromaticity of cyclobutadiene? To derive the DSTP, one should divide this number by 4 for the four electrons of the "offending" central four-membered ring. The resulting value, 6 kcal (25 kJ), is large but not unreasonably so and may be compared with the one given earlier for cyclooctatetraene, 7.6 kcal (33 kJ). Our intuition suggests the quoted value for cyclobutadiene is significantly lower than the "real" one because in three of the four resonance structures we customarily draw for biphenylene, no cyclobutadiene appears. This intuition can be put on a more quantitative footing by recognizing that in the four resonance structures for biphenylene, the central ring appears twice as 3,4-dimethylenecyclobutene (**19**) and once each as [4]-radialene (**33**) and as cyclobutadiene. Let us now assume that **19** and **33** are nonaromatic (ie, their

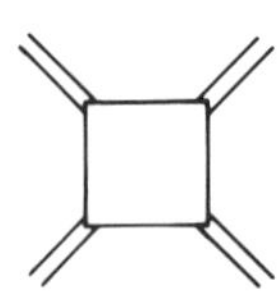

33

DSTP values are approximately that of butadiene) and that in biphenylene all four resonance structures contribute equally.

≪To the extent that either 3,4-dimethylenecyclobutene or [4]radialene are at all aromatic and/or that the resonance structures containing these substructures contribute more than the statistical $\frac{3}{4}$ or $\frac{1}{4}$, respectively, the degree of antiaromaticity of cyclobutadiene is increased.≫

The earlier obtained value of 6 kcal (25 kJ) may thus be set equal to $\frac{1}{4}$ [DSTP(cyclobutadiene) + 3DSTP(butadiene)]. The DSTP of butadiene is *not* zero but rather is 1.8 kcal (7.5 kJ)—remember, the nonconjugated ethylene and the aromatic benzene are our reference states. For cyclobutadiene, the standard methods of this chapter yield a DSTP equaling an astonishing 18.3 kcal (76 kJ) for cyclobutadiene. As destabilized as we normally believe cyclobutadiene to be, this last value nonetheless seems too high. It thus appears reasonable to divide it by 2, noting that there is but one formal single bond for four π electrons in the acyclic butadiene, but twice as many in cyclobutadiene (two such bonds for four electrons), in 3,4-dimethylenecyclobutene (three bonds for six electrons), and in [4]radialene (four bonds for eight electrons).

≪It can be verified that no such miscount arises in finding the DSTP for the other species in this section by observing that $=CH_2$ has no such formal single bond while $=CH-$ and $=C<$ have $\frac{1}{2}$ and $\frac{2}{2}$, respectively. The factors of 2 in the denominators of the last two groups arise from the necessity to avoid double counting of bonds in the formal "synthesis" of the compounds of interest from the component groups.≫

So dividing results in 9.2 kcal (38 kJ), a "formidable" but much more reasonable value for the antiaromaticity of cyclobutadiene.

Consolidating of our numerical results for the DSTP of the annulenes, we find: [4]annulene, $\geq$9.2 kcal (38 kJ); benzene = [6]annulene, 0 kcal (0 kJ); cyclooctatetraene = [8]annulene, 7.6 kcal (33 kJ); and [10]annulene, $\leq$4.4 kcal (18 kJ). We find an encouraging consistency with each other and with standard preconceptions about the decreasing magnitude of aromaticity and antiaromaticity with increasing ring sizes in the annulenes. We so close this section rather contentedly.

6. THE AROMATICITY OF THE "POLYHEDRAL" *CLOSO*-BORANE IONS $[(BH)_n]^{2-}$

We now turn to a discussion of the structure and energetics of the *closo*-borane ions, a class of inorganic compunds with the general formula $[(BH)_n]^{2-}$. As is suggested by the adjective "polyhedral," these species are manifestly three-dimensional. For example, the molecular geometry of the n = 6, 10, and 12 species are an octahedron (alternatively labeled a bicapped square or a trigonal antiprism), a bicapped square antiprism, and an icosahedron (or bicapped pentagonal antiprism), structures **34**, **35**, and **36**, respectively. While a more thorough discussion of the structural and

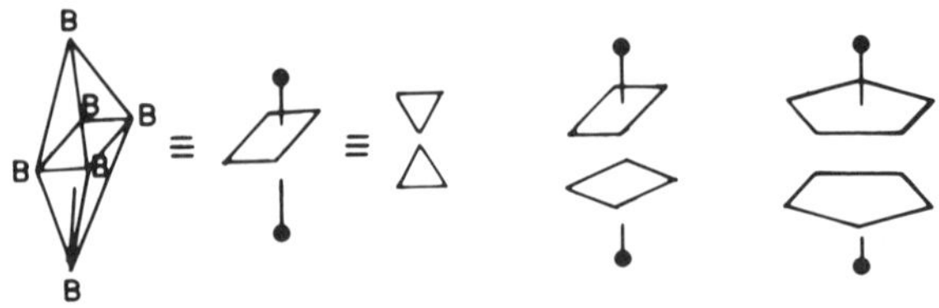

34 **35** **36**

energetics features of *closo*-borane ions is found in Volume 1 (Chapters 5 on chemical topology and 6 on general cluster compounds, respectively), it suffices to say here that the particular stabilization found in this class of compounds is a mainstay of the chemistry of boron hydrides.[42–44]

It is important to note that while experimental evidence of this aromatic stabilization has generally come from reaction chemistry,[42] such as substitution or thermal decomposition reactions, as opposed to calorimetry, this generally presumed thermodynamic effect is also indicated by both calculational and topologically based theory .[42–44] More precisely, the experimental data consist of observations such as that $[(BH)_{12}]^{2-}$ may be perhalogenated by Cl_2, Br_2, and I_2 without destruction of the "cage" and that it reacts with aqueous, elemental F_2 to give $[B_{12}F_{11}OH]^{2-}$ somewhat in excess of 30% yield.[45] Also the disproportionation of the cesium salt of $[(BH)_{11}]^{2-}$ proceeds[46] above 600°C quantitatively into a equimolar mixture of $[(BH)_{10}]^{2-}$ and $[(BH)_{12}]^{2-}$ ions. Direct thermochemical data on any of these compounds or of other salts of the "polyhedral" *closo*-borane ions seemingly are absent. A fortiori, the desired data on unsolvated, unassociated forms of *closo*-borane ions are also absent. That (in principle) one can deduce this thermochemistry from data of the former type is shown by consideration of the formally related, though planar, $[(CH)_n]^{2-}$ ions. A discussion for these carbon-containing anions with n = 6, 8, and 16 is made in Reference 47 and in Chapter 2 of Volume 3. Suffice it to say that the quantitative assignments corroborate conclusions drawn from nonthermochemical methods of deducing the relative degrees of aromaticity for these and other hydrocarbons. Thus we feel that these approaches should be applicable to boron hydrides and other aromatic species, whether they be inorganic or organic.

Nonetheless, we limit our attention to *closo*-borane ions in this section and use the diagonal reference states approach to analyze and to understand the relative aromaticity of this class of compounds. Literature studies have often characterized these species by the number of borons with four, five, and six boron neighbors.[43] This parallels the characterization earlier in this chapter of strained alicyclic hydrocarbons in terms of the number of carbons with two, three, and four carbon neighbors. As befits this boron hydride–hydrocarbon analogy, a $[(BH)_n]^{2-}$ ion will accordingly be described by

(ν_4, ν_5, ν_6), where ν_4 is the (necessarily integer) number of borons with four boron neighbors, and so on. For example, the n = 6, 10, and 12 $[(BH)_n]^{2-}$ ions above have the number triplets (6,0,0), (2,10,0), and (0,12,0) (**34**, **35**, **36**), while the yet unknown "supericosahedral"[48] species **37**, *closo*-borane ion $[(BH)_{14}]^{2-}$, has the number triplet (0,12,2).

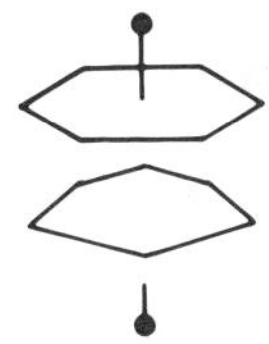

37

The stabilization energy, STE, of a $[(BH)_n]^{2-}$ ion may then be defined as follows:

$$\text{STE} = \Sigma_i^{\text{l}} \; {}_i\epsilon_i - E = \boldsymbol{\nu} \cdot \boldsymbol{\varepsilon} - E \tag{8-11}$$

where the ϵ_i are the energies of the structural groups or components, suitably defined, and E is the total energy of the ion.

≪We use the term "total energy" instead of "heat of formation" because all the input numerical quantities in this section are derived from Reference 42. These authors used partial retention of diatomic differential overlap (PRDDO) calculations, an approximate quantum chemical procedure that mimics the ab initio STO-3G minimum basis set.[49] Only if these numbers were derived from experiment *or* from the exacting level of theory described by Del Bene in Chapter 9 of Volume 1 do we feel the term "heat of formation" is applicable.≫

The stabilization energy per boron, STE_B, is likewise defined as follows:

$$\text{STE}_\text{B} = \frac{\text{STE}}{n} \tag{8-12}$$

Here we used the term "stabilization energy" instead of "resonance energy" to allow for the existence of strain energy in the three-membered rings or triangular faces implicit in *closo*-boranes. (Are *closo*-boranes strained? Consider Chapter 7 in Volume 1.) The use of this term also allows for the presence of resonance energy in the reference state. (Notice that had ethylene, as opposed to graphite, been used as a reference state for the π-bonding in aromatic hydrocarbons, all the resulting stabilization energies would have quantitatively, although not qualitatively, changed. See Sections 4 and 5 of this chapter for a related approach and also Chapter 2 for a general discussion of the energetics of graphite.)

The major problem remaining is to determine the values for the individual ϵ_i. Our earlier analysis shows there are but three sets of ϵ_i (ϵ_4, ϵ_5, ϵ_6) that may be independently defined. Perhaps not surprisingly, we opt to use numbers derived from diagonal reference states. For the cases of the four- and five-neighbored borons, the obvious choice of reference species consists of the well-characterized and highly symmetric $[(BH)_6]^{2-}$ and $[(BH)_{12}]^{2-}$ which have **v** components of (6,0,0) and (0,12,0). For these ions, we set STE equal to zero and so derive:

$$\epsilon_4 = \tfrac{1}{6}\{E[(BH)_6]^{2-}\} = -25.215 \text{ au} \qquad (8\text{-}13)$$

and

$$\epsilon_5 = \tfrac{1}{12}\{E[(BH)_{12}]^{2-}\} = -25.275 \text{ au} \qquad (8\text{-}14)$$

where 1 au = 627.5 kcal = 2625 kJ.

≪While the use of the literature PRDDO calculations (Reference 42) no doubt incurred errors, the difference between ϵ_4 and ϵ_5 is large (~40 kcal (170 kJ)), and thus we believe that meaningful conclusions may be drawn from our analysis using them and the value of ϵ_6 that follows. We cannot believe that the entire dependence on n shown in References 43 and in what follows in this section is an artifact of using an admittedly inexact quantum chemical methodology.≫

Evaluation of ϵ_6 is the only real problem remaining, since it may be proven[43b] that there is no possible diagonal reference species for boron bound to six other borons. More precisely, no possible n exists for which $[(BH)_n]^{2-}$ is describable by the number triplet $(0,0,n)$, since Euler's theorem (Equations 8-15) would have to be violated:

$$\text{Number of faces} = \text{number of B—B bonds} - \text{number of B atoms} + 2 \qquad (8\text{-}15a)$$

$$\text{Number of faces} = \text{number of edges} - \text{number of vertices} + 2 \qquad (8\text{-}15b)$$

In the inherent absence of such a diagonal reference state, we opt to use an "asymptotically" diagonal species $[(BH)_n]^{2-}$ with suitably large n, which minimizes the relative numbers of four- and five-neighbored borons. Thus we define:

$$\epsilon_6 = \lim_{\substack{(\nu_4+\nu_5)/n \to 0 \\ n \to \infty}} \frac{E\{[(BH)_n]^{2-}\} - \nu_4\epsilon_4 - \nu_5\epsilon_5}{n - \nu_4 - \nu_5} \qquad (8\text{-}16)$$

Literature investigations on *closo*-borane ions cease with $[(BH)_{24}]^{2-}$. Of those studied in the literature, this species also is the closest to the desired limit. Thus we choose it to represent the "asymptotically" diagonal species despite its associated triplet (0,12,12). This results in $\epsilon_6 = -25.238$ au and a derived associated ordering of energies $\epsilon_5 < \epsilon_6 < \epsilon_4$.

That ϵ_5 is the largest (ie, that five-neighbored boron is preferred over four- and six-neighbored boron) is in accord with the experimental structure of

TABLE 8-1. Values[a] for the Closo-borane Ions $[(BH)_n]^{2-}$ $E_B = E/n$ and $STE_B = STE/n$

n	Symmetry	ν_4	ν_5	ν_6	Energies (au)	
					E_B	STE_B[b]
6		6	0	0	−25.215	0
7		5	2	0	−25.232	0.000
8		4	4	0	−25.238	−0.008
9		3	6	0	−25.249	−0.007
10		2	8	0	−25.259	−0.005
11		2	8	1	−25.257	−0.004
12		0	12	0	−25.276	0
13	(C_2)	0	12	1	−25.260	−0.005
14	(D_{6d})	0	12	2	−25.266	−0.005
15	(D_{3h})	0	12	3	−25.260	−0.008
16	(D_2)	0	12	4	−25.262	−0.005
17	(D_{5h})	0	12	5	−25.267	+0.002
18	(D_{3d})	0	12	6	−25.262	−0.007
19	(C_s)	0	12	7	−25.254	−0.008
20	(D_{3h})	0	12	8	−25.262	+0.001
21	(D_3)	0	12	9	−25.261	+0.001
22	(D_{5d})	0	12	10	−25.261	+0.002
23	(D_3)	0	12	11	−25.255	−0.003
24	(T)	0	12	12	−25.257	0

[a]All values except STE_B from References 43a and 43b.
[b]Writing STE_B as zero implies that this value is zero by definition; if positive, the species is more stable than the reference states would suggest, while if negative, likewise less.

elemental boron[50] and with the cap–ring orbital overlaps analysis discussed in Reference 51. Admittedly ϵ_4 and ϵ_6 are in a different order in this work of Jemmis and Schleyer. Is this discrepancy due to the increased diffuseness of orbitals in the borane ions compared to the organometallics and carboranes discussed more explicitly therein? Is the discrepancy due to the increased intramolecular electrostatic destabilization of the smaller dianions, which could cause ϵ_4 and ϵ_5 to be too small? We cannot now answer these questions. Nonetheless, Table 8-1 shows the stabilization energies (as defined by this chapter's scheme) per boron of the *closo*-borane ions with the accompanying point group description of the molecular symmetry for $n \geq 13$, and demonstrates that most of these species are less stable than the highly symmetric, diagonal reference states. Indeed, it may be argued that since the largest stabilization energy found is +0.002 au per boron (ie, ~1 kcal (5 kJ)), a slightly more "stable" ϵ_6 would result in net destabilization of all nondiagonal borane ions. Surely these conclusions overshadow any monotonic trends with increasing n, such as the increasing number of resonance structures or the decreasing number of bonding electron pairs per skeletal atom.

We conclude this section by explicitly encouraging the direct measurement, and subsequent disentanglement, of thermochemical data on *closo*-

borane ions, as well as the synthesis of the related "supericosahedral" species so that the nature and magnitude of aromaticity of the *closo*-borane ions may be rigorously understood.

7. EXOTHERMICITY VERSUS ENDOTHERMICITY OF SIMPLE INORGANIC REACTIONS

We now digress briefly from comparatively complicated organic compounds and consider an application of the conceptual vector-type logic to rather simple, inorganic species. This section considers a rather restricted universe—only compounds that are composed of hydrogen, sodium, and chlorine, and no other element. Since there are three independent chemical elements, this suggests treating the individual types of atoms as groups, and indeed the dichotomy of elements and groups is discussed in Section 10. We recognize, however, that because there are three groups, there are three possible choices of independent reference states.

We opt not to discuss any of them immediately, but rather to commence with a question that could be taken from an examination in introductory chemistry. "Given the reaction of sodium and hydrogen chloride to form sodium chloride and hydrogen, does sodium or hydrogen have the greater 'affinity' for chlorine? Is this reaction of sodium and hydrogen chloride endothermic or exothermic, and why?"

From various sources of chemical insight and intuition derived from our understanding of the "real" universe, the answer "exothermic" may be offered with the following interrelated reasons:

1. Sodium is more metallic, or electropositive, than hydrogen and so sodium displaces hydrogen.
2. It is easier to ionize sodium than hydrogen—salts of Na^+ are easier to form than of H^+.
3. The sodium–chlorine bond is more polar (ionic) than the hydrogen–chlorine bond and so "enjoys" greater Coulombic or electrostatic stabilization; sodium chloride is stabilized over hydrogen chloride.
4. Sodium reacts exothermically, indeed nearly violently and explosively, with water. Should not a strong acid, HCl, react with sodium ever so more exothermically than a weak acid, H_2O?

These reasons are all mutually compatible and also compatible with numerous other experimental observations and conceptual models involving the other species that compose the "real" universe. But is the answer "exothermic" correct?

The reader may have noticed that in the question we did not write the symbols of the elements and instead used words. This was a conscious choice. Translating to symbols, do we now use H or H_2 for hydrogen, Cl or Cl_2 for chlorine? And what about sodium? Does Na mean solid, metallic, or gaseous atomic sodium? Or should we consider the diatomic (dimeric) gaseous Na_2 form? Hydrogen chloride is less ambiguous because we can't mean hydrochloric acid, given the exclusion of oxygen for the water of aqueous solutions from the ingredients of our imaginary universe. Sodium chloride, however, can be the ionic salt, $Na^+ Cl^-(s)$, or the diatomic molecule, NaCl(g). Which species among all the above do we mean? In addition, depending on how we define "the reaction"—for example, as a two-body collision in "free space" or as everything that happens between mixing the starting materials and analyzing the products—the question of its exothermicity may or may not be equivalent to that of the relative affinity of sodium and hydrogen for chlorine. Which clarification for all these ambiguities did we make?

Our failure to respond to the foregoing question does not mean that there is no standard state available to be chosen as the reference states for the species described. Quoting from what we take as the major archive of the energetics of inorganic species[52]:

> (a) For a pure solid the standard thermodynamic properties are for the pure substance in the solid phase under a pressure of 0.1 MPa (1 bar).
> (b) For a gaseous substance the standard thermodynamic properties are for the pure substance at a pressure of 0.1 MPa (1 bar) and in a (hypothetical) state in which it exhibits ideal gas behavior. Its . . . enthalpy [values] are those of the real gas extrapolated to zero pressure.
> (c) The reference phase at 298.15 K for each element . . . has been chosen to be the phase that is thermodynamically stable at that temperature and 0.1 MPa.

These quite elaborately described conditions are nearly identical to the conventionally stated conditions of standard temperature and pressure, 25°C and 1 atm pressure, under which conditions the heat of formation of the elements is taken as exactly 0. This means we consider gaseous H_2, Cl_2, and HCl, and solid Na and NaCl, and as reference states choose the set [Na(s), $H_2(g)$, $Cl_2(g)$]. Given all these preliminaries, we are not at all surprised that reaction 8-17, so written to be in accord with those most standard of standard conditions, is markedly exothermic.

$$Na(s) + HCl(g) \rightarrow NaCl(s) + \tfrac{1}{2}H_2(g) \qquad (8\text{-}17)$$

Numerically, 76 kcal (319 kJ) is liberated per mole of solid Na and gaseous HCl reacting. But notice that we have $\tfrac{1}{2}H_2(g)$ on the right-hand side of reaction 8-17. This implicitly means that the second definition of the "reaction" is being used. Only in the unlikely case that the total transformation occurs in a single mechanistic step does the derived exothermicity answer the affinity question.

But what about the other choices of reference states? Another, equally valid though less practical, is to take the reference state of all species as gases under the conditions in item (b) of the foregoing list and set the heat of formation of elemental gases as zero. Equivalently, the reference state set is [Na_2(g), H_2(g), Cl_2(g)]. If this choice of reference states were to be considered seriously by the outside world, one would have to choose either Na or Na_2 as the form of gaseous sodium. For now, we choose Na_2 to parallel H_2 in reaction 8-18:

$$\tfrac{1}{2}Na_2(g) + HCl(g) \rightarrow NaCl(g) + \tfrac{1}{2}H_2(g) \qquad (8\text{-}18)$$

This reaction is also exothermic, though by a significantly smaller value: 37 kcal (156 kJ) per mol of HCl. Once again, however, the presence of H_2 on the right-hand side of reaction 8-18 means that the derived exothermicity does not answer the affinity question.

Our third possible choice of reference states continues to utilize gaseous species and differs from the second choice by taking the elements to be atomic with an associated heat of formation of zero by definition. This reference state set is thus [Na(g), H(g), Cl(g)]. Moreover, we choose atomic H to parallel Na in reaction 8-19:

$$Na(g) + HCl(g) \rightarrow NaCl(g) + H(g) \qquad (8\text{-}19)$$

This reaction represents the conceptually simplest, mechanistically most elementary representation of the process of interest that one can write and, unlike reactions 8-17 and 8-18, it compares nothing but the relative affinity of sodium and hydrogen for chlorine. Experimentally, it is *endothermic* by 6 kcal (26 kJ)! That is, atomic hydrogen has a greater affinity for atomic chlorine than does atomic sodium. Indeed, it is the strongly exothermic *secondary* reaction of two atoms of hydrogen to form a molecule that drives reactions 8-17 and 8-18 to the right by keeping the concentration of atomic hydrogen so low that the reverse of reaction 8-19 may be neglected. We must remember that the H—H bond in H_2 is among the strongest of all single bonds, so that reactions that must break this bond are hindered and those that form it are driven.

The energetics of reaction 8-19 may be translated into a comparison of the competing reactions 8-20 and 8-21.

$$Cl(g) + NaH(g) \rightarrow NaCl(g) + H(g) \qquad (8\text{-}20)$$
$$Cl(g) + NaH(g) \rightarrow HCl(g) + Na(g) \qquad (8\text{-}21)$$

While both reactions are exothermic, 8-21 is more so, again suggesting that hydrogen has a stronger affinity for chlorine than does sodium. Moreover, reaction 8-22 is exothermic for X = Br and I, but endothermic for X = F:

$$Cl(g) + HX(g) \rightarrow HCl(g) + X(g) \qquad (8\text{-}22)$$

As such, there appears to be a new trend: exothermicity appears to parallel the electronegativities of the elements X and so the polarities of the HX bonds. The result for X = Na is consistent with this trend (see exam answer 3). However, there is an expensive conceptual price. Do we really want to have the HX and the NaH bonds "polarized" the same way, whether they be with negatively charged sodium and X, or with positively charged X and sodium?

We shall not ask, "Which of the three diagonal reference states 'should' be used: [Na(s), H_2(g), Cl_2(g)], [Na_2(g), H_2(g), Cl_2(g)], or [Na(g), H(g), Cl(g)]?" This question is too narrow. Indeed, as the foregoing discussion illustrates, the choice of reference states could be tailored to the details of the use to which they are to be put, most especially how the reactions are defined. One should actually consider other sets of reference states that may be written by defining the structural groups differently, for example, sodium and its salts [Na(s), NaH(s), NaCl(s)], and a set of stable binary compounds [NaH(s), NaCl(s), HCl(g)]. That these latter choices are equally fundamental may be shown by rewriting the various "old" reference states in terms of the "new" ones; thus Cl may alternatively be defined as (NaCl − Na) and ½[(NaCl + HCl) − NaH]. The choice of reference state defines only which species are assigned a heat of formation of precisely zero—the independence of the total heat of reaction on the individual reaction steps (also called Hess' law) prevents this assignment from affecting the heat of reaction corresponding to any specified process we care to consider. The vectorial logic says we are free to use our own criteria—the authors of this chapter opt for generality, conceptual intelligibility, and experimental accuracy. Generality as well as familiarity suggests using the elements—not all elements form salts, much less isolable compounds. Use of the elements generates at least an "atom-diagonal" reference state that has many, though not all, of the earlier enunciated advantages. Since heats of vaporization and sublimation are not reliably known for all elements, no less all substances, condensed phase data will have to do occasionally. Standard temperature and pressure are quite achievable experimentally. The elements in their most stable form under these specified conditons are thus chosen—inorganic thermochemistry convention and intuition concur.

≪Many of the same ambiguities and seeming resolutions apply to the analysis of related reactions involving alkali metals other than sodium and/or halogens other than chlorine. However, no such unifying picture has arisen for general reactions of elements with binary compounds.≫

8. "ENERGY COMPONENT ANALYSES"

We now turn to a relatively "orthodox" theoretical application of our vectorial analysis. The total nonrelativistic energy of a molecule is usually computed using the Born–Oppenheimer approximation (see Reference 53 and Chapter 8 of Volume 1). Implicit in this approximation is the assumption that the nuclei are fixed and the electrons are mobile, that the nuclei are points and the electrons are "smeared out." Equivalently, one may speak of well-defined positions of the nuclei constituting the molecular framework or structure while necessarily speaking of electron "clouds" or distributions such as are connoted by orbitals. Indeed, the Born–Oppenheimer approximation corresponds to much of the chemist's image of molecular "reality." It does not limit one to time-independent phenomena of a hypothetical frozen state "at a temperature below 0 K": by calculating molecular energies for a variety of relative nuclei positions, one derives "quantum thermochemistry" and potential energy surfaces and hypersurfaces that are providing increasingly rich insights for organic and inorganic chemists (see Chapter 9 of Volume 1). Rather, the Born–Oppenheimer approximation allows us to express molecular energies (ie, the energy associated with some specified number of electrons and an arbitrary set of nuclear positions) as the sum of the following four intelligible energy components. These four components are: V_{nn} = nuclear–nuclear repulsion, V_{ne} = nuclear–electron attraction, V_{ee} = electron–electron repulsion, and T = the electronic kinetic energy. (Note that the nuclei are assumed at rest.) Accordingly, the total energy difference associated with any chemical process is expressed as the sum of the differences of the four individual energy components:

$$\Delta E = \Delta V_{nn} + \Delta V_{ne} + \Delta V_{ee} + \Delta T \quad \text{(8-23a)}$$

So far, no new errors have been introduced, nor any new insights. However, in practice numerous and numerically large errors often beset computational approaches to solving "chemical problems." To compensate for these errors, energy component analysis was developed. This approach consisted of recombining these four individual terms into more suggestive, chemically meaningful and interpretable combinations. In addition, there was the hope that these new combinations would result in calculations that were less sensitive to basis set limitations, artifacts, and other defects. Of all the possible combinations of these four terms, those suggested most cogently in the literature were:

Zero- and one-electron versus two-electron[54]:

$$\Delta V_{nn} + \Delta V_{ne} + \Delta T \text{ versus } \Delta V_{ee} \quad \text{(8-23b)}$$

Repulsive versus attractive[55]:

$$\Delta V_{nn} + \Delta V_{ee} + \Delta T \text{ versus } \Delta V_{ne} \quad \text{(8-23c)}$$

Repulsive versus one-electron[56]:

$$\Delta V_{nn} + \Delta V_{ee} \text{ versus } \Delta V_{ne} + \Delta T \quad \text{(8-23d)}$$

Antiphase analysis[57]:

$$\Delta V_{ne} + \Delta V_{ee} + \Delta T \text{ versus } \Delta V_{nn} \quad \text{(8-23e)}$$

These combinations have been successfully applied to numerous chemical problems, most notably conformational changes, as demonstrated in References 54–57 and other papers by the same authors. However, it appeared to us that knowledge gained on one class of compounds and/or problems (eg, inversion barriers of amines) has rarely been transferred to another (eg, rotational barriers in substituted ethanes). What follows is a demonstration that this seemingly a posteriori nature of energy component analyses is an inherent, hence unavoidable aspect of this approach.

We commence with the observation that the individual energy components are independent of each other. One might object that the virial theorem relates the kinetic energy T to the total potential energy V $(= V_{nn} + V_{ne} + V_{ee})$. However, the commonly cited scaling ratio, $-V/2T$, must equal exactly 1 *only* when the calculated geometry corresponds to one for which there are no intramolecular forces or torques. While true for extrema (relative and absolute maxima and minima) on the multidimensional potential energy hypersurfaces, this condition is generally not true, and so all four components remain independent. Thus an arbitrary energy component analysis approach may be written as follows:

$$c_{nn}\,\Delta V_{nn} + c_{ne}\,\Delta V_{ne} + c_{ee}\,\Delta V_{ee} + c_{T}\,\Delta T = \sum_i c_i \mathrm{A}_i = \mathbf{c}\cdot\mathbf{A} \qquad (8\text{-}24)$$

where the subscripts $i = 1, 2, 3$, and 4 correspond to the nuclear–nuclear repulsion, nuclear–electron attraction, electron–electron repulsion, and kinetic energy, respectively; and $\mathbf{c}$ and $\mathbf{A}$ are 4-vectors. We identify $\mathbf{c}$ as a row vector of coefficients and $\mathbf{A}$ as the "universal" column 4-vector col $(\Delta V_{nn}, \Delta V_{ne}, \Delta V_{ee}, \Delta T)$. It is thus the row vector $\mathbf{c}$ that differentiates the individual energy component analyses from each other and from the change in the total energy. The change in the total energy corresponds to the 4-vector $(+1, +1, +1, +1)$, while the energy component approaches as defined by expressions 8-23b through 8-23e correspond to the vectors $(+1, +1, -1, +1)$, $(+1, -1, +1, +1)$, $(+1, -1, +1, -1)$, and $(+1, -1, -1, -1)$, respectively.

From the properties of general vectors, it is clear that the foregoing total set of five 4-vectors *cannot* be linearly independent. However, any two or even three chosen from that set of five are. More important, the final value of the total $\mathbf{c}\cdot\mathbf{A}$ expression from a single energy component analysis for a given compound does not yield the value of $\mathbf{c}\cdot\mathbf{A}$ for the same compound for any other approach unless at least three of the individual $\mathbf{A}_i$ (ie, ΔV_{nn}, ΔV_{ne}, ΔV_{ee} and ΔT) are known. In that quantum chemical calculations are geared to reliably computing directly the total energy (eg, via the variational principle), techniques have not yet been developed to independently calculate the appropriate sum for any of the general energy component analysis approaches except by separately calculating and adding the individual $\mathbf{A}_i$. In that case, all four terms would be available separately, and all five $\mathbf{c}\cdot\mathbf{A}$ values could be determined unambiguously. That is, calculationally we know either all or none of the values for the energy component analyses. Thus the hopes that any of these analyses would be less sensitive to errors in the components and that conclusions derived from one molecule would make predictions of any other more facile are groundless. In summary, the four energy component analyses of Equations 8-23b through e can give the energy difference that is fundamentally the quantity of greatest chemical interest only in a posteriori manner.

9. DIFFERENCES AND EQUIVALENCES OF THE "GROUP," "EXCHANGE," AND "MACRO-INCREMENTATION" ESTIMATION APPROACHES

In this section we briefly discuss the inherent similarities, differences, and assumptions of the methodologies of "group increment contributions,"[58–60] "group substitution (or conversion)"[60,61] (or "exchange increments,") and "macroincrementation."[62] Before proceeding, let us briefly define the three methods. We commence with the well-characterized group increment approach. This methodology describes the molecule as the set of polyvalent, multicoordinated atoms and their associated bonds and bonding atoms, which constitute the groups chemists would classically assert to compose the molecule. Such groups include:

$C-(H)_3(C)$, a carbon bonded to three hydrogens and to another carbon and so recognized as a methyl group.

$C-(H)_2(C)_2$, a carbon bonded to two hydrogens and two other carbons and so recognized as a methylene group.

$C_B-(H)$, an aromatic or benzenoid carbon bonded to a hydrogen and implicitly bonded to two other benzenoid carbons.

$C_B-(Cl)$, a benzenoid carbon bonded to a chlorine and implicitly bonded to two other benzenoid carbons.

To first order, each type of bonded unit or group with associated bonds for joining to other groups is assigned a numerical value for its linearly additive contribution to the total molecular property of interest. For these four groups and the property taken as heat of formation, the appropriate values are −10.20, −4.93, 3.30, and −3.8 kcal (−42.7, −20.6, 13.8, and −15.9 kJ). To derive the heats of formation of species of interest, say the three isomeric dichlorobenzenes, **38**, one adds the appropriate values for the

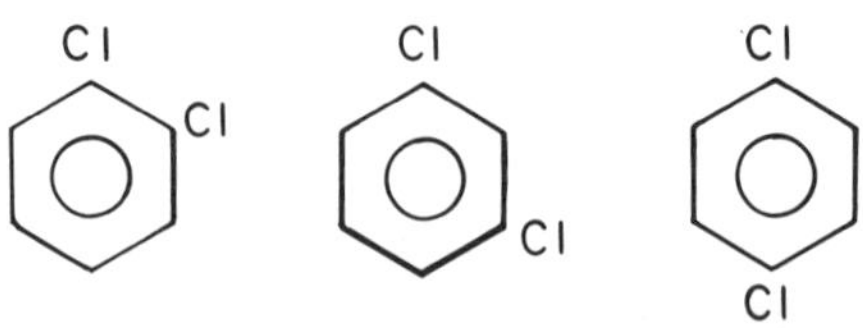

38-o **38-m** **38-p**

groups, in this case four of $C_B-(H)$, and two of $C_B-(Cl)$. This results in 5.6 kcal (23.4 kJ) for the predicted heat of formation of the any of the dichlorobenzenes, species **38.** For comparison, the experimental values for the ortho, meta, and para-isomers are 7.2(0.4), 6.1(0.5) and 5.4(0.4) kcal (30.2(1.8), 25.7(2.1), and 22.5(1.6) kJ), respectively. Theory and experiment are in agreement for the *m*- and *p*-isomers and adequately so for the *o*-isomer.

Can one do better for the *o*-isomer? Considering the total number of compounds that can be studied with group increments, the user "should" not be distressed by the additional presence of a comparatively small set of numerical correction terms. These are included to take explicitly into account additional factors such as orthosubstitution of benzene rings, ring strain, resonance of conjugated double bonds, gauche-methyl interactions, and geminal or vicinal dihalogen substitution of aliphatic or alicyclic hydrocarbons. Implicitly, properties for simply bonded, minimally substituted, and unstrained molecules are computed and then modified to accommodate the effects of "molecular reality." In practice, the values for a given group are generally derived either from a statistical fit to the properties of a collection of "normal" species or from the often few or even single species with that component for which the desired property is known experimentally. Alternatively, by making explicit assumptions about the magnitude of intramolecular interactions (eg, setting the unknown quantity of one species equal to that found experimentally for another), appropriate values are obtained. Correction terms are likewise derived. For the current problem, a correction term is needed for the ortho(Cl)(Cl) substitution. Numerically, this term is 2.2 kcal (9.2 kJ), its use resulting in the new and more accurate prediction of 7.8 kcal (32.6 kJ) for the heat of formation of *o*-dichlorobenzene. In principle, many correction terms could be made obsolete by the use of larger groups such as the isomeric phenylenes $-C_6H_4-$, but this is not done in current practice.

≪For most groups, current practice also usually employs the values of heat of formation, internal entropy, and heat capacity presented in considerable detail in the appendices of Benson's book.[59b] We have not done so here.≫

Exchange incrementation takes as its starting point the formulation of a molecule of interest as a derivative of another species that is better understood—both qualitatively and quantitatively. Correction terms are used to reflect the conceptual interconversion of the two species. Ideally, the steric and electronic differences between the two species are small and are essentially independent of whatever changes are needed to change one molecule into another. An example from the recent thermochemical literature[63] is the prediction of the heat of formation of an aliphatic azoxy compound from the related (E)-olefin with the same substituents flanking the double bond:

$$\Delta H_f(\mathrm{R{-}N{=}N(O'){-}R'}) = \Delta H_f(\mathrm{R{-}CH{=}CH{-}R'}) + 14.6\ \mathrm{kcal}\ (61\ \mathrm{kJ}) \qquad (8\text{-}25)$$

The 14.6-kcal (61-kJ) additive correction term arose from averaging the difference of "adjusted" experimental heats of formation of R−N-N(O)−R′ and (*E*)−R−CH=CH−R′ for R = R′ = *t*-butyl and *n*-propyl. There are, however, no thermochemical data from which to deduce whether the same additive correction is also applicable to acyclic (*E*)−R−N=N(R′)−O and (*Z*)−olefinic species or to conjugated species such as pyridazine-*N*-oxide (**39**) and benzene, (**7**).

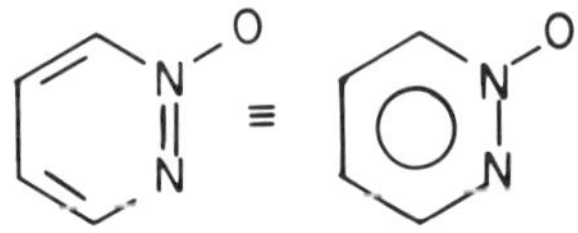

39

The third approach, macroincrementation, explicitly asserts that if for each of two sets of molecules the total numbers and types of atoms, bonds, and more general structural types (eg, methyl groups, benzene rings) are the same, then the totals or sums of the numerical values for a given property for the two sets are the same. As such, all macroincrementation reactions are presumed to be thermoneutral as written. Equivalently, the desired property of the molecule of interest is written as the arithmetic sum and difference of those of other, better understood species. In either formulation of macroincrementation reactions, simple arithmetic allows one to estimate a missing number if the values for all the others are available.

Correction terms such as those employed with group increments are less universally but more explicitly incorporated by using known species that contain these features. Alternatively, the predicted results from established rules and regularities (eg, direct use of symmetry number corrections) may immediately be used. One may even use "alphanumeric corrections," a rather fancy way of saying a numerical value obtained from some rule of thumb, piece of folklore, or more qualitative relationship. For example, the heat of vaporization of a hydrocarbon is almost independent of its molecular structure and instead depends almost completely on its total number of carbons.[64] One benefit is the ability to maximally use the most "trusted" experimental data, theoretical results, or intuitive understanding without any apprehension for the effects of the value of one experimentally measured quantity on another. This contrasts with the group increment approach, where "jiggling" one increment will modify the parameters for all the others. For example, a macroincrementation reaction equivalent of expression 8-28 explicitly replaces the 14.6 kcal (61 kJ) by the difference of the heats of formation of R−N=N(O)−R′ and of (*E*)−R−CH=CH−R′ where either R = R′ = *t*-butyl and/or *n*-propyl is explicitly stated. That is, one writes macroincrementation reactions 8-26 and 8-27.

$$\text{R}-\text{N}=\text{N(O)}-\text{R}' = t\text{-C}_4\text{H}_9-\text{N}=\text{N(O)}-t\text{-C}_4\text{H}_9 + (E)-\text{R}-\text{CH}=\text{CH}-\text{R}' - t\text{-C}_4\text{H}_9-\text{CH}=\text{CH}-t\text{-C}_4\text{H}_9 \qquad (8\text{-}26)$$

$$\mathrm{R{-}N{=}N(O){-}R' = \textit{n-}C_3H_7{-}N{=}N(O){-}\textit{n-}C_3H_7 + (\textit{E}){-}R{-}CH{=}CH{-}R' - \textit{n-}C_3H_7{-}CH{=}CH{-}\textit{n-}C_3H_7} \qquad (8\text{-}27)$$

Furthermore, since the macroincrementation description of a molecule is rarely unique (eg, the two reactions above to decide between are rather different), several estimates may usually be made and compared. To the extent they agree, one is reassured that the estimate may be trusted.

≪Data on azoxy compounds are insufficient to test whether Equation (8-26) or (8-27) is preferable. However, an internal check may be performed by setting R = R′ = n-C_3H_7 in expression 8-26 and R = R′ = t-C_4H_9 in 8-27. Combining these two equations and canceling out common terms as in (8-28) "should" result in an equality. Does it?

$$\mathrm{\textit{t-}C_4H_9{-}N{=}N(O){-}\textit{t-}C_4H_9 + \textit{n-}C_3H_7\text{-}CH{=}CH{-}\textit{n-}C_3H_7 = \textit{n-}C_3H_7{-}N{=}N(O){-}\textit{n-}C_3H_7 + \textit{t-}C_4H_9{-}CH{=}CH{-}\textit{t-}C_4H_9} \qquad (8\text{-}28)$$

Using experimental heats of formation of all four species in this macroincrementation reaction taken from Reference 63, Equation (8-28) is shown valid to approximately 1 kcal (4 kJ).≫

From the foregoing descriptions of the three methods, the objectivity, ease of computerization, immediate usefulness, and uniqueness of predictions decrease in the following order: group increments, exchange incrementation, macroincrementation. Conversely, the subjectivity, relevance of new quantitative data, and ability to incorporate one's own prejudices, biases, or intuition of chemical phenomena increase in the same order. Which should one use? As S. W. Benson, a pioneer and major practitioner of group increment methodologies, has pointed out, "Were they flawless, they would agree with each other and with experiment."[65]

We will now use our vectorial logic to show the fundamental interrelationships and equivalence of these approaches other than by the tautology or even wishful thinking implied by the quotation above. A suitably simple set of molecules can demonstrate this. In particular, consider the n-alkyl chlorides, $CH_3(CH_2)_nCl$, and *no* other class of compounds. To simplify the notation, we will abstain from the orthodox symbolism such as $C{-}(H)_3(C)$ and instead use the more commonplace, abbreviated, $CH_3{-}$ in discussions using group increments. In this modified group increment language, the property of an arbitrary n-alkyl chloride may be determined from:

$$P[\mathrm{CH_3(CH_2)_nCl}] = P(\mathrm{CH_3{-}}) + nP(\mathrm{{-}CH_2{-}}) + P(\mathrm{{-}Cl}) \qquad (8\text{-}29)$$

Expression 8-29 may be recast in terms of vectors:

$$P[\mathrm{CH_3(CH_2)_nCl}] = \mathbf{v} \cdot \mathbf{P} \qquad (8\text{-}30)$$

where the $\mathbf{v}$ (row) vector (ν_1, ν_2, ν_3) gives sequentially the number of $CH_3{-}$, ${-}CH_2{-}$, and ${-}Cl$ groups and the $\mathbf{P}$ (column) vector col(P_1, P_2, P_3) gives the properties of the same groups in the same order. To find the **P** vector, a least-squares fit is performed to minimize the discrepancy or deviation:

$$\sum_{n} \{P[CH_3(CH_2)_nCl] - \mathbf{v} \cdot P\}^2 \qquad (8\text{-}31)$$

for a set of compounds wherein only n, the number of CH_2 units, changes.

Exchange incrementation can also be discussed in this vectorial language. Consider expressions 8-32, 8-33, and 8-34:

$$P[CH_3(CH_2)_nCl] = P[CH_3Cl] + n\{P[CH_3CH_2Cl] - P[CH_3Cl]\} \qquad (8\text{-}32)$$

$$P[CH_3(CH_2)_nCl] = P[CH_3CH_2Cl] + (n - 1)\{P[CH_3CH_2Cl] - P[CH_3Cl]\} \qquad (8\text{-}33)$$

$$P[CH_3(CH_2)_nCl] = P[CH_3CH_2Cl] + (n - 1)\{P[CH_3(CH_2)_2Cl] - P[CH_3CH_2Cl]\} \qquad (8\text{-}34)$$

Expression 8-32 approximates the n CH_2 compound by starting with the methyl compound, $n = 0$, and adding n times the difference of ethyl and methyl chlorides, $n = 1$ and 0, respectively. Expression 8-33 likewise approximates the n CH_2 compound by starting with the ethyl compound, $n = 1$ and adding but $n - 1$ times the difference of ethyl and methyl chlorides. Expression 8-34, like 8-33, starts with the ethyl compound and adds $n - 1$ times the difference of the propyl and ethyl chlorides, with $n = 2$ and 1, respectively. (Quite clearly, in practice the terms enclosed in braces and the names of the compounds above would be replaced by the appropriate numerical values.) The predictions made by these three expressions 8-32, 8-33, and 8-34 are not, fundamentally, particularly conceptually distinct. That is, their logical structures are identical and indeed they can all be written as follows:

$$P[CH_3(CH_2)_nCl] = \mathbf{v} \cdot \mathbf{P} \qquad (8\text{-}35)$$

where **v** in Equation 8-32 is the row 2-vector $(1, n)$ and **P** is the column 2-vector col($P[CH_3Cl]$, $\{P[CH_3CH_2Cl] - P[CH_3Cl]\}$). In Equation 8-33, however, **v** is now the row vector $(1, n - 1)$ and **P** is the column vector col ($P[CH_3CH_2Cl]$, $\{P[CH_3CH_2Cl] - P[CH_3Cl]\}$). Likewise Equation 8-34 may so be expressed wherein **v** is the row vector $(1, n - 1)$ and P is now the column vector col($P[CH_3CH_2Cl]$, $\{P[CH_3(CH_2)_2Cl] - P[CH_3CH_2Cl]\}$). By the properties of vectors, these three 2-vector approaches cannot be linearly independent. Indeed, approaches 8-32 and 8-33 make identical predictions because they are identical: expression 8-36 is an identity for all n and can be used to interconvert them.

$$CH_3Cl + n(CH_3CH_2Cl - CH_3Cl) = CH_3CH_2Cl + (n - 1)(CH_3CH_2Cl - CH_3Cl) \qquad (8\text{-}36)$$

TABLE 8-2. Heats of Formation for Gaseous *n*-Alkyl Chlorides Containing N_C Carbons

N_C	ΔH_f (kcal/mol)		
	Experimental[a]	Predicted by Equation 8-32[a,b]	Predicted by Equation 8-34[a,b]
1	−19.6(0.1)	−19.6(0.1)*	−22.0(0.2)
2	−26.8(0.1)	−26.8(0.1)*	−26.8(0.1)*
3	−31.6(0.1)	−34.0(0.3)	−31.6(0.1)*
4	−36.9(0.9)	−41.2(0.4)	−36.4(0.3)
5	−41.8(0.7)	−48.4(0.6)	−41.2(0.4)
8	−57.1(0.4)	−70.0(1.0)	−55.6(0.9)
12	−76.9(0.6)	−98.7(1.6)	−74.8(1.4)
18	−106.6(0.7)	−141.2(2.4)	−103.6(2.3)

[a]Error bar given in parentheses.
[b]Asterisks indicate input numbers to the appropriate equation.

Because the exchange incrementation vectors are two-dimensional, vectorial analysis asserts that there are two fundamentally distinct models and we have seemingly found two such models. Which one do we choose? Using input numbers from our archive, Table 8-2 shows our "predictions" for the heat of formation of the gaseous *n*-alkyl chlorides.

The results from Equation 8-32 are surprisingly poor for such a simple class of compounds. Contrariwise, the results from Equation 8-34 and from experiment are much closer, except for the 1-carbon case of CH_3Cl—indeed, within experimental and calculated error bars, the two sets of results are indistinguishable. The error bars for the predicted numbers for Equations 8-32 and 8-34 are found using the intuitively "appealing" expressions

$$e(N_C) = \{e^2(1) + (N_C - 1)^2\,[e^2(2) + e^2(1)]\}^{1/2} \tag{8-37}$$

and

$$e(N_C) = \{e^2(2) + (N_C - 2)^2[e^2(3) + e^2(2)]\}^{1/2} \tag{8-38}$$

respectively where $e(N_C)$ is the error bar for the N_C case. Thus while vectorial analysis cannot decide, in this case the model implicit in Equation 8-34 is quite obviously better than that in Equations 8-32 and 8-33. Notice the consistency of the failure of Equation 8-32 based on its use of CH_3Cl as an input species and the erroneous "prediction" for this species in Equation 8-34. This suggests that CH_3Cl is itself anomalous and indeed, it is a standard "folklore" observation in the thermochemistry of organic compounds that anomaly is customary with methyl derivatives.[66] We have thus learned, or at least confirmed, a basic assumption to be explicitly employed in constructing exchange and macroincrementation reactions. This also suggests that it is best to avoid methyl derivatives, if possible, in generating group increments, and so the least-squares fit of Equation 8-31 should not include input numbers from CH_3Cl. This last suggestion and its generalizations admittedly represent subjective decisions, which nevertheless appear to be simultaneously unavoidable, educational, and successful.

It appears that we can adequately describe *n*-alkyl chlorides using 2-vectors when using exchange incrementation but require 3-vectors for the group increment approach. Upon first glance, this discrepancy in dimension is surprising. Does it not violate vector logic to have 2- and 3-vectors equally valid in describing "reality"? Yes and no. If all we care about are *n*-alkyl chlorides, then 2-vectors in fact also suffice for the group increment approach. To see this, note that expression 8-31 can be rewritten as follows:

$$P[CH_3(CH_2)_nCl] = \mathbf{v}' \cdot \mathbf{P}' \qquad (8\text{-}39)$$

where the new **v**′ (row) vector (ν_1, ν_2) gives sequentially 1 (the universally common number of CH_3- and $Cl-$ groups alike for all *n*-alkyl chlorides) and the number of $-CH_2-$ groups. The first element of the new **P**′ (column) vector col(P_1, P_2) is the sum of the properties of CH_3- and $Cl-$ groups; owing to the anomalies cited earlier for CH_3Cl, it should not be confused with any property of this substance. The second element of **P**′ is the value for the property of the $-CH_2-$ group.

We would have expected 3-vectors to contain more information than 2-vectors. Indeed, this is the case. The original group increments devised for *n*-alkyl chlorides in terms of 3-vectors are equally applicable for *n*-alkanes, a class of compounds unexpressible and inexplicable in terms of the 2-vectors in Equation 8-39. Do we wish to understand these additional species? If so, 3-vectors are needed, and in the earlier least-squares analysis (8-31) for $P[CH_3(CH_2)_nCl]$, one should have included the properties for this "new" class of molecules, $P[CH_3(CH_2)_nCH_3]$. These new data on *n*-alkanes do not affect the dimensionality of the vectors to be discussed, and so we must decide what to use as the reference 3-vectors to allow for the consideration of *n*-alkyl chlorides and *n*-alkanes alike. Consider two typical examples of exchange incrementation reactions that incorporate explicit, quantitative consideration of *n*-alkanes:

$$\begin{aligned} CH_3(CH_2)_3Cl = {} & CH_3(CH_2)_2CH_3 - (CH_3CH_2CH_2CH_3 \\ & - CH_3CH_2Cl) + 2(CH_3(CH_2)_2Cl - CH_3CH_2Cl) \end{aligned} \qquad (8\text{-}40)$$

$$CH_3(CH_2)_6CH_3 = CH_3(CH_2)_2CH_3 + 4(CH_3(CH_2)_2Cl - CH_3CH_2Cl) \qquad (8\text{-}41)$$

Rewriting the right-hand sides of Equations 8-40 and 8-41 as 8-42 and 8-43, respectively:

$$\begin{aligned} & (1)CH_3(CH_2)_2CH_3 + (-1)(CH_3CH_2CH_2CH_3 - CH_3CH_2Cl) \\ & \qquad + (2)(CH_3(CH_2)_2Cl - CH_3CH_2Cl) \end{aligned} \qquad (8\text{-}42)$$

$$\begin{aligned} & (1)CH_3(CH_2)_2CH_3 + (0)(CH_3CH_2CH_2CH_3 - CH_3CH_2Cl) \\ & \qquad + (4)(CH_3(CH_2)_2Cl - CH_3CH_2Cl) \end{aligned} \qquad (8\text{-}43)$$

shows that a universally applicable **P** vector we seek may be written as follows:

$$\text{col}[P(CH_3(CH_2)_2CH_3, \{P[CH_3CH_2CH_2CH_3] - P[CH_3CH_2Cl]\}, \{P[CH_3(CH_2)_2Cl] - P[CH_3CH_2Cl]\}]$$

with **v**'s for the cases 8-42 and 8-43 of (1, −1, 2) and (1, 0, 4), respectively. Numerically the property of the heat of formation of gaseous species, the **P** vector, is (−30.2, −3.4, −4.9) kcal [(−126.5, −14.4, −20.3) kJ], while for the property of the heat of vaporization the **P** vector is (5.0, −0.9, 1.0) kcal [(21.0, −3.7, 4.2) kJ]. Our "predictions" using Equation 8-40 for the heats of formation and of vaporization of *n*-butyl chloride are −36.4 and 6.9 kcal (−153 and 29 kJ), while the experimental values are −36.9 and 8.1 kcal (−155 and 34 kJ). For *n*-octane using Equation 8-39, we find −49.6 and 9.0 kcal (−208 and 38 kJ), while the experimental values are −49.8 and 9.9 kcal (−209 and 42 kJ). All four sets comparing theory and experimental results are encouraging.

As written, the exchange incrementation reactions are indistinguishable from macroincrementation reactions, although in practice one might have taken a suitably averaged collection of *n*-alkyl chlorides and *n*-alkanes from which to derive the second and third components of the **P** vector. Likewise, other qualitative, intuitive awarenesses might result in new macroincrementation reactions, although often none are apparent unless one can go beyond the current classes of compounds. For example, one can get a handle on the anomalous heat of formation of CH_3Cl by considering the whole class of methyl derivatives bearing comparably electronegative substituents. In particular, consider the related macroincrementation reactions 8-44 and 8-45:

$$CH_3Cl = CH_3NH_2 + CH_3CH_2Cl - CH_3CH_2NH_2 \qquad (8\text{-}44)$$

$$CH_3Cl = CH_3OH + CH_3CH_2Cl - CH_3CH_2OH \qquad (8\text{-}45)$$

From reaction 8-44, the predicted heat of formation is −5.5 + (−26.8) − (−11.4) = −20.9 kcal (−88 kJ), while from reaction 8-45 the prediction is −48.2 + (−26.8) − (−56.8) = −18.2 kcal (−79 kJ). We remind the reader that the experimental value is −19.6 kcal (−82 kJ). The agreement is encouraging.

It should be noted that direct application of any of the three approaches, group, exchange or macroincrementation, could yield properties of 1,*n*-dichloro-*n*-alkanes, $Cl(CH_2)_nCl$. That is, one stumbles into, rather naturally, such effects as vicinal or 1,2-dihalosubstitution. When using group increments, one can create de novo correction terms for such dichlorosubstitution. This can be directly incorporated into the vectorial formulation by increasing the dimensionality of the vectors and including components in the **v** and **P** vectors corresponding to the number and contributions of these features. In the particular, one can write the new **v** for $ClCH_2CH_2Cl$ as the 4-vector:

[number (CH_3), number (Cl), number (CH_2), number (1,2-dihalo interactions)] = (0, 2, 2, 1)

a related 4-vector **P** in the group increment methods. Exchange and macroincrementation approaches would more naturally, but admittedly subjectively, incorporate some species with analogous substitution patterns. For example, macroincrementation reactions analogous to 8-44 and 8-45 predict heats of formation of this species as −35.1 and −34.0 kcal (−147 and −142 kJ), respectively, to be compared with the recent experimental value[67] of −32.0(0.2) kcal (−134(1) kJ). Yes, it would appear that there is a special vicinal dihalo effect, a subclass of the extensively studied[68] "gauche effect." An "alphanumeric" correction may be automatically incorporated into any of these approaches: group increments, exchange incrementation, and macroincrementation as part of the **P** vector and/or as part of the "folklore" of the energetics of organic species. Vectorial logic cannot be used to decide which approach is the best used. However, by providing numerical limitations on the number and nature of estimation methodologies, this approach suggests a greater degree of conceptual unity and flexibility than normally is supposed.

10. UNIVERSALITY

The thorough reader may be questioning whether this chapter has been written from a consistent viewpoint. Admittedly in Sections 2 and 3 we spoke about strain energy in the context of unsubstituted alicyclic hydrocarbons; we introduced a 3-vector for the number of $>C<$, $>CH-$, and $-CH_2-$ groups in the species. In Sections 4 and 5 we spoke about aromaticity and antiaromaticity in the context of a different class of hydrocarbons; we introduced another 3-vector for the number of groups $=CH_2$, $=CH-$, and $>C=$ in the species. These choices of groups seemed to arise naturally from the structure of the compounds when one adopts an organic chemist's viewpoint. An inorganic chemist's viewpoint dominated Section 7, where, in the context of species containing solely Na, H, and Cl, we spoke about the energetics of inorganic compounds and suggested yet another 3-vector for the number of Na, H, and Cl "groups." This last 3-vector arose quite naturally from molecular and crystal structures because there were three distinct chemical elements being considered. However there is not really any contradiction involved in these various approaches, both of which use standard groupings of atoms that are deemed to be well understood by the community using them. Indeed, that inorganic chemists often in fact consider polyatomic groups is indicated by the conventional assumption that SiO_4, $H_2NCH_2CH_2NH_2$, CH_3COO, and C_5H_5 are useful conceptual units

≪noting that many of the carbon-containing groups are given short, elementlike abbreviations (eg, en, AcO, Cp).≫

and by our suggestion that the salts of Na could be used as the reference states for a sodium–hydrogen–chlorine "universe." Indeed the main difference between the organic and inorganic logic seems to involve whether the same elements appear in more than one different group in a given compound—they usually do in organic chemistry and but rarely in inorganic.

We have demonstrated the usefulness of at least six different groups containing nothing but carbon and hydrogen in studying various hydrocarbons. It is instructive to consider what can be learned about the same class of compounds by considering only the monoatomic groups. We immediately note that the analysis is simpler, since there are only two groups rather than six to consider. Thermochemical convention suggests that the choice of reference states be [C(s, graphite), ½ H_2(g)] in accord with our pragmatic choice in the case of Na, H, and Cl. Now note that assigning the value zero for the heat of formation of both solid graphite and diatomic hydrogen gas (which have **v** vectors (∞, 0) and (0, 2)) is precisely the basis for *all* the heats of formation of *all* the organic compounds described in this chapter.

≪That the value of zero is given for the heat of formation of graphite means that it makes no numerical difference whether we consider its **v** vector as (∞, 0) or (1, 0). The latter, though clearly chemically incorrect, is more generally useful in that one normally refers to the energy per mole-C.≫

This arises because the heat of formation of any compound is defined to be the difference between the rest (total) energy of the compound and the energies of its constituent elements in their standard states. Thus the elements must always have zero heat of formation, even though such constituent elements never have zero rest energies in their standard states. We do not use the rest energies directly: they are large numbers, and this makes their use inconvenient because the chemically important numbers are the small changes in these energies during a reaction. Because the number of atoms of each element is conserved in any chemical reaction, the changes in the heat of formation exactly equal the changes in the rest energies of the compounds. Thus when we cite the heat of formation of ethylene, C_2H_4, as 12.5 kcal (52.2 kJ), we mean that ethylene lies 12.5 kcal (52.2 kJ) per mole higher in energy than 2 mols of solid graphite and $\frac{4}{2}$ mols of gaseous molecular hydrogen under the well-defined conditions quoted in Section 7. Additionally, independent of the types of group we are using, the heat of interconversion of two isomers is precisely the difference of the heats of formation.

The use of groups other than atoms or elements leads to other numerical quantities. We recall that the strain or resonance energy of a molecule equals its heat of formation from a set of preformed components (>C<, >CH−, $-CH_2-$) for the former and ($=CH_2$, =CH−, >C=) for the latter. These new components can be said to correspond to new elements, or perhaps better said "quasi-elements," and they play precisely the same role in discussions of strain or resonance energies as conventional elements do in discussions of heats of formation and general thermochemistry. Indeed the mathematics of the definitions of strain and resonance energies suggest that these concepts are actually generalizations of that of the heat of formation. One of the few distinctions is that all the species discussed are either explicitly or implicitly taken as gases in the cases of studying strain and resonance energies, while the elements are usually taken in whatever phase they are most stable in at standard temperature and pressure.

≪Analogous logic using conveniently defined groups and the energetics of their combinations is found in polymer chemistry, biochemistry and solid state chemistry. Appendix B documents some examples.≫

How to go beyond the intuitive, though generally successful, understanding of the average chemist of how groups and atoms are interrelated to rigorous theory is not obvious. Nonetheless, we comment that while atoms compose groups in the same fashion as groups compose molecules and groups can be combined to form molecules, groups cannot be *uniquely* "subtracted" to yield atoms. Recall that there are often two or more groups with the same atomic composition but different bonding arrangements. Furthermore, it is only because groups are essentially unchanged from species to species that one is allowed to consider them as chemically significant entities. By speaking of larger molecules in terms of hypothetical groups such as the "isomeric" $-CH=$ and $>CH-$, the largest numerical contributions to *total* molecular energies are automatically subtracted out.

≪Note that quantum chemists (see Section 7 above and Chapter 9 of Volume 1) refer to a different reference state, that of the component nuclei and electrons infinitely far apart. For example, the energetics of hydrocarbons is referenced to C^{6+}, H^+, and e^-, and so the total energies in theoretical papers are much larger than atomization energies in thermochemical papers. The "orthodox" nuclear physicist likewise speaks of energies binding nuclei relative to protons and neutrons, and the particle physicist speaks of protons and neutrons relative to quarks. Most chemists ignore all this. Is ignorance bliss? See Appendix C.≫

The resulting strain and resonance energies are but small fractions of the atomization energy required to dismember a molecule into its component atoms and are even small compared to the total energy required to separate a molecule into the "real" group (ie, into observable species with the same formula as the groups of interest, such as the diatomic molecule CH).

Nevertheless, the chemical behavior is strongly predicted by the strain and resonance energies—recall the comparison of antiaromatic cyclobutadiene $(CH)_4$ and aromatic benzene $(CH)_6$, in Section 4. However nothing about the magnitudes of the strain energy of prismane and the resonance energy of benzene tells us that prismane is thermodynamically unstable with respect to spontaneous decomposition to benzene. This admittedly valuable information is lost when one considers strain and resonance energies rather than heats of formation. However, it is fortunately rare that one must consider strain and resonance energies simultaneously to estimate molecular properties. The sole case in this chapter was that of 1, 6-methano[10]annulene. One can, of course, invoke a 6-vector, the "direct sum" of the two 3-vectors for strain and resonance energies ($>C<$, $>CH-$, $-CH_2-$, $=CH_2$, $=CH-$, $>C=$), but few people will have their intuitive understanding enhanced by such a construct.

There are additional more subtle, though equally important relationships between atoms and groups. These include numerous isoelectronic equivalences (see Chapter 2 in Volume 1) between the quasi- and real chemical elements, between groups and atoms. That is, one finds atoms bonded in the standard way that mimic the groups discussed. Starting with the reference species for the study of strained species, the energetics of the $>C<$ component utilizes the energetics of diamond, a higher energy form of carbon than graphite. The cases of the isoelectronic tetracoordinate, tetravalent $>Si<$ and $>C<$ are automatically parallel, since the most stable form of elemental silicon has the diamond structure—indeed, the graphitelike form of Si remains unknown. Moreover, the study of species containing the $>CH-$ group proceeded despite the admitted lack of a diagonal, strainless species containing this group. The thermochemistry of species containing phosphorus, isoelectronic with CH, have been referenced to either the solid white or the red form of the element. Both allotropes contain various assemblages composed solely of $>P-$ units, and so correspond with $>CH-$, and are at least "atom-diagonal" in both cases. Species containing $-CH_2-$ and $-S-$ are parallel in that one uses $1/N_C$ $[(CH_2)_{N_C}]$ and $1/N_S$ $[S_{N_S}]$ as reference species in the respective thermochemistries. This is true even though for purely "idiosyncratic" reasons, the molecules choose $N_C = 6$ and $N_S = 8$ to minimize their total energies. For the study of aromatic species, $>C=$ is graphite, the standard form of elemental carbon. The next component, $=CH-$, is not mimicked by any chemical element: apparently the second-row, group V element nitrogen prefers triple bonding over double or single and so forms N_2, while the third-row, group V element phosphorus prefers single bonding over double or triple. It appears that neither N_6 nor P_6 is stable, nor is the all-nitrogen or phosphorus analogue of any other [*n*]annulene.

≪Indeed not even P_3N_3 or any other alternating PN species is a known species, although another phosphorus nitride P_3N_5 has been studied. The related $\lambda^4P(V)$ species $(PR_2)_3N_3$, the so-called triphosphazenes or phosphonitrilic derivatives, are well known for numerous R. While many species of type $(PR_2)_nN_n$ for *n* other than 3 have also been synthesized, very few species are known with differing numbers of phosphorus and nitrogen atoms.≫

Likewise note the existence of the dimers of the isoelectronic O and CH_2, both O_2 and $(CH_2)_2$, elemental oxygen and ethylene, that serve as input quantities for oxygen and $=CH_2$ chemistry, respectively.

We close by noting that these isoelectronic analogies provide useful insights to the inorganic and organic chemist alike and indeed bridge these fields. For example, white phosphorus with its tetrahedral assemblages of four $>P-$ atoms antedated by several *centuries any* species containing an analogous tetrahedral assemblage of four carbon atoms, regardless of whether they be $>CH-$ or $>CR-$ for some "special" R. We admit that the "real" ground state of O_2—chosen by the thermochemist as the oxygen reference state and as a generally used oxidant in calorimetry (and breathing!)—has two unpaired electrons and $(CH_2)_2$ has nonc.

≪The reader should thus note that O_2 and $(CH_2)_2$ are not strictly isoelectronic.≫

However, the lowest lying excited state of each species corresponds to the ground state of the other, and the reaction chemistries of excited oxygen with ground state ethylene and its derivatives and of ground state oxygen with excited ethylene and its derivatives are related, interesting, and important.

Thus when inorganic chemists consider heats of formation and organic chemists consider strain and resonance energies, they use exactly the same logical structure and use as large structural units as they consider well understood to focus on the difference of two large terms that quantifies the less-well-understood effect they study. In this chapter we have tried to clarify this common thought pattern and to make more explicit the presumptions of, differences, similarities, and equivalences between these approaches as they relate to the understanding of chemical "reality." As such, we hope to have helped the chemist to better understand why and how the practices already in use may be "justified."

ACKNOWLEDGMENTS

One of us (D.V.V.) thanks the National Research Council for her NRC–NRL Postdoctoral Research Associateship while the other (J.F.L.) thanks Sharon G. Lias, Rhoda D. Levin, and Eugene S. Domalski of the National Bureau of Standards for their hospitality while he was at the NBS on sabbatical.

In addition to the credits given in References 32 and 33, we make the following acknowledgments: We thank our co-workers Prof. Ivan Kramer and Noah P. Adams for their preliminary studies in the aromaticity and antiaromaticity of benzene versus cyclobutadiene, as covered in Section 4, M. Scott Kramer and James L. Tanko for their preliminary studies of the aromaticity of "polyhedral" *closo*-borane ions, as discussed in Section 6, and Trudy K. Holt for her preliminary studies in the exothermicity/endothermicity area treated in Section 7.

REFERENCES

1. Van Vechten, D.; Liebman, J.F. *Isr. J. Chem.* **1981,** *21,* 105.
2. While there have been many studies of the effect of but one singled-out substituent on numerous strained rings and studies probing the properties of various substituents on but one or a small, restricted set of ring compounds, there have been few systematic studies of the effects of a large number of substituents on likewise a large number of strained species. Two such studies are: Dill, J.D.; Greenberg, A.; Liebman, J.F. *J. Am. Chem. Soc.* **1979,** *101,* 6814 and Chapter 5 of Volume 3 in this series, by Greenberg and Stevenson.
3. See Greenberg, A.; Liebman, J.F. "Strained Organic Molecules". Academic Press: New York, 1978, and numerous articles cited therein.
4. Unless explicitly said otherwise, the heats of formation of all organic compounds in this chapter are taken from: Pedley, J.B.; Rylance, J. "Sussex-N.P.L. Computer Analysed Thermochemical Data: Organic and Organometallic Compounds". University of Sussex, Brighton, UK, 1977. In the cases of condensed phase species, the additional reference given will be for the enthalpy of vaporization or sublimation.
5. See Greenberg, A.; Liebman, J.F., Reference 3, pp. 38–42.
6. It is not obvious whether to consider diamond an inorganic or organic compound or both. In any case, the thermochemical data for diamond—and for all inorganic compounds in this chapter—are from: Wagman, D.D.; Evans, W.H.; Parker, V.B.; Schumm, R.H.; Halow, I.; Bailey, S.M.; Churney, K.L.; Nuttall, R.L. "The NBS Tables of Chemical Thermodynamic Properties: Selected Values for Inorganic and C_1 and C_2 Organic Substances in SI Units". *J. Phys. Chem. Ref. Data* **1982,** *11,* Suppl. 2.
7. Two of the more extensively documented discussions are: Greenberg, A.; Liebman, J.F., Reference 3, pp. 55–58; and Dewar, M.J.S. *J. Am. Chem. Soc.* **1984,** *106,* 669.
8. See Greenberg, A.; Liebman, J.F. Reference 3, pp. 222–224.
9. Rosenstock, H.M.; Dannacher, J.; Liebman, J.F. *Radiat. Phys. Chem.* **1981,** *20,* 7.
10. Greenberg, A.; Liebman, J.F. *J. Am. Chem. Soc.* **1981,** *103,* 44.
11. See Greenberg, A.; Liebman, J.F. Reference 3, pp. 178–202.
12. Katz, T.J.; Acton, N. *J. Am. Chem. Soc.* **1973,** *95,* 2738.
13. Woodward, R.B.; Hoffmann, R. *Angew. Chem. Int. Ed. Engl.* **1969,** *8,* 789.
14. Subach, D.J.; Zwolinski, B.J. *J. Chem. Thermodyn.* **1975,** *7,* 443; *J. Chem. Eng. Data* **1975,** *20,* 232.
15. (a) Lin, C.Y.; Krantz, A. *J. Chem. Soc. Chem. Commun.* **1972,** 1111. (b) Masamune, S.; Suda, M.; Ono, H.; Leichter, L.M. ibid, **1972,** 1268. (c) Chapman, O.L.; McIntosh, C.L.; Pacansky, J. *J. Am. Chem. Soc.* **1973,** *95,* 1337. (d) Maier, G.; Hoppe, B. *Tetrahedron Lett.* **1973,** 881.
16. (a) Maier, G.; Alzerreca, A. *Angew. Chem. Int. Ed. Engl.* **1973,** *12,* 1015. (b) Lauer, G.; Muller, C.; Schulte, K.W.; Schweig, A.; Krebs, A. ibid, **1974,** *13,* 544.
17. (a)A naphthocyclobutadiene: Cava, M.P.; Chang, B.Y.; Van Meter, J.P. *J. Am. Chem. Soc.* **1963,** *85,* 4032. (b) A benzocyclobutadiene: Straub, H. *Angew. Chem. Int. Ed. Engl.* **1974,** *13,* 405. (c)A benzo-1,2:4,5-cyclobutadiene: Toda, F.; Ohi, M. *J. Chem. Soc. Chem. Commun.* **1975,** 506.
18. See Clar, E. "The Aromatic Sextet". Wiley: New York, 1972, and Chapter 2 by Stein and Brown, this volume.
19. (a) Breslow, R. *Chem. Eng. News* June 28, 1965, p. 90. (b) Dewar, M.J.S. *Adv. Chem. Phys.* **1965,** *8,* 121.
20. Clar, E. Reference 18, pp. 13–14. Clar discusses briefly the case of an infinite number of rings (see below) but regrettably, he does not distinguish between thermodynamic stability and kinetic reactivity.
21. Herndon, W.C. *Thermochim. Acta* **1974,** *8,* 225. This paper presents new experimental data and a theoretical analysis of numerous polynuclear hydrocarbons.

22. (a) Breslow, R.; Napierski, J.; Clarke, T. *J. Am. Chem. Soc.* **1975,** *97,* 6275. (b) For the differences and similarities of butalene and *p*-benzyne, 1,4-benzenediyl and 1,2,4,5-cyclohexatetraene, see Chapter 3 in Volume 3 of this series, by Johnson.
23. Fritch, J.R.; Vollhardt, K.P.C. *J. Am. Chem. Soc.* **1978,** *100,* 3643.
24. Wong, H.N.C.; Garratt, P.J.; Sondheimer, *J. Am. Chem. Soc.* **1974,** *96,* 5604.
25. (a) Roberts, J.D., Streitwieser, A., Jr; Regan, C.M. *J. Am. Chem. Soc.* **1952,** *74,* 4579. (b) Dewar, M.J.S.; Kohn, M.C.; Trinajastic, N. ibid, **1971,** *93,* 3432. (c) Herndon, W.C.; Ellzey, M.L., Jr. *J. Am. Chem. Soc.* **1974,** *96,* 6631. (d) Randic, M. *J. Am. Chem. Soc.* **1977,** *99,* 444.
26. (a) Halford, J.O. *J. Am. Chem. Soc.* **1967,** *89,* 5338. (b) Kramer, I.; Adams, N.P. Unpublished results.
27. (a) Streitwieser, A., Jr. "Molecular Orbital Theory for Organic Chemists". Wiley: New York, 1961. (b) Roberts, J.D. "Molecular Orbital Calculations". Benjamin: New York, 1962. (c) Coulson, C.A.; Streitwieser, A.; Jr. "Dictionary of π-Electron Calculations". Freeman: San Francisco, 1965.
28. For a largely qualitative discussion of the interplay of strain and conjugative effects in aromatic and antiaromatic species, see Greenberg, A.; Liebman, J.F. Reference 3, pp. 100–105. For a quantitative discussion of these effects limited almost exclusively to diphenyl cyclopropenone, see Greenberg, A.; Tomkins, R.P.T.; Dobrovolny, M.; Liebman, J.F. *J. Am. Chem. Soc.* **1983,** *105,* 6855.
29. Coulson, C.A.; Rushbrooke, G.S. *Proc. R. Soc. (Edinburgh) Ser. A* **1948,** *62,* 350.
30. The list is small of "infinite" systems for which literature HMO numbers are available: Gutman, I. *Naturwissenschaften* **1974,** *61,* 216.
31. Hoffmann, R.; Alder, R.W.; Wilcox, C.F., Jr. *J. Am. Chem. Soc.* **1970,** *92* 4992.
32. We thank Prof. L.A. Schaad of Vanderbilt University for making this number of his and Prof. B.A. Hess available to us. To determine this, a double integral had to be numerically evaluated. Using a smaller number of points, Coulson and Rushbrooke[29], earlier obtained 1.576 β.
33. We thank the University of Maryland Computing Center for a gift of computer time that made these calculations possible.
34. Donohue, J. "The Structure of the Elements". Wiley: New York, 1974, Chapter 7.
35. Stein, S.E.; Golden, D.M.; Benson, S.W. *J. Phys. Chem.* **1977,** *81,* 314.
36. Smith, N.K.; Stewart, R.C., Jr.; Osborn, A.G.; Scott, D.W. *J. Chem. Thermodyn.* **1980,** *12,* 919.
37. For benz[*a*]anthracene, the heat of formation of the solid was taken from our archival source, while the heat of sublimation was taken from De Kruif, C.G. *J. Chem. Thermodyn.* **1980,** *12,* 243. For triphenylene, a new heat of formation of the solid (Good, W.D. *J. Chem. Thermodyn.* **1978,** *10,* 553) was combined with the heat of sublimation taken from the work of De Kruif.
38. Dewar, M.J.S.; Dougherty, R.C. "The PMO Theory of Organic Chemistry". Plenum Press: New York, 1975, pp. 93–94.
39. For an explicit discussion of the conflicting effects of strain and aromaticity in 1,6-methano[10]annulene, see Bremser, W.; Hagen, R.; Heilbronner, E.; Vogel, E. *Helv. Chim. Acta* **1969,** *52,* 418. For a simple documentation of the strain in azulene, note that the fully saturated decahydroazulene is about 12 kcal (50 kJ) less stable than the corresponding decahydronaphthalene.
40. See Greenberg, A.; Liebman, J.F. Reference 3, pp. 66 and 94.
41. This value was obtained by combining the most recent heat of formation of the solid, 79.7(0.4) kcal (333.4(1.7) kJ (Good, W.D. *J. Chem. Thermodyn.* **1978,** *10,* 553) with a heat of sublimation measured some years before, 20.0(0.1) kcal (89.2(0.4) kJ (Morawetz, E. *J. Chem. Thermodyn.* **1972,** *4,* 455).
42. (a) Lipscomb, W.N. "Boron Hydrides". Benjamin: New York, 1963. (b) See the chapters by Muetterties, E.L., Lipscomb, W.N.; and Middaugh, R.L. In "Boron Hydride Chemistry," Muetterties, E.L., Ed.; Academic Press: New York, 1975.
43. (a) Brown, L.D.; Lipscomb, W.N. *Inorg. Chem.* **1977,** *16,* 2989. (b) Bicerano, J.; Marynick, D.S.; Lipscomb, W.N. ibid, **1978,** *17,* 2041, 3443.
44. (a) Aihara, J.-I. *J. Am. Chem. Soc.* **1978,** *100,* 3339. (b) Housecroft, C.E.; Snaith, R.; Wade, K. *Inorg. Nucl. Chem. Lett.* **1979,** *15,* 343.
45. Knoth, W.N.; Miller, H.C.; Sauer, J.C.; Balthis, J.H.; Chia, Y.T.; Muetterties, E.L. *Inorg. Chem.* **1964,** *3,* 159.

46. (a) Klanberg, F.; Muetterties, E.L. *Inorg. Chem.* **1966,** *5,* 1955. (b) Klanberg, F.; Eaton, D.R.; Guggenberger, L.J.; Muetterties, E.L. ibid, **1967,** *6,* 1271.
47. Stevenson, G.R.; Zigler, S.S.; Reiter, R.C. *J. Am. Chem. Soc.* **1981,** *103,* 6057.
48. See, for example, Hawthorne, M.F. *J. Organomet. Chem.* **1975,** *100,* 97.
49. (a) Halgren, T.A.; Lipscomb, W.N. *J. Chem. Phys.* **1973,** *58,* 1569. (b) Zyubin, A.S., Charkin, O.P. *J. Struct. Chem.* (Engl. transl.) **1980,** *21,* 403, 407.
50. Donohue, J. Reference 34, Chapter 5.
51. (a) Jemmis, E.D.; Schleyer, P.v.R. *J. Am. Chem. Soc.* **1982,** *104,* 4781. (b) Jemmis, E.D. *J. Am. Chem. Soc.* **1982,** *104,* 7017.
52. Wagman et al, Reference 6.
53. See, for example, Levine, I.N. "Quantum Chemistry", 3rd ed.; Allyn & Bacon: Boston, 1983, pp. 313–316.
54. Fink, W.H.; Allen, L.C. *J. Chem. Phys.* **1967,** *46,* 2261, 2276.
55. Allen, L.C. *Chem. Phys. Lett.* **1968,** *2,* 597.
56. Liberles, A.; O'Leary, B.; Eilers, J.E.; Whitman, D.R. *J. Am. Chem. Soc.* **1972,** *94,* 6894.
57. Csizmadia, I.G.; Theodorakopoulos, G.; Schlegel, H.B.; Whangbo, M.-H., Wolfe, S. *Can. J. Chem.* **1977,** *55,* 986.
58. Franklin, J.L. *Ind. Eng. Chem.* **1949,** *41,* 1070.
59. (a) Benson, S.W.; Buss, J.H. *J. Chem. Phys.* **1958,** *29,* 546. (b) Benson, S.W. "Thermochemical Kinetics", 2nd ed.; Wiley: New York, 1976.
60. This method has been used in many chapters in this series to estimate heats of formation of numerous species. It was used and described, extensively in Chapter 2 of this volume, by Stein and Brown, and in Chapter 4 of Volume 3, by Smart. The reader is also addressed to Chapter 4 in this volume, by Wiberg, wherein "exchange incrementation" is discussed.
61. This method has been used "informally" for many years by numerous investigators of molecular properties. It is used extensively, in both research and pedagogical ways, in: Stull, D.R.; Westrum, E.F., Jr.; Sinke, G.C. "The Chemical Thermodynamics of Organic Compounds". Wiley: New York, 1969. Various classes of compounds are so treated (eg, pp. 151–153), and the authors discuss the application to aliphatic thiols and sulfides (pp. 460–461) to primary, secondary, and tertiary amines, nitriles, nitro compounds, and to mono- and polyhalogenated aliphatic and aromatic hydrocarbons (pp. 488–493).
62. This method, though not by this name, has been used "informally" for many years by numerous investigators of molecular properties. Three studies by one of the current authors in which it was used systematically are: Liebman, J.F.; Greenberg, A. *J. Org. Chem.* **1974,** *39,* 123; Rosenstock, H.M.; Dannacher, J.; Liebman, J.F. *Radiat. Phys. Chem.* 1982, **20,** 7; Liebman, J.F., This series, Vol. 3, Chapter 6.
63. Bystrom, K. *J. Chem. Thermodyn.* **1982,** *14,* 865.
64. Chickos, J.S.; Ladon, L.H.; Hyman, A.S.; Liebman, J.F. *J. Org. Chem.* **1981,** *46,* 4294.
65. Benson, S.W. Personal communication.
66. (a) Montgomery, R.L.; Rossini, F.D. *J. Chem. Thermodyn.* ***1977,*** *10,* 471. (b) Sellers, P.; Stridh, G.; Sunner, S. *J. Chem. Eng. Data* **1978,** *23,* 250.
67. In an analysis of the thermochemistry of haloethanes, this new value was explicitly suggested: Kolesov, V.P.; Papina, T.S. *Russ. Chem. Rev.* (Engl. transl.) **1983,** *52,* 425. This value is meaningfully different from that reported in our organic archive: −31.0(0.4 kcal (−130(2) kJ).
68. (a) Wolfe, S. *Acc. Chem. Res.* **1972,** *5,* 102. (b) Kirby, A.J. "The Anomeric Effect and Related Stereoelectronic Effects at Oxygen". Springer-Verlag: Berlin, 1983, pp. 33–36, 75–77.

APPENDIX A: AN EXTENSION OF THE CONCEPT OF DIAGONAL STRUCTURAL UNITS

In this chapter we have made extensive use of the diagonal reference states for the understanding of the energetics of the groups composing numerous classes of compounds. In hindsight, in Reference 1 and thus in Section 10, it would have been better to call compounds composed of a single type of group "quasi-atom-diagonal" or "group-diagonal," rather than "atom-diagonal," to emphasize our freedom to define structural groups in the most convenient, chemically plausible way. We clearly do not mean to restrict consideration of possible reference states to just the elements. Notice that our discussion has been limited to the atoms that constitute the framework of the molecule and their affixed hydrogen atoms (or substituents) and associated bonds to other framework atoms. As such, we recognize benzene (**7**) and prismane (**4**), both formally $(CH)_6$, and cyclooctatetraene (**27**) and cubane (**1**), both formally $(CH)_8$, to be "quasi-atom-diagonal."

We have also spoken of a species as "bond-diagonal," one in which all the bonds are equivalent. Again, quite naturally, we meant the framework bonds as opposed to those to affixed hydrogens or substituents. A species may be quasi-atom-diagonal but not bond-diagonal—of the species above, prismane and cyclooctatetraene are only quasi-atom-diagonal, while benzene and cubane are both quasi-atom and bond-diagonal. We recognize all prismanes (eg, pentaprismane, **A-1**), as well as species with antiprismatic geometry (eg, B_8Cl_8, **A-2**), as quasi-atom-diagonal. A species may be bond-diagonal and not atom-diagonal. In the text we explicitly mentioned adamantane (**2**) as such an example. Other examples included are bicyclo(1.1.1)pentane (**A-3**), borazine (**A-4**), and 1,3,5-trioxane (**A-5**).

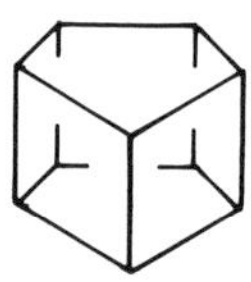

A-1

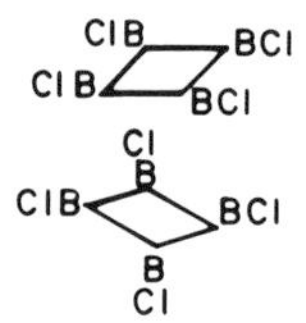

A-2

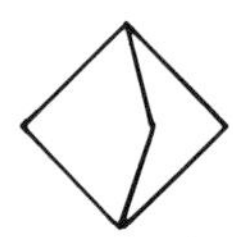

A-3

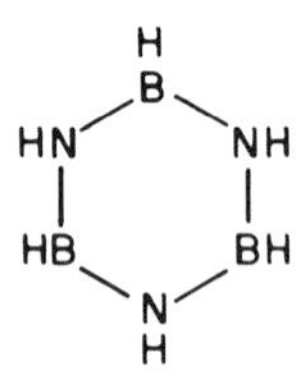

A-4

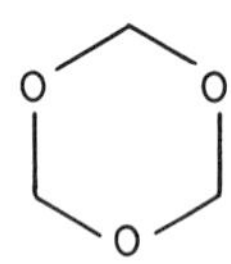

A-5

Noting the conceptual simplicity and utility of species that are quasi-atom-diagonal in the understanding of carbocyclic species, we are confident that bond-diagonal species should likewise allow for a better understanding of the thermochemistry of heterocyclic as well as general carbocyclic species. Indeed, the heat of formation of trioxane was used by Liebman in Chapter 6 of Volume 3 as an important input number for the understanding of the chemical energetics of oxygen-containing ring systems such as the likewise bond-diagonal tetroxane **A-6.** It is to be noted that all these bond-diagonal, non-quasi-atom-diagonal species are "synthesizable" by replacing all the

TABLE A-1. Interrelated Bond-Diagonal and Quasi-atom-diagonal Species

Bond-diagonal species	Quasi-atom-diagonal species	−X−	Number
Adamantane (**2**)	Tetrahedrane (**A-7**)	$-CH_2-$	6
Bicyclo(1.1.1)pentane (**A-3**)	Acetylene (**A-8**)	$-CH_2-$	3
Borazine (**A-4**)	Triazirane (**A-9a**)	−NH−	3
	or Triborirane (**A-9b**)	−BH−	3
Trioxane (**A-5**)	Trioxirane (**A-10a**)	$-O_2-$	3
	or Cyclopropane (**A-10b**)	−CH−	3
Tetroxane (**A-6**)	Tetroxetane (**A-11a**)	$-O_2-$	4
	or Cyclobutane (**A-11b**)	−CH−	4

bonds in an atom-diagonal species with −X−. In the particular, species **2** and **A-3** through **A-6** arise by so modifying the appropriate species **A-7** through **A-11** as indicated in Table A-1. (Note acetylene has been somewhat more tellingly drawn as bicyclo(0.0.0)ethane (**A-3**). Interestingly, bicyclo(1.1.1)pentane (**A-3**) may further be described as arising from the quasi-atom- and bond-diagonal cyclopropane (**A-10b**) wherein all intraring bonds totaling six electrons are replaced by a pair of >CH− groups. Analogous replacement of the six intraring π electrons of benzene by a pair of >CH− groups results in cubane.

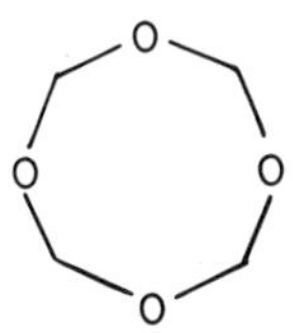

A-6

A-7

A-8

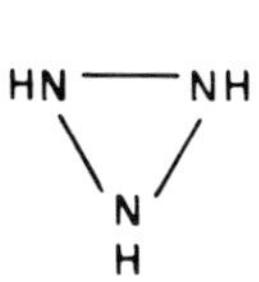

A-9a

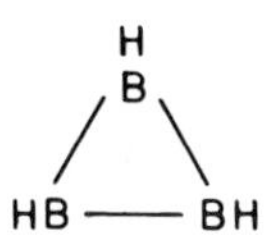

A-9b

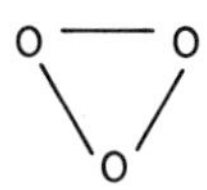

A-10a

A-10b

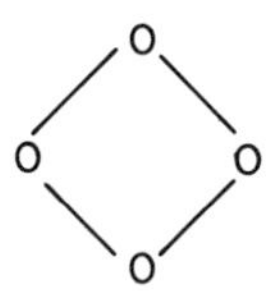

A-11a

A-11b

There is yet one more aspect to "diagonality" to be discussed here. Having spoken about atoms and bonds, we now speak more explicitly about rings. One may observe that in tetrahedrane, bicyclo(1.1.1)pentane, cubane, and adamantane all the component rings are equivalent: three cyclopropanes, two cyclobutanes, five cyclobutanes, and three cyclohexanes, respectively. These species may thus be said to be "ring-diagonal." In contrast, prismane—and most other carbocycles—are not. Adamantane documents that a bond-diagonal and ring-diagonal species need not be quasi-atom-diagonal as well. The earlier mentioned infinite quadragonal ladder (**17**) is quasi-atom-diagonal and ring-diagonal but is not bond-diagonal, while the infinite hexagonal ladder (**16**) is ring-diagonal but neither quasi-atom-diagonal nor bond-diagonal, a description shared by the various cata- and pericondensed polynuclear hydrocarbons discussed in the chapter. Very few species are simultaneously quasi-atom-, bond-, and ring-diagonal. Indeed, this class is limited to those whose framework is that of the regular polyhedra. Thus they include tetrahedral $(C-(t\text{-Bu}))_4$ (**A-12**) and $(BCl)_4$ (**A-13**), the octahedral *closo*-borane $(BH)_6^{2-}$ (**34**); cubane, reappearing now as the "representative" of the cube (**1**); the icosahedral *closo*-borane $(BH)_{12}^{2-}$ (**36**); and (pentagonal) dodecahedral $(CH)_{20}$ (**A-14**). In almost a reductio ad absurdum way, the infinite hexagonal net chemically realized as graphite sheets (**20**), also qualifies, as do some more unavoidably three-dimensional and so solid arrangements of atoms such as the diamond lattice (**A-15**). Will diagonal reasoning provide insights into the energetics of solids as it has for gaseous species? Noting the successes going in the other direction by employing data from both solid graphite and diamond for the understanding of gaseous hydrocarbons, we answer this question with enthusiastic optimism. On to the future!

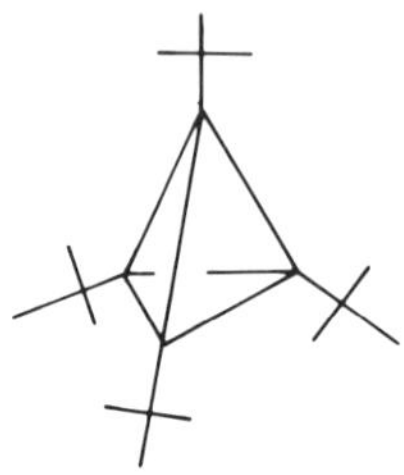

A-12

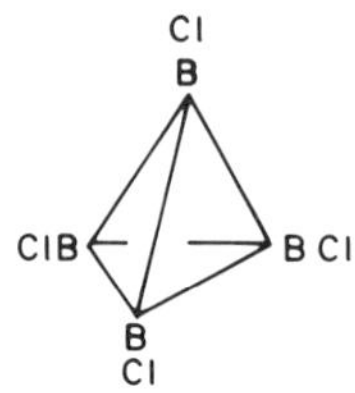

A-13

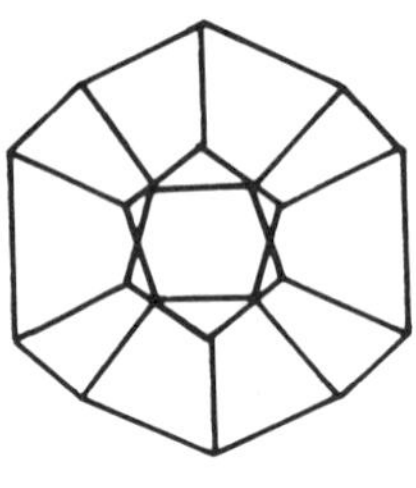

A-14

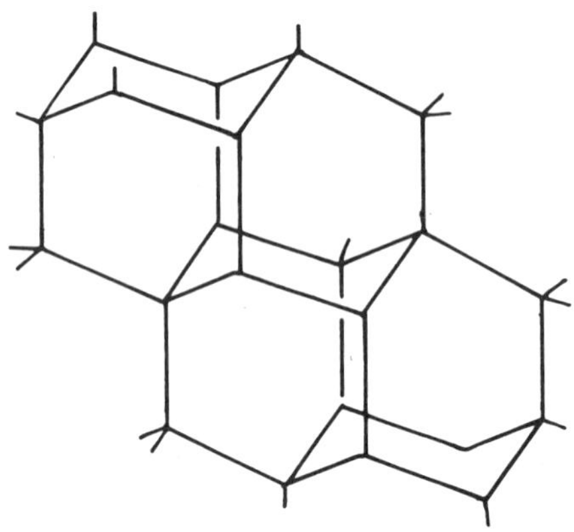

A-15

APPENDIX B: THE LOGIC OF GROUPS AND ENERGIES AS APPLIED TO POLYMER CHEMISTRY, BIOCHEMISTRY, AND SOLID STATE CHEMISTRY

Analogous grouplike constructs are also useful in understanding the energetics considerations of numerous other areas of physical science. For example, almost by definition, the polymer chemist speaks of both the monomer and the repeating unit in the polymer. The former is almost always chemically realizable, but the latter almost never is. For example, polyethylene is generally synthesized from ethylene (CH_2CH_2) and is conceptually composed of an assemblage of hypothetical $-CH_2-CH_2-$ groups. In the case of polyethylene, the polymer can also be conceptually derived from the earlier mentioned $-CH_2-$ group and indeed from $-(CH_2)_n-$ groups with any positive integer value of n. However, while the word "polymethylene" corresponding to $n = 1$ occasionally is used as a synonym of polyethylene in the polymer literature, we know of no mention of "polycyclopropane" or "polypropane-1,3-diyl" for $n = 3$ or any other "poly . . . alkane" reference for $n > 3$. It would thus appear that the group $-CH_2-CH_2-$ is treated as a divalent quasi-atom. Analogous considerations apply to other vinyl monomers, CH_2CHX (eg, X = CH_3, C_6H_5, Cl, $OCOCH_3$, CN, $COOCH_3$), and their corresponding polymers in which $-CH_2-CHX-$ likewise behaves as a divalent quasi-atomic.

Earlier we discussed cases of two or more groups that have the same atomic composition but different bonding arrangements. This applies to general vinyl monomers in that one talks about head-to-tail, head-to-head, or tail-to-tail polymerization. The two arrangements, $-CH_2-CHX-CH_2-CHX-$ (head-to-tail) and $-CHX-CH_2-CH_2-CHX-$ (tail-to-tail) are distinct, and since the derived polymers are different, it is convenient to view $-CHX-CH_2-$ and $-CH_2-CHX-$ as groups of separate types. In addition, both groups are also chiral. Noting the fundamental differences of isotactic, syndiotactic, and atactic polymers, it is therefore convenient to consider the chirality as a further distinguishing feature. There are thus a total of four possible divalent groups derived from any vinyl monomer.

The various types of polymer derived from the same monomer may be distinguished by numerous physical properties, among them the heat of formation of the polymer. In the polymer literature, it is more common to use the term "heat of polymerization," which would precisely equal the heat of formation *if* the heat of formation of the monomers were taken as zero. Indeed, the concept of "monomer" and "the heat of polymerization per unit monomer" play logically the same role in polymer chemistry as "group" and SE_C and DSTP do in general organic chemistry.

In other types of polymer such as the polyamides, there are analogous distinctions between the monomers and the resultant polymers that relate to group properties. For example, consider the polyamide polymers. Among them are the commercially important $[NH(CH_2)_5CO-]_n$ and $[-CO(CH_2)_4CONH(CH_2)_6NH-]_{n'}$. Given $-CH_2-$, $-CO-$, and $-NH-$ groups, one can derive numerous new and important larger groups that are relevant in a polymer chemistry context. The specific polymers of interest are composed of three so-derived new and important groups, $-NH(CH_2)_5CO-$, $-CO(CH_2)_4CO-$, and $-NH(CH_2)_6NH-$. Both these polymers, $[NH(CH_2)_5CO-]_n$ and $[-CO(CH_2)_4CONH(CH_2)_6NH-]_{n'}$, can be made by loss of water from appropriate monomers, $H_2N(CH_2)_5COOH$ for the former and $HOOC(CH_2)_4COOH + H_2N(CH_2)_6NH_2$ for the latter. The term "heat of polymerization" in these cases explicitly refers to the process involving such loss of water. Since this loss of water is stoichiometric, the heat of formation of water can be directly incorporated into the relationships between heat of formation and heat of polymerization. That is, the heat of polymerization associated with $[NH(CH_2)_5CO-]_n$ formation is defined as follows:

$$\frac{\Delta H_f\{[-NH(CH_2)_5CO-]_n\}}{n} + \Delta H_f(H_2O) - (\Delta H_f(H_2N(CH_2)_5COOH)$$

where it is additionally specified whether the species are to be taken in their standard form at STP or as ideal gases.

The reader should be warned that in some cases a given polymer can be formed by more than one method. For example, direct polymerization of a suitable monomer, ϵ-caprolactam = cyclo-$(NH(CH_2)_5CO)$, will result in $[-NH(CH_2)_5CO-]_n$. Clearly there is a heat of formation from the monomer associated with this process as well, and thus there is a chance for ambiguity if one just speaks of "the" heat of polymerization. That is, there is the thermochemical quantity:

$$\frac{\Delta H_f\{[-NH(CH_2)_5CO-]_n\}}{n} - \Delta H_f\{\text{cyclo-}(NH(CH_2)_5CO)\}$$

and certainly, one cannot set the heats of formation of both $(NH_2(CH_2)_5COOH - H_2O)$ and ϵ-caprolactam equal to zero. However, as with vinyl monomers and their derived polymers, as long as one defines carefully and consistently the reference species, and the starting and final products (including water when appropriate), no problem need arise. That is, one "should" not mix discussion of the energetics of the direct polymerization of lactams with that of the condensation polymerization of amines with carboxylic acids, regardless of whether it be intra- or intermolecular. However, once one is clear which process is being discussed, the polymer chemist's concept of "heat of polymerization per unit monomer" is again logically equivalent to the organic chemist's DSTP and SE_C.

The biochemist also generally speaks of big groups under the guise of small elementlike labels. For example, in the study of nucleic acids, biochemists often write their macromolecules as a string of monomers from which the polymer is formed by loss of water. However, even though the primary chemical components, the nucleotides, are even bigger than most monomers of the polymer chemist, they are generally given one-letter symbols (A, G, C, T, etc.). Limiting our attention to double-stranded DNA, there are four chemical components derived from the four bases, and so we may say that there are four groups. The strands are paired by hydrogen bonds between the groups. Our analysis is eased by the fact that rather than forming $4^2 = 16$ pairs—effectively new groups (pairs), by random pairwise interactions—only four new groups are ever observed. These are AT, TA, GC, and CG, where the position of the letter in the group designates the DNA strand to which the base belongs.

In aqueous solution, double-stranded DNA is in thermal equilibrium with the single-stranded form; that is, both are present in temperature-dependent concentrations. It is customary to speak of a "melting" temperature for DNA that corresponds to the temperature at which ΔG for the double/single-stranded reaction changes from positive to negative.

≪Why (except for historical reasons) the term "melting" rather than "vaporization" is used is not obvious to the authors, since the double-stranded chain is still isolated from others in the dilute solution and thus lacks the long-range order characteristic of nonamorphous solids. What the double-stranded chain does have is short-range order: all the bases have predictable interstrand neighbors, and the azimuthal angle of each base relative to the chain is a predictable function of position along the chain. When the strands separate, both these elements of order disappear and the only order that remains is the definite sequence of bases. Thus the phase transition corresponds to a change in the short-range order, as is usually the case for vaporization. Moreover, the temperature dependence of the relative concentration of the single strand-species is fully analogous to the saturated vapor pressure above its liquid.≫

For a given polymer size (ie, number of nucleotides), the melting point depends intimately on the relative numbers of GC (and CG) pairs as opposed to AT (and TA) pairs. Since this "melting" physically corresponds to the rupturing of the interstrand hydrogen bonds and the formation of additional hydrogen bonds to the solvent water molecules, microscopically it is clear that this phase transition is truly a chemical reaction.

≪Indeed, all phase transitions involving latent heat (ie, first-order transitions) involve changes in such chemical properties as the average coordination number or bond characteristics and so may be viewed as chemical reactions. This is one source of the entropy differences between the two phases.≫

Thus we are not surprised that the latent heat of the reaction—distinguished from the usual "heat of reaction" by referring to processes at the melting temperature rather than under STP conditions—likewise depends on the composition of the polymer (ie, on the number and nature of the groups present). Indeed the estimation of this latent heat would seem to be relatively straightforward if the sequence of bases is accurately known. (We expect there to be some dependence of the interstrand bond strength on the identity of the intrastrand neighbors.) Because any composition of DNA is allowed—subject to the pairing condition—the biochemist's situation is more difficult than that of the organic or inorganic chemist in that all the different combinations (sequences) go by the same general name and isolation of a particular sequence (combination) is distinctly less trivial. Yet the underlying logical structure, the relationship of heat of melting to nucleotide composition, parallels that of the heat of formation of a compound from its component groups.

In the context of "small molecules" and strained rings and aromaticity, we are not used to properties and group compositions varying essentially continuously. However, this continuous variation is seen in certain natural and synthetic silicates in which tetrahedral SiO_4 units may be quite freely replaced by AlO_4 (and some accompanying cation for electroneutrality) and in some alloys such as $Zn_{1-x}Cu_x$. What are the systematics of these species? We recognize that alloys may be stoichiometric and some are even saltlike (eg, CsAu, wherein the 1:1 intermetallic ratio is as circumscribed as the 1:1 elemental ratio in CsCl. It would appear that the concept of "heat of alloying" relates in a logical sense to that of "heat of formation" as practiced by the inorganic chemist. We find these connections between the inorganic solid state and biologically important macromolecules quite remarkable and certainly encouraging. We hope that the reader has shared our enthusiasm in the study of the regularities of chemical phenomena.

APPENDIX C: THE CHEMIST'S UNDERSTANDING OF GROUPS VERSUS ATOMS AND THE FORMAL LANGUAGE OF QUANTUM MECHANICS

This chapter records some of the progress that has been made translating the chemist's understanding of groups versus atoms into the formal language of quantum mechanics. The central conceptual element is the realization that the eigenstates of the Hamiltonian for a system are states which describe *all* of the members of that system concurrently. Moreover, when dealing with explicitly multi-particle systems, the total system's wavefunction is only approximated by the product of the separately (anti)symmetrized and normalized product of the effective single-particle wavefunctions of the constituent identical particles. For example, letting ψ_{n1} and ψ_{n2}, ψ_{p1} and ψ_{p2}, and ψ_{e1} and ψ_{e2}, be the lowest lying single-particle eigenstates of neutrons, protons and electrons respectively, a complete description of an atom of ^{4}He is *not*

$$\Psi \equiv 2^{-3/2}[\psi_{n1}(\mathbf{r}_1)\psi_{n2}(\mathbf{r}_2) - \psi_{n2}(\mathbf{r}_1)\psi_{n1}(\mathbf{r}_2)][\psi_{p1}(\mathbf{r}_3)\psi_{p2}(\mathbf{r}_4) - \psi_{p2}(\mathbf{r}_3)\psi_{p1}(\mathbf{r}_4)][\psi_{e1}(\mathbf{r}_5)\psi_{e2}(\mathbf{r}_6) - \psi_{e2}(\mathbf{r}_5)\psi_{e1}(\mathbf{r}_6)].$$

However Ψ so defined is a good first approximation to the true ground state wave function. Moreover it is convenient to consider since the properties of the total system can be shown to be the sum of those of the individual particles ($\psi_i(\mathbf{r}_i)$). For example, the total energy is approximated by

$$E^{(0)} = \epsilon_{n1} + \epsilon_{n2} + \epsilon_{p1} + \epsilon_{p2} + \epsilon_{e1} + \epsilon_{e2}$$

Discrepancies, both here and in more complicated problems, between the prediction $E^{(0)}$ and experiment are ascribable to errors in the calculation of the single particle wavefunctions, ψ_i and their energies ϵ_i, to interactions omitted or misdescribed within a given class of identical components, and to interactions between the various classes of components. For many problems, one may safely ignore one class of components while explicitly studying another and thus the first two sources of errors are more frequently explicitly considered than the last. For example, in most studies of atomic ^{4}He (and likewise of most atomic and molecular phenomena as well) the nucleus is approximated with little error by a point charge of infinite mass and the neutrons are totally ignored, the protons' existence being reflected only by a time independent potential energy term in the Hamiltonian for the electrons. Notice, however, that when one studies the He nucleus (and nuclear excited states and shell structure in general), it is rare to consider the effects of the electrons because the interaction of the electrons with the nucleons is so much weaker than that among the nucleons that these effects are small. Notice that it is well understood that it is unreasonable to expect agreement between theory and experiment if one ignores the motion of the neutrons when considering the protons. Protons and neutrons are too similar in their masses, energy levels, time scales and strength of interaction to consider one type of constituent while ignoring the other.

Similarly, in the chemical case, the image of a complex molecule as an assemblage of one or more types of identical particles (the different groups) which do not interact with one another is obviously a considerable oversimplification. However, it is a very useful one since it allows us to calculate a good first approximation to the heat of formation ($\mathbf{v} \cdot \boldsymbol{\eta}$). The discrepancies which arise (eg, strain and resonance energy) reflect the fact that the groups are actually too similar in size and the inter- and intra-group energies too similar for the conceptual separation into groups to correspond to truly noninteracting pieces. In addition, it is sometimes the case that the assemblage labelled as a group does not truly correspond to the archetypical group because it is characterized by considerable structural rearrangement and/or electronic promotion, both generically labelled excitation. For example, in cyclopropane, the CCC angles are 60° whereas in the reference species for the $>CH_2$ group, cyclohexane, they are ca. 112°. Intuitively, we recognize the former $>CH_2$ group as strained, distorted, and destabilized relative to the latter, the former as an excited state of the latter. Thus we were not surprised that the heat of formation of the former compound is ca. 9 kcal/mol per group higher than the latter. This destabilization arises from both electronic excitation and structural rearrangement of the $>CH_2$ group as well as the interactions among $>CH_2$ groups. Indeed, most other compounds containing the $>CH_2$ group are likewise destabilized, although to a less extent because the degree of excitation and rearrangement in them is comparatively small.

One should be careful not to equate either weak bonding between groups or small strain or resonance energies with only a small degree of structural rearrangement or electronic excitation. This is dramatically shown by the energetics of elemental phosphorus, an unequivocally atom-diagonal species. Upon sublimation, the resultant gas consists of tetrahedral P_4 molecules. The rather low heat of sublimation[6], ca. 3.5 kcal/mol-P (15 kJ), might seem to suggest that distinct P_4 molecules exist in the solid. In fact, the most common form of solid element, white phosphorus, has these preformed tetrahedra. However, red and black phosphorus are more stable by ca. 1 and 2 kcal/mol-P respectively and consist of phosphorus atoms in more highly interconnected networks, though still having 3 neighboring phosphorus atoms each. It is unobvious why the three solid forms should be so close in energy. When the red and black forms are sublimed, they too produce P_4 groups. That the entropy of sublimation for the red (and presumably the black) form is higher than that of the white reflects the need for rearrangement prior to detachment. (It is not surprising that P_4 groups are preferred over atomic P in the vapor since P—P bonds are each worth ca. 50 kcal (200 kJ).) Analogously, the very stable ^{4}He units may be exothermically combined to form the small "nuclear polymers" ^{12}C, ^{16}O, ^{20}Ne, ^{24}Mg, ^{28}Si, ^{32}S, ^{36}Ar, and ^{40}Ca, although we do not mean to imply a linear arrangement of the components. We are not surprised radioactive decay by loss of ^{4}He is a common decay mode while loss of either single proton or neutron is not. However, one does not usually assert that these or any other nuclei are composed of ^{4}He or α particles as opposed to a more complicated collection of interacting protons and neutrons.

In summary the rarity of zero values for strain and resonance energy does not arise from some obscure quantum mechanical feature such as the distinction between mixtures and superpositions. Rather this rarity arises from the fact that the structural units into which we conceptually divide molecules are too similar in size and properties and not sufficiently differentiated physically to be expected to be non-interacting.

Addendum

The authors of three chapters in this volume have elected to submit material to elaborate on and extend points made in their chapter. Recall that Chapter 2, written by Stein and Brown, deals with the energetics of graphite and Liebman and Van Vechten studied it in Chapter 8 as well. Herein Stein and Brown derive approximate formulas for total π-electron energy in the graphite limit. Ciosłowski[1] has tabulated several semi-empirical formulas for calculating the total Hückel π energy, E, of benzenoid hydrocarbons. It is interesting to see what these predict for the energy per carbon atom in this graphite limit. Two of these formulas use as input the number of carbon atoms, N, bonds, M, and Kékulé structures, K. The first, due to Hall,[2] is

$$E = 0.442N + 0.788M + 0.34K(0.632)^{M-N} \qquad (1)$$

This gives $E/N = 0.442 + 0.778(3/2) = 1.624\beta$ as N goes to infinity, appropriate for graphite. (Remember $1\beta \cong 18$ kcal/mol.)

The second formula, derived by Gutman and coworkers,[3] is

$$E = 1.804(Mq)^{1/2} + 0.421K(q/M)^{q/2} + 1.686;\ q = N/2 - 1 \qquad (2)$$

This gives $E/N = 1.804(3/4)^{1/2} = 1.562\beta$ as N goes to infinity. The third method, developed by Ciosłowski,[1] expresses E as the sum of 12 types of empirical nodal increments. These nodes are the vertices of a dual graph constructed from the molecular graph of the hydrocarbon by replacing all of the hexagons by dots. In the graphite limit, only one nodal type occurs; that surrounded by six other nodes. As a result we get $E/N = 3.094499\beta/2 = 1.5472495\beta$ as N goes to infinity.

These three limiting values can be compared to the true graphite limit of E/N of 1.57460β, obtained by reevaluating an earlier derived double integral of Coulson and Taylor[4] by either using a finer numerical grid[5] or a literature polynomial approximation[6] for the requisite elliptic integral.[7] Only minor adjustments of the numerical parameters in expressions (1) and (2) are required to bring these expressions into agreement with the graphite limit. It is hoped that this could be done without seriously compromising their accuracy for finite molecules.

Chapter 3 written by Chickos deals in large part with the properties of conglomerates and estimation methods for heats of sublimation. With regard to the former, racemic 2,3-dibromobutane-1,4-diol is an example of a system known to form a conglomerate in the solid state.[8] The vapor pressures, heats of fusion and heats of sublimation of d and dl-2,3-dibromobutane-1.2-diol have recently been measured.[9] Values of 28.9 ($\pm$0.8), 33.5 ($\pm$0.8), 111.7 and 118.4 kJ/mol were measured for ΔH_{fus} and ΔH_S for dl and d crystals respectively. (Remember 1 kcal $\equiv$ 4.184 kJ). In both cases the higher melting pure enantiomer gave the larger value for ΔH_{fus} and ΔH_S. The racemic form, however, was the most volatile. The ratio of vapor pressures for the dl/d forms varied from 2.6 to 2.26 over the temperature range investigated (348-361 K, respectively). This ratio is above 2 as predicted by theory and provides experimental evidence supporting the theoretical basis of the double solubility rule.

With regard to estimations of the heat of sublimation, it is noted that most methods have been either very qualitative or highly restricted, unlike the variety of successful methods developed to estimate H_V. Recently a successful semiempirical method has been developed[10] for estimating ΔH_S of hydrocarbons that starts with the approximation earlier given in the text

$$\Delta H_S(298) = \Delta H_{fus}(T_M) + \Delta H_V(T) \tag{3}$$

Heats of vaporization were estimated from the relationship:[11]

$$\Delta H_V(298) = (1.29 \pm 0.21)n_Q + (4.69 \pm 0.04)\bar{n}_C + (2.97 \pm 0.63) \tag{4}$$

where $\bar{n}_C \equiv n_C - n_Q$, the difference of the total number of carbon atoms and the number of quaternary carbon atoms. Experimental values[12] were used for ΔH_{fus}, the most difficult term to predict accurately. The calculated values of heat of sublimation, ΔH_S^c, were compared for the 45 hydrocarbons (of diverse structure) for which both the experimental heats of sublimation and fusion, ΔH_S and ΔH_{fus}, were available. The accuracy of calculated values for heat of sublimation were generally within 5% of the experimental.[13] Some typical results follow in the accompanying Table A3-7. (see next page)
The diadamantane entry serves as an example of the usefulness of this estimation technique. Two discordant ΔH_S results have been reported, 95.8 and 117.2 kJ/mol. That the former is more accurate of the two values can be readily concluded by this estimation technique. It thus allows for a rational section of the more probable value.

A third point about chapter 3 is that the conscious omission of heats of vaporization data has been remedied with the publication[14] of a data compendium and associated discussion in this important area.

Chapter 8, written by Liebman and Van Vechten, deals in part with the aromaticity of the higher annulenes, thermochemistry of azoxy compounds, and studies of P_6 and P_8. Commencing here with the higher annulenes, a class of diagonal compounds earlier discussed in the section on polynuclear hydrocarbons, note that from indirect thermochemical measurements and some semiempirical estimates, Oth and coworkers ascertained[15] the heat of formation of [18] annulene to be 125 ($\pm$6) kcal/mol. This corresponds to a

TABLE A3-7. Experimental and Estimated Heats of Sublimation of Structurally Diverse Hydrocarbons

Compound	ΔH_V	ΔH_{fus}	$\Delta H^C{}_S$	ΔH_S	$\Delta H_S - \Delta H^C{}_S/\Delta H_S$
C_8					
bicyclooctane	40.6	8.4	49.0	47.7	−0.03
2,2,3,3-tetramethyl butane	33.9	9.2	43.1	43.5	0.01
C_{10}					
naphthalene	49.8	18.8	68.6	72.4	0.05
1,2,4,5-tetramethyl benzene	43.1	20.9	64.0	71.1	0.10
C_{12}					
acenaphthene	59.4	20.1	79.5	83.3	0.05
C_{14}					
1,2-diphenylethane	69.9	23.0	92.9	91.2	−0.05
diadamantane	69.9	22.2	92.1	95.8	0.04
				117.2	0.27
C_{18}					
o-terphenyl	87.4	17.2	104.6	97.1	−0.08
C_{19}					
triphenylmethane	92.0	20.9	112.9	100.0	−0.13

DSTP of 3.7 (±0.3) kcal/mol, a result consistent with a recent high accuracy quantum chemical calculation by Haddon and Raghavachari[16] from which a value of 3.3 kcal/mol may be derived. These latter authors also report and discuss quantum chemical calculations on [10] annulene and on the [∞] annulene (trans-polyacetylene), from which DSTP values of 2.4 and 4.8 kcal/mol can be deduced. The earlier logic on the DSTP of annulenes is seen to be self-consistent.

It is also noted a recent study by Byström[17] of cyclic (E)-azoxy compounds complements the earlier discussion of the related, but acyclic, (Z)-azoxy species. From data on these new species applicable group increment and macroincrementation reactions may be deduced. The literature study also compares related azo and azoxy compounds.

The addendum is completed by a brief discussion of studies of P_6 and P_8. While hexagonal P_6 is as experimentally unknown as ever, a "triple decker sandwich" complex containing it, $[\eta^5(CH_3)_5C_5Mo]_2(\mu,\eta^6\text{-}P_6)$, has been synthesized in good yield[18] and shown to have local 6-fold symmetry. Finally, the reader should note a collection[19] of high quality ab-initio quantum chemical calculations on the unknown atom-, bond- and face-diagonal form of elemental phosphorus, cubic P_8. "Natural" and "reasonable" assumptions suggest that this species is comparatively strainless. It would thus be expected to be more stable than the likewise diagonal tetrahedral P_4 that is found in the white solid and in the vapor. However, these studies unequivocally document the surprising conclusion that P_8 is considerably less stable (ca. 30-50 kcal/mol-P_8) than $2P_4$. The reversal of the relative stability of P_4

and P_8, compared to that of their carbon analogs, tetrahedrane $(CH)_4$ and cubane $(CH)_8$, is worthy of further investigation.

ADDENDUM REFERENCES

1. Ciosłowski, J. *Theoret. Chim. Acta* **1985,** *68,* 315.
2. Hall, G.G. *Inst. Math. Appl.* **1981,** *17,* 70.
3. Gutman, I.; Teodorovič, A.V.; Nedeljkovič, Lj. *Theoret. Chim. Acta* **1984,** *65,* 23.
4. Coulson, C.A.; Taylor, R. *Proc. Phys. Soc.* **1952,** *A65,* 815.
5. Schaad, L.A.; Hess, A.B. Unpublished results cited by Liebman and Van Vechten in chapter 8 in this volume.
6. Abramowitz, M.; Stegun, I.A. "Handbook of Mathematical Functions," U.S. National Bureau of Standards, Applied Mathematics Series, 55, 1964, p. 592.
7. Brown, R.L.; Stein, S.E. Unpublished calculation.
8. (a) Collet, A.; Brienne, M.J.; Jacques, J. *Bull. Soc. Chim. Fr.* **1972,** 127; (b) Luttringhaus, A.; Berrer, D. *Tetrahedron Lett.* **1959,** 10 and *Angew. Chem.* **1958,** *70,* 439.
9. Chickos, J.S.; Annunziata, R. Unpublished results.
10. Chickos, J.S.; Annunziata, R.; Ladon, L.H.; Hyman, A.S.; Liebman, J.F. Unpublished results, and *J. Org Chem.* In press.
11. Ref. 70 of chapter 3 (Chickos, J.S.; Ladon, L.H.; Hyman, A.S.; Liebman, J.F. *J. Org. Chem.* **1981,** *46,* 4294.)
12. All ΔH_{fus} values are taken from Domalski, E.S.; Evans, W.H.; Hearing, E.D. *J. Phys. Chem. Ref. Data* **1984,** 13 (Supplement 1).
13. Citations to the experimental values of ΔH_S may be found in Tables 5 and 6 of Chapter 3.
14. "Enthalpies of Vaporization of Organic Compounds: A Critical Review and Data Compilation" (eds. Majer, V.; Svoboda, V. Blackwell Scientific Publications, Inc., Palo Alto, 1985).
15. Oth, J.F.M.; Bunzli, J.-C.; de Zelicourt, Y. de J. *Helv. Chim. Acta* **1977,** *57,* 2276.
16. Haddon, R.C.; Raghavachari, K. *J. Am. Chem. Soc.* **1985,** *107,* 289.
17. Byström, K. *J. Comput. Chem.* **1983,** *4,* 308, and unpublished work cited of hers with Ehrbar, W.; Månsson, M.; Snyder, J.P.
18. Scherer, O.J.; Sitzmann, H.; Wolmershäuser, G. *Angew. Chem. Int. Ed.* **1985,** *24,* 351.
19. (a) Trinquier, G.; Daudey, J.-P.; Komiha, N. *J. Am. Chem. Soc.* **1985,** *107,* 7210; (b) Ahlrichs, R.; Brode, S.; Ehrhardt, J. *J. Am. Chem. Soc.* **1985,** *107,* 7260; (c) Raghavachari, K.; Haddon, R.C.; Binkley, J.S. *Chem. Phys. Lett.* **1985,** *122,* 219; (d) Schmidt, M.W.; Gordon, M.S. *Inorg. Chem.* **1985,** *24,* 4503.

CUMULATIVE CONTENTS: VOLUMES 1 AND 2

(Article title followed by volume number with chapter in parentheses)

GENERAL INDEX

D

E

I

N

O

P

V

W

X

Z